Microbial Growth on C₁ Compounds

Microbial Growth on C_1 Compounds

edited by

J.C. MURRELL
Department of Biological Sciences
University of Warwick, Coventry, UK

D.P. KELLY
Natural Environment Research Council
Swindon, UK

Intercept Ltd
Andover

Copyright © Intercept Ltd. 1993

British Library CIP Data available
ISBN 0 946707 54 5

All rights reserved. No part of this publication may be reproduced (including photocopying), stored in a retrieval system of any kind, or transmitted by any means without the written permission of the Publishers. Permission to copy is granted to libraries and other users on the condition that the appropriate fee is paid directly to the Copyright Clearance Center, 21 Congress Street, Salem, MA 01970, USA. For *Microbial Growth on C_1, Compounds* the copying fee per chapter is $20.00

Published by Intercept Limited, P.O. Box 716, Andover, Hampshire SP10 1YG, England

Typeset in Times by Techset Ltd. Salisbury, Wilts.
Printed in Great Britain by Athenaeum Press, Newcastle upon Tyne

List of Contributors

B. E. Alber
Department of Anaerobic Microbiology
Virginia Polytechnic Institute
 and State University
Blacksburg
VA 24061
USA

M.J. Alperin
Curriculum in Marine Sciences
University of North Carolina
Chapel Hill
North Carolina 27599-3300
USA

L. Alvarez-Cohen
Department of Civil Engineering
726 Davis Hall
University of California
Berkeley
CA 94720
USA

K.K. Andersson
4-225 Millard Hall
Department of Biochemistry
University of Minnesota
Minneapolis
MN 55455
USA

C. Anthony
SERC Centre for Molecular Recognition
Biochemistry Department
University of Southampton
Southampton
SO9 3TU
UK

N. Arfman
Department of Microbiology
University of Groningen
Kerklaan 30
9751 NN Haren
THE NETHERLANDS

R. Bader
Mikrobiologisches Institut ETH
ETH-Zentrum
CH-8092
Zurich
SWITZERLAND

C. Balny
Institut National de la Sante et de la
 Recherche Medicale
Unite 128
BP 5051
34033 Montpellier Cedex
FRANCE

R. Bednarski
Institut fur Mikrobiologie
Georg-August-Universitat Gottingen
Grisebachstrasse 8
D-W-3400 Gottingen
GERMANY

B. Becher
Institut fur Mikrobiologie der
 Georg-August-Universitat Gottingen
Grisebachstr. 8
W-3400 Gottingen
GERMANY

E. Bellion
Department of Chemistry and Biochemistry
The University of Texas at Arlington
Arlington
TX 76019-0066
USA

G.W. Black
Department of Agricultural Biochemistry
University of Newcastle-upon-Tyne
Newcastle-upon-Tyne
SR1 3SD
UK

M. Blaut
Institut fur Mikrobiologie der
 Georg-August-Universitat Gottingen
Grisebachstr. 8
W-3400 Gottingen
GERMANY

E.S. Boulygina
Gray Freshwater Biological Institute
University of Minnesota
P.O. Box 100
Navarre
MN 55392
USA

B. Bowien
Institut fur Mikrobiologie
Georg-August-Universitat Gottingen
Grisebachstr. 8
D-W-3400 Gottingen
GERMANY

B.J. Bratina
Department of Microbiology
University of Minnesota
P.O. Box 100
Navarre
MN 55392
USA

List of contributors

S. Braus-Stromeyer
Mikrobiologisches Institut ETH
ETH-Zentrum
CH-8092
Zurich
SWITZERLAND

G.A. Brusseau
Department of Microbiology
University of Minnesota
P.O. Box 100
Navarre
MN 55392
USA

L.V. Bystrykh
Department of Microbiology
University of Groningen
Kerklaan 30
9751 NN Haren
THE NETHERLANDS

M.A. Carver
ICI Bio Products
Billingham
Cleveland
TS23 1LB
UK

C.M. Cavanaugh
Department of Organismic and
 Evolutionary Biology
Harvard University
16 Divinity Avenue
Cambridge
MA 02138
USA

H.T.C. Chan
SERC Centre for Molecular Recognition
Biochemistry Department
University of Southampton
Southampton
SO9 3TU
UK

S.I. Chan
A.A. Noyes Laboratory of Chemical Physics
California Institute of Technology
Pasadena
CA 91125
USA

A.Y. Chistoserdov
Environmental Engineering Science
W.M. Keck Laboratories 138-78
California Institute of Technology
Pasadena
CA 91125
USA

K.M. Chumakov
FDA Center for Biologics Evaluation
 and Research
8800 Rockville Pike
Bldg. 29A
Bethesda
MD 20892
USA

A.P. Clements
Department of Anaerobic Microbiology
Virginia Polytechnic Institute
 and State University
Blacksburg
VA 24061
USA

J. Colby
School of Health Sciences
University of Sunderland
Sunderland
SR1 3SD
UK

A.M. Cook
Mikrobiologisches Institut ETH
ETH-Zentrum
CH-8092
Zurich
SWITZERLAND

J.M. Cox
SERC Centre for Molecular Recognition
Biochemistry Department
University of Southampton
Southampton
SO9 3TU
UK

H. Dalton
Department of Biological Sciences
University of Warwick
Coventry
CV4 7AL
UK

U. Deppenmeier
Institut fur Mikrobiologie der
 Georg-August-Universitat Gottingen
Grisebachstr. 8
W-3400 Gottingen
GERMANY

L. Dijkhuizen
Department of Microbiology
University of Groningen
Kerklaan 30
9751 NN Haren
THE NETHERLANDS

H.L. Drake
Lehrstuhl fur Okologische
 Mikrobiologie
BITOK
Universitat Bayreuth

List of contributors

Postfach 10 12 51
D-8580 Bayreuth
GERMANY

J.A. Duine
Department of Microbiology and Enzymology
Delft University of Technology
Julianalaan 67
2628 BC Delft
THE NETHERLANDS

G.J.W. Euverink
Department of Microbiology
University of Groningen
Kerklaan 30
9751 NN Haren
THE NETHERLANDS

D.L. Falcone
Department of Microbiology and the
   Biotechnology Center
The Ohio State University
484 West 12th Avenue
Columbus
OH 43210–1292
USA

J.G. Ferry
Department of Anaerobic Microbiology
Virginia Polytechnic Institute
   and State University
Blacksburg
VA 24061
USA

J. Frank
Department of Microbiology and Enzymology
Delft University of Technology
Julianalaan 67
NL-2628 BC Delft
THE NETHERLANDS

A. Freter
Institut fur Mikrobiologie
Georg-August-Universitat Gottingen
Grisebachstr. 8
D-W-3400 Gottingen
GERMANY

W.A. Froland
4-225 Millard Hall
Department of Biochemistry
University of Minnesota
Minneapolis
MN 55455
USA

K. Frunzke
Lehrstuhl fur Mikrobiologie
Universitat Bayreuth
Universitatsstraße 30
8580 Bayreuth
GERMANY

F. Gasser
Unite de Regulation de l'Expression
   Genetique
Department de Biochimie et Genetique
   Moleculaire
Institut Pasteur
28 rue du Dr. Roux
75724 Paris Cedex 15
FRANCE

J.L. Gibson
Department of Microbiology and the
   Biotechnology Center
The Ohio State University
484 West 12th Avenue
Columbus
OH 43210–1292
USA

P.N. Green
NCIMB Ltd
23 St Machar Drive
Aberdeen
AB2 1RY
Scotland
UK

G. Gottschalk
Institut fur Mikrobiologie der
   Georg-August-Universitat Gottingen
Grisebachstr. 8
W-3400 Gottingen
GERMANY

R.S. Hanson
Department of Microbiology &
Gray Freshwater Biological Institute
University of Minnesota
P.O. Box 100
Navarre
MN 55392
USA

W. Harder
Biological Center
University of Groningen
Kerklaan 30
9751 NN Haren
THE NETHERLANDS

N. Harms
Department of Microbial Physiology
Biological Laboratory
Vrije Universiteit
De Boelelaan 1087
1081HV Amsterdam
THE NETHERLANDS

I. Heiber-Langer
Institut National de la Sante et de la
   Recherche Medicale
Unite 128
BP 5051
34033 Montpellier Cedex
FRANCE

viii List of contributors

A.N. Hennigan
Department of Microbiology
The Ohio State University
483 W. 12th Avenue
Columbus
OH 43210-1292
USA

G.I. Hessels
Department of Microbiology
University of Groningen
Kerklaan 30
9751 NN Haren
THE NETHERLANDS

P.E. Jablonski
Department of Anaerobic Microbiology
Virginia Polytechnic Institute
  and State University
Blacksburg
VA 24061
USA

M. Janvier
Unite de Regulation de l'Expression
  Genetique
Department de Biochimie et Genetique
  Moleculaire
Institut Pasteur
28 rue du Dr. Roux
75724 Paris Cedex 15
FRANCE

D.B. Janssen
Department of Biochemistry
University of Groningen
Nijenborgh 4
9747 AG Groningen
THE NETHERLANDS

Y. Jiang
Department of Biological Sciences
University of Warwick
Coventry
CV4 7AL
UK

C.W. Jones
Department of Biochemistry
University of Leicester
Leicester
LE1 7RH
UK

J.G. Jones
Department of Radiology
The University of Texas
  Southwestern Medical Center at Dallas
Dallas
TX 75235
USA

B. Kamlage
Institut fur Mikrobiologie der
  Georg-August-Universitat Gottingen
Grisebachstr. 8
W-3400 Gottingen
GERMANY

D.P. Kelly
Natural Environment Research Council
Polaris House
Swindon
SN2 1EU
UK

J.T. Keltjens
Department of Microbiology
Faculty of Science
University of Nijmegen
Toernooiveld
NL-6525 ED Nijmegen
THE NETHERLANDS

R.P. Kiene
University of Georgia Marine Institute
Sapelo Island
Georgia 31327
USA

G.M. King
Darling Marine Center
University of Maine
Walpole
ME 04573
USA

B. Kusian
Institut fur Mikrobiologie
Georg-August-Universitat Gottingen
Grisebachstr. 8
D-W-3400 Gottingen
GERMANY

S. La Roche
Frederick Cancer Research and
  Development Center
Frederick
MD 21701
USA

M.T. Latimer
Department of Anaerobic Microbiology
Virginia Polytechnic Institute
  and State University
Blacksburg
VA 24061
USA

S. Lee
4-225 Millard Hall
Department of Biochemistry
University of Minnesota
Minneapolis
MN 55455
USA

T. Leisinger
Mikrobiologisches Institut ETH
ETH-Zentrum

List of contributors   ix

CH-8092
Zurich
SWITZERLAND

L. Li
Department of Microbiology and the
  Biotechnology Center
The Ohio State University
484 West 12th Avenue
Columbus
OH 43210–1292
USA

M.E. Lidstrom
Environmental Engineering Science
W.M. Keck Laboratories 138–78
California Institute of Technology
Pasadena
CA 91125
USA

J.D. Lipscomb
4-225 Millard Hall
Department of Biochemistry
University of Minnesota
Minneapolis
MN 55455
USA

Y. Liu
4-225 Millard Hall
Department of Biochemistry
University of Minnesota
Minneapolis
MN 55455
USA

G. Malin
School of Environmental Sciences
University of East Anglia
Norwich
NR4 7TJ
UK

O. Meyer
Lehrstuhl fur Mikrobiologie
Universitat Bayreuth
Universitatsstraße 30
8580 Bayreuth
GERMANY

G. Morsdorf
Lehrstuhl fur Mikrobiologie
Universitat Bayreuth
Universitatsstraße 30
8580 Bayreuth
GERMANY

V. Muller
Institut fur Mikrobiologie der
  Georg-August-Universitat Gottingen
Grisebachstr. 8
W-3400 Gottingen
GERMANY

J.C. Murrell
Department of Biological Sciences
University of Warwick
Coventry
CV4 7AL
UK

A.I. Netrusov
Microbiology Department
Moscow State University
Moscow 119899
RUSSIA

H.T. Nguyen
A.A. Noyes Laboratory of Chemical Physics
California Institute of Technology
Pasadena
CA 91125
USA

R. Oldenhuis
Department of Biochemistry
University of Groningen
Nijenborgh 4
9747 AG Groningen
THE NETHERLANDS

C. O'Reilly
School of Health Sciences
University of Sunderland
Sunderland
SR1 3SD
UK

G.C. Paoli
Department of Microbiology and the
  Biotechnology Center
The Ohio State University
484 West 12th Avenue
OH 43210–1292
USA

D.M. Pearson
School of Health Sciences
University of Sunderland
Sunderland
SR1 3SD
UK

T.D. Pihl
Department of Microbiology
The Ohio State University
483 W. 12th Avenue
Columbus
OH 43210–1292
USA

P.C. Raemakers-Franken
Department of Microbiology
Faculty of Science
University of Nijmegen
Toernooiveld
NL-6525 ED Nijmegen
THE NETHERLANDS

List of contributors

B.A. Read
Department of Microbiology and the
  Biotechnology Center
The Ohio State University
484 West 12th Avenue
Columbus
OH 43210-1292
USA

W.S. Reeburgh
The Institute of Marine Science
University of Alaska
Fairbanks
Alaska 99775-1080
USA

J.N. Reeve
Department of Microbiology
The Ohio State University
483 W. 12th Avenue
Columbus
OH 43210-1292
USA

I.W. Richardson
SERC Centre for Molecular Recognition
Biochemistry Department
University of Southampton
Southampton
SO9 3TU
UK

J. Schaferjohann
Institut fur Mikrobiologie
Georg-August-Universitat Gottingen
Grisebachstr. 8
D-W-3400 Gottingen
GERMANY

M. Schmid-Appert
Mikrobiologisches Institut ETH
ETH-Zentrum
CH-8092
Zurich
SWITZERLAND

B. Schworer
Laboratorium fur Mikrobiologie
Fachbereich Biologie
Philipps-Universitat
Karl-von-Frisch-Straße
D-3550 Marburg/L.
GERMANY

A.K. Shiemke
A.A. Noyes Laboratory of Chemical Physics
California Institute of Technology
Pasadena
CA 91125
USA

V.J. Steigerwald
Department of Microbiology
The Ohio State University
483 W. 12th Avenue
Columbus
OH 43210-1292
USA

G. Sulter
Biological Center
University of Groningen
Kerklaan 30
9751 NN Haren
THE NETHERLANDS

F.R. Tabita
Department of Microbiology and the
  Biotechnology Center
The Ohio State University
484 West 12th Avenue
Columbus
OH 43210-1292
USA

K.C. Terlesky
Department of Microbiology and the
  Biotechnology Center
The Ohio State University
484 West 12th Avenue
Columbus
OH 43210-1292
USA

R.K. Thauer
Laboratorium fur Mikrobiologie
Fachbereich Biologie
Philipps-Universitat
Karl-von-Frisch-Straße
D-3550 Marburg/L.
GERMANY

M. Veehuis
Biological Center
University of Groningen
Kerklaan 30
9751 NN Haren
THE NETHERLANDS

J.W. Vrijbloed
Department of Microbiology
University of Groningen
Kerklaan 30
9751 NN Haren
THE NETHERLANDS

G.D. Vogels
Department of Microbiology
Faculty of Science
University of Nijmegen
Toernooiveld
NL-625 ED Nijmegen
THE NETHERLANDS

X. Wang
Department of Microbiology and the
  Biotechnology Center
The Ohio State University
484 West 12th Avenue

Columbus
OH 43210–1292
USA

D. Westenberg
Institut fur Mikrobiologie der
 Georg-August-Universitat Gottingen
Grisebachstr. 8
W-3400 Gottingen
GERMANY

R. Wever
E.C. Slater Institute for Biochemical Research
Plantage Muidergracht 12
1018 TV Amsterdam
THE NETHERLANDS

S.C. Whalen
The Institute of Marine Science
University of Alaska
Fairbanks
Alaska 99775–1080
USA

P.C. Wilkins
Department of Biological Sciences
University of Warwick
Coventry
CV4 7AL
UK

U. Windhovel
Institut fur Mikrobiologie
Georg-August-Universitat Gottingen
Grisebachstr. 8
D-W-3400 Gottingen
GERMANY

A.P. Wood
Division of Life Sciences
King's College London
London
W8 7AH
UK

J.-G. Yoo
Institut fur Mikrobiologie
Georg-August-Universitat Gottingen
Grisebachstr. 8
D-W-3400 Gottingen
GERMANY

C. Zirngibl
Laboratorium fur Mikrobiologie
Fachbereich Biologie
Philipps-Universitat
Karl-von-Frisch-Straße
D-3550 Marburg/L.
GERMANY

List of Participants

A.P.S. Adamsen
B. Alber
L. Alvarez-Cohen
D. Andrews
C. Anthony
N. Arfman
A. Avezoux
R. Bader
M. Baev
S. Baker
T.M. Barta
E. Bellion
F. Biville
M. Blaut
G. Borjesson
H. Botte
E. Boulygina
B. Bowien
J. Bowman
L. Bystrykh
H.N. Carlsen
M.A. Carver
C. Cavanaugh
C. Chan
S. Chan
A. Chistoserdov
L. Chistoserdova
M.L. Collins
A. Cornish
J. Cox
P. Daas
S. Dales
H. Dalton
S. Daniel
M. Davey
P. De Marco
L. Dijkhuizen
G. Gilworth
A. DiSpirito
C.S. Dow
H.L. Drake
J.A. Duine
E. Eisenstadt
R. J. Ellis
E.A. Elmorsi
P. Fareleira
J. Ferry
K. Fiebig
L. Florencio dos Santos
J. Frank
F. Gasser
J.L. Gibson
F. Girio
P.N. Green
P. Goodwin

R. Hanson
N. Harms
A. Hart
S. Heising
H. Hektor
T. Higgins
Y. Igarashi
N. Iversen
Y. Izumi
D.B. Janssen
Y. Jiang
L. Joergensen
C.W. Jones
S. Jordon
G. Jud
T. Kanagawa
D. Kelly
J. Keltjens
E. Kenna
R. Kiene
D. Kightley
Y. M. Kim
G. King
C.R. Klein
T. Kodama
J.-U. Kreft
D.A. Kunz
K. Lachtchev
M. Latimer
Th. Leisinger
S. Li
J.D. Lipscombe
M. Maclean
Gill Malin
G. Marchenko
H. Martin
I. McDonald
A. McEwan
W.G. Meijer
O. Meyer
J. Mills
K. Miniami
A. Miura
C. Morris
A. Moss
H. Motoyama
V. Muller
J. C. Murrell
D. Nedwell
J. Nesvera
A.I. Netrusov
J. Newbold
H-H. Nguyen
A. Nozhevnikova
E. Odintsova

List of participants

C. O'Reilly
M. Ostafin
D. Pearson
P. Peltola
R. Quayle
W.S. Reeburgh
J. Reeve
P. Roslev
Y. Sakai
G.P.C. Salmond
M. Schmid
S. Schnell
A. Schwartz
J. Semrau
S. Shima
M. Shimoda
D. Skladnev
L. Smith
B. Speer
A. Stainthorpe
D. Stax
R. Stettler
A. Stouthamer
E. Stupperich
G. Sulter
I. Sundh
B. Svensson

F.R. Tabita
S. Tate
R.K. Thauer
A. Thompson
J. Trickett
Y.A. Trotsenko
Y. Tsygankov
E. Turlin
H. Uchiyama
G-J. van Alebeek
R. van den Bergh
C. van der Drift
P. van Ophem
R. van Spanning
F. Vellieux
G. Vogels
A. Wasserfallen
D. Westenberg
R. Wever
P. Wilkins
C. Wirenfeldt Klaus
A. Wood
N. Woods
N. Wyborn
H. Xu
L. Zatman

Preface

The 7th International Symposium on Microbial Growth on C_1 Compounds was held at the University of Warwick, 15–20 August 1992. The meeting attracted some 165 delegates from 18 countries. The eight Scientific Sessions, running from Sunday 16 August to Wednesday 20 August, saw the presentation of 40 papers, along with 139 posters which were displayed and discussed by their authors in three separate sessions.

The Symposium marked nearly twenty years since 'Symposium zero' was convened in Edinburgh University in 1973. From that gathering of about 50 participants grew the 3-yearly symposium series, starting in Tokyo in 1974. A number of the 'founder members' of that Edinburgh meeting were still going strong at the 7th Symposium, demonstrating the addictive nature of the field! 1992 was also a particularly appropriate year for a symposium in which much of the interest focused on the metabolic transformations of methyl compounds, as it was the centenary year of Loew's isolation of '*Bacillus methylicus*' [1], an organism growing methylotrophically on methanol, formaldehyde and other now familiar C_1 substrates.

The 7th Symposium was agreed by its participants to be scientifically and socially a great success, and to have met the aim of providing a forum in which the stage of the art of one-carbon microbiology could be presented to a diverse audience. As in earlier meetings, all aspects of 'C_1' were covered, from aerobes to anaerobes, basic and applied, global and regional ecosystem impacts, molecular biology, taxonomy, energy conservation and carbon assimilation.

These ongoing symposia provide a useful historical record of the development of understanding of all aspects of C_1 microbiology and biochemistry, and the current one is no exception in providing the most up-to-date compilation of all the researcher and advanced student could wish to know about this field – until the next symposium, scheduled for San Diego, California, USA in 1995.

Some readers may be surprised at the use of the terms 'carboxidotrophy' and 'carboxidobacteria' to describe those bacteria able to grow on carbon monoxide as sole energy substrate, in the chapters by Meyer and O'Reilly. The more familiar prefix is 'carboxydo-' in use for many years (and indeed in specific names such as *carboxydovorans*), but this has no strict etymological justification. Its unsuitability for bacteria growing specifically on **carbo**-n mono-**xid**-e was recognized by Hans Schlegel, who used the 'i' rather than 'y' spelling in the 5th and 6th editions of his book [2–4]. It is notable that he was able to avoid this issue in the 4th edition [5] in which he referred only to Kohlenmonoxidoxidation, as the biochemistry and microbiology of these organisms was only put on a firm modern footing by the work of Zavarzin and Nozhevnikova [6] and

Meyer and Schlegel [7]. Interestingly, Schlegel and Meyer used 'carboxido-' as the prefix not only for the metabolic mode, but also for all the organism names in their contribution to the Third C_1 Symposium [8], but this had reverted to the original 'y' form in the chapters by Meyer and his colleagues in the Fourth and Sixth Symposia [9,10].

ACKNOWLEDGEMENTS

The 7th Symposium owed its successful organization and comprehensive programme to the combined efforts of its Organizing Committee and an international Scientific Advisory Panel. The Committee members were Howard Dalton (Chairman), Colin Murrell (Secretary and Symposium Editor), Chris Anthony, Crawford Dow, Don Kelly (Symposium Editor), Colin Jones, George Salmond, Carol Howes, and members of the Department of Biological Sciences of Warwick University. The Panel members were Gerhard Gottschalk, Dick Hanson, Wim Harder, K. Komagata, Mary Lidstrom, Ortwin Meyer, Yuri Trotsenko, and Rod Quayle.

The Symposium was greatly helped by the finanical support provided by the following organizations: Amoco Technology Co; Celgene Corporation; Dansk BioProtein A/S; DuPont Central Research & Development; General Electric Central Research & Development; ICI Pharmaceuticals; Idemitsu Kosan Co Ltd; International Union of Biochemistry and Molecular Biology; Nippon Mining Co Ltd; Phillips Petroleum Co; Shell Research Ltd; The Royal Society; The Society for General Microbiology; The Wellcome Trust; US Office of Naval Research and the University of Warwick.

REFERENCES

[1] Loew, I. (1892) Z. Bakteriol. 12, 462. Cited by Quayle, J.R. (1987) An eightieth anniversary of the study of microbial C_1 metabolism. In: Microbial Growth on C_1 Compounds (van Verseveld, H.W. and Duine, J.A., Eds), pp. 1–5, Nijhoff, Dordrecht, The Netherlands.
[2] Schlegel, H.G. (1981) Allgemeine Mikrobiologie, 5. Auflage, p. 351. Thieme, Stuttgart, Germany.
[3] Schlegel, H.G. (1985) Allgemeine Mikrobiologie, 6. Auflage, p. 363. Thieme, Stuttgart, Germany.
[4] Schlegel, H.G. (1986) General Microbiology, 6th edition (translated by Kogut, M.), p. 363, Cambridge University Press, Cambridge, England.
[5] Schlegel, H.G. (1976) Allgemeine Mikrobiologie, 4. Auflage, p. 303. Thieme, Stuttgart, Germany.
[6] Zavarzin, G.A. and Nozhevnikova, A.N. (1977) Aerobic carboxydobacteria. Microb. Ecol. 3, 305–326.
[7] Meyer, O. and Schlegel, H.G. (1983) Biology of the carbon monoxide oxidizing bacteria. Ann. Rev. Microbiol. 37, 277–310.

[8] Schlegel, H.G. and Meyer, O. (1981) Microbial growth on carbon monoxide, formate and hydrogen: a biochemical assessment. In: Microbial Growth on C_1 Compounds (Dalton, H., Ed.), pp. 105–115, Heyden, London, England.
[9] Meyer, O. and Rohde, M. (1984) Enzymology and bioenergetics of carbon monoxide-oxidizing bacteria. In: Microbial Growth on C_1 Compounds (Crawford, R.L. and Hanson, R.S., Eds), pp. 26–33, American Society for Microbiology, Washington, DC.
[10] Meyer, O., Frunzke, K., Gadkari, D., Jacobitz, S., Hugendieck, I. and Kraut, M. (1990) Utilization of carbon monoxide by aerobes: recent advances. *FEMS Microbiol. Rev.* **87**, 253–260.

October 1992

J. COLIN MURRELL
DON P. KELLY

Contents

List of contributors	v
List of participants	xiii
Preface	xv

1
The Role of Methylotrophy in the Global Methane Budget
W.S. Reeburgh, S.C. Whalen and M.J. Alperin 1

2
Microbial Sources and Sinks for Methylated Sulfur Compounds in the Marine Environment
R.P. Kiene 15

3
Sources and Sinks of Halogenated Methanes in Nature
R. Wever 35

4
Microbial Transformations and Biogeochemical Cycling of One-carbon Substrates Containing Sulphur, Nitrogen or Halogens
D.P. Kelly, G. Malin and A.P. Wood 47

5
Structure and Mechanism of Action of the Hydroxylase of Soluble Methane Monooxygenase
H. Dalton, P.C. Wilkins and Y. Jiang 65

6
The Catalytic Cycle of Methane Monooxygenase and the Novel Roles Played by Protein Component Complexes During Turnover
W.A. Froland, K.K. Andersson, S-K. Lee, Y. Liu and
J.D. Lipscombe   81

7
Biochemical and Biophysical Studies Toward Characterization of the Membrane-associated Methane Monooxygenase
S.I. Chan, H.T. Nguyen, A.K. Shiemke and M.E. Lidstrom   93

8
Molecular Biology of Methane Oxidation
J.C. Murrell   109

9
Degradation of Trichloroethylene by Methanotrophic Bacteria
R. Oldenhuis and D.B. Janssen   121

10
Methanopterin, its Structural Diversity and Functional Uniqueness
J.T. Keltjens, P.C. Raemakers-Franken and G.D. Vogels   135

11
Enzymes Involved in Methanogenesis from CO_2
R.K. Thauer, B. Schwörer and C. Zirngibl   151

12
Enzymology of the Methanogenic Fermentation of Acetate by *Methanosarcina thermophila*
B.E. Alber, A.P. Clements, P.E. Jablonski, M.T. Latimer and
J.G. Ferry   163

13
Mechanisms of Energy Conservation in Methanogenic Bacteria
M. Blaut, U. Deppenmeier, B. Kamlage, D. Westenberg,
B. Becher, V. Müller and G. Gottschalk   171

14
Genes Encoding the Methyl Viologen-reducing Hydrogenase, Polyferredoxin and Methyl Coenzyme M Reductase II are Adjacent in the Genomes of *Methanobacterium thermoautotrophicum* and *Methanothermus fervidus*
V.J. Steigerwald, A.N. Hennigan, T.D. Pihl and J.N. Reeve   181

15
A Molecular Analysis of Peroxisome Biogenesis and Function in the Methylotrophic Yeast *Hansenula polymorpha*
G. Sulter, M. Veenhuis and W. Harder   193

16
Structural Aspects of Methanol Oxidation in Gram-negative Bacteria
J. Frank, M. Janvier, I. Heiber-Langer, J.A. Duine, F. Gasser and
C. Balny   209

17
Methanol Dehydrogenase and Cytochrome Interactions
C. Anthony, H.T.C. Chan, J.M. Cox and I.W. Richardson   221

18
Genetics of Methanol Oxidation in *Paracoccus denitrificans*
N. Harms   235

19
The Methanol-oxidizing Enzyme Systems in Gram-positive Methylotrophic Bacteria
L.V. Bystrykh, N. Arfman and L. Dijkhuizen   245

20
Overview of the Current State of Methylotroph Taxonomy
P.N. Green   253

21
Taxonomy of Thermotolerant Methylotrophic Bacilli
N. Arfman and L. Dijkhuizen   267

22
Systematics of Gram-negative Methylotrophic Bacteria Based on 5S rRNA Sequences
E.S. Boulygina, K.M. Chumakov and A.I. Netrusov   275

23
Phylogeny and Ecology of Methylotrophic Bacteria
R.S. Hanson, B.J. Bratina and G.A. Brusseau   285

24
Ecophysiological Characteristics of Obligate Methanotrophic Bacteria and Methane Oxidation *in situ*
G.M. King   303

25
Methanotroph–Invertebrate Symbioses in the Marine Environment: Ultrastructural, Biochemical and Molecular Studies
C.M. Cavanaugh   315

26
L-Phenylalanine Synthesis by the Facultative RuMP Cycle Methylotroph *Amycolatopsis methanolica*
L. Dijkhuizen, G.J.W. Euverink, G.I. Hessels and J.W. Vrijbloed   329

27
Application of Methanotrophic Oxidations for the Bioremediation of Chlorinated Organics
L. Alvarez-Cohen   337

28
Chlorinated Methanes as Carbon Sources for Aerobic and Anaerobic Bacteria
T. Leisinger, S. La Roche, R. Bader, M. Schmid-Appert, S. Braus-Stromeyer and A.M. Cook   351

29
Production and Applications of Amidase from *Methylophilus methylotrophus*
M.A. Carver and C.W. Jones   365

30
Molecular Biology and Genetics of Methylamine Dehydrogenases
M.E. Lidstrom and A.Y. Chistoserdov   381

31
New Developments in the Biochemistry of Quinoproteins in Methylotrophs
J.A. Duine   401

32
Methylamine Utilization in Yeast and Bacteria: Studies Using *in vivo* NMR
E. Bellion and J.G. Jones   415

33
Biochemistry of the Aerobic Utilization of Carbon Monoxide
O. Meyer, K. Frunzke and G. Mörsdorf   433

34
Molecular Genetics of Carbon Monoxide Dehydrogenase
C. O'Reilly, J. Colby, D.M. Pearson and G.W. Black   461

35
Current Studies on the Molecular Biology and Biochemistry of CO_2 Fixation in Phototrophic Bacteria
F.R. Tabita, J.L. Gibson, D.L. Falcone, X. Wang, L-A. Li, B.A. Read, K.C. Terlesky and G.C. Paoli   469

36
Genetic Regulation of CO_2 Assimilation in Chemoautotrophs
B. Bowien, R. Bednarski, B. Kusian, U. Windhövel, A. Freter, J. Schäferjohann and J-G. Yoo   481

37
CO_2, Reductant, and the Autotrophic Acetyl-CoA Pathway: Alternative Origins and Destinations
H.L. Drake   493

Species Index   509

Subject Index   513

1

The Role of Methylotrophy in the Global Methane Budget

W.S. Reeburgh[1,2], S.C. Whalen[1] and M.J. Alperin[3]

[1]*Institute of Marine Science, University of Alaska, Fairbanks, Alaska 99775-1080, USA;* [2]*Department of Geosciences, University of California, Irvine, CA 92717, USA;* [3]*Curriculum in Marine Sciences, University of North Carolina, Chapel Hill, North Carolina 27599-3300, USA*

1 SUMMARY

Studies dealing with the present atmospheric CH_4 increase and its climate consequences have focused almost entirely on the atmospheric CH_4 budget. The atmospheric CH_4 budget involves a framework of atmospheric burden, residence time, isotope, and model constraints. Net CH_4 emission from a variety of sources is nearly balanced by photochemical oxidation in the atmosphere.

Microbially-mediated CH_4 oxidation occurs in soil and aquatic environments, where it modulates emission and is a possible negative feedback on atmospheric CH_4 increases. We extend existing CH_4 oxidation rate measurements to an estimate of global CH_4 oxidation and find that microbially-mediated oxidation in major atmospheric CH_4 budget source terms is about 700 Tg CH_4 yr^{-1}, 200 Tg yr^{-1} larger than the 500 Tg yr^{-1} emitted to the atmosphere. Soil processes account for approximately 80% of the global CH_4 oxidation.

2 INTRODUCTION

Microbially-mediated CH_4 oxidation has been studied in environments ranging from marine and wetland sediments to desert soils. The process has been recognized and suggested as globally-important [1,2], but there have been no

---

Microbial Growth on C_1 Compounds
© Intercept Ltd, PO Box 716, Andover, Hampshire SP10 1YG, UK

attempts at consolidating the diverse data into a global estimate of CH_4 oxidation. Recent reports dealing with the atmospheric CH_4 increase [3-8], its climate consequences [9,10], and strategies for decreasing CH_4 emission [11,12] have focused almost entirely on the atmospheric CH_4 budget (actually an emission budget nearly balanced by photochemical oxidation). Recent atmospheric CH_4 budgets consider photochemical oxidation as the major sink [6-8], and are based on constraints involving the atmospheric burden, residence time, and isotope (^{14}C, ^{13}C) distributions of CH_4. New information on source strengths can be confirmed and new photochemical reaction rates [13] are easily applied within this framework. Since the observed CH_4 concentration increase is occurring in the atmosphere, the locus of greenhouse warming, and since most information on atmospheric CH_4 sources results from direct measurements of net emission, the current emphasis on the atmospheric budget is appropriate. The overall atmospheric CH_4 budget is reasonably well-constrained; individual terms in the budget, however, are subject to wide ($\sim$2-fold or more) ranges [7].

Understanding the atmospheric CH_4 budget under present conditions and predicting changes in various terms under altered climate requires information in processes that affect CH_4 prior to and after emission to the atmosphere. The present emphasis on the atmospheric CH_4 budget must be extended to the global gross budget. Global gross production is the quantity of CH_4 entering the atmosphere (net emission) plus global oxidation (photochemical and microbial).

Unfortunately, the budget constraints used in the atmospheric CH_4 budget do not apply for estimates of gross CH_4 production and CH_4 oxidation; neither can be constrained or checked independently. *A priori* estimates of gross global CH_4 production are not possible; measurements of gross CH_4 production are possible only for sources that introduce CH_4 directly to the atmosphere, such as biomass burning, natural gas flaring, coal production, and enteric production. Gross CH_4 production has been estimated indirectly from measurements of potential production [14-16]. These involve determining the accumulation rate of CH_4 in systems isolated from consumption and emission. However, it is unclear whether experimental manipulation biases these estimates. A specific inhibitor for methane oxidation, methylfluoride (CH_3F), has been described recently [17], and has the potential of providing more direct estimates of CH_4 oxidation.

3 METHANE OXIDATION RATE MEASUREMENTS

Microbially-mediated CH_4 oxidation rates have been measured for several important source terms in the atmospheric CH_4 budget. Microbially-mediated aerobic and anaerobic CH_4 oxidation occur largely at interfaces and in zones adjacent to interfaces [18,19], and thus play important roles as 'biofilters' [20] in modulating CH_4 emission. Aerobic CH_4 oxidizers are reasonably well-understood and characterized, but their global importance is not well known [21–23]. Anaerobic CH_4 oxidation is important in anoxic marine environments [19,24–27], but the responsible organism has not been cultured and the mechanism is unknown [28]. Direct measurement of CH_4 oxidation rates is possible using $^{14}CH_4$ or $C^{13}H_4$ tracers [24,27,28]. Static flux chamber measurements and jar experiments may be used to estimate net CH_4 consumption [18,29] in instances where production is zero or consumption exceeds production.

4 GLOBAL ESTIMATE OF MICROBIALLY-MEDIATED CH_4 OXIDATION

The above CH_4 oxidation rate measurements are few and have not been consolidated or extended to an estimate of global oxidation. This review summarizes available microbial CH_4 oxidation rate measurements and makes a first attempt at extending them to a global oxidation estimate, highlighting information gaps. We use terms in the atmospheric CH_4 budget [6,8] as a framework for compiling and extending available CH_4 oxidation rate measurements. Using the atmospheric CH_4 budget as a classifying framework for summing independent consumption estimates highlights uncertainties and provides insight for future research. We acknowledge that this approach propagates uncertainties in individual terms of the atmospheric budget to the gross budget.

We estimate global gross CH_4 production by adding global CH_4 emission (net production) to measurements and estimates of global CH_4 oxidation (consumption) in categories in the current atmospheric CH_4 budget. Table 1 summarizes recent estimates of CH_4 oxidation in atmospheric budget source terms, dividing entries into natural and man-influenced sinks. The range and average are reported when possible. Consumption is zero for sources that introduce CH_4 directly to the atmosphere, namely, biomass burning, natural gas flaring, coal production and enteric production.

5 NATURAL CH_4 SINKS

5.1 *Wetlands*

Vascular plants play an important role in facilitating CH_4 transport from waterlogged anoxic soils to the atmosphere [30–33]. While most of the emphasis has been on CH_4 transport to the atmosphere, vascular plants also play an important and simultaneous role in transporting O_2 and other oxidants to the root zone in soils and sediments, maintaining oxic conditions within the roots and promoting oxidation of reduced species [32–33]. Several mechanisms for gas transport in vascular plants have been identified: diffusion along concentration gradients and active mass transport driven by pressure differences

Table 1 Microbially-mediated CH$_4$ oxidation

Sink	CH$_4$ Oxidation, Tg yr^{-1}	Comment, Reference
Natural		
Wetlands	27	30% of production oxidized; weighted value from [34]
Bogs/tundra (boreal)	15	85% of emission by bubbles [36]; Assume oxidation equals diffusive loss
Swamps/alluvial (tropical)	12	
Termites	24	55% of production oxidized [37]
Ocean	70.3 (35.5–105)	
Shelf sediments	55–99.5	>99% of production oxidized anaerobically [24]
Water column	30–52	Sulfate reduction-based mode [25]
Anoxic basins	0.5	[27,42]
Aerobic waters	5	[85]
Freshwaters	5?	Assume oxidation equals emission [49]
Hydrates	40	Assume consumption equals production
Soils		
Model estimate	40	[8]
Global estimate (vegetation)	5.6–58.2	[57]
Global estimate (soil texture)	28.7	[64]
Regional estimates		[29,57]
Boreal forest	0.8; 1–15.6	[57,61,62,73]
Temperate forest	0.4–7.3; 0.9–1.3	

Temperate grassland	0.5–5.6; 0.9–1.3	[57,60]
Savannah	21; 2.9–10.5	[37,57]
Woodland/shrubland	0.6–10.4	[57]
Tropical forest	0–1.5	[63]
Cultivated land	0.3–2.8	[57]
Desert	7	[74]
Total Natural	171.3	
Man-influenced		
Animals	0	Direct release to atmosphere
Rice Culture	477 (43–910)	
	247–900	45–90% of production oxidized [12]
	138	58% of production oxidized [11]
	400	80% of production oxidized [10]
	33	25% of production oxidized (K. Inubushi, unpub.)
	10	Assumption based on soil and wetland rates
Fallow field oxidation		
Gas production	18	Assume 60% of amount leaked is oxidized
Distribution leaks	0	Direct release to atmosphere
Flaring	22 (4–40)	10–50% (30% av.) of production is oxidized [65,70]
Landfills	0	Direct release to atmosphere
Coal production	0	Direct release to atmosphere
Biomass burning		
Total man-influenced	517	
Total oxidation	688.3	

(thermo-osmosis, negative pressures resulting from root O_2 consumption, and positive pressures from equilibration with soil CH_4, resulting in effusion) [32]. These mechanisms are specific to particular plants and stages of development. The effectiveness of these transport mechanisms varies widely with plant structure, so that broad generalizations are not possible. Information on CH_4 oxidation in the root zone is quite limited, but it is clear that large and unquantified amounts of CH_4 are oxidized as a result of vascular transport of O_2. Half of the CH_4 production in moss areas of the tundra is oxidized [34] and potential CH_4 oxidation rates adjacent to small roots in wetland sites are 2 to 10-fold greater than emission rates [35]. We consider that 30% of the CH_4 production is oxidized in high latitude wetlands. A large fraction of the CH_4 production in tropical systems enters the atmosphere by ebullition [36], and thus escapes oxidation. The tropical wetland oxidation entry in Table 1 reflects these bubble losses.

5.2   Termites

Methane oxidation has been observed in tropical soils adjacent to termite colonies [37]. The termite emission term is one of the most uncertain in the atmospheric budget [7]; we use 55% of the termite emission term as an estimate of oxidation.

5.3   Ocean

The ocean is a large potential source of CH_4, but is a relatively small term in the net CH_4 budget [7,8]. Anaerobic oxidation in organic-rich high deposition shelf sediments [25,26] and aerobic oxidation at the pycnocline [25,38] effectively consume CH_4 generated within and transported through the ocean. Anaerobic CH_4 oxidation occurs in a subsurface zone in shelf sediments and consumes virtually all CH_4 generated within the sediments [25]. Rate measurements from widely separated shelf environments [24,39,40] agree within a factor of two. Water column anaerobic oxidation is the dominant sink term in permanently anoxic marine basins like the Black Sea [27] and the Cariaco Trench [41,42]. Aerobic CH_4 oxidation rates in the ocean water column range over a factor of 10^7 [23]. We consider that CH_4 emitted from continental shelf seeps [43] is oxidized by *in situ* aerobic processes, based on the occurrence of isotopically light (CH_4-derived) carbonate cements and the presence of CH_4-consuming communities adjacent to vents [44,45]. The ocean's large capacity for CH_4 oxidation and the presence of both anaerobic and aerobic CH_4 oxidizing zones can be expected to effectively control internal CH_4 additions. The ocean should play a negligible role in consuming atmospheric CH_4 because of slow atmosphere:ocean gas exchange rates and atmosphere:ocean concentration gradients that will decrease as atmospheric concentrations increase. Rising sea levels should suppress CH_4 emissions from coastal wetlands due to decreased CH_4 production resulting from competition by sulfate reducers [46] and/or anaerobic oxidation [19].

5.4 Freshwaters

Methane oxidation in freshwaters occurs at the sediment:water interface, at the oxycline in stratified lakes, and throughout the water column during seasonal overturn [47,48]. It also occurs in sediments adjacent to vascular plants along lake margins [32]. A recent review [49, Table 3] suggests that CH_4 oxidation in freshwaters equals emission.

5.5 Hydrates

Decomposition of CH_4 hydrates (clathrates) has been suggested as a potentially significant atmospheric CH_4 source. Large ($\sim 10,000$ Gt C) deposits of CH_4 hydrates occur beneath terrestrial permafrost regions, in sediments of the Arctic continental shelves, and in ocean sediments at the base of the continental slope [50]. A hydrate decomposition term appears as a 'placeholder' [17] in most net CH_4 budgets because of its uncertainty. Hydrate decomposition is frequently cited as a possible positive feedback to global warming [7,50] and scenarios suggesting release of large quantities of CH_4 through gradual [50] or sudden hydrate decomposition as a result of warmer climate have been proposed [51-55]. Methane release by decomposing hydrates will be controlled by hydrate melting rates, which are controlled by heat transport, not temperature alone. Hydrates occur in sites well-removed from potential rapid temperature changes and are subject to small temperature gradients. The hydrates most susceptible to decomposition are those in permafrost areas inundated by the post-glacial sea-level rise (Arctic shelves). Heat transport (sensible plus latent heat of fusion) to most hydrates will be so slow that rapid decomposition and large CH_4 releases are unlikely. Methane oxidation by microbial communities [44,45] adjacent to ocean vents should keep pace with releases. No radiatively active gas additions to the atmosphere will result, as the product of CH_4 oxidation, CO_2, will dissolve in waters adjacent to oxidation sites. We assume that oxidation equals production for the hydrate source.

5.6 Soils

Methane oxidation by soils has been regarded as a negligible to small term in the net global CH_4 budget [6,7]. This view resulted from laboratory kinetic studies on pure cultures that indicate atmospheric CH_4 concentrations are too low for microbial growth with CH_4 as the sole substrate [56] and the notion that resistance limits transport of CH_4 into soils. However, recent measurements from a wide variety of environments [18,37,56-63,73,74] indicate that soils may be an atmospheric CH_4 sink of 20-60 Tg yr^{-1}. A global soil sink estimate (wetlands excluded) of 28.7 Tg yr^{-1} results from parameterizing soil CH_4 oxidation with soil texture [64]. Soil CH_4 oxidizing communities respond rapidly to CH_4 additions with no lag time, and have low thresholds (~ 0.2 ppm) and high capacities (uptake saturation at concentrations >3000 ppm) for CH_4 oxidation [18,64]. Transport and reaction kinetics both limit CH_4 consumption in moist soils. The rate constant for CH_4 equilibration (or transport) in moist soils is some 100-fold faster than the rate constant for microbial oxidation at atmospheric concentrations [18]. Methane oxidation in soils occurs over a texture-dependent moisture range. Oxidation ceases under dry conditions [65] and is diffusion-limited in waterlogged soils [65,66]. Small decreases (~ 10–20 cm) in wetland water table levels are sufficient to balance CH_4 emission and oxidation [34,66]. Recent reports [60,61,63] indicate that disturbance (cultivation, fertilization) suppresses soil CH_4 oxidation by some 30%, and that these effects may persist for decades to centuries.

6 MAN-INFLUENCED CH_4 SINKS

6.1 *Rice culture*

Oxidation in rice paddies is the largest and most uncertain CH_4 oxidation term, in part because it is expressed as a fraction of a large and uncertain net budget term. Estimates of CH_4 consumption in rice paddies range from 25% to 90% of production. It is not clear whether this large range results from seasonal variability, from differences in measurement of CH_4 oxidation rates or agricultural practices (cultivar selection, management of fertilizer and water). Rice plants may have lower CH_4 emission capacity than other wetland plants because of their hard and less permeable root and rhizome cortex tissues [32]. Direct tracer measurements of CH_4 oxidation and CH_4 inhibition experiments are needed in rice paddies to confirm the present indirect (potential production minus emission) estimates. Year-round measurements of CH_4 oxidation are also needed, as the rice paddy emission estimates are derived from growing season data only; fallow-season CH_4 consumption measurements in unflooded paddy soils are unavailable. Methane oxidation associated with rice culture should be a high priority for future study.

6.2 *Natural gas production*

The emission term for natural gas distribution leaks is an indirect estimate based on 'unaccounted for' natural gas [8], and should be considered a gross production term. Methane from leaks in buried transmission lines is effectively oxidized in soils adjacent to the leaks [67]; we estimate that all but the largest leaks are oxidized and use 60% of the overall term for an oxidation estimate in Table 1.

Table 2 Net atmospheric CH$_4$ emission, global consumption and gross global production, Tg CH$_4$yr^{-1}

Source/sink term	E Net atmospheric budget[a]	+	C Global consumption[b]	=	P Global production
Animals	80		0		80
Wetlands	115		27		142
bogs/tundra (boreal)	35		15		50
swamps/alluvial	80		12		92
Rice Production	100		477		577
Biomass burning	55		0		55
Termites	20		24		44
Landfills	40		22		62
Oceans, freshwaters	10		75.3		85.3
Hydrates	5?		5		10
Coal production	35		0		35
Gas production	40				40
venting, flaring	10		0		10
distribution leaks[c]	30		18		58
Total sources	500[d]				
Chemical destruction	−450				
Soil consumption	−10		40		40[e]
Total sinks	−460[d]		688.3		1188.3
			Total production		

[a] Ref. 8, scenario 7
[b] From Table 1
[c] Should be considered P
[d] 500−460 = 40 Tg CH$_4$ yr^{-1} = annual atmospheric (0.9% yr^{-1}) increment
[e] Soil consumption of atmospheric CH$_4$ added to the gross budget as an equivalent production term

6.3 Landfills

Landfills have received limited study as a CH_4 budget term [68]. Extensive oxidation occurs in landfill cover soils [69], consuming 10 to 50% of gross landfill CH_4 production [65,70]. Landfills are an uncertain, large source term, and should also be a high priority for study.

7 SUMMARY

Table 2 gives a recent net atmospheric budget [8], summarizes the consumption terms from Table 1 and gives a gross CH_4 budget, which results from adding consumption (Table 1) to the corresponding net atmospheric CH_4 budget terms. Since moist soils consume atmospheric CH_4 which escaped microbial oxidation as it entered the atmosphere, this term is entered as an equivalent production term in the gross budget. Table 2 shows that global microbially-mediated CH_4 oxidation is about 200 Tg yr^{-1} larger than emission to the atmosphere; the present emphasis on net sources neglects approximately 60% of the gross CH_4 budget. The view that CH_4 oxidation by soils is negligible [6–8] is largely a classification artifact resulting from the choice of the source terms employed in the net atmospheric CH_4 budget. Methane oxidation associated with wetlands, rice production, termites, landfills and gas production – all soil processes – accounts for over 80% of global CH_4 consumption.

The CH_4 oxidation estimates compiled here are based on limited data. Although the values cannot be constrained, they are conservative and the terms whose values are guesses are small. Methane oxidation is capable not only of modulating net emission, but may also provide an important negative feedback on future CH_4 increases in wetland and soil environment [18,34]. Indications that wetland CH_4 releases are more sensitive to changes in water table level than temperature [34,66] make predictions of future CH_4 budgets based on temperature increases alone [71,72] questionable. Seasonal CH_4 oxidation measurements in rice paddies and wetlands should be incorporated in future budget studies to improve our ability to predict the effects of climate change on the CH_4 budget. Attention to CH_4 oxidation in rice paddies, wetlands, landfills and soils adjacent gas transmission leaks may provide fruitful means of mitigating future atmospheric CH_4 increases.

ACKNOWLEDGEMENTS

This work was supported by NSF, NASA and EPA.

REFERENCES

[1] Higgins, I.J., Best, D.J., Hammond, R.C. and Scott, D. (1980) New findings in methane-utilizing bacteria highlight their importance in the biosphere and their commercial potential. *Nature* **286**, 561–564.

[2] Large, P.J. (1983) Methylotrophy and Methanogenesis. pp. 1–10, American Society for Microbiology, Washington, DC.
[3] Blake, D.R. and Rowland, F.S. (1988) Continuing worldwide increase in tropospheric methane. *Science* **239**, 1129–1131.
[4] Khalil, M.A.K. and Rasmussen, R.A. (1983) Sources, sinks, and seasonal cycles of atmospheric methane. *J. Geophys Res.* **88**, 5131–5144.
[5] Steele, L.P., Fraser, P.J., Rasmussen, R.A., Khalil, M.A.K., Conway, T.J., Crawford, A.J., Masaie, R.H. and Thoning, K.W. (1987) The global distribution of methane in the troposphere. *J. Atmos. Chem.* **5**, 125–171.
[6] Ehhalt, D. (1974) The atmospheric cycle of methane. *Tellus* **26**, 58–70.
[7] Cicerone, R.J. and Oremland, R.S. (1988) Biogeochemical aspects of atmospheric methane. *Global Biogeochem. Cycles* **2**, 299–327.
[8] Fung, I., John, J., Lerner, J., Matthews, E., Prather, M., Steele, L.P. and Fraser, P.J. (1991) Three-dimensional model synthesis of the global methane cycle. *J. Geophys. Res.* **96**, 13,033–13,065.
[9] Mitchell, J.F.B. (1989) The greenhouse effect and climate. *Rev. Geophys.* **27**, 115–139.
[10] Ramanathan, V., Callis, L., Cess, R., Hansen, J., Isaksen, I., Kuhn, W., Lacis, A., Luther, F., Mahlman, R., Reck, R., and Schliesinger, M. (1987) Climate-chemical interactions and the effect of changing atmospheric trace gases. *Rev. Geophys.* **26**, 1441–1482.
[11] Watson, R.T., Rodhe, H., Oeschger, H. and Siegenthaler, U. (1990) Greenhouse Gases and Aerosols in Climate Change: The IPCC Scientific Assessment. (Houghton, J.T., Jenkins, G.J. and Ephraums, J.J., Eds) pp. 1–40, Cambridge University Press.
[12] Hogan, K.B., Hoffman, J.S. and Thompson, A.M. (1991) Methane on the greenhouse agenda. *Nature* **354**, 181–182.
[13] Vaghjiani, G.L. and Ravishankara, A.R. (1991) New measurement of the rate coefficient for the reaction of OH with methane. *Nature* **350**, 406–409.
[14] Conrad, R. and Rothfuss, F. (1991) Methane oxidation in the soil surface layer of a flooded rice field and the effect of ammonium. *Biol. Fertil. Soils* **12**, 28–32.
[15] Sass, R.L., Fisher, F.M., Harcombe, P.A. and Turner, F.T. (1990) Methane producion and emission in a Texas rice field. *Global Biogeochem. Cycles* **4**, 47–68.
[16] Schütz, H., Seiler, W. and Conrad, R. (1989) Processes involved in formation and emission of methane in rice paddies. *Biogeochem* **7**, 33–53.
[17] Oremland, R.S. and Culbertson, C.W. (1992) Importance of methane-oxidizing bacteria in the methane budget as revealed by the use of a specific inhibitor. *Nature* **356**, 421–423.
[18] Whalen, S.C. and Reeburgh, W.S. (1990) Consumption of atmospheric methane by tundra soils. *Nature* **346**, 160–162.
[19] Reeburgh, W.S. (1980) Anaerobic methane oxidation: Rates and rate depth distributions in Skan Bay sediments. *Earth. Planet. Sci. Lett.* **47**, 345–352.
[20] Galchenko. V.F., Lenin, A. and Ivanov, M. (1989) Biological sinks of methane. In: Exchange of Trace Gases between Terrestrial Ecosystems and the Atmosphere (Andreae, M.O. and Schimel, D.S., Eds) pp. 59–71, John Wiley, New York.
[21] Hanson, R.S. (1980) Ecology and diversity of methylotrophic organisms. *Adv. App. Microbiol.* **26**, 3–39.
[22] Higgins, I.J., Best, D.J., Hammond, R.C. and Scott, D. (1981) Methane-oxidizing microorganisms. *Microbiol. Rev.* **45**, 556–590.
[23] Topp, E. and Hanson, R.S. (1991) Metabolism of radiatively active trace gases by methane-oxidizing bacteria. In: Microbial Production and Consumption of Greenhouse Gases: Methane, Nitrogen Oxides and Halomethanes (Rogers, J.E. and Whitman, W.E., Eds) pp. 71–90, American Society for Microbiology, Washington, DC.

[24] Alperin, M.J. and Reeburgh, W.S. (1984) Geochemical observations supporting anaerobic methane oxidation. In: Microbial Growth on C-1 Compounds (Crawford, R. and Hanson, R., Eds) pp. 282–289, American Society for Microbiology, Washington, DC.
[25] Henrichs, S.M. and Reeburgh, W.S. (1987) Anaerobic mineralization of marine sediment organic matter: Rates and the role of anaerobic processes in the oceanic carbon economy. *Geomicrobiol. J.* **5**, 191–237.
[26] Reeburgh, W.S. (1983) Rates of biogeochemical processes in anoxic sediments. *Ann. Rev. Earth Planet. Sci.* **11**, 269–98.
[27] Reeburgh, W.S., Ward, B.B., Whalen, S.C., Sandbeck, K.A., Kilpatrick, K.A. and Kerkhof, L.J. (1991) Black Sea methane geochemistry. *Deep-Sea Res.* **38**, S1189–S1210.
[28] Alperin, M.J. and Reeburgh, W.S. (1984) Inhibition experiments on anaerobic methane oxidation. *Appl. Environ. Microbiol.* **50**, 940–945.
[29] Whalen, S.C., Reeburgh, W.S. and Kizer, K. (1991) Methane consumption and emission from taiga sites. *Global Biogeochem. Cycles* **5**, 261–274.
[30] Sebacher, D.I., Harriss, R.C. and Bartlett, K.B. (1985) Methane emission to the atmosphere through aquatic plants. *J. Environ. Qual.* **14**, 40–46.
[31] Aslemann, I. and Crutzen, P.J. (1989) Freshwater wetlands: Global distribution of natural wetlands and rice paddies, their net primary productivity, seasonality and possible methane emissions. *J. Atmos. Chem.* **8**, 307–358.
[32] Schütz, H., Schröder, P. and Rennenberg, H. (1991) Role of plants in regulating the methane flux to the atmosphere. In: Gas Emissions from Plants (Mooney, H., Holland, E. and Sharkey, T., Eds) pp. 29–63, Academic Press, San Diego.
[33] Chanton, J.P. and Dacey, J.W.H. (1991) Effect of vegetation on methane flux, reservoirs and isotopic composition. In: Gas Emissions from Plants (Mooney, H., Holland, E. and Sharkey, T., Eds) pp. 65–92, Academic Press, San Diego.
[34] Whalen, S.C., Reeburgh, W.S. and Reimers, C.E. (1992) Processes controlling methane fluxes from tundra environments. In: Landscape Function: Implications for Ecosystem Response to Disturbance. A Case Study in Arctic Tundra (Reynolds, J.F. and Tenhunen, J.D., Eds), Springer-Verlag, New York, in press.
[35] Gerard, G. (1992) Methane oxidation associated with aquatic macrophytes, M.S. Thesis, Florida State University, Tallahassee, FL.
[36] Devol, A.H., Richey, J.E., Clark, W.A., King, S.A. and Martinelli, L.A. (1988) Methane emissions to the troposphere from the Amazonian floodplain. *J. Geophys. Res.* **93**, 1583–1592.
[37] Seiler, W., Conrad, R. and Scharffe, D. (1984) Field studies of methane emission from termite nests into the atmosphere and measurements or methane uptake by tropical soils. *J. Atmos. Chem.* **1**, 171–186.
[38] Ward, B.B., Kilpatrick, K.K., Wopat, A.E., Minnich, E.C. and Lidstrom, M.E. (1989) Methane oxidation in Saanich Inlet during summer stratification. *Cont. Shelf Res.* **9**, 65–75.
[39] Devol, A.H. (1983) Methane oxidation rates in the anaerobic sediments of Saanich Inlet. *Limnol. Oceanogr.* **28**, 738–742.
[40] Iversen, N. and Jørgensen, B.B. (1985) Anaerobic methane oxidation rates at the sulfate-methane transition in marine sediments from the Kattegat and Skaggerrak (Denmark). *Limnol. Oceanogr.* **30**, 944–955.
[41] Scranton, M.I. (1988) Temporal variations in the methane content of the Cariaco Trench. *Deep-Sea Res.* **35**, 1511–1523.
[42] Ward, B.B., Kilpatrick, K.A., Novelli, P.C. and Scranton, M.I. (1987) Methane oxidation and methane fluxes in the ocean surface layer and in deep anoxic waters. *Nature* **327**, 226–229.
[43] Hovland, M. and Judd, A.G. (1988) Seabed Pockmarks and Seepages: Impact on Geology, Biology and the Marine Environment, Graham & Trotman, London.

[44] Martens, C.S., Chanton, J.P. and Paull, C.K. (1991) Biogenic methane from abyssal brine seeps at the base of the Florida escarpment. *Geology* **19**, 851–854.
[45] Childress, J.J., Fischer, C.R., Brooks, J.M., Kennicutt, M.C., II, Bidigare, R. and Anderson, A.E. (1986) A methanotrophic marine molluscan (Bivalvia: Mytilidae) symbiosis: Mussels fueled by gas. *Science* **233**, 1306–1308.
[46] Bartlett, K.B., Bartlett, D.S., Harriss, R.C. and Sebacher, D.I. (1987) Methane emissions along a salt marsh gradient. *Biogeochemistry* **4**, 183–202.
[47] Lidstrom, M.E. and Somers, L. (1984) Seasonal study of methane oxidation in Lake Washington. *Appl. Environ. Microbiol.* **47**, 1255–1260.
[48] Rudd, J.W.M. and Taylor, C.D. (1980) Methane cycling in aquatic environments. *Adv. Aquatic Microbiol.* **2**, 77–150.
[49] Kiene, R.P. (1991) Production and consumption of methane in aquatic systems. In: Microbial Production and Consumption of Greenhouse Gases: Methane, Nitrogen Oxides and Halomethanes (Rogers, J.E. and Whitman, W.E., Eds) pp. 111–146, American Society for Microbiology, Washington, DC.
[50] Kvenvolden, K.A. (1988) Methane hydrates and global climate. *Global Biogeochem. Cycles* **2**, 221–230.
[51] Lashof, D.A. (1989) The dynamic greenhouse: feedback processes that may influence future concentrations of atmospheric trace gases and climatic change. *Climatic Change* **14**, 213–242.
[52] Leggett, J. (1990) The nature of the greenhouse threat. In: Global Warming, The Greenpeace Report (Leggett, J., Ed.) pp. 14–43, Oxford University Press, Oxford.
[53] MacDonald, G.J. (1990) The role of methane clathrates in past and future climates. *Climatic Change* **16**, 247–281.
[54] Nisbet, E.G. (1989) Some northern sources of atmospheric methane: production, history, and future implications. *Can. J. Earth Sci.* **26**, 1603–1611.
[55] Revelle, R.R. (1983) Methane hydrates in continental slope sediments and increasing carbon dioxide. In: Changing Climate, pp. 252–261, National Academy of Science Press, Washington, DC.
[56] Conrad, R. (1984) Capacity of aerobic microorganisms to utilize and grow on atmospheric trace gases (H_2, CO, CH_4). In: Current Perspectives in Microbial Ecology (Klug, M.J. and Reddy, C.A., Eds) pp. 461–467, American Society for Microbiology, Washington, DC.
[57] Born, M., Dörr, H. and Levin, I. (1990) Methane consumption in aerated soils of the temperate zone. *Tellus* **42B**, 2–8.
[58] Harriss, R.C., Sebacher, D.I. and Day, F.P., Jr. (1982) Methane flux in the Great Dismal Swamp. *Nature* **297**, 673–674.
[59] Keller, M., Goreau, T.J., Wofsey, S.C., Kaplan, W.A. and McElroy, M.B. (1983) Production of nitrous oxide and consumption of methane by forest soils. *Geophys. Res. Lett.* **10**, 1156–1159.
[60] Mosier. A., Schimel, D., Valentine, D., Bronson, K. and Parton, W. (1991) Methane and nitrous oxide in native, fertilized and cultivated grasslands. *Nature* **350**, 330–332.
[61] Steudler, P.A., Bowden, R.D., Melillo, J.M. and Aber, J.D. (1989) Influence of nitrogen fertilization on methane uptake in temperate forest soils. *Nature* **341**, 314–316.
[62] Yavitt, J.B., Downey, D.M., Lang, G.E. and Sextone, A.J. (1990) Methane consumption in two temperate forest soils. *Biogeochemistry* **9**, 39–52.
[63] Keller, M., Mitre, M.E. and Stallard, R.F. (1990) Consumption of methane in soils of central Panama: Effects of agricultural development. *Global Biogeochem. Cycles* **4**, 21–27.
[64] Dörr, H., Katruff, L. and Levin, I. (1992) Soil texture parameterization of the methane uptake in aerated soils. *Chemosphere*, in press.
[65] Whalen, S.C., Reeburgh, W.S. and Sandbeck, K.A. (1990) Rapid methane oxidation in a landfill cover soil. *Appl. Environ. Microbiol.* **56**, 3405–3411.

[66] Roulet, N., Moore, T., Bubier, J. and Lafleur, B. (1992) Northern fens: methane flux and climatic change. *Tellus* **44B**, 100–105.
[67] Hoeks, J. (1972) Changes in composition of soil air near leaks in natural gas mains. *Soil Science* **113**, 46–54.
[68] Bingemer, H.G. and Crutzen, P.J. (1987) The production of methane from solid wastes. *J. Geophys. Res.* **92**, 2181–2187.
[69] Jones. H.A. and Nedwell, D.B. (1990) Soil atmosphere concentration profiles and methane emission rates in the restoration covers above landfill sites: Equipment and preliminary results. *Waste Management and Research* **8**, 21–31.
[70] Mancinelli, R.L. and McKay, C.P. (1985) Methane oxidizing bacteria in sanitary landfills. In: Proc 1st Symp. on Biotechnological Advances in Processing Municipal Wastes for Fuels and Chemicals (A.A. Antonopoulos, Ed.), Argonne National Laboratory Report ANL/CNSV-TM-167.
[71] Hameed, S. and Cess, R.D. (1983) Impact of global warming in biospheric sources of methane and its climate consequences. *Tellus* **35**, 1–7.
[72] Khalil, M.A.K. and Rasmussen, R.A. (1989) Climate-induced feedbacks for the global cycles of methane and nitrous oxide. *Tellus* **41B**, 554–559.
[73] Crill, P.M. (1991) Seasonal patterns of methane uptake and carbon dioxide release by a temperate woodland soil. *Global Biogeochem. Cycles* **4**, 319–334.
[74] Striegl, R.G., McConnaughey, T.A., Thorstensen, D.C., Weeks, E.P. and Woodward, J.C. (1992) Consumption of atmospheric methane by desert soils. *Nature* **357**, 145–147

2
Microbial Sources and Sinks for Methylated Sulfur Compounds in the Marine Environment

Ronald P. Kiene

University of Georgia Marine Institute, Sapelo Island, Georgia 31327, USA

1 SUMMARY

Methylated sulfur compounds, primarily dimethyl sulfide (DMS), contribute >50% of the global biogenic sulfur input to the atmosphere and play important roles in atmospheric chemistry, aerosol formation and possibly cloud/climate dynamics. There is considerable evidence that DMS in seawater is derived primarily from the degradation of β-dimethylsulfoniopropionate (DMSP), an osmolyte and compatible solute produced by marine plants and some heterotrophic flagellates. The biogeochemical cycle of DMS in surface seawater is complex and involves the interactions of many different processes and organisms. While synthesis of DMSP appears to be restricted mainly to the primary producers, the exact mechanisms of its release and/or degradation to DMS are not well understood. Laboratory studies suggest that grazing on phytoplankton promotes the release of DMS and that bacterial lyase degradation (cleavage) of the dissolved DMSP pool may be the main route to the formation of DMS. A substantial fraction of the dissolved DMSP in seawater appears to be metabolized by a demethylation pathway which does not yield DMS. This pathway yields 3-methiolpropionate and may be carried out by methylotrophic bacteria which compete with organisms utilizing the lyase pathway. Once produced, DMS is degraded by methylotrophic pathways, but is also lost by air-sea exchange and photochemical degradation. Biological consumption appears to be a major sink for DMS in most marine environments. Both DMSP and DMS are among the many potential methylotrophic substrates which

---

Microbial Growth on C_1 Compounds
© Intercept Ltd, PO Box 716, Andover, Hampshire SP10 1YG, UK

contribute to C_1 cycling in the marine environment. I speculate on the role of methylated sulfur compounds in the marine C_1 cycle and I discuss some new information on the potential interactions among methylated sulfur and nitrogen compounds and their effects on DMS cycling.

2 INTRODUCTION

> It has been shown that some multicellular marine algae evolve volatile sulfur compounds which have "iso-no-kaori", a unique smell of the sea
> *Yuzaburo Ishida, 1968.*

Research on the cycling of methylated sulfur compounds in marine environments has been stimulated by recognition of the important role played by volatile organosulfur compounds in global biogeochemical cycles [1,2] and climate regulation [3,4]. While the 'global geochemical' aspects of volatile sulfur cycling have received a great deal of attention, methylated sulfur compounds have proven to be fascinating subjects from the perspectives of microbiology, microbial ecology, biochemistry and biogeochemistry [5–8]. Much of the recent research has focused on the study of dimethyl sulfide (DMS) because it is the principal form of volatile sulfur emitted from coastal marine and oceanic areas [9,10] and because DMS plays a central role in the biogeochemical cycling of a variety of related methylated sulfur compounds including β-dimethylsulfoniopropionate (DMSP), dimethylsulfoxide (DMSO), methionine, dimethyldisulfide (DMDS) and methanethiol (MeSH) [6–10].

Here I briefly describe the microbial processes involved in the cycling of DMS in marine environments and I discuss some of the recent developments in the field of DMS biogeochemistry. Because DMS, its precursors and degradation products are potential C_1 substrates, and our understanding of C_1 cycling in the oceans is poor, I speculate on the role of methylated sulfur compounds in the C_1 cycle of seawater. Finally, I discuss some new information describing how several C_1 substrates might interact to affect the DMS cycle.

3 THE ORIGIN OF DMS IN THE MARINE ENVIRONMENT

3.1 *Identification of the plant osmolyte DMSP as a major precursor of DMS*

Haas [11] first identified DMS in the emissions from the marine rhodophyte alga *Polysiphonia fastigiata*. Challenger [12] later showed that *P. fastigiata* contained dimethyl-2-carboxyethyl sulfonium salts (later known as DMPT or DMSP) which was the precursor of DMS. Subsequent studies showed that DMS and its precursors were found in a wide variety of marine macroalgae, phytoplankton and a heterotrophic flagellate [13–18] which accounts for the ubiquitous occurrence of DMSP and DMS in marine systems.

DMSP functions as an osmolyte and compatible solute in certain marine algae and rooted halophytes [18–21]. Its presence at significant intracellular concentrations (>10 mM) is highly species specific although some taxonomic generalizations can be made [14,22]. The dinoflagellates and prymnesiophytes have a large incidence of significant DMSP production whereas the chlorophytes and cyanobacteria have a much lower incidence. It is interesting to note that while the cyanobacteria isolated from the marine water column such as *Synecococcus* sp., *Synechocystis* and *Trichodesmium* sp. do not produce DMSP [14], the filamentous mat-forming species *Microcoleous cthonoplastes* does [23,24]. The mat habitat of *M. cthonoplastes* is characterized by extremes of salinity, temperature, sulfide and other physicochemical parameters [24] which might account for the production of a compatible solute/osmolyte such as DMSP. It is important to recognize, however, that not all algae which produce DMSP live in environments with fluctuating salinities. Many oceanic phytoplankton living in the relatively stable salinities of the open ocean synthesize DMSP [14,22] presumably for its osmotic and compatible solute functions.

3.2 Production of DMS from DMSP

DMSP can be degraded via an enzymatic elimination reaction which yields DMS and acrylic acid according to the following equation [25,26]:

$$(CH_3)_2S^+CH_2CH_2COO^- \xrightarrow{\text{lyase}} (CH_3)_2S + CH_2{=}CHCOO^- + H^+$$

DMSP lyase activity has been found in algal extracts [13,25] and bacteria [26–28]. In seawater, DMSP is found in a particulate pool ($DMSP_{part}$), which is presumably algae, grazers of algae, fecal pellets and detritus; and in a dissolved or 'free' pool ($DMSP_{diss}$) [29, 30]. Grazing by zooplankton on DMSP-containing phytoplankton is thought to play an important role in DMS production [31–34]. The role of micrograzers such as flagellates and ciliates could also play an important role in release of DMS as well as forming a reservoir of particulate DMSP which can be degraded or transferred through the food web [34]. However, the processes involved in conversion between $DMSP_{part}$ and $DMSP_{diss}$ and the production of DMS from these pools are still not well understood. In particular, the release of dissolved DMSP during the grazing process has not been demonstrated but is likely to be important. Recent studies indicate that DMS is formed rapidly from $DMSP_{diss}$ and that the turnover time of $DMSP_{diss}$ may be on the order of hours [28,35,36].

It is clear that DMSP is enzymatically degraded to DMS in seawater [28,37] and in sediments [23,38] but there are no definitive studies to prove that DMSP is the only significant precursor of the DMS in seawater. Other possible precursors such as sulfocholine [39], methionine, DMSO and others have not been ruled out. In anoxic sediment environments it is possible that methylation of sulfide [40] or degradation of methoxy compounds such as syringic acid could also contribute to DMS production[41]. Finster et al. [41] found that bacteria in freshwater and marine anoxic sediments used the methyl groups of lignin-derived methoxyaromatic compounds to methylate HS-. The MeSH formed was methylated further to yield DMS. This type of mechanism may explain the occurrence of DMS in habitats with little or no DMSP.

3.3 Relationship between DMS, algal biomass and DMSP in seawater

The extensive data sets of Andreae and his co-workers [1,2,42–44] showed that the distribution of DMS in the oceans is largely restricted to the euphotic zone and is loosely correlated with the biomass of phytoplankton (Chl *a*). Only recently has the distribution of DMS been related to the distribution of DMSP and DMSP-producing algae [30,45–48]. The relationship between the numbers or biomass of individual algal species and DMS or DMSP is often more significant than that with a general indicator of algal biomass such as Chl *a* [29,45]. However, in these cases the relationships, while statistically significant, typically do not explain more than 50% of the variability in DMS. Close examination of data in the literature indicates that it is only when algal biomass is extremely high, such as in bloom situations, that correlations between algae, Chl *a* and DMSP with DMS are highly significant [29,45,47–49]. These studies have found significant correlations between DMS and DMSP in nearly monospecific blooms of the prymnesiophyte *Phaeocystis pouchetii* as well as the coccolithophore *Emiliania huxleyi*. Likewise in sediment environments, very high DMSP and DMS concentrations are found in the upper few mm of microbial mats where dense populations of the DMSP-producing cyanobacterium *M. cthonoplastes* reside [23,24]. It is interesting to note that Iverson et al. [30] reported a significant increase in the ratio of DMSP to Chl *a* in transects from estuarine to oceanic waters (Fig. 1). It is not clear why this is so but similar high DMSP:Chl *a* ratios have been observed in oceanic waters by other investigators [10,48]. We can only speculate at present, but greater DMSP production by oceanic phytoplankton may be related to the low nitrogen status of oceanic regions because DMSP may substitute for N-containing osmolytes [10].

In many non-bloom situations the relationships between DMS, DMSP and algal biomass are not very strong [30,33,45,47,50]. While a consensus has emerged among researchers that DMSP is the principal precursor of DMS in seawater and most other marine habitats including sediments, a relationships between DMSP and DMS in seawater is not always clear. For example, a depth profile of DMSP and DMS obtained approximately 250 km off the coast of Seattle, WA, USA in the Northeastern Pacific Ocean showed DMS concentrations which were very low (<1 nM) and uniform with depth, whereas particulate and dissolved pools of DMSP were much higher (25–30 nM total(Σ)DMSP) and showed significant variations with depth (Fig. 2). This pattern was relatively

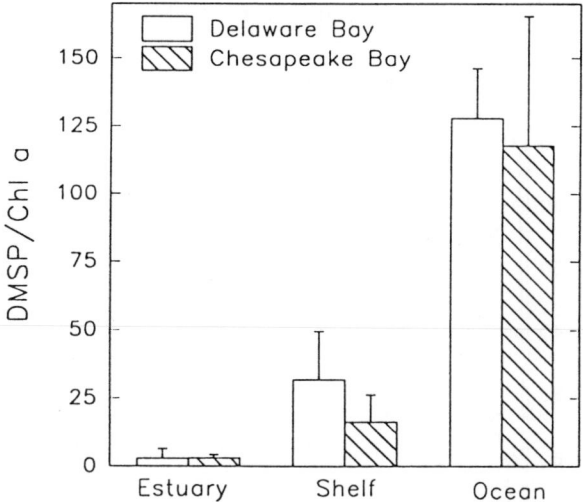

Figure 1 Average values of DMSP/Chl a (nmol·μg^{-1}) obtained in estuarine, shelf and oceanic zones along transects in Delaware and Chesapeake Bays. The bars indicate one standard deviation. The data are taken from Table 2 of Iverson et al. [30].

constant over a 2-week period during which the water mass was tracked with a drogue. Another example of how DMS concentrations can be uncoupled from DMSP pools comes from incubations of Georgia estuarine waters in the light (Fig. 3). The concentration of $DMSP_{part}$ doubled from 70 to 140 nM during the 11 h incubation due to growth of phytoplankton. The $DMSP_{diss}$ rose quickly from 1.5 nM to 6 nM in a few hours and then declined to approach the initial value. During these substantial changes in DMSP pools, the DMS concentration was virtually invariant (Fig. 3). Therefore the distribution of DMSP is certainly not the only factor which affects the concentration of DMS in seawater. Other factors such as the rates of DMSP release and degradation to DMS as well as the removal of DMS must also be considered (see below).

3.4. Food web interactions and the production of DMS

Work with laboratory cultures of phytoplankton has shown that little DMS or DMSP is released by healthy growing cells [22,29]. Dacey and Wakeham [32] were the first to illustrate that grazing by zooplankton on DMSP-producing phytoplankton greatly accelerated DMS release in algal cultures. They also showed that copepods which had fed on DMSP-containing algae continued to release DMS after being removed from the food source. This suggested that DMSP degradation occurred in the zooplankton or in their fecal pellets. In addition it is likely that dissolved DMSP is released during the grazing process, although this has not yet been demonstrated. Because

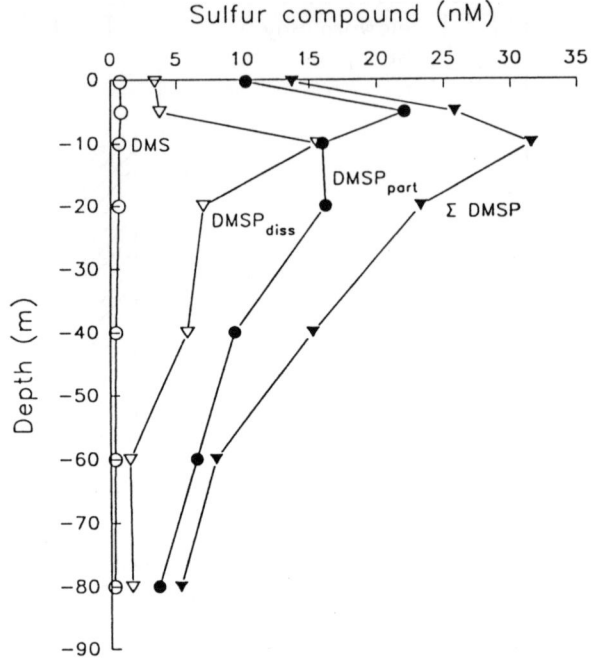

Figure 2 Depth profiles of DMS, dissolved DMSP (DMSP$_{diss}$), particulate DMSP (DMSP$_{part}$) and total DMSP (ΣDMSP) obtained during PSI-3 (Pacific Sulfur/Stratus Investigation) in April 1991. The station was located 250 km west of the Washington, USA coast near the mouth of Puget Sound. The water temperature was 9°C. Total DMSP was obtained from the sum of separate dissolved and particulate measurements. (Previously unpublished data of R.P. Kiene.)

DMSP$_{diss}$ is consumed rapidly in seawater, it may be difficult to detect an increase in this pool as a function of grazing. Bellviso et al. [34] found that microzooplankton grazers including ciliates and flagellates were important agents in the release of DMS. Further evidence for the role of grazers in DMS production comes from Leck et al. [33] who found that zooplankton biomass was significantly correlated with DMS concentration in the Baltic Sea. They also found that high DMS concentrations were associated with low inorganic nutrient concentrations, particularly nitrate. The mechanisms of DMS production are still unknown and because of the complex interactions involved, considerably more work will be required to gain an understanding of DMS production in surface seawater.

3.5 *DMSP degradation to products other than DMS*

Implicit in much of the research which has been done on ocean DMS/DMSP cycling is the idea that DMS is the only sulfur product of DMSP degradation.

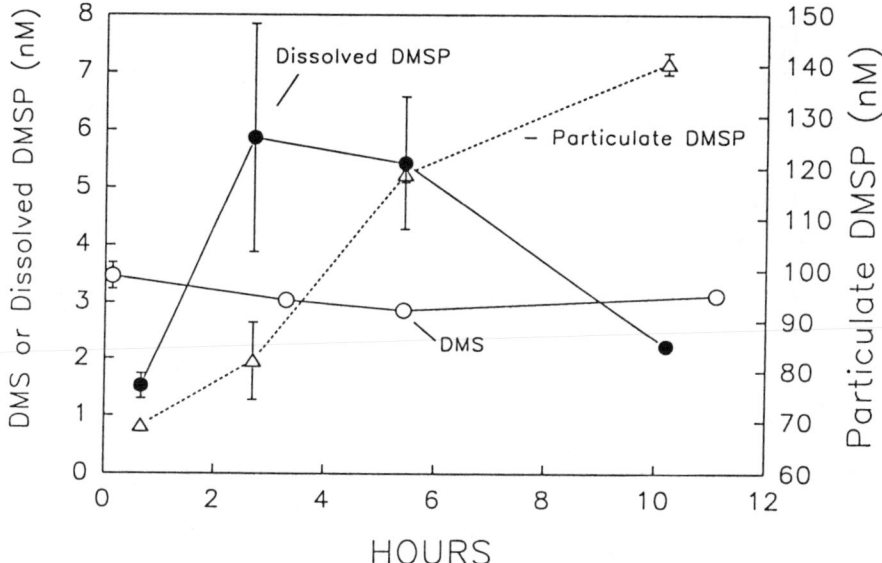

Figure 3 Time courses of dissolved and particulate DMSP as well as DMS concentrations in estuarine water samples incubated in the light. Data values are the mean of duplicate bottles and the bars indicate the range. (Previously unpublished data of R.P. Kiene.)

Recent evidence from work with field samples and with bacterial cultures has called this idea into question. In experiments designed to estimate how much of the DMSP degraded in water samples was converted to DMS, Kiene and Service [36] found that <30% of the ΣDMSP lost during the incubations was recovered as DMS; even when DMS consumption was inhibited by the inclusion of chloroform (Fig. 4). A similar conclusion was reached by Kiene [37] who used oceanic water samples from the Equatorial Pacific. Photochemical [51] and other chemical losses have been ruled out as explanations for the low yield of DMS, leaving the conclusion that a major fraction of DMSP must be degraded by a pathway which does not yield DMS.

It now appears that the alternative $DMSP_{diss}$ degradation pathway involves demethylation. A demethylation pathway for the degradation of DMSP was first proposed by Mopper and Taylor [52] and later shown to occur in anoxic sediments by Kiene and Taylor [53,54]. DMSP demethylation leads to the production of 3-methiolpropionate which can be further demethylated to 3-mercaptopropionate [53]. Taylor and Gilchrist [55] have isolated aerobic marine bacteria which metabolize 3-methiolpropionate, an intermediate in the demethylation of DMSP. Recently Diaz et al. [56] isolated and studied a marine bacterium which grew methylotrophically on glycine betaine but which cleaved DMSP to DMS and acrylate rather than using it as a methyl substrate.

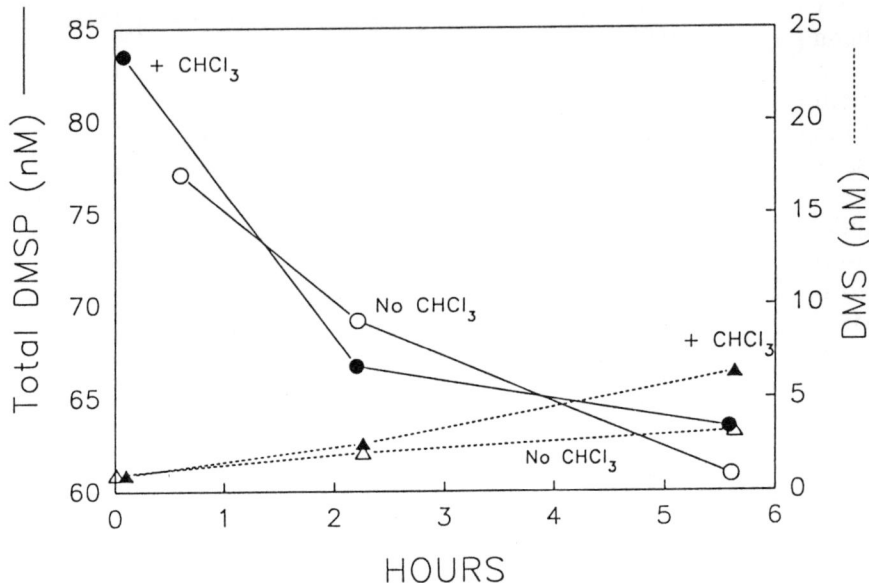

Figure 4 Time courses of total DMSP and DMS in incubations of estuarine water from Sapelo Island Georgia showing a large decrease in DMSP but only a small increase in DMS. Approximately 40 nM dissolved DMSP was added to the samples. Chloroform (CHCl₃) was added to one set of samples to inhibit biological consumption of DMS. Data are taken from Fig. 4 of Kiene and Service [36].

Chemical demethylation of DMSP by reaction with iodide is possible [17,57], but chemical losses of DMSP in seawater are insignificant compared to biological losses [36]. It seems likely that two groups of organisms, utilizing two different degradation pathways, compete for $DMSP_{diss}$ in seawater. The outcome of competition for DMSP by the different pathways (organisms) may be a major factor governing the amount of DMS formed in seawater.

4 DEGRADATION OF DMS IN SEAWATER

4.1 *Biological degradation of DMS*

DMS may be used as a substrate by aerobic bacteria including chemolithotrophs such as *Thiobacilus thioparus* [6,58] and methylotrophs such as *Hyphomicrobium* sp. [59,60]. DMS is also degraded anaerobically by phototrophs such as *Thiocapsa roseopersicina* [61] and by methanogens [62,63]. The aerobes and methanogens initially demethylate DMS to produce methanethiol whereas the phototrophs utilize DMS as an electron donor and form DMSO. The organisms involved and the pathways used are covered in detail in several recent reviews [5,6]. In marine environments DMS consumption has been observed in anoxic and aerobic sediments [23,64], as well as in the water column [35,36]. Measurements of DMS concentrations in incubated water samples often show that DMS is consumed at significant rates [35-37]. Inhibitor studies with chloroform or antibiotics suggest that DMS consumption is biologically mediated. Further evidence for enzymatic degradation of DMS in seawater comes from observations that DMS consumption rates are concentration dependent and may be saturated at elevated concentrations [37]. In addition, a temperature optimum for DMS consumption rates was observed between 30 and 50°C [36]. Recent investigations by Wolfe [46] using low nM levels of ^{14}C-DMS with seawater samples indicated that the methyl groups of DMS are converted to both CO_2 and cell material, suggesting use of DMS by methylotrophic organisms.

There is every indication that biological degradation of DMS in seawater and sediments is a major sink for DMS. Kiene [64] found that the flux of DMS out of sediment cores was very small unless biological inhibitors were added to the cores. Similarly, Dacey *et al.* [65] concluded that salt marsh sediments were likely to be a sink for DMS rather than a source. Wakeham *et al.* [66] and Leck *et al.* [33] used estimates of DMS and DMSP pools and air-sea exchange rates to conclude that microbiological or chemical sinks in the water columns of a coastal salt pond and the Baltic Sea were the major loss mechanisms for DMS. Using a chloroform inhibition technique Kiene and Bates [35] estimated that DMS turnover times in the Equatorial Pacific Ocean were of the order of 1-2 days and that rates of biological consumption of DMS in the surface waters exceeded the rates of DMS lost to the atmosphere by a factor of 10 in most cases.

5 ROLE OF DMSO IN THE DMS CYCLE

5.1 *Formation of DMSO: biological and photochemical reactions*

DMS is oxidized to DMSO by both biological and photochemical processes. Certain phototrophic bacteria utilize DMS as an electron donor with the production of DMSO [24,67]. The reaction is strictly light dependent and DMSO was the only sulfur product of DMS oxidation by *Thiocapsa roseopersicina* [61]. Recent findings by Zhang *et al.* [68] have shown that DMSO may be produced from DMS by a heterotroph, *Pseudomonas acidovorans*, and that this may be a co-metabolic reaction rather than use as a growth substrate. Photochemical degradation of DMS in seawater was observed by Brimblecombe and Shooter [51] and they found evidence that DMSO was the product of the photosensitized reaction. Only recently has DMSO been measured as the product of DMS photodegradation (D.J. Kieber and R.P. Kiene, in preparation) and the conversion of DMS to DMSO appears to be stoichiometric.

5.2 Distribution and degradation of DMSO

DMSO appears to be a major methylated sulfur compound in seawater. Andreae [69] measured concentrations as high as 200 nM and similarly high concentrations were observed in Antarctic waters [49]. It should be noted, however, that in the former study, DMSP may have been measured along with the DMSO [70], and in the latter study few details on the methodology used were provided. More recent measurements from the Equatorial Pacific indicate that DMSO concentrations are lower (ca. 20 nM) but are still greater than the combined pools of DMS and ΣDMSP (R.P. Kiene, unpublished data). DMSO is a potential substrate for methylotrophic bacteria with the known pathways for its metabolism involving initial reduction to DMS followed by demethylation to MeSH and subsequently HS^- [59,71] (see also reviews by Taylor and Kiene [5] and Kelly and Smith [6]). Nothing is known about the fate of DMSO in surface seawater yet it represents a significant pool of methylated sulfur which could be used by methylotrophs.

Under reducing conditions in marine sediments, DMSO is readily reduced to DMS [54] presumably by bacteria which can use it as an electron acceptor [72,73]. It is interesting to note that Zhang et al. [68] observed that an aerobic DMS-oxidizing bacterium would also reduce DMSO back to DMS under certain conditions. Little is known of the importance of these reactions or whether DMSO is a significant precursor of DMS in natural environments.

6 LIMITATIONS ON THE FLUX OF DMS FROM THE OCEANS

The biogeochemical cycle of DMS in seawater has emerged as a much more complex cycle than earlier believed with many potential pathways for production and consumption (Fig. 5). From laboratory experiments [36] and from field data such as in Fig. 2, it is becoming increasingly clear that only a small fraction of the DMSP-S in seawater may be converted to DMS and an even smaller fraction may be emitted from the oceans. Similar conclusions have been reached by other investigators [33,46,47,66]. I have constructed a simple conceptual model, based on recent findings, to illustrate this point (Fig. 6). In this scheme I assume that all of the algal DMSP in the surface mixed layer turns over in one day, but the time chosen has little impact on the conclusions drawn. If all of the algal DMSP which is turned over in a given period of time passes through the $DMSP_{diss}$ pool, then <30% of this sulfur will be converted to DMS because the predominant fate of $DMSP_{diss}$ appears to be demethylation [36,55]. Of the DMS formed, only 10% or so ($\sim 3\%$ of the original DMSP-S) is likely to escape to the atmosphere because the biological turnover time of DMS in surface seawater is generally 10 times faster than DMS exchange with the atmosphere [35]. If phytoplankton degraded all of their intracellular DMSP to DMS (no $DMSP_{diss}$ formed) a somewhat greater fraction DMSP-S would escape to the atmosphere, but even with this scenario, 90% of the DMS is likely

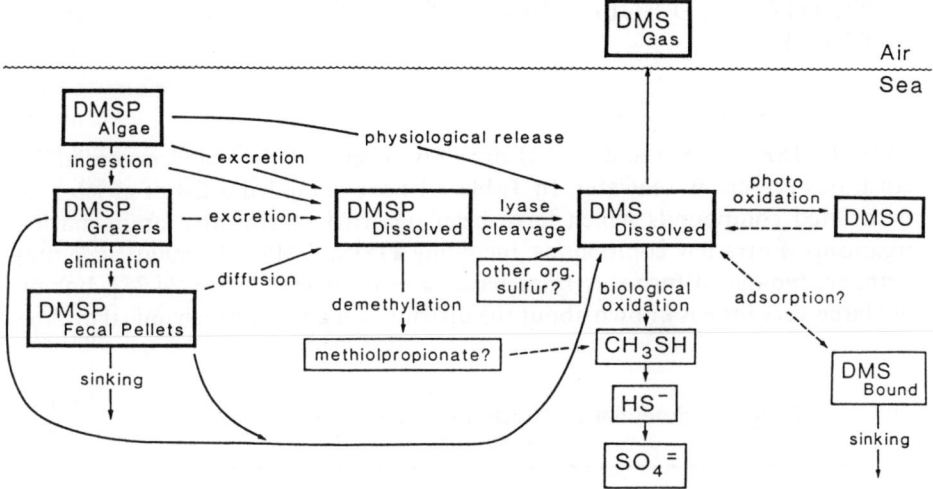

Figure 5 Scheme of the sources and sinks of DMS in surface seawater.

to be consumed within the water column [35]. Obviously, this conceptual model is overly simplistic, given our poor understanding of the complex processes involved. It does, however, serve to illustrate that the emission of DMS may only be a small leak from a large cycle.

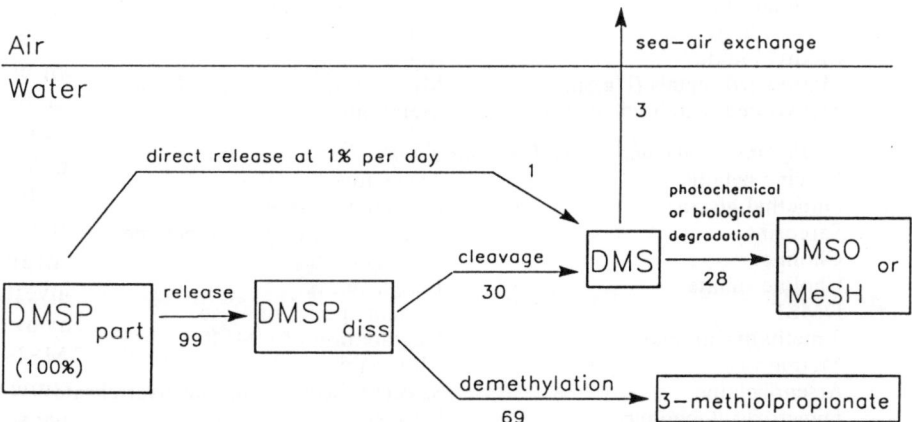

Figure 6 Idealized daily DMSP/DMS budget in surface seawater assuming that 100% of particulate DMSP is released and turned over per day. The pools of $DMSP_{part}$, $DMSP_{diss}$ and DMS are assumed to be in steady state.

7 METHYLATED SULFUR COMPOUNDS AND THE MARINE C_1 CYCLE

7.1 C_1 substrates in seawater

DMS, DMSP, DMSO and MeSH are only several of the many potential C_1 substrates present in seawater. In Table 1 I have compiled a list of at least 28 methylated compounds which have been reported in seawater or in marine organisms. For a few compounds, including DMS, DMSP, formaldehyde and methane, biological turnover rates have been estimated [35,36,74,75], but by and large very little is known about the distribution and biogeochemical cycling

Table 1 Potential C_1 substrates found in seawater and their sources

Substrate	Sources
C_1 compounds	
Methane	Sediments, seeps, microbiological?
Methanol	Other C_1 compounds
Formaldehyde (HCHO)	Photochemical source
Formate	Oxidation of HCHO
Trimethylamine	Algae, glycine betaine, TMAO
Dimethylamine	TMA, TMAO
Methylamine	TMA, TMAO
Trimethylamine oxide (TMAO)	Algae, fish
Dimethyl sulfoxide	Photochemical source from DMS
Dimethyl, sulfide (DMS)	Alghae, DMSP, methionine
Methane thiol	DMS, methionine, methiolpropionate
Dimethyl disulfide (DMDS)	Methanethiol, methionine
Methane sulfonate[a]	DMS, methanethiol
Carbon monoxide	Photochemical source, DOC
Methyl iodide	Algae
Methylated metals (Hg,Sn)	Metabolites?
Methylated metalloids (Ge,As)	Metabolites?
Methylated compounds capable of being utilized as C_1 substrates	
Glycine betaine	Algae, invertebrates
Dimethyl glycine	Degradation of glycine betaine
Sarcosine	Degradation of glycine betaine
Choline	All organisms
Choline sulfate	Fungi, yeasts?
DMSP	Certain species of algae
3-methiolpropionate	Degradation of DMSP
Methionine	All organisms
Arsenobetaine	Selected species of marine invertebrates
Dimethylsulfocholine	Selected diatoms
S-methylmentionine	Possibly algae

[a] Methane sulfonate has not been measured in seawater but it is present in rain and aerosols and should occur in seawater.

of these potential C_1 substrates. Are DMS and DMSP metabolized as part of the collective pool of C_1 compounds by generalist methylotrophs? Are different organisms specialized to use specific compounds? Does methylotrophy contribute significantly to bacterial production and carbon turnover in seawater as has been suggested by Sieburth? [76,77] These are just a few of the many interesting questions concerning C_1 cycling in the ocean. Certainly, it is not likely that C_1 compounds support a *major* fraction of bacterial growth in the sea but it might not be a trivial fraction either. As an example, one can estimate from DMSP:Chl *a* ratios reported in the literature [30] (see also Fig. 1) that particulate DMSP alone makes up 0.3–9% of the carbon in living phytoplankton biomass. I have assumed that most of the DMSP is in living phytoplankton and that the organic carbon to Chl *a* ratio is 80 [78]. The methyl groups of DMSP appear to be very labile substrates whether in the form of DMSP or its degradation products DMS and 3-methiolpropionate [36,79] therefore this compound could conceivably support methylotrophs in seawater. As mentioned above, DMS and DMSP are only part of the C_1 pool. Of great interest with respect to C_1 cycling in the ocean is the fact that formaldehyde, which is a central intermediate in all forms of methylotrophy, is produced in seawater by photochemical degradation of refractory organic matter. Significant production rates (50 $nM \cdot d^{-1}$) of formaldehyde as well as biological turnover rates (turnover times as short as 0.3 d) have been measured in surface waters of the Black Sea [75]. As some of the methylated substrates in Table 1 are also potential heterotrophic substrates (e.g. DMSP, methionine and glycine betaine), the competitive interactions between methylotrophic and heterotrophic microorganisms could potentially influence the cycling of C_1 sulfur compounds (see below).

7.2 Interactions among DMSP, glycine betaine and DMS

Several recent studies have pointed to some potentially interesting interactions between DMS and the functionally-similar osmolytes DMSP and glycine betaine (GBT) in seawater and in bacterial cultures. Kiene and Service [80] observed that DMS concentrations increased in seawater which was spiked with GBT (125–5000 nM). Paradoxically they also observed that $DMSP_{diss}$ consumption and DMS production from DMSP were significantly inhibited by 500 nM GBT. They concluded that inhibition of $DMSP_{diss}$ consumption, perhaps due to competitive uptake of GBT, led to an increase in $DMSP_{diss}$. The inhibition of DMSP consumption was transient and as consumption resumed, DMS was produced in greater amounts than in non-GBT treatments. Diaz *et al.* have recently found that a novel marine bacterium MD-14-50 could take up and utilize both DMSP and GBT as osmolytes when subjected to osmotic stress [56]. This organism metabolized GBT methylotrophically but cleaved DMSP to DMS and acrylate with only the acrylate being used as a substrate. Kiene and Service [36] also reported that addition of $5\mu M$ GBT to estuarine water samples resulted in lower production of $DMSP_{part}$ by a growing

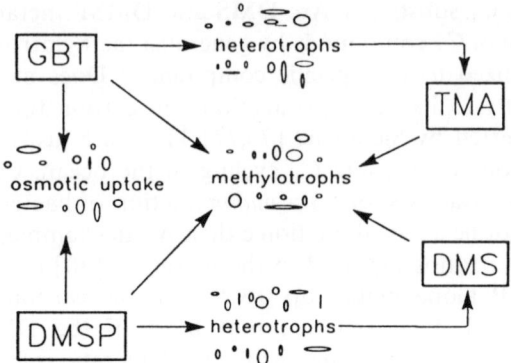

Figure 7 Potential microbial interactions in the competition for methylated sulfur (DMSP & DMS) and nitrogen (glycine betaine(GBT) and trimethylamine (TMA)) compounds in seawater.

phytoplankton community. These authors suggested that phytoplankton may have taken up GBT and therefore required less DMSP. These recent findings support the notion that several naturally occurring C_1 substrates may interact with each other in interesting and complex ways. Some of the potential microbial interactions involving uptake and degradation of GBT, DMSP and DMS are illustrated in Fig. 7. Dissolved pools of DMSP and GBT may be taken up by bacteria, algae or microzooplankton for osmotic purposes. So far this has only been demonstrated with bacteria [56,81] but is likely to occur with other organisms as well because DMSP and GBT are valuable osmolytes and compatible solutes. Organisms which take up DMSP and GBT for osmotic purposes will have to compete with other microorganisms which use the compounds as energy substrates either heterotrophically or methylotrophically. Heterotrophs utilizing the carbon chain of GBT might produce trimethylamine (TMA) [82] which is a substrate for methylotrophs. Likewise, heterotrophs degrading DMSP by the lyase pathway would utilize acrylate [28] and make DMS available to methylotrophs. Both DMSP and GBT can be utilized directly by methylotrophic bacteria [55,83] which employ demethylation pathways. Competition no doubt occurs within and among groups of microorganisms for the methylated compounds and this undoubtedly affects the DMS concentrations in marine environments.

8 CONCLUSION

Methylated sulfur compounds play key roles in the marine environment, serving as 1. energy substrates for a variety of microorganisms, 2. osmolytes and compatible solutes and 3. agents of transfer of sulfur to the atmosphere. In these roles methylated sulfur compounds have an impact on the global atmospheric chemistry, climate and the structure of microbial communities. DMS and DMSP are the most important of the methylated sulfur compounds in marine systems and considerable work still needs to be done for a better understanding of the processes involved in the production and consumption of these compounds. Practically nothing is known about the role of methylated sulfur compounds in C_1 cycling in seawater or of the marine C_1 cycle in general. The information presented here suggests that DMS and DMSP are perhaps important substrates for marine methylotrophs and that other C_1 compound such as GBT may affect DMS(P) cycling. Studies of DMS and DMSP, therefore are likely to shed light on the more general picture of C_1 cycling in the ocean.

ACKNOWLEDGEMENT

I thank Gordon Wolfe for stimulating discussions during the preparation of this manuscript and for commenting on the text. This work was supported by the National Science Foundation (Grants OCE-8817442 and OCE-9203728) and The University of Georgia Marine Institute. I am also indebted to Susan Service for assistance in the laboratory. Contribution 714 of the University of Georgia Marine Institute.

REFERENCES

[1] Andreae, M.O. and Raemdonck, H. (1983) Dimethyl sulfide in the surface ocean and the marine atmosphere: a global view. *Science* **221**, 744–747.
[2] Andreae, M.O. (1985), The emission of sulfur to the remote atmosphere: background paper. In: The Biogeochemical Cycling of Sulfur and Nitrogen in the Remote Atmosphere (J.N. Galloway *et al.*, Eds), pp. 5–25. D. Reidel Publishing Co., Boston, MA, USA.
[3] Bates, T.S., Charlson, R.J. and Gammon, R.H. (1987) Evidence for the climatic role of marine biogenic sulphur. *Nature* **329**, 319–321.
[4] Charlson, R.J., Lovelock, J.E., Andreae, M.O. and Warren, S.G. (1987) Oceanic phytoplankton, atmospheric sulfur, cloud albedo and climate. *Nature* **326**, 655–661.
[5] Taylor, B.F. and Kiene, R.P. (1989) Microbial metabolism of dimethyl sulfide. In: Biogenic Sulfur in the Environment (E.S. Saltzman and W.J. Cooper, Eds) pp. 202–221. American Chemical Society, Washington, DC.
[6] Kelly, D.P. and Smith, N.A. (1990) Organic sulfur compounds in the environment: Biogeochemistry, microbiology, and ecological aspects. *Adv. Microb. Ecol.* **11**, 345–385.
[7] Kelly, D.P. and Baker, S.C. (1990) The organosulfur cycle: aerobic and anaerobic processes leading to turnover of C_1-sulphur compounds. *FEMS Microbiol. Rev.* **87**, 241–246.
[8] Kiene, R.P. (1992), Microbial cycling of organosulfur gases in marine and freshwater environments. In: Cycling of Reduced Gases in the Hydrosphere (D. Adams, S. Seitzinger and P. Crill, Eds), E. Schweizerbart'sche Verlagsbuchhandlung (Naglele U. Obermiller), Stuttgart, in press.
[9] Bates, T.S., Lamb, B.K., Guenther, A., Dignon, J. and Stoiber, R.E. (1992) Sulfur emissions to the atmosphere from natural sources. *J. Atmos. Chem.* **14**, 315–337.

[10] Andreae, M.O. (1990) Ocean-atmosphere interactions in the global biogeochemical sulfur cycle. *Mar. Chem.* **30**, 1–29.
[11] Haas, P. (1935), CLVII. The liberation of methyl sulphide by seaweed. *Biochem. J.* **29**, 1297–1299.
[12] Challenger, F. and Simpson, M.I. (1948) Studies on biological methylation, Part XII. A precursor of the dimethyl sulfide evolved by *Polysiphonia fastigiata*. Dimethyl-2-carboxyethyl sulfonium hydroxide and its salts. *J. Chem. Soc.* **1948**, 1591–1597.
[13] Ishida, Y. (1968) Physiological studies on evolution of dimethylsulfide. *Mem. Coll. Agric. Kyoto Univ.* **94**, 47–82.
[14] Keller, M.D., Bellows, W.K. and Guillard, R.R.L. (1989), Dimethyl sulfide production in marine phytoplankton. In: Biogenic Sulfur in the Environment. (E. Saltzman and W.J. Cooper, Eds), pp. 167–182. American Chemical Society, Washington, DC.
[15] Blunden, G. and Gordon, S.M. (1986), Betaines and their sulphonio analogues in marine algae. In: Progress in Phycological Research (F.E. Round and D.J. Chapman, Eds), pp.39–80. Biopress Ltd., Bristol.
[16] Ackman, R.G., Tocher, C.S. and McLachlan, J. (1966) Occurrence of dimethyl-β-propiothetin in marine phytoplankton. *J. Fish. Res. Bd. Canada* **23**, 357–364.
[17] White, R.H. (1982) Analysis of dimethyl sulfonium compounds in marine algae. *J. Mar. Res.* **40**, 529–535.
[18] Reed, R.H. (1983) Measurement and osmotic significance of β-dimethylsulfoniopropionate in marine macroalgae. *Mar. Biol. Lett.* **4**, 173–181.
[19] Grone, T. and Kirst, G.O. (1991) Aspects of dimethylsulfoniopropionate effects on enzymes isolated from the marine phytoplankter *Tetraselmis subcordiformis* (Stein). *J. Plant Physiol.* **138**, 85–91.
[20] Edwards, D.M., Reed, R.H. and Stewart, W.D.P. (1988) Osmoacclimation in *Enteromorpha intestinalis*: long-term effects of osmotic stress on organic solute accumulation. *Marine Biology* **98**, 467–476.
[21] Vairavamurthy, A., Andreae, M.O. and Iverson, R.L. (1985) Biosynthesis of dimethylsulfide and dimethylpropiothetin by *Hymenomonas carterae* in relation to sulfur source and salinity variations. *Limnol. Oceanogr.* **30**, 59–70.
[22] Keller, M.D. (1991) Dimethyl sulfide production and marine phytoplankton: The importance of species composition and cell size. *Biol. Oceanogr.* **6**, 375–382.
[23] Visscher, P. and Van Gemerden, H. (1991) Production and consumption of dimethylsulfoniopropionate in marine microbial mats. *Appl. Environ. Microbiol.* **57**, 3237–3242.
[24] Visscher, P.T. (1992) Microbial sulfur cycling in laminated marine ecosystems, Ph.D Thesis, University of Groningen, The Netherlands.
[25] Cantoni, G.L. and Anderson, D.G. (1956) Enzymatic cleavage of dimethylpropiothetin by *Polysiphonia* leaves. *J. Biol. Chem.* **222**, 171–177.
[26] Dacey, J.W.H. and Blough, N. (1987) Hydroxide decomposition of DMSP to form DMS. *J. Geophys. Res. Lett.* **14**, 1246–1249.
[27] Ledyard, K. and Dacey, J.W.H. (1990) Production of DMS from DMSP by a marine bacterium. *EOS* **71**, 104.
[28] Kiene, R.P. (1990) Dimethyl sulfide production from dimethylsulfoniopropionate in coastal seawater samples and bacterial cultures. *Appl. Environ. Microbiol.* **56**, 3292–3297.
[29] Turner, S.M., Malin, G. and Liss, P.S. (1988) The seasonal variation of dimethyl sulfide and dimethylsulfoniopropionate concentrations in nearshore waters. *Limnol. Oceanogr.* **33**, 364–375.
[30] Iverson, R.L., Nearhoof, F.L. and Andreae, M.O. (1989) Production of dimethylsulfonium propionate and dimethylsulfide by phytoplankton in estuarine and coastal waters. *Limnol. Oceanogr.* **34**, 53–67.

[31] Wakeham, S.G. and Dacey, J.W.H. (1989), Biogeochemical cycling of dimethyl sulfide in marine environments. In: Biogenic Sulfur in the Environment (E.S. Saltzman and W.J. Cooper, Eds), pp. 152–166. American Chemical Society, Washington, DC.

[32] Dacey, J.W.H. and Wakeham, S.G. (1986) Oceanic dimethylsulfide: production during zooplankton grazing on phytoplankton. *Science* **233**, 1314–1316.

[33] Leck, C., Larsson, U., Bagender, L.E., Johansson, S., and Hajdu, S. (1990) Dimethyl sulfide in the Baltic Sea: Annual variability in relation to biological activity. *J. Geophys. Res.* **95**, 3353–3364.

[34] Belviso, S., Kim, S.K., Rassoulzadegan, F., Krajka, B., Nguyen, B.C., Mihalopoulos, N. and Buat-Menard, P. (1990) Production of dimethylsulfonium propionate (DMSP) and dimethylsulfide (DMS) by a microbial food web. *Limnol. Oceanogr.* **35**, 1810–1821.

[35] Kiene, R.P. and Bates, T.S. (1990) Biological removal of dimethyl sulphide from seawater. *Nature* **345**, 702–705.

[36] Kiene, R.P. and Service, S.K. (1991) Decomposition of dissolved DMSP and DMS in estuarine waters: dependence on temperature and substrate concentration. *Mar. Ecol. Prog. Ser.* **76**, 1–11.

[37] Kiene, R. (1992) Dynamics of dimethyl sulfide and dimethylsulfoniopropionate in oceanic seawater samples. *Mar. Chem.* **37**, 29–52.

[38] Kiene, R.P. and Visscher, P.T. (1987) Production and fate of methylated sulfur compounds from methionine and dimethylsulfoniopropionate in anoxic salt marsh sediments. *Applied and Environmental Microbiology* **53**, 2426–2434.

[39] Bisseret, P., Ito, S., Tremblay, P.A., Volcani, B.E., Dessort, D. and Kates, M. (1984) Occurrence of phosphatidylsulfocholine, the sulfonium analog of phosphatidylcholine in some diatoms and algae. *Bichimica et Biophysica Acta* **796**, 320–327.

[40] Drotar, A., Burton, G.A., Taverier, J.E. and Fall, R. (1987) Widespread occurrence of bacterial thiol methyltransferase and the biogenic emission of methylated sulfur gases. *Appl. Environ. Microbiol.* **53**, 1626–1631.

[41] Finster, K., King, G.M. and Bak, F. (1990) Formation of methylmercaptan and dimethylsulfide from methoxylated aromatic compounds in anoxic marine and fresh water sediments. *FEMS Microbiol. Ecol.* **74**, 295–302.

[42] Andreae, M.O. and Barnard, W.R. (1984) The marine chemistry of dimethylsulfide. *Marine Chem.* **14**, 267–279.

[43] Andreae, M.O. (1985) Dimethylsulfide in the water column and the sediment porewaters of the Peru upwelling area. *Limnol. Oceanogr.* **30**, 1208–1218.

[44] Barnard, W.R., Andreae, M.O., and Iverson, R.L. (1983) Dimethylsulfide and *Phaeocystis pouchetii* in the south-eastern Bering Sea. Continental Shelf Res. **3**, 103–113.

[45] Burgermeister, S., Zimmermann, R.L., Georgii, H.W., Bingemer, H.G., Kirst, G.O., Janssen, M. and Ernst, W. (1990) On the biogenic origin of dimethylsulfide: relation between chlorophyll, ATP, organismic DMSP, phytoplankton species, and DMS distribution in Atlantic surface water and atmosphere. *J. Geophys. Res.* **95**, 20607–20615.

[46] Wolfe, G.V. (1992) The cycling of climatically active dimethyl sulfide (DMS) in the marine euphotic zone: Biological and chemical constraints on the flux to the atmosphere, Ph.D., University of Washington.

[47] Matrai, P.A. and Keller, M.D. (1992) Dimethylsulfide in a large-scale cocolithophore bloom in the Gulf of Maine. *Cont. Shelf Res.*, in press.

[48] Malin, G., Turner, S., Liss, P., Holligan, P. and Harbour, D. (1992) Production of dimethyl sulfide and dimethylsulphoniopropionate in the North East Atlantic during the summer coccolithophore bloom. *Deep-Sea Res.*, in press.

[49] Gibson, J.A.E., Garrick, R.C., Burton, H.R., and McTaggart, A.R. (1990) Dimethylsulfide and the alga *Phaeocystis pouchetii* in antarctic coastal waters. *Mar. Biol.* **104**, 339–246.

[50] Cline, J.D. and Bates, T.S. (1983) Dimethyl sulfide in the equatorial Pacific ocean: A natural source of sulfur to the atmosphere. *Geophys. Res. Lett.* **10**, 949–952.

[51] Brimblecombe, P. and Shooter, D. (1986) Photo-oxidation of dimethylsulphide in aqueous solution. *Mar. Chem.* **19**, 343–353.

[52] Mopper, K. and Taylor, B.F. (1986), Biogeochemical cycling of sulfur. In: Organic Marine Geochemistry (M.L. Sohn, Ed.), pp. 324–339. American Chemical Society, Washington, DC.

[53] Kiene, R.P. and Taylor, B.F. (1988) Demethylation of dimethylsulfoniopropionate and production of thiols in anoxic marine sediments. *Appl. Environ. Microbiol.* **54**, 2208–2212.

[54] Kiene, R.P. and Taylor, B.F. (1988) Biotransformations of organosulphur compounds in sediments via 3- mercaptopropionate. *Nature* **332**, 148–150.

[55] Taylor, B.F. and Gilchrist, D.C. (1991) New routes for aerobic biodegradation if dimethylsulfoniopropionate. *Appl. Environ. Microbiol.* **57**, 3581–3584.

[56] Diaz, M.R., Visscher, P.T., and Taylor, B.F. (1992) Metabolism of dimethylsulfoniopropionate and glycine betaine by a marine bacterium. *FEMS Microbiol. Lett.* **96**, 61–66.

[57] Brinckman, F.E., Olson, G.J. and Thayer, J.S. (1985), Biological mediation of marine metal cycles: the case of methyl iodide. In: Marine and Estuarine Geochemistry (A.C. Sigleo and A. Hattori, Eds), pp. 227–238. Lewis Publishers, Chelsea.

[58] Kanagawa, T. and Kelly, D.P. (1986) Breakdown of dimethyl sulphide by mixed cultures and by *Thiobacillus thioparus*. *FEMS Microbiol Lett* **34**, 13–19.

[59] De Bont, J.A.M., van Dijken, J.P. and Harder, W. (1981) Dimethyl sulphoxide and dimethyl sulfide as a carbon, sulphur and energy source for growth of *Hyphomicrobium* S. *J. Gen. Microbiol.* **127**, 315–323.

[60] Suylen, G.M.H. and Kuenen, J.G. (1986) Chemostat enrichment and isolation of *Hyphomicrobium* EG, a dimethyl sulphide oxidizing methylotroph and reevaluation of *Thiobacillus* MS1. *Antonie van Leeuwenhoek* **52**, 281–293.

[61] Visscher, P.T. (1991) Photo-autotrophic growth of *Thiocapsa roseopersicina* on dimethyl sulfide. *FEMS Microbiol. Lett.* **81**, 247–250.

[62] Oremland, R.S., Kiene, R.P., Mathrani, I., Whiticar, M.J. and Boone, D.R. (1989) Description of an estuarine methylotrophic methanogen which grows on dimethyl sulfide. *Appl. Environ. Microbiol.* **55**, 994–1002.

[63] Ni, S. and Boone, D.R. (1991) Isolation and characterization of a dimethyl sulfide-degrading methanogen, *Methanolobus siciliae* HI350, from and oil well, characterization of *M. siciliae* T4/M^T, and emendation of *M. siciliae*. *Int. J. Syst. Bacteriol.* **41**, 410–416.

[64] Kiene, R.P. (1988) Dimethyl sulfide metabolism in salt marsh sediments. *FEMS Microbiol. Ecol.* **53**, 71–78.

[65] Dacey, J.W.H., King, G.M. and Wakeham, S.G. (1987) Factors controlling emission of dimethylsulphide from salt marshes. *Nature* **330**, 643–645.

[66] Wakeham, S.G., Howes, B.L., Dacey, J.W.H., Schwarzenbach, R.P. and Zeyer, J. (1987) Biogeochemistry of dimethylsulfide in a seasonally stratified coastal salt pond. *Geochim. Cosmochim. Acta* **51**, 1675–1684.

[67] Zeyer, J., Eicher, P., Wakeham, S.G. and Schwartzenbach, R.P. (1987) Oxidation of dimethyl sulfide to dimethyl sulfoxide by phototrophic purple bacteria. *Appl. Environ. Micro.* **53**, 2026–2032.

[68] Zhang, L., Kuniyoshi, I., Hirai, M. and Shoda, M. (1991) Oxidation of dimethylsulfide by *Pseudomonas acidovorans* DMR-11 isolated from peat biofilter. *Biotechnol. Lett.* **13**, 223–228.

[69] Andreae, M.O. (1980) Dimethylsulfoxide in marine and freshwaters. *Limnol. Oceanogr.* **25**, 1054–1063.
[70] Andreae, M.O. (1980) Determination of trace quantities of dimethylsulfoxide in aqueous solutions. *Anal. Chem.* **52**, 150–153.
[71] Suylen, G.M.H., Stefess, G.C., and Kuenen, J.G. (1986) Chemolithotrophic potential of a *Hyphomicrobium* species, capable of growth on methylated sulfur compounds. *Arch. Microbiol.* **146**, 192–198.
[72] Zinder, S.H. and Brock, T.D. (1978) Dimethyl sulfoxide as an electron acceptor for anaerobic growth. *Arch. Microbiol.* **116**, 35–40.
[73] Zinder, S.H. and Brock, T.D. (1978) Dimethyl sulfoxide reduction by microorganisms. *J. Gen. Microbiol.* **105**, 335–342.
[74] Ward, B.B., Kilpatrick, K.A., Novelli, P.C. and Scranton, M.I. (1987) Methane oxidation and methane fluxes in the ocean surface layer and deep anoxic waters. *Nature* **327**, 226–229.
[75] Mopper, K. and Kieber, D.J. (1991) Distribution and biological turnover of dissolved organic compounds in the water column of the Black Sea. *Deep-Sea Res.* **38**, S1021–S1047.
[76] Sieburth, J.M. and Keller, M.D. (1991) Methylaminotrophic bacteria in xenic nanoalgal cultures: incidence, significance, and role of methylated algal osmoprotectants. *Biol. Oceanogr.* **6**, 383–395.
[77] Sieburth, J.McN. (1988), The nanoalgal peak in the dim oceanic pycnocline: is photosynthesis augmented by microparticulates and their bacterial consortia. In: Biogeochemical Cycling and Fluxes Between the Deep Euphotic Zone and Other Oceanic Realms (C.R. Agegian, Ed.), pp. 101–130. NURP Report 88-1,
[78] Banse, K. (1977) Determining the carbon-to-chlorophyll ratio of natural phytoplankton. *Marine Biol.* **41**, 199–212.
[79] Taylor, B.F. and Gilchrist, D.C. (1991) New routes for aerobic biodegradation of dimethylsulfoniopropionate. *Appl. Environ. Microbiol.* **57**, 3581–3584.
[80] Kiene, R.P. and Service, S.K. (1992) The influence of glycine betaine on dimethyl sulfide and dimethylsulfoniopropionate concentrations in seawater. In: The Biogeochemistry of Global Change: Radiatively Important Trace Gases (R.S. Oremland, Ed.), Chapman and Hall, New York, in press.
[81] Chambers, S.T., Kunin, C.M., Miller, D. and Hamada, A. (1987) Dimethylthetin can substitute for glycine betaine as a osmoprotectant molecule for *Escherichia coli. J. Bacteriol.* **169**, 4845–4847.
[82] King, G.M. (1984) Metabolism of trimethylamine, choline, and glycine betaine by sulfate-reducing and methanogenic bacteria in marine sediments. *Appl. Environ. Microbiol.* **48**, 719–725.
[83] Shieh, H.S. (1966) Further studies on the oxidation of betaine by a marine bacterium, *Achromobacter cholinophagum. Can. J. Microbiol.* **12**, 299–302.

3

Sources and Sinks of Halogenated Methanes in Nature

R. Wever

E.C. Slater Institute for Biochemical Research, Plantage Muidergracht 12, 1018 TV Amsterdam, The Netherlands

1 SUMMARY

The release of halogenated volatile organic compounds such as bromoform, dibromomethane and methyl chloride by natural sources and the biochemical pathways by which they are synthesized are discussed. It will be shown that seaweeds and phytoplankton are involved in the production of the brominated compounds found in oceans and in which haloperoxidases appear to play a key role. The biosynthesis of methyl chloride in oceans is not well understood, but in contrast its formation in the terrestrial environment by white rot fungi is known in some detail and a methyl transferase is involved in its production. These halogenated compounds are produced in substantial amounts and after reaching the atmosphere will be photolyzed and participate in chemical processes in the atmosphere. In particular the interaction of bromoform with ozone in the troposphere is discussed in some detail.

2 INTRODUCTION

Since the discovery of a dramatic thinning of the stratospheric ozone layer over Antarctica caused by man-made halocarbons and possible global climate changes, there is an increasing interest in atmospheric chemical processes. Most studies are focused on the chlorofluorocompounds entering the stratosphere after their release at the surface of the Earth [1,2]. The realization that natural processes contribute to the halocarbon load is also triggering research to assess the sources, to determine the amounts involved, their possible sinks and interaction with other biochemical processes. Unlike the chlorofluorocompounds, with the exception of methyl chloride, most of the natural halogenated volatiles, due to their short life time in the troposphere after photolysis, will not reach the stratosphere. It should be realized, however, that bromine in the stratosphere has a much larger impact on ozone destruction that does chlorine [2,3]. Thus, even a small amount of a compound like methyl bromide may still be involved in stratospheric ozone destruction because of its high ozone depletion potential. This review focuses on the sources of bromoform, its biosynthesis and its sinks in the atmosphere. Also the biogenic formation of methyl chloride is discussed. The other volatile halocarbons (methyl bromide and methyl iodide) will not be treated here since they have been reviewed recently [4].

3 BROMOMETHANES

3.1 *Bromoform in oceans*

Several studies have shown that in the Arctic Ocean halocarbons are present in the ng/l range. The data from Dyrssen and Fogelqvist [5,6] demonstrate that the source of bromoform is biogenic and that algal belts in the coastal area of Svalbard are partly responsible. This conclusion was reached from depth profiles and from the bromoform concentration in surface water collected as a function of the distance from the island. Some sampling sites in and outside fjords showed the presence of very high concentrations of bromoform (up to 300 ng/l). High surface water concentrations (up to 12 ng/l) were found for stations situated far from the coast. This biogenic production of bromoform in the Arctic Ocean causes an over-saturation in open sea surface waters when compared with the atmospheric background and this suggests that the flux of bromoform to the air is the major source of the elevated concentrations of atmospheric bromine as will be discussed later. The distribution pattern of bromoform in the Idefjorden on the border of Norway and Sweden and the Skagerrak also shows [7] that bromoform in sea water has a natural source and is derived from algal belts outside the fjords.

The studies by Fogelqvist [6,8] show that there is another natural source. Phytoplankton in the Arctic Ocean are apparently involved in the production of bromoform. The more detailed study of bromoform measured throughout the water column on a transection across the Nansen Basin in the Arctic Ocean by Krysell [9] confirmed this. There is a bromoform maximum (3–7 ng/l) which is indicative of production very close to the surface (5–10 m) and this suggests that pelagic marine algae are the source of bromoform in the northern Arctic Ocean. As pointed out, part of the bromoform will be released into the atmosphere.

Polyhalomethanes ($CHBr$, $CHBr_2Cl$, CH_2Br_2, $CHBr_3$, CH_2I_2) have also been detected in surface waters of the North and South Atlantic and in the North Atlantic surface water samples in a quite high concentration (>6ng/l) [10] and exceptionally high values of bromo- and bromo-chloromethanes were detected in marine air samples of highly bioactive tropical regions by Class et al. [10]. A correlation exists between high concentrations of these compounds in air and water and the occurrence of algae at the coast lines of various islands (the Azores, Bermuda, Tenerife) and in a region of high bioactivity in the Atlantic Ocean near the West Africa Coast [11].

Systematic data concerning the bromoform concentrations in Antartic waters are scarce. The concentration of bromoform was reported to be less than 2 ng/l (M. Krysell, personal communication, quoted in [12]). Unfortunately, further details of this investigation are lacking. A recent analysis of the surface sea water near the Antarctic Peninsula between October and December has been carried out [13] and bromoform was the dominant brominated methane in the South Polar Sea with a mean concentration of 6.2 ng/l. Interestingly, the bromoform concentration and also the concentration of the other brominated methanes is clearly higher at the coastline of the Antarctic Peninsula and the different Antarctic Islands than in the open sea. High concentrations of $CHBr_3$ of up to 50 ng/l were found at the coast of Elephant Island and King George Island. These values are close to those reported [5–8] for the Arctic Ocean, though slightly lower. The high concentrations close to the islands suggest again that seaweeds are involved in the production. However, a detailed study of a number of seaweeds collected near Signy Island [12] failed to demonstrate the presence of bromoperoxidase, an enzyme which is supposed, as will be discussed below, to be involved in the biosynthesis of brominated methanes. The presence of vanadium bromoperoxidases in some seaweeds, however, is a function of the season [14,15] and the highest activity is found in some seaweeds in winter in the northern hemisphere. The plants investigated [12] and which are dominating in the Antarctic, were collected in the Weddell Sea in November/December and the bromoperoxidase may have been absent in this season.

The sea water concentration of the different brominated methanes shows a good correlation and this suggests [13] a common biosynthetic pathway. In contrast methyl iodide does not show any correlation with the presence of the brominated methanes, suggesting a different origin.

3.2 Biosynthesis of the brominated methanes

Direct measurements [16,17] have been carried out on the release of halogenated compounds by the brown seaweeds *Ascophyllum nodosum*, *Fucus vesiculosis*, the green seaweed *Ulva lacta* and the red seaweed *Gigertina stellata*. Various halogenated compounds have been detected, the three major compounds being bromoform, dibromochloromethane and dibromomethane. In the study by Class et al. [10] it was reported that the brown alga *Fucus sargassum* collected near the Bermudian Islands and *Laminaria laminaria* collected at the Cape of Good Hope showed a specific emission pattern of volatile organohalides into the air of bromoform, dibromomethane, dibromochloromethane and bromodichloromethane. Iodinated compounds have also been detected. Brominated methanes and other brominated compounds have also been found in the marine red algae *Bonnemaisonia hamifera* [18,19] and the red algae *Asparagopsis taxiformis* [20,21] although release in the sea water has not been reported.

There are at least two pathways by which brominated methanes are formed by seaweeds. One mechanism which has been proposed [18] is the bromination of keto-acids present in the seaweed which yields a number of unstable intermediates which decay *via* the haloform reaction to bromomethane and bromoform. This bromination is catalyzed by a bromoperoxidase and up to now two classes of haloperoxidases have been detected in seaweed: the classical peroxidases which contain a heme as a prosthetic group [22] and the novel vanadium-containing bromoperoxidases (for a review see [23]). These vanadium bromoperoxidases catalyze bromination reactions by formation of HOBr (Eqn. 1) which in a subsequent reaction with a nucleophilic acceptor (AH) gives rise to the formation of the brominated product ABr (Eqn. 2).

$$H_2O_2 + Br^- + H^+ \rightarrow HOBr + H_2O \quad (1)$$

$$HOBr + AH \rightarrow ABr + H_2O \quad (2)$$

Recently it has been shown that the vanadium bromoperoxidases [24] are also able to carry out chlorination reactions albeit at a slow rate compared to bromination. This enzymic activity may be responsible for the formation of the mixed chloro-bromomethanes which are also released by some seaweeds in the sea water.

An alternative pathway for the formation of bromoform has been proposed by Wever et al. [25]. They showed that in some macro-algae, bromoperoxidases are located on the surface of the plants and are able to release HOBr in sea water. HOBr will rapidly react with organic matter [26,27], consisting mainly of fulvic and humic acids, to give rise to compounds that upon decay lead to formation of bromoform. However, the pathway by which these volatile brominated compounds are formed remains to be characterized. It is of interest to note that upon chlorination of fulvic and humic acid in aqueous solution, chloroform is found. The corresponding bromine-substituted products are thought to be the result of parallel bromination reactions [28].

Whatever the pathway may be, it is clear that bromoperoxidases in seaweeds are involved in the biosynthesis of the volatile brominated compounds and bromoperoxidases are present in a great number of seaweeds as found by Hewson and Hager [29] in a survey of the species along the Caribbean Coast of Central America. Some of these enzymes are hemoproteins [22], others contain vanadium. In [12] a summary is given of the various seaweed species which up to now have been shown to contain the novel vanadium bromoperoxidases. These species come from all parts of the world including the North Atlantic Ocean. Some of the species like *Laminaria saccharina, L. digitata, L. hyperborea* are also found in the Arctic Ocean and it is likely that these also contain the vanadium bromoperoxidases. The vegetation along the coastal shores of this Ocean is luxuriously developed and is dominated [30,31] by these and other brown seaweeds and their biomass is high.

3.3 Amounts of brominated methanes formed

Direct measurements on the amounts of organohalides produced by seaweeds are still scarce. The only data available are those of Gschwend *et al.* [16,17] who measured the rate of release of volatile halogenated compounds. Values of 1–10 microgram bromine in the form of organobromides per gram seaweed per day were found. Assuming that all seaweeds would release these compounds at the same rate and that the total global biomass [32] of algae is 10^{13} g, an annual input of 10^{10} g per year was estimated. Wever *et al.* [25] showed that the brown macro-alga *A. nodosum* was able to release HOBr directly in the seawater when the plant was exposed to light. The amount was 68 nmol HOBr/g/h. Although the biomass of this seaweed is not known, it grows in large quantities in the North Atlantic and Western Russian polar seas and a biomass of 10^{11} g is certainly an acceptable value. From this an input of 2×10^9 g of HOBr in the biosphere, produced annually by this seaweed alone was estimated [25]. However, the phenomenon that light was able to induce HOBr formation was seen only with this seaweed and not with some others tested. Nevertheless, in these macro-algae the bromoperoxidase is accessible to added substrates and addition of hydrogen peroxide to the intact plant leads to formation of HOBr [25]. The reason why these algae form HOBr and brominated compounds is not clear. HOBr is a strong biocidal agent and it is likely that formation of these compounds is part of a host defense system and may prevent fouling by micro-organisms or act as an antifeeding system.

3.4 Sinks and atmospheric bromine

As discussed earlier, the Arctic Ocean is saturated with bromoform and ventilation to the atmosphere will occur. For the Arctic Ocean, on the basis of a mean surface concentration of 5.5 ng/l, an annual flux of 10^9 g of bromoform from the water to the atmosphere is estimated [9]. It should be noted that this value is an upper limit and may be affected by seasonal variation and factors like ice cover and light intensity. This flux should be compared to the global yearly flux of bromoform [33] produced by all oceans of about 10^{12} g. An estimate of the life time of bromoform has also been made. In cold seawater the hydrolysis of bromoform is a very slow process [9], however, the atmospheric lifetime is much shorter and may be as low as 2 weeks [33]. Because of this short lifetime it is unlikely that any significant amount of this brominated compound will reach the stratosphere and will thus contribute little to stratospheric bromine. As a result of absorption of solar radiation in the lower troposphere, bromoform will be photolyzed [34,35] and a reaction with ozone or OH radicals may occur [11,35,36].

High levels of brominated organic species, of which bromoform was the main component, have been detected in the Arctic atmosphere [36]. Further, the bromine content of Arctic aerosol at Point Barrow (Alaska) and Alert (Canadian High Arctic) shows an annual sharp maximum between February and May [37,38] which was also observed in the aerosol over Spitsbergen at the same time [36]. When compared to the natural background level these results show [36] that the bromine concentrations in the Arctic atmosphere are extraordinarily high, the highest found anywhere in the world. Direct measurements of atmospheric methyl bromide and bromoform at Barrow also showed a large seasonal variation with maximal bromoform concentrations in winter and minimal in summer [39].

There is considerable interest in this phenomenon since Barrie et al. [34] and Bottenheim et al. [40] showed that at polar sunrise ozone destruction occurs in the lower Arctic atmosphere at Alert and that this phenomenon strongly anticorrelated with the presence of filterable bromine in aerosols. Organobromides like bromoform are easily photolyzed and are precursors of filterable bromine in aerosols [34]. Both bromoform production and ozone destruction show a seasonal cycle and when there is a decrease in ozone there is a concurrent increase in bromine compounds collected on filters [34,40]. However, it has been questioned [40] whether sufficient Br atoms may be formed from bromoform photolysis to cause substantial ozone depletion. Further, there is a dispute [41] concerning the source of bromine on filters e.g. it has been suggested to result from the interaction of NOx compounds with NaBr from sea salt in aerosols to yield $BrNO_2$. This compound will rapidly photolyse at sunrise forming ozone-destroying bromine atoms. Alternatively, a mechanism has been proposed [42] involving the conversion of sea salt derived Br^-, which is accumulated in the dark polar night on aerosols and snow pack surface, to gas-phase Br_2 which is driven off the snow as the solar flux becomes more intense [42]. Photolysis of the liberated Br_2 could yield Br which can then destroy ozone through the cycle:

$$Br + O_3 \longrightarrow BrO + O_2 \qquad (3)$$

$$BrO + BrO \longrightarrow 2\,Br + O_2 \qquad (4)$$

Normally in the gas-phase the Br_2 would be converted to HBr or brominated organic compounds and this would shut off further depletion of O_3. However, particles (aerosols and ice crystals) are efficient scavengers of HOBr and the brominated organic compounds. Subsequent photo-induced heterogeneous conversion to Br_2 in the gas-phase may occur, restoring the cycle. Although part of the chemistry occurring in the atmosphere can be explained by this photochemical bromine production it ignores the input of volatile bromomethanes from the marine environment to the atmosphere and also the fact that bromoform in the Arctic atmosphere is maximal in the dark polar night [39].

A very likely possibility is that the source of bromine-containing particles and bromoform in the atmosphere is biogenic and is produced in the oceans in the northern part of the northern hemisphere. It has been proposed [15] that the vanadium bromoperoxidases present in seaweeds in the Arctic Ocean and North Atlantic are responsible for the production of bromoform and other halogenated compounds and that this enzymatic activity is linked to the observed ozone destruction at ground level in the Arctic. It is not clear yet whether the seasonal effects observed in the Arctic atmosphere are due to triggering of chemical processes in the atmosphere by sunlight or that the biogenic source shows a seasonal activity. If the latter is the case, peroxidases in seaweeds and their products have had a much greater impact on the chemical processes in the atmosphere than anticipated.

4 METHYL CHLORIDE

This halocarbon is produced in higher amounts than the others. According to the estimates [43–45] about $2.5–5 \times 10^{12}$ g/yr is released which is substantially more than the industrial source. Methyl chloride accounted for about 20% of the total atmospheric chlorine for the year 1985, the major part coming from chlorofluorocompounds [2]. Part of its biosynthetic pathway has been resolved. Wuosmaa and Hager [46] showed that marine algae produced methyl chloride probably via a methyl transferase that catalyzes the methylation of chloride in the presence of S-adenosyl methionine. This work follows that of Turner et al. [47] and Harper et al. [48-49] who showed that mushrooms and many species in the Hymenochaetaceae, a wide spread family of white-rot fungi produce methyl chloride as a secondary metabolite in a high yield. In a survey amongst polypores [50], of the 63 species investigated in the family of Hymenochaetaceae more than 50% were able to synthesize methyl chloride. This suggests that forest habitats could be a major source. Although quantitative measurements have been made [49] concerning the formation of the compound in culture media, no data or estimates are available concerning the amounts of methyl chloride produced by these fungi at natural sites and the release into the atmosphere. Harper and Hamilton [52] were able to demonstrate that methionine is a highly effective precursor of methyl chloride. Further, when NaCl in the growth medium of *P. pomaceus* was substituted by NaBr no significant difference was found [50] in the incorporation of the label into CH_3Br from that into CH_3Cl. This shows that these halomethanes are almost certainly derived from the action of the same enzyme from the same precursor pool. Cell extracts of *P. pomaceus* indeed catalyze the enzymatic synthesis of methyl halides using S-adenosyl methionine as a methyl donor [46]. The physiological role of the compound has been obscure for some time. However, Harper et al. [53–55] demonstrated that methyl chloride may act as a methyl donor in the biosynthesis by *P. pomaceus* of methyl esthers of benzoic and furoic acid, and the methylation of phenol and butyrate. Also methylation via methyl chloride was observed in a non-$CHCl_3$ releasing fungus. This participation suggested that this compound may have a role as an intermediate in the biosynthesis of nonhalogenated natural products. Indeed, as was shown [56], it also acts as a methyl donor in the veratryl alcohol synthesis in the lignin degrading fungi *Phanerochaete chrysosporium*, *Phlebia radiata* and *Coriolus versicolor*. Veratryl alcohol has a key role in the degradation of lignin, a major component of biomass on Earth, and its biosynthesis may be of critical importance for these lignin degrading fungi. It was proposed [56] that the lignin degrading fungi possess a tightly coupled multi-enzyme system in which methyl chloride biosynthesis is coupled to methyl chloride utilization for methylation of veratryl alcohol precursor without any significant release of methyl chloride to the external environment. Recent results [57] support this very interesting idea. It was shown that the biosynthesis of veratryl alcohol by the fungus was initiated much earlier in fungal cultures supplemented with methyl chloride. A parallel effect of the compound was noted on lignin peroxidase activity.

5 CONCLUSIONS

From the literature a picture emerges that explains at least part of the bromoform formation in the biosphere. An important issue remains whether the bromine in particles collected on filters and which peaks every year at polar sunrise in the Arctic troposphere, is of biogenic origin. Clearly, data are still lacking on the seasonal dependence of bromoform in sea water and possible relation to activity of seaweeds and phytoplankton. It should be realized that the biomass of phytoplankton in the Arctic Ocean is high, in particular in spring during blooming and it may lead to a short but large flux of bromoform to the atmosphere. It would be of great interest to see which plankton species are responsible for its formation and whether peroxidases are involved in the biosynthesis. Also, the biochemical pathway of methyl chloride formation is being unravelled. This gas is especially interesting because it can reach the stratosphere and may be involved in a natural cycle of ozone destruction. This cycle may be affected by changes in the natural flux of this gas. For example it should also be noted that during biomass burning, methyl chloride is released [58]. This additional release may, however, be compensated by the deforestation which will affect the habitat of various fungi involved in wood degradation and may lead to lower natural terrestrial emissions.

ACKNOWLEDGEMENTS

This work is supported in part by the Netherlands Foundation for Chemical Research (SON) with financial aid from the Netherlands Organisation for Scientific Research (NWO).

REFERENCES

[1] Brune, W.H., Anderson, J.G., Toohey, D.W., Fahey, D.W., Kawa, S.R., Jones, R.L., McKenna, D.S. and Pole, L.R. (1991) The potential for ozone depletion in the Arctic polar stratosphere. *Science* **252**, 1260–1266.

[2] Prather, M.J. and Watson, R.T. (1990) Stratospheric ozone depletion and future levels of atmospheric chlorine and bromine. *Nature* **344**, 729–734.

[3] Solomon, S. and Akbritton, D.L. (1992) Time-dependent ozone depletion: potentials for short- and long-term forecasts. *Nature* **357**, 33–37.

[4] Wever, R. (1991) Formation of halogenated gases by natural sources. In: Microbial Production and Consumption of Greenhouse Gasses: Methane, Nitrogen Oxides, and Halomethanes (Rogers, J.E. and Whitman, W.B., Eds) pp. 277–285, American Society for Microbiology, Washington, DC.

[5] Dyrssen, D. and Fogelqvist, E. (1981) Bromoform concentrations of the Arctic Ocean in the Svalbard area. *Oceanol. Acta* **4**, 313–317.

[6] Fogelqvist, E. (1985) Carbon tetrachloride, tetrachloroethane, 1,1,1,-trichloroethane and bromoform in Arctic sea water. *J. Geophys. Res.* **90**, 9181–9193.

[7] Fogelqvist, E. and Krysell, M. (1988) The anthropogenic and biogenic origin of low molecular weight halocarbons in a polluted fjord, the Idefjorden. *Mar. Poll. Bull.* **17**, 378–382.

[8] Fogelqvist, E., Josefsson, B. and Roos, C. (1982) Halocarbons as tracer substances in studies of the distribution pattern of chlorinated waters in coastal areas. *Environ. Sci. Technol.* **16**, 479–482.

[9] Krysell, M. (1991) Bromoform in the Nansen Basin in the Arctic Ocean. *Mar. Chem.* **33**, 187–197.

[10] Class, Th., Kohnle, R. and Ballschmiter, K. (1986) Chemistry of organic traces in air VII: bromo- and bromochloromethanes in air over the Atlantic Ocean. *Chemosphere* **4**, 429–436.

[11] Class, Th. and Ballschmiter, K. (1988) Chemistry of organic traces in air. VIII: Sources and distribution of bromo- and bromochloromethanes in marine air and surface water of the Atlantic Ocean. *J. At. Chem.* **6**, 35–46.

[12] Wever, R., Tromp, M.G.M., Van Schijndel, J.W.P.M., Vollenbroek, E.M., Olsen, R.L. and Fogelqvist, E. (1992) Bromoperoxidases, their role in the formation of HOBr and bromoform by seaweed. In: The Biogeochemistry of Global Change: Radiative Trace Gases (Oremland, R.S., Ed.), Chapman and Hall, New York, in press.

[13] Reifenhauser, W. and Heumann, K.G. (1992) Bromo- and chloromethanes in the Antarctic atmosphere and the south polar sea. *Chemosphere* **24**, 1293–1300.

[14] Vilter, H., Glombitza, K.-W. and Grawe, A. (1983) Peroxidases from Phaeophyceae I: extraction and detection of the peroxidases. *Bot. Mar.* **26**, 331–340.

[15] Wever, R. (1988) Ozone destruction by algae in the Arctic atmosphere. *Nature* **335**, 501.

[16] Gschwend, P.M., MacFarlane, J.K. and Newman, K.A. (1985) Volatile halogenated organic compounds released to sea water from temperate marine macroalgae. *Science* **227**, 1033–1036.

[17] Gschwend, P.M. and MacFarlane, J.K. (1986) Polybromomethanes, a year round study of their release to seawater from *Ascophyllum nodosum* and *Fucus vesiculosis*. In: Organic Marine Geochemistry (Sohn, M.L., Ed.) pp. 314–322, ACS Symposium Series 305.

[18] Theiler, R., Cook, J.C., Hager, L.P. and Siuda, J.F. (1978) Halohydrocarbon synthesis by bromoperoxidase. *Science* **202**, 1094–1096.

[19] Fenical, W. (1982) Molecular aspects of halogen-based biosynthesis of marine natural products. *Recent Adv. Phytochem.* **13**, 219–239.

[20] Burreson, J.A., Moore, R.E. and Rohler, P.P. (1976) Volatile halogen compounds in the alga *Asparagopsis taxiforms* (Rhodophyta). *J. Agric. Food Chem.* **24**, 856–861.

[21] Fenical, W. (1974) Polyhaloketons from the red seaweed *Asparagopsis taxiformis*. *Tetrahedron Lett.* **51**, 4463–4466.

[22] Manthey, J.A. and Hager, L.P. (1989) Characterization of the catalytic properties of bromoperoxidase. *Biochemistry* **28**, 3052–3057.

[23] Wever, R. and Kustin, K. (1990) Vanadium: A biologically relevant element. *Adv. Inorg. Chem.* **35**, 81-115.

[24] Soedjak, H.S. and Butler, A. (1990) Chlorination catalyzed by vanadium bromoperoxidases. *Inorg. Chem* **29**, 5015–5017.

[25] Wever, R., Tromp, M.G.M., Krenn, B.E., Marjani, A. and Van Tol, M. (1991) Brominating activity of the seaweed *A. nodosum*: Impact on the biosphere. *Environ. Sci. Technol.* **25**, 446–449.

[26] Helz, G.R. and Hsu, R.Y. (1970) Volatile chloro- and bromocarbons in coastal waters. *Limnol. Oceanogr.* **23**, 859–869.

[27] Jaworske, D.A. and Helz, G.R. (1985) Rapid consumption of bromine oxidants in river and estuarine waters. *Environ. Sci. Technol.* **19**, 1188–1191.

[28] Boyce, S.D. and Hornig, J.F. (1983) Reaction pathways of trihalomethane formation from the halogenation of dihydroxy aromatic model compounds from humic acid. *Environ. Sci. Technol.* **17**, 202–211.

[29] Hewson, W.D. and Hager, L.P. (1980) Bromoperoxidases and halogenated lipids in marine algae. *J. Phycol.* **16**, 340–345.

[30] Kjellman, F.R. (1883) The algae of the Arctic Sea. Kongl. Svenska Vetenskaps Akademiens Handlingar 20, 1–61. Reprinted (1971) by Otto Koeltz Antiquariat Koenigstein-Taunus, BRD.

[31] Svendson, P. (1959) The algal vegetation of Spitsbergen. Norsk Polar Institutt Skrifter **116**, 3–52.

[32] Waaland, R.J. (1981) Commercial utilization. In: The Biology of Seaweeds (Lobban, C.S. and Wynne, M.J., Eds) pp. 726–741, University of California Press, Berkeley and Los Angeles.

[33] Penkett, S.A., Jones, B.M.R., Rycrofft, M.J. and Simons, D.A. (1985) An interhemispheric comparison of the concentrations of bromine compounds in the atmosphere. *Nature* **318**, 550–553.

[34] Barrie, L.A., Bottenheim, J.W., Schnell, R.C., Crutzen, P.J. and Rasmussen, R.A. (1988) Ozone destruction and photochemical reactions at polar sunrise in the lower Arctic atmosphere. *Nature* **334**, 138–141.

[35] Cicerone, R.J. (1981) Halogens in the atmosphere. *Rev. Geophysics Space Physics* **19**, 123–139.

[36] Berg, W.W., Heidt, L.E., Pollock, W.H., Sperry, P.D. and Cicerone, R.J. (1984) Brominated organic species in the Arctic atmosphere. *Geophys. Res. Lett.* **11**, 429–432.

[37] Berg, W.W., Sperry, P.D., Rahn, K.A. and Gladney, E.S. (1983) Atmospheric bromine in the Arctic. *J. Geophys. Res.* **88**, 6719–6736.

[38] Sturges, W.T. and Barrie, L.A. (1988) Chlorine, bromine and iodine in Arctic aerosols. *Atmos. Environment* **22**, 1179–1194.

[39] Cicerone, R.J., Heidt, L.E. and Pollock, W.H. (1988) Measurements of atmospheric methyl bromide and bromoform. *J. Geophys. Res.* **93**, 3745–3749.

[40] Bottenheim, J.W., Barrie, L.A., Atlas, E., Heidt, L.E., Niki, H., Rasmussen, R.A. and Shepson, P.B. (1990) Depletion of lower tropospheric ozone during Arctic spring: The Polar Sunrise experiment 1988. *J. Geophys. Res.* **95**, D11, 18555–18568.

[41] Finlayson-Pitts, B. L., Livingstone, F.E. and Berko, H.N. (1990) Ozone destruction and bromine photochemistry at ground level in the Arctic spring. *Nature* **343**, 622–625.

[42] McConnel, J.C., Henderson, G.S., Barrie, L.A., Bottenheim, J.W., Niki, H., Langford, C.H. and Templeton, E.M.J. (1992). Photochemical bromine production implicated in Arctic boundary-layer ozone depletion. *Nature* **355**, 150–152.
[43] Rasmussen, R.A., Rasmussen, L.E., Khalil, M.A.K. and Dalluge. R.W. (1980) Concentration distribution of methyl chloride in the atmosphere. *J. Geophys. Res.* **85**, 7350–7356.
[44] Singh, H.B., Salas, L.J., Shigeishi, H. and Scribner. E. (1979) Atmospheric halocarbons, hydrocarbons and SF_6: global distributions, sources and sinks. *Science* **203**, 899–903.
[45] Singh, H.B., Salas, L.J. and Stiles, R.E. (1983) Methyl halides in and over the eastern Pacific. *J. Geophys. Res.* **88**, 3684–3690.
[46] Wuosmaa, A.M. and Hager, L.P. (1990) Methyl chloride transferase: a carbocation route for biosynthesis of halometabolites. *Science* **249**, 160–162.
[47] Turner, E.M., Wright, M., Ward, T., Osborne, D.J. and Self, R. (1975) Production of ethylene and other volatiles and changes in cellulase and lactase activities during the life cycle of the cultivated mushroom. *J. Gen. Microbiol.* **91**, 167–176.
[48] Harper, D.B. (1985) Halomethane from halide ion – a highly efficient fungal conversion of environmental significance. *Nature* **315**, 55–57.
[49] Harper, D.B. and Kennedy, J.T. (1986) Effect of growth conditions of halomethane production by *Phellinus* species: biological and environmental implications. *J. Gen. Microbiol.* **132**, 1231–1246.
[50] Harper, D.B., Kennedy, J.T. and Hamilton, J.T.G. (1988) Chloromethane biosynthesis in poroid fungi. *Phytochemistry* **27**, 3147–3153.
[51] White, R.N. (1982) Biosynthesis of methyl chloride in the fungus *Phellinus pomaceus*. *Arch. Microbiol.* **132**, 100–102.
[52] Harper, D.B. and Hamilton, J.T.G. (1988) Biosynthesis of chloromethane in *Phellinus pomaceus*. *J. Gen. Microbiol.* **134**, 2831–2838.
[53] Harper, D.B., Hamilton, J.T.G., Kennedy, J.T. and McNally, K.J. (1989) Chloromethane, a novel methyl donor for biosynthesis of esters and anisoles in *Phellinus pomaceus*. *Appl. Env. Microbiol.* **55**, 1981–1989.
[54] McNally, K.J., Hamilton, J.T.G. and Harper, D.B. (1990) The methylation of benzoic and n-butyric acids by chloromethane in *Phellinus pomaceus*. *J. Gen. Microbiol.* **136**, 1509–1515.
[55] McNally, K.J. and Harper, D.B. (1991) Methylation of phenyl by chloromethane in the fungus *Phellinus pomaceus*. *J. Gen. Microbiol.* **137**, 1029–1032.
[56] Harper, D.B., Buswell, J.A., Kennedey, J.T. and Hamilton, J.T.G. (1990) Chloromethane, methyl donor in veratryl alcohol biosynthesis in *Phanerochaete chrysosporium* and other liginin-degrading fungi. *Appl. Env. Microbiol.* **56**, 3450–3457.
[57] Harper, D.B., Buswell, J.A. and Kennedy, J.T. (1991) Effect or chloromethane on veratryl alcohol and lignin peroxidase production by the fungus *Phanerochaete chrysosporium*. *J. Gen. Microbiol.* **137**, 2867–2872.
[58] Anderson, C. (1990) Methyl chloride gas implicated. *Nature* **348**, 377.

4

Microbial Transformations and Biogeochemical Cycling of One-carbon Substrates Containing Sulphur, Nitrogen or Halogens

Don P. Kelly, Gill Malin[1] and Ann P. Wood[2]

Natural Environment Research Council, Polaris House, Swindon SN2 1EU, UK;
[1]*School of Environmental Sciences, University of East Anglia, Norwich NR4 7TJ, UK;* [2]*Division of Life Sciences, King's College London, W8 7AH, UK*

1 SUMMARY

Over the past twenty years the predominant gaseous component of the sulphur cycle has been incontrovertibly shown to be dimethyl sulphide (DMS). This is known to be degraded by diverse bacteria as a methylotrophic and chemolithotrophic substrate. More recently DMS has been shown to be converted to methanesulphonic acid (MSA) in the atmosphere, and MSA shown to be degraded by some methylotrophic bacteria. We review the present understanding of MSA transformation by a pure culture of a novel methylotroph, in which the MSA is initially cleaved to formaldehyde and sulphite by a MSA monooxygenase. Lesser contributors to the biogeochemical carbon cycle are the halomethanes, carbon disulphide (CS_2), thiocyanate (SCN) and carbonyl sulphide (COS), all of which arise in the environment from natural biological sources. Carbon disulphide is produced by some plants, and novel gram positive endospore-forming bacteria able to use it as a sole substrate for growth have recently been isolated from the leaf surface of the oak, *Quercus lobata*. Another plant product, thiocyanate, is degraded by some heterotrophs and by some thiobacilli. The probable central role of carbonyl sulphide as an intermediate in the microbiological transformation of both carbon disulphide and thiocyanate is considered. COS is the most abundant of the atmospheric sulphur trace gases, arising from natural sources and from the atmospheric oxidation of CS_2. Halogenated methanes represent a small but significant component in the atmospheric cycle of one-carbon gases, and microbial conversion in the marine environment may be a biologically significant process involving methanes containing chlorine, bromine or iodine substitutions, even to the extent represented by CCl_4, CH_2I_2 and $CHBr_2Cl$. Microbiological and biochemical understanding of most of the lesser one-carbon sulphur and halogen compounds are in early stages of elucidation, and much remains to be done to study this field. We also consider the sources, and microbiological and biogeochemical significance of the methylamines, amino-carbon compounds with one or more methyl substitutions, about which considerably more is known in relation to their biochemical transformation.

2 INTRODUCTION

Second only to carbon dioxide, methane is well-established as the most important gaseous input of carbon to the atmosphere [W.S. Reeburgh *et al.*, this volume]. There is a diverse collection of other one-carbon compounds that also enter the atmosphere, and cumulatively contribute a small but significant proportion of the total atmospheric carbon cycling. More importantly, these include the major gaseous inputs to the atmosphere of sulphur, halogens, and possibly of combined nitrogen. Predominant among these is dimethyl sulphide (DMS), with lesser amounts of other methylated sulphides [1–3; R.P. Kiene, this volume], contributing some 40–50 Tg S annually to the atmosphere. Lesser contributors to the sulphur cycle are carbon disulphide (CS_2) and carbonyl sulphide (COS), mainly of biological origin, and thiocyanates of anthropogenic and plant origin. All these are used as sources of energy and carbon by several bacterial groups, as are the methylamines. Chlorinated, brominated and iodinated methanes occur particularly in seawater, presumably mainly of algal origin, and are substrates for diverse bacteria, although the metabolism of the mono- and di-chloromethanes have received most attention.

We now review these one-carbon compounds individually, devoting most attention to those aspects not considered in more depth by other contributors to this volume.

3 METHANESULPHONIC ACID (MSA)

3.1 *Natural sources of MSA*

MSA is produced both in the light (by reaction with OH or NO_x) and at night time by dark reactions with NO_3 radical [4–9], as summarized below:

(a) Photochemical (hydroxyl radical) [Daytime]

$$CH_3SCH_3 + OH \longrightarrow CH_3(OH)SCH_3$$
$$CH_3(OH)SCH_3 + 2O_2 \longrightarrow CH_3SO_3H + CH_3O_2$$

(b) Dark reaction with nitrate radical [Night time]

$$CH_3SCH_3 + NO_3 \longrightarrow CH_3SCH_2 + HNO_3$$
$$SCH_3SCH_2 + O_2 \longrightarrow CH_3SCH_2O_2\cdot$$
$$\xrightarrow{NO} CH_3SCH_2O\cdot \longrightarrow CH_3S\cdot$$
$$CH_3S\cdot + NO_2 \longrightarrow CH_3SO\cdot + NO$$
$$CH_3SO\cdot \to \to \to \to CH_3SO_3H + SO_2$$
$$[O_2, NO, H, H_2O][\text{Ratio } 3:1]$$

The absolute amount of MSA generated in the atmosphere is unknown, and the proportion of DMS converted to MSA is believed to differ at different latitudes. Conversion of only 50% of atmospheric DMS could produce about 70 Tg MSA annually, all of which would ultimately return to the earth's surface in rain or by dry deposition [10].

3.2 Microbiological conversions of MSA

Until recently, the only reports of MSA metabolism by microorganisms were of its use as a sulphur source [1]. It is now known that diverse soil and marine bacteria are able to use MSA as a substrate for methylotrophic growth [1,10–12; A.S. Thompson, N.J.P. Owens and J.C. Murrell, unpublished]. One of these, the gram negative strain M2 from soil, has been studied in some detail [11; J. Trickett, M. Davey, R. Finch and J.C. Murrell, unpublished] and is now known to cleave MSA to formaldehyde and sulphite by means of a NADH-dependent mono-oxygenase. Formaldehyde is subsequently oxidised *via* formate to carbon dioxide and NADH and energy thereby generated for growth and to sustain the oxygenase reaction (Fig. 1). Carbon assimilation is by means of the serine pathway [11, and in preparation], and no role for ribulose bisphosphate carboxylase has yet been proved for this organism even for growth on formate as sole substrate. Suspensions of strain M2 effect the complete oxidation of MSA, formaldehyde, formate and methanol, consistent with the equations [J. Trickett, personal communication]:

$$CH_3SO_3H + 2O_2 = CO_2 + H_2SO_4 + H_2O$$
$$HCHO + O_2 = CO_2 + H_2O$$
$$HCOOH + 0.5O_2 = CO_2 + H_2O$$
$$CH_3OH + 1.5O_2 = CO_2 + 2H_2O$$

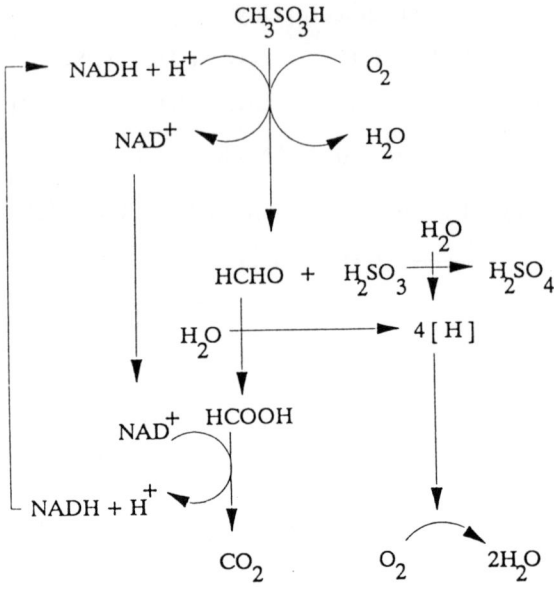

Figure 1 Oxidation of methanesulphonic acid by Strain M2.

Interestingly, suspensions of strain M2 previously grown on MSA could also oxidize amino-MSA, ethanesulphonate, monomethylsulphate, dichloromethane, ethanol and methanol at 24–114% of the rate of MSA oxidation (each supplied at 1.7 mM), and showed slower but measurable oxidation of formate, propanesulphonate and ethane 1,2-disulphonate. While M2 also grew on amino-MSA, ethanesulphonate, formaldehyde, formate, nitromethane, methanol, methylammoniums and (weakly) on propane-, butane- and pentanesulphonates (as well as amino acids, aliphatic acids and sugars), it could not grow on methane, methanephosphonate, hydrogen + carbon dioxide, monomethylsulphate, DMS, COS, CS_2, methyl-MSA or ethanol. The taxonomic position of this organism has not yet been established, except as a fairly typical facultative methylotroph (61 mol% G+C), but it is clearly only one of a range of novel methylotrophs able to use MSA, and demonstrates that microbial degradation completes the 'DMS cycle', whereby global carbon cycling proceeds through dimethylsulphonium propionate, DMSP (and other DMS precursors), via MSA back to carbon dioxide. These bacteria have evolved enzyme systems enabling them to use MSA with high efficiency, distinguishing them from many methylotrophs unable to use MSA, but can also degrade more familiar one-carbon substrates such as methanol and methylamine, as well as being able to oxidize dichloromethane, nitromethane and other less studied compounds. Such organisms can clearly play an important overall role in one-carbon cycling.

4 CARBON DISULPHIDE

4.1 Biological sources of CS_2

Production by soils, higher plants, salt marshes, marine sediments and anaerobic muds give rise to about 80% of the estimated 5Tg CS_2 entering the atmosphere annually [1]. Little is yet known of the biochemistry of CS_2 generation or the total global turnover of this compound.

4.2 Bacterial degradation of CS_2

Until recently the only bacteria known to use CS_2 as the sole source of energy for growth were specialized strains of *Thiobacillus thioparus* [1,10,13]. There is no biochemical evidence for the nature of the initial cleavage of CS_2, although the ability of anaerobic suspensions to produce COS and H_2S suggested that hydrolytic splitting was a probability [1,13]:

$$CS_2 + H_2O = COS + H_2S$$
$$COS + H_2O = CO_2 + H_2S$$

Such a degradative route would dictate that the only way that energy could possibly be conserved by the organism would be from the chemolithotrophic oxidation of H_2S to sulphate:

$$H_2S + 2O_2 = H_2SO_4$$

Such an oxidation route would be typical in thiobacilli but of course also dictates that the only source of carbon for biosynthesis must be carbon dioxide, fixed by an autotrophic process such as the Calvin Cycle. This would be consistent with the only organisms described as using CS_2 as sole substrate being obligately chemolitho-autotrophic *Thiobacillus* strains. If this is the case, the amount of growth on the three substrates CS_2, COS, H_2S (and thiosulphate, which is energetically equivalent to H_2S) should be determined only by their content of reduced sulphur, and hence be in the ratio 2:1:1. Experimentally, *T. thioparus* strain TK-m gave relative yields (g dry wt. mol^{-1}) of 1.35:1:1. The 'shortfall' in yield on CS_2 might indicate that initial cleavage of CS_2 requires a mono-oxygenase [12] (as reported for rat liver [1,14]):

$$CS_2 + 0.5O_2 + 2[H] = COS + H_2S$$

This would require molecular oxygen and a source of reducing equivalents (such as NAD(P)H) which could lower the 'expected' growth yield from 2 (relative to COS) to about 1.25 if one assumes that of the 16[H] made available from oxidising $2H_2S$, to $2H_2SO_4$, 2[H] are consumed by a monooxygenase, 4[H] might be required to provide energy for 'reverse election flow' from cytochromes [15,16] to provide NAD(P)H for the monooxygenase, and would thus leave only 10[H] to support autotrophic growth. These questions invite resolution, and there is the likelihood that more than one cleavage pathway exists. To date no anaerobic CS_2-using strain of *T. denitrificans* has been isolated, in which a hydrolytic cleavage pathway (and hence higher growth yield) would be expected.

4.3 Novel CS_2-utilizing bacteria

Recently, several new strains of bacteria have been isolated and shown to be capable of growth on CS_2 as sole substrate. S. Jordan and A.P. Wood [unpublished data] have obtained such bacteria from the soil in which the oak, *Quercus lobata* was growing, and from the leaf surfaces of this tree. *Q. lobata* is known to be a producer (through the leaves) of CS_2 [17]. Soil isolates were gram negative rods, capable of autotrophic growths on sulphide, thiosulphate or thiocyanate, and might therefore be strains of *Thiobacillus*. They are, however, facultatively heterotrophic and also grow on DMS, diethylsulphide, thioacetic acid, methylamine, formate, glucose and complex media, indicating them not to be *T. thioparus*, and indeed not being immediately attributable to any of the well-defined *Thiobacillus* species [18]. The bacteria isolated from the *Q. lobata* leaf surface were gram positive, endospore-forming rods, initially classifiable as '*Bacillus*'. To date [S. Jordan and A.P. Wood, personal communication] these bacteria have proved impossible to culture autotrophically on thiosulphate or tetrathionate, but do grow on nutrient broth or yeast extract, and on methanethiol, thiophene 2- or 3-carboxylate, and thiophene-2-acetate. It is clear that terrestrial habitats in which CS_2 is available are likely to yield a previously unexpected diversity of more or less specialised bacteria able to use this compound as sole growth substrate.

4.4 CS_2 utilization by Thiothrix ramosa

Recent work by E. Odintsova and A.P. Wood [unpublished observations] has demonstrated a remarkable and previously unsuspected versatility in the metabolism of this filamentous, rosette-forming bacterium [19,20]. Not only has it been shown to grow wholly autotrophically on thiosulphate as sole energy source (with growth yields in the chemostat typical of chemolithotrophs such as *Thiobacillus neapolitanus*), but also to grow on CS_2 both mixotrophically and as sole growth substrate. These observations further extend the range of bacteria involved in the cycling of one-carbon sulphur compounds, as well as extending the capacity for chemolithoautotrophy further among the 'filamentous sulphur bacteria', the study of some of which helped lead Winogradsky to the concept of autotrophy [21]. *T. ramosa* was also able to grow twice as rapidly mixotrophically with lactate and methanethiol as with lactate alone [E. Odintsova and A.P. Wood, unpublished data], indicating such bacteria also to assist in turnover of methylated sulphides.

5 CARBONYL SULPHIDE (COS)

COS is the major product of the photochemical breakdown of CS_2 in the atmosphere and also arises from marine, salt marsh and terrestrial sources [1]. Some reports show it to be the major sulphur trace gas arising from soils, and to be present in the atmosphere at about 510 pg/l, compared to a mean value of 100 pg/l for CS_2 [1,21a–c]. Pine forest canopy appears to be a source of COS (as well as CS_2 and DMS) [22]. Plants have been reported to be a sink for atmospheric COS, but bacterial oxidation is likely to account for the bulk of reoxidation of COS generated in terrestrial and aquatic environments. Much of the turnover of biogenic COS is likely to be by closed cycling in those environments in which it arises. Only strains of *Thiobacillus thioparus* have been shown to grown on COS, which can only be a substrate for chemolithoautotrophy, as it can presumably only provide energy through sulphide oxidation and carbon as carbon dioxide. It is noteworthy that the strains of bacteria able to grow on COS also grow on thiocyanate, but not necessarily on CS_2.

6 THIOCYANATE (SCN)

The thiocyanate ion (S=C=N⁻) is produced both as an anthropogenic pollutant and from natural sources. The best known biological origin is from the enzymatic hydrolysis of the glucosinolates produced by Cruciferous plants [10,23]:

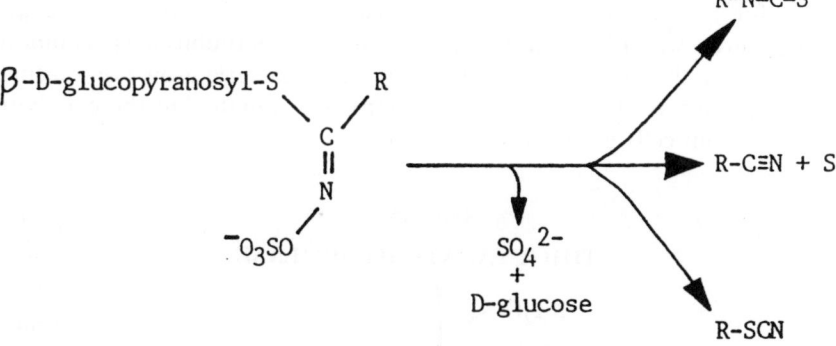

These give rise also to organic isothiocyanates and cyanides, the former also being possible sources of carbon disulphide in soils [24].

Thiocyanate has been shown to be a nitrogen source for some heterotrophic bacteria, in which it is initially hydrolysed to cyanate and sulphide, and the cyanate broken down by cyanase to liberate ammonia [10,25]:

$$SCN^- + H_2O = CNO^- + H_2S$$
$$CNO^- + 2H_2O = CO_2 + NH_3 + OH^-$$

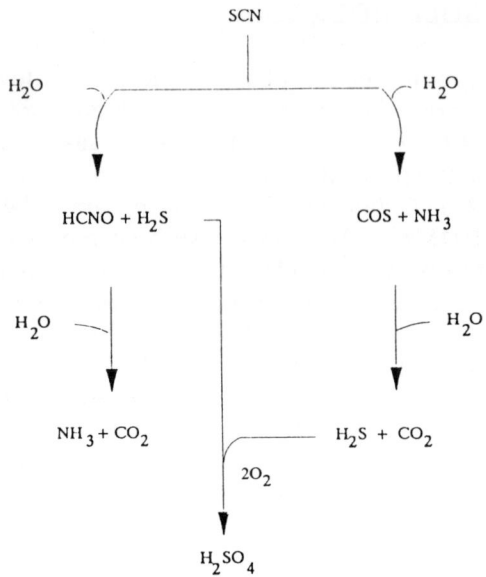

Figure 2 'Cyanate pathway' and 'Carbonyl sulphide pathway' as alternative routes for the degradation of thiocyanate [10].

Evidence for a thiocyanase in *Pseudomonas putida* using thiocyanate as a nitrogen source was obtained by A. Gil Aguirre [unpublished]. Ammonia production from thiocyanate by cell-free extracts of bacteria grown on succinate and thiocyanate was at a rate similar to that from cyanate, but these activities were absent from cells grown with ammonium ion.

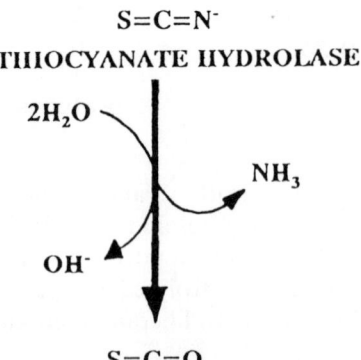

Figure 3 Formation of carbonyl sulphide by the thiocyanate hydrolase from *Thiobacillus thioparus* [28].

Thiocyanate is used as sole autotrophic growth substrate by aerobic and anaerobic *Thiobacillus* strains, and early work suggested that cyanate was an intermediate, but enzymatic evidence was not presented [26,27]. The possibility of COS as an intermediate in SCN breakdown by thiobacilli was subsequently postulated [10]. These possible alternative routes are illustrated in Fig. 2. Evidence for a 'COS pathway' of SCN breakdown has recently been presented for *T. thioparus* [28], in which Katayama et al. have shown a thiocyanate hydrolase (Fig. 3). This enzyme was induced by growth on thiocyanate and while it has a seemingly poor affinity for the substrate (K_m, 11mM), it is the first demonstration of an enzymic specific to thiocyanate oxidation in a *Thiobacillus*.

7 A CENTRAL ROLE FOR COS

COS seems to occupy a linking position in the biogeochemical and photochemical turnover of CS_2 and the breakdown of SCN (and isothiocyanates). Certainly COS now seems to play an essential role as a central intermediate in the oxidation of CS_2 and SCN by thiobacilli. The evidence that plants may both produce and consume COS, that soils and marine environments are exporters of COS, and that much atmospheric CS_2 is converted to COS, all leads to the view that bacteria using COS are likely to be ubiquitous, and that the amount of COS turnover in the natural environment is much greater than that seen to flow through the atmosphere (Fig. 4).

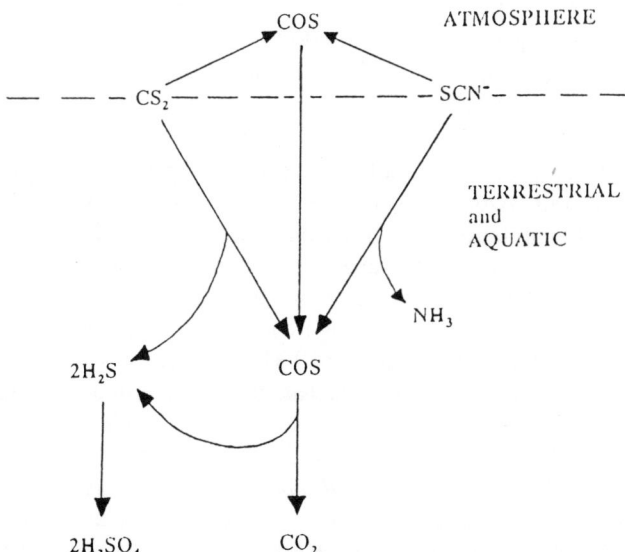

Figure 4 Carbonyl sulphide (COS) as the central intermediate in the biogeochemical oxidation of both carbon disulphide (CS_2) and thiocyanate (SCN^-).

8 METHYLAMINES

The metabolism of methylamines, especially monomethylamine, has been much studied in methylotrophic bacteria and is considered elsewhere in this volume. Methylamines can be the energy-yielding substrates for both methylotrophic and autotrophic biosynthetic pathways, and are also used as nitrogen sources by some heterotrophs. Our purpose is to review here the more general biogeochemical and microbiological roles of the methylamines.

The methylamines (MA's) mono-, di- and trimethylamine (MMA, DMA, TMA) are volatile one-carbon compounds which are derived from quaternary amines (QAs) such as choline, glycine-betaine (GBT), β-alanine-betaine and trimethylamine-N-oxide (TMAO). QAs are abundant in marine animals, plants and macroalgae and are thought to function as osmoregulatory compounds [29,30]. They have also been detected in cyanobacteria [31] and the freshwater microalgae *Ochromonas* and *Nitzchia* [32]. GBT has been found in several unicellular marine algae [33,34] and although a comprehensive survey has not been carried out, it is thought that GBT will prove to be as widely distributed in marine phytoplankton [35] as its sulphur analogue dimethylsulphonium propionate (DMSP) [36]. R. Kiene [this volume] discusses how GBT might influence the cycling of DMSP and DMS.

There have been many studies on the metabolism of MAs and QAs in bacteria and yeasts. MMA, DMA, TMA and TMAO can serve as sole carbon or nitrogen sources for methylotrophic or methazotrophic microorganisms respectively [37]. King [38] found that TMA, GBT and choline simultaneously stimulated methanogenesis and sulphate reduction in intertidal surface sediments. In addition, MMA is a substrate for glutamine synthetase and can therefore be incorporated by non-methylo- and non-methazo-trophs. This property has been exploited in transport studies where MMA is used as an ammonium analogue [39,40].

Several pathways for the dissimilation of MAs have been identified in methylotrophs. In a study of 12 cultures of methylotrophic bacteria Colby and Zatman [41] found two pathways for oxidation of TMA to DMA and formaldehyde. The obligate methylotrophs used TMA dehydrogenae and the facultative methylotrophs a two-enzyme system consisting of TMA monooxygenase and TMAO aldolase which produces DMA and formadelhyde via a non-oxidative reaction. Since the obligate methylotrophs produced a molecule of reduced co-factor for each molecule of TMA oxidized, whereas the facultative species required one molecule of co-factor, they concluded that the metabolic route used by the obligate methylotrophs was more efficient for production of reducing power and harnessing metabolic energy. Many bacteria, including *Alteromonas*, *Escherichia coli* and *Rhodopseudomonas capsulatus*, can use TMAO as an electron acceptor to support growth under anaerobic conditions [42]. DMA is converted to MMA by DMA mono-oxygenase in most methylotrophic bacteria. An exception is *Hyphomicrobium X* which uses DMA dehydrogenase when grown under low partial pressures or oxygen [43]. Finally, MMA can be oxidized to formaldehyde and ammonia by three enzymes: a stable primary-amine dehydrogenase, membrane bound MMA oxidase and a route involving methylated amino acids the most significant being N-methylglutamate dehydrogenase [see 37]. Methazotrophic bacteria utilize MAs via the same pathways as methylotrophs with MMA oxidation representing a key step since the nitrogen atom of methylamine is released at this stage.

There is considerable interest in the role of organic compounds in the global nitrogen cycle, including the relationships between metabolism of QAs and fluxes of volatile MAs and methane to the atmosphere, and in methods for quantifying MAs. Techniques are available for quantifying MAs and their precursors in samples where relatively high concentrations are present [see 30]. A method where samples were prepared by vacuum distillation and analysed using a gas chromatograph with a chemiluminescent detector was used to detect micromolar concentrations of MAs in coastal sediments [44], coastal surface seawater and atmospheric samples [45]. However, method precisions were in the range 22 to 47% for seawater, atmospheric and rainwater samples [45]. A recently developed flow injection extraction − ion chromatography method which uses a comductivity detector shows promise for precise and convenient analysis of nanomolar levels of MAs in seawater and atmospheric samples [S.W. Gibb, personal communication]. Flux calculations based on existing data suggest that the oceans are a source of MMA and TMA [45]. The emission of MAs to the atmosphere from the ocean is potentially highly significant since they would be sources of base to marine aerosols and therefore likely to play a role in the acid-base balance of remote marine aerosols and rainwater. It is clear that further studies are required to define the biogeochemical cycle of MA's in seawater, in particular how these compounds and/or their precursors are produced by marine phytoplankton and whether transformation or utilisation by marine bacteria might influence the air-sea exchange of MMA, DMA and TMA.

9 HALOMETHANES

Some of the C_1-halocarbons found in the environment are listed in Table 1. Many of these may be of biological origin (as well as having anthropogenic sources), and chloro-, bromo-, and chlorobromomethanes in seawater and marine air samples are likely to arise primarily from algal metabolism. Even the polychloromethanes, $CHCl_3$ and CCl_4, are produced and degraded biologically [46], although their importance is not well quantified. Bromoform ($CHBr_3$) is also a significant macroalgal and cyanobacterial product. Probably the major natural halomethane is methyl chloride (CH_3Cl), whose natural production is around 5Tg per year, compared to industrial production of about 2Tg annually [see chapters by Leisinger *et al.* and Wever in this volume for detailed reviews].

Some halocarbons are manufactured for use as intermediates and solvents in the chemical industry, and fumigants for pest control in horticulture and

Table 1 C_1-halocarbons found in the environment

Compound	Formula	Common name
Chloromethane	CH_3Cl	Methyl chloride
Bromomethane	CH_3Br	Methyl bromide
Iodomethane	CH_3I	Methyl iodide
Dichloromethane	CH_2Cl_2	Methylene chloride
Dibromomethane	CH_2Br_2	Methylene bromide
Diiodomethane	CH_2I_2	Methylene iodide
Trichloromethane	$CHCl_3$	Chloroform
Tribromomethane	$CHBr_3$	Bromoform
Triiodomethane	CHI_3	Iodoform
Tetrachloromethane	CCl_4	Carbon tetrachloride
Bromodichloromethane	$CHBrCl_2$	
Bromochloromethane	CH_2BrCl	
Dibromochloromethane	$CHBr_2Cl$	
Chloroiodomethane	CH_2ClI	

food storage. They can also be produced when drinking water and waters used as coolants for power plants are chlorinated. As many of these compounds are known or suspect carcinogens and/or mutagens, they are designated as priority pollutants and may have an impact on public health [47]. C_1-halocarbon compounds are also produced by macroalgae, fungi, various bacteria and an actinomycete (see Wever, this volume). There is also a recent report of halocarbon production by microalgae in Arctic ice [48]. There is great interest in the release of compounds containing chlorine and bromine into the atmosphere because of their role in ozone depletion [49]. There is growing evidence of the use of halomethanes as substrates by bacteria. A bacterium thought to be of the genus *Pseudomonas* was found to degrade dichloromethane (methylene chloride) *via* dehalogenation to formaldehyde [50]. The isolate was a facultative anaerobe but utilization required aerobic conditions. Mikesell and Boyd [51] found that two *Methanosarcina* strains dehalogenated tetrachloromethane (carbon tetrachloride), trichloromethane (chloroform) and tribromomethane (bromoform) while generating methane from methanol, methylamines or acetate. Bouwer and McCarty [52] observed trichloromethane and tetrachloromethane removal by bio-oxidation to CO_2 in the presence of a mixed methanogenic culture with acetate as the primary substrate. Anaerobic microbiological processes were also significant in the transformation of bromodichloromethane, dibromochloromethane and tribromomethane, although chemical degradation also occurred. Using anaerobic enrichment cultures and inhibitors, Freedman and Gossett [53] found that non-methanogenic organisms oxidized dichloromethane to CO_2 and acetate, and methanogens converted dichloromethane degradation products to methane. This study also showed utilisation of dichloromethane as a sole carbon and energy source, but it was necessary to keep the dichloromethane concentration low to prevent inhibition of methanogenesis. Bouwer and McCarty [54] demonstrated transformation of tetrachloromethane, tribromomethane, bromodichloromethane and dibromochloromethane after 8 weeks under anoxic conditions in the presence of denitrifying bacteria with nitrate as an electron acceptor [54]. The reaction pathway was not determined, but production of trichloromethane from tetrachloromethane suggested reductive dechlorination. Chloromethane (methyl chloride) is cometabolized by methylotrophic bacteria and can be used as a sole carbon and energy source by a strain of *Hyphomicrobium* isolated from industrial sewage [55]. Chloromethane, dichloromethane and trichloromethane are substrates for the broad-range enzyme soluble methane mono-oxygenase [see 37]. The reaction requires NADH or an additional readily oxidizable co-substrate such as formaldehyde. In all of these investigations the halocarbon compounds were supplied at microgram per litre concentrations. We are not aware of any studies which have considered whether bacteria can degrade the nanogram per litre concentrations of C_1 halocarbon compounds typically found in seawater [e.g. 56,57]. Several C_1-halocarbons are believed to possess significant oceanic sources [58], hence further research is warranted in order to determine the role of bacteria in production and consumption of these compounds in the natural environment.

10 EXCHANGE OF TRACE GASES TO THE ATMOSPHERE, GLOBAL EFFECTS AND GLOBAL CLIMATE CHANGE

The flux to the air of any trace gas is dependent upon its production and consumption in the underlying aquatic or terrestrial environment and its rate of transfer to the atmosphere. Whereas transport processes for most biogeochemically important gases are governed by physical processes, e.g. air-sea exchange [59], production and consumption processes are primarily microbial or due to photochemical conversions of organic matter. Recently there has been increasing awareness of how uptake and emissions of trace gases by the biosphere can influence the chemical composition and physical properties of the global atmosphere [60,61]. Furthermore, C-, S-, N- and halogen-containing biogenic trace gases can potentially affect the acid/base and redox chemistry and/or radiative balance of the atmosphere (and hence climate) either directly or via aerosol/cloud formation [60]. However, the specific microbial production, transformation and consumption processes, whether these processes control the emissions of atmospherically significant gases, and the magnitudes of fluxes from different ecosystems are poorly understood.

The question of whether global climate change will affect biogeochemical C_1 cycles is difficult to answer. Average global temperatures are expected to increase as greenhouse gases build up, and precipitation patterns (and hence soil moisture content) are also likely to change [63]. Extreme temperature and soil moisture episodes due to long hot summers could greatly influence trace gas emissions. Depletion of stratospheric ozone continues to increase the amount of UV-B radiation reaching the Earth, potentially decreasing primary productivity and possibly influencing community structure in terrestrial and aquatic ecosystems [64]. Climate changes such as these could alter microbial production and consumption of volatile C_1 compounds, physical transfer processes, and additionally create a feedback effect on climate change [65].

ACKNOWLEDGEMENTS

Original work reported by the authors in this review was funded by the Natural Environment Research Council, Science and Engineering Research Council, and a UFC Initiative in Environmental Biotechnology. We wish to thank our colleagues, whose original work is cited in our text, Colin Murrell, Simon Baker, Margaret Davey, Stuart Gibb, Sarah Jordan, Elena Odintsova, Nicholas Owens, Andrew Thompson and Jim Trickett.

REFERENCES

[1] Kelly, D.P. and Smith, N.A. (1990) Organic sulfur compounds in the environment: biogeochemistry, microbiology, and ecological aspects. *Adv. Microb. Ecol.* **11**, 345–385.
[2] Andreae, M.O. (1986) The ocean as a source of atmospheric sulfur compounds. In: The Role of Air-Sea Exchange in Geochemical Cycling (Buat-Menard, P., Ed.), pp. 331–362, Reidel, New York.
[3] Watts, S.F., Brimblecombe, P. and Watson, A.J. (1990) Methanesulphonic acid, dimethyl sulphoxide and dimethyl sulphone in aerosols. *Atmos. Environ.* **24A**, 353–359.
[4] Yin, F., Grosjean, D. and Seinfeld, J.H. (1990) Photooxidation of dimethyl sulfide and dimethyl disulfide. I: Mechanism development. *J. Atmos. Chem.* **11**, 309–364.
[5] Yin, F., Grosjean, D., Flagan, R.C. and Seinfeld, J.H. (1990) Photooxidation of dimethyl sulfide and dimethyl disulfide. II: Mechanism evaluation. *J. Atmos. Chem.* **11**, 365–399.
[6] Jensen, N.R, Hjorth, J., Lohse, C., Skov, H. and Restelli, G. (1991) Products and mechanism of the reaction between NO_3 and dimethylsulphide in air. *Atmos. Environ.* **25A**, 1897–1904.
[7] Jensen, N.R., Hjorth, J., Lohse, C., Skov, H. and Restelli, G. (1992) Products and mechanism of the gas phase reactions of NO_3 with CH_3SCH_3, CD_3SCD_3, CH_3SH and CH_3SSCH_3. *J. Atmos. Chem.*, in press.
[8] Berresheim, H., Andreae, M.O., Ayers, G.P., Gillett, R.W., Merrill, J.T., Davis, V.J. and Chameides, W.L. (1990) Airborne measurements of dimethylsulfide, sulfur dioxide, and aerosol ions over the Southern Ocean south of Australia. *J. Atmos. Chem.* **10**, 341–370.
[9] Cocks, A. and Kallend, T. (1988) The chemistry of atmospheric pollution. *Chem. Brit.* **24**, 884–888.
[10] Kelly, D.P. (1991) Environmental significance and biological turnover of dimethyl sulphide, carbon disulphide and other volatile sulphur compounds. In: International Symposium: Environmental Biotechnology (Verachtert, H. and Verstraete, W., Eds), pp. 107–114, Koninklijke Vlaamse Ingenieursvereiniging vzw.
[11] Baker, S.C., Kelly, D.P. and Murrell, J.C. (1991) Microbial degradation of methanesulphonic acid: a missing link in the biogeochemical sulphur cycle. *Nature* **350**, 627–628.

[12] Kelly, D.P. and Baker, S.C. (1990) The organosulphur cycle: aerobic and anaerobic processes leading to turnover of C_1-sulphur compounds. *FEMS Microbiol. Rev.* **87**, 241–246.
[13] Smith, N.A. and Kelly, D.P. (1988) Oxidation of carbon disulphide as the sole source of energy for the autotrophic growth of *Thiobacillus thioparus* strain TK-m. *J. Gen. Microbiol.* **134**, 3041–3048.
[14] Chengelis, C.P. and Neal, R.A. (1987) Oxidative metabolism of carbon disulfide by isolated rat liver hepatocytes and microsomes. *Biochem. Pharmacol.* **36**, 363–368.
[15] Kelly, D.P. (1982) Biochemistry of the chemolithotrophic oxidation of inorganic sulphur. *Phil. Trans. R. Soc. London Sect. B.* **298**, 499–528.
[16] Kelly, D.P. (1990) Energetics of chemolithotrophs. In: The Bacteria, Vol. 12 (Krulwich, T.A., Ed.), pp. 479–503, Academic Press, San Diego, CA.
[17] Haines, B., Black, M. and Bayer, C. (1989) Sulfur emissions from roots of the rain forest tree *Stryphnodendron excelsum*. In: Biogenic Sulfur in the Environment (Saltzman, E.S. and Cooper, W.J., Eds), pp. 58–69, American Chemical Society (Symposium Series 393), Washington, DC.
[18] Kelly, D.P. and Harrison, A.P. (1989) Genus *Thiobacillus*. In: Bergey's Manual of Systematic Bacteriology, Vol. 3 (Staley, J.T., Bryant, M.P., Pfennig, N. and Holt, J.G., Eds), pp. 1842–1858, Williams and Wilkins, Baltimore, MD.
[19] Odintsova, E.V. and Dubinina, G.A. (1990) New filamentous colourless sulfur bacteria *Thiothrix ramosa* nov. sp. *Microbiologiya* **59**, 637.
[20] Larkin, J.M. (1989) Genus 11. *Thiothrix*. In: Bergey's Manual of Systematic Bacteriology, Vol. 3 (Staley, J.T., Bryant, M.P., Pfennig, N. and Holt, J.G., Eds), pp. 2098–2101, Williams and Wilkins, Baltimore, MD.
[21] Zavarzin. G.A. (1989) Sergei N. Winogradsky and the discovery of chemosynthesis. In: Autotrophic Bacteria (Schlegel, H.G. and Bowien, B., Eds), pp. 17–32, Science Tech, Madison, WI.
[21a] Mihalopoulos, N., Nguyen, B.C., Putaud, J.P. and Belviso, S. (1992) The oceanic source of carbonyl sulfide (COS). *Atmos. Environ.* **26A**, 1383–1394.
[21b] Bandy, A.R., Thornton, D.C., Scott, D.L., Lalevic, M., Lewin, E.E. and Driedger, A.R. (1992) A time series for carbonyl sulfide in the northern hemisphere. *J. Atmos. Chem.* **14**, 527–534.
[21c] Mihalopoulos, N., Putaud, J.P., Nguyen, B.C. and Belviso, S. (1992) Annual variation of atmospheric carbonyl sulfide in the marine atmosphere in the southern Indian Ocean. *J. Atmos. Chem.* **13**, 73–82.
[22] Berresheim, H. and Vulcan, V.D. (1992) Vertical distributions of COS, CS_2, DMS and other sulfur compounds in a loblolly pine forest. *Atmos. Environ.* **26A**, 2031–2036.
[23] Drobnica, L., Kristian, P. and Augustin, J. (1977) The chemistry of the NCS group. In: The Chemistry of Cyanates and their Derivatives (Patai, S., Ed), pp. 1003–1221, Wiley, New York, NY.
[24] Challenger, F. (1959) Aspects of the Organic Chemistry of Sulphur, pp. 132–133. Butterworths, London.
[25] Kunz, D.A. and Nagappan, O. (1989) Cyanase-mediated utilization of cyanate in *Pseudomonas fluorescens* NCIB 11764. *Appl. Environ. Microbiol.* **55**, 256–258.
[26] Youatt, J.B. (1954) Studies on the metabolism of *Thiobacillus thiocyanoxidans*. *J. Gen. Microbiol.* **11**, 139–149.
[27] De Kruyff, C.D., van der Walt, J.P. and Schwartz, J.M. (1957) The utilization of thiocyanate and nitrate by thiobacilli. *Antonie v. Leeuwenhoek J. Microbiol. Serol.* **23**, 305–316.
[28] Katayama, Y., Narahara, U., Inoue, U., Amano, F., Kanagawa, T. and Kuraishi, H. (1992) A thiocyanate hydrolase of *Thiobacillus thioparus*: a novel enzyme catalyzing the formation of carbonyl sulfide from thiocyanate. *J. Biol. Chem.* **267**, 9170–9175.

[29] Yancey, P.H., Clark, M.E., Hand, S.C., Bowlus, R.D. and Somero, G.N. (1982) Living with water stress: evolution of osmolyte systems. *Science* **217**, 1214–1222.

[30] King, G.M. (1988) Distribution and metabolism of quaternary amines in marine sediments. In: Nitrogen Cycling in Coastal Marine Environments (Blackburn, T.H. and Sorenson, J. Eds), pp. 143–173, Wiley, London.

[31] Borowitzka, L.J. (1988) Osmoregulation in blue-green algae. *Progr. Phycol. Res.* **4**, 243–256.

[32] Herrmann, V. and Jüttner, F. (1977) Excretion productions of algae. Identification of biogenic amines by gas-liquid chromatography and mass spectrometry of their trifluoroacetamides. *Anal. Biochem.* **78**, 365–373.

[33] Dickson, D.M.J. and Kirst, G.O. (1987) Osmotic adjustment in marine eukaryotic algae: the role of organic ions, quarternary ammonium, tertiary sulphonium and carbohydrate solutes. I Diatoms and a rhodophyte. *New Phytol.* **106**, 645–655.

[34] Dickson, D.M.J. and Kirst, G.O. (1987) Osmotic adjustment in marine eukaryotic algae: the role of inorganic ions, quaternary ammonium, tertiary sulphonium and carbohydrate solutes. II Prasinophytes and haptophytes. *New Phytol.* **106**, 657–666.

[35] Sieburth, J. McN. and Keller, M.D. (1989) Methylaminotrophic bacteria in xenic nanoalgal cultures: incidence, significance, and role of methylated algal osmoprotectants. *Biol. Oceanog.* **6**, 383–395.

[36] Keller, M.D., Bellows, W.K. and Guillard, R.R.L. (1989) Dimethyl sulfide production in marine phytoplankton. In: Biogenic Sulfur in the Environment (Saltzman, E.S. and Cooper, W.J., Eds), pp. 167–182, ACS Symposium Series No. 393.

[37] Large, P.J. and Bamforth, C.W. (1988) Methylotrophy and Biotechnology, Longman Scientific and Technical, Harlow.

[38] King, G.M. (1984) Metabolism of trimethylamine, choline, and glycine betaine by sulfate-reducing and methanogenic bacteria in marine sediments. *Appl. Environ. Microbiol.* **48**, 719–725.

[39] Wheeler, P.A. and McCarthy, J.J. (1982) Methylammonium uptake by Chesapeake Bay phytoplankton: evaluation of the use of ammonium analogue for field uptake measurements. *Limnol. Oceanog.* **27**, 1129–1140.

[40] Kleiner, D. (1985) Bacterial ammonium transport. *FEMS Microbiol. Rev.* **32**, 87–100.

[41] Colby, J. and Zatman, L.J. (1973) Trimethylamine metabolism in obligate and faculatative methylotrophs. *Biochem. J.* **132**, 101–112.

[42] Barrett, E.L. and Kwan, H.S. (1985) Bacterial reduction of trimethylamine oxide. *Ann. Rev. Microbiol.* **39**, 131–149.

[43] Meiberg, J.B.M and Harder, W. (1979) Dimethylamine dehydrogenase from *Hyphomicrobium X*: purification and some properties of a new enzyme that oxidises secondary amines. *J. Gen. Microbiol.* **115**, 49–58.

[44] Lee, C. and Olsen, B.L. (1984) Dissolved exchangeable and bound aliphatic amines in marine sediments, initial results. *Org. Geochem.* **6**, 259–263.

[45] Van Neste, A., Duce, R.A. and Lee, C. (1987) Methylamines in the remote marine atmosphere. *Geophys. Res. Lett.* **14**, 711–714.

[46] Gschwend, P.M., McFarlan, J.K. and Newman, K.A. (1985) Volatile halogenated organic compounds released to seawater from temperate marine macroalgae. *Science* **227**, 1033–1035.

[47] Fawell, J.K. and Hunt, S. (1988) Environmental Toxicology: Organic Pollutants (Series in Water and Wastewater Technology) Ellis Harwood.

[48] Sturges, W.T. and Cota, G.F. (1991) Ice algal production of volatile organic bromine compounds: release in seawater, the atmosphere, and potential influence on surface ozone. *Amer. Geophys. Union* **27** (44 suppl.) p. 237, Abstract O12C-10.

[49] Russell-Jones, R. and Wigley, T. (eds) (1989) Ozone Depletion: Health and Environmental Consequences, John Wiley and Sons, Chichester, UK.

[50] Brunner, W., Staub, D. and Leisinger, T. (1980) Bacterial degradation of dichloromethane. *Appl. Environ. Microbiol.* **40**, 950–958.

[51] Mikesell, M.D. and Boyd, S.A. (1990) Dechlorination of chloroform by *Methanosarcina* strains. *Appl. Environ. Microbiol.* **56**, 1198–1201.

[52] Bouwer, E.J. and McCarty, P.L. (1983a) Transformations of 1- and 2-carbon halogenated aliphatic organic compounds under methanogenic conditions. *Appl. Environ. Microbiol.* **45**, 1286–1294.

[53] Freedman, D.L. and Gossett, J.M. (1991) Biodegradation of dichloromethane and its utilization as a growth substrate under methanogenic conditions. *Appl. Environ. Microbiol.* **57**, 2847–2857.

[54] Bouwer, E.J. and McCarty, P.L. (1983b) Transformations of halogenated organic compounds under denitrification conditions. *Appl. Environ. Microbiol.* **45**, 1295–1299.

[55] Hartmans, S., Schmuckle, A., Cook, A.M. and Leisinger, T. (1986) Methyl chloride: naturally occurring toxicant and C-1 growth substrate. *J. Gen. Microbiol.* **132**, 1139–1142.

[56] Singh, H.B., Salas, L.J. and Stiles, R.E. (1983) Methyl halides in and over the eastern Pacific (40°N-32°S). *J. Geophys. Res.* **88C**, 3684–3690.

[57] Nightingale, P.D. (1991) Low molecular weight halocarbons in seawater. PhD Thesis, University of East Anglia, Norwich, UK.

[58] Liss, P.S. (1986) The air-sea exchange of low molecular weight halocarbon gases. In: The Role of Air-Sea Exchange in Geochemical Cycling (Buat-Menard, P., Ed.), pp. 283–294, Reidel, Dordrecht, Netherlands.

[59] Liss, P.S. and Merlivat, L. (1986) Air-sea gas exchange rates: introduction and synthesis. In: The Role of Air-Sea Exchange in Geochemical Cycling (Buat-Menard, P., Ed.), pp. 113–127, Reidel, Dordrecht, Netherlands.

[60] Andreae, M.O. and Schimel, D.S. (Eds) (1989) Exchange of Trace Gases Between Terrestrial Ecosystems and the Atmosphere, John Wiley and Sons, Chichester, UK.

[61] Liss, P.S. and Galloway, J.N. (1992) Air-sea exchange of sulphur and nitrogen and their interactions in the marine atmosphere. In: Interactions of C, N, P and S Biogeochemical Cycles (Wollast, R. *et al.*, Eds), Springer Verlag, in press.

[62] Wolfe, G.V., Bates, T.S. and Charlson, R.J. (1991) Climatic and environmental implications of biogas exchange at the sea surface: modeling DMS and the marine biologic sulfur cycle. In: Ocean Margin Processes in Global Change (Mantoura, R.F.C., Martin, J.-M. and Wollast, R., Eds), pp. 383–400, John Wiley and Sons, Chichester, UK.

[63] Rosswall, T. and 12 others (1989) Group report. What regulates production and consumption of trace gases in ecosystems: biology or physicochemistry. In: Exchange of Trace Gases Between Terrestrial Ecosystems and the Atmosphere (Andreae, M.O. and Schimel, D.S., Eds), pp. 73–95, John Wiley and Sons, Chichester, UK.

[64] Worrest, R.C. and Grant, L.D. (1989) Effects of ultraviolet-B radiation on terrestrial plants and marine organisms. In: Ozone Depletion: Health and Environmental Consequences (Russell-Jones, R. and Wigley, T., Eds), pp. 197–206, John Wiley and Sons, Chichester, UK.

[65] Malin, G., Turner, S.M. and Liss, P.S. (1992) Sulfur: the plankton/climate connection. *J. Phycol.* **28**, 590–597.

5

Structure and Mechanism of Action of the Hydroxylase of Soluble Methane Monooxygenase

H. Dalton, P. C. Wilkins and Y. Jiang

Department of Biological Sciences, University of Warwick, Coventry CV4 7AL, UK

1 SUMMARY

Results from spin-trapping and H_2O_2/hydroxylase experiments indicate that there are at least three pathways in the sMMO catalytic cycle by which substrate oxidation can occur. These include: 1. hydrogen atom abstraction from substrate by a ferryl species; 2. H atom abstraction by hydroxyl radical; and 3. direct oxygen insertion. Evidence is presented for participation by each of the pathways. Thus, sMMO catalysis may be kinetically controlled with the chosen route dictated by substrate. From the alignment of the sequence of the α subunit of the hydroxylase with the crystallographically determined structure of the R2 protein of ribonucleotide reductase we have been able to propose a rational model of the active site of sMMO, including the substrate binding pocket.

2 INTRODUCTION

Methanotrophs possess the enzyme methane monooxygenase (MMO) which allows them to grow on methane as their sole carbon source. This is the first of four enzymes required for the conversion of methane to carbon dioxide. There are soluble-(sMMO) and membrane-associated particulate (pMMO) forms of the enzyme, the nature of which species is expressed depending on the Cu(II):biomass ratio under which the organism is grown [1]. The substrate specificity of pMMO is not nearly as broad as that of sMMO which catalyses in addition to methane, oxygen insertion into a wide variety of alkanes, alkenes, ethers, alicyclic, aromatic and heterocyclic compounds [2,3]. Little is known about the structure and mechanism of the particulate enzyme but soluble MMO's from four organisms have been well characterized. The enzyme systems from *Methylococcus capsulatus* (Bath) [4], *Methylosinus trichosporium* OB3b [5] and *Methylosinus sporium* [6] all require three protein components for activity: the hydroxylase (previously designated component A), the reductase (previously component C) and component B. Enzyme from *Methylobacterium* CRL-26 may lack protein B, although it could remain tightly bound to the hydroxylase throughout purification [7, 8]. The hydroxylase (Mr = 250 kDa) is comprised of three subunits in an $\alpha_2, \beta_2, \gamma_2$ arrangement and has been shown to be the site of catalytic activity [5]. It contains a bridged diiron site similar to those in hemerythrin [9], rubrerythrin [10], ribonucleotide reductase R2 protein [11], purple acid phosphatase [12] and the initial stages of iron accumulation in apoferritin [13]. The reductase (Mr = 38.5 kDa) is a single polypeptide containing an FAD and an Fe_2S_2 cluster which accepts electrons from NAD(P)H and transfers them, one at a time, to the diiron site of the hydroxylase [4, 14]. Component B (Mr = 16 kDa) is a single polypeptide containing no metal ions or cofactors and functions as a regulatory protein [15–17]. The molecular masses for the enzyme components are based on the gene sequence for *M. capsulatus* (Bath) soluble methane monooxygenase [18].

At the 1989 C_1-Symposium we discussed the roles of the three components in catalysis showing how kinetic studies, spectroscopy and substrate specificity had enabled us to propose a mechanistic cycle for sMMO [3, 19]. The P450-like mechanism [20] involved formation of a high valent iron intermediate which could be generated either via homolysis or heterolysis of the O-O bond in what is formally a ferric peroxo species. We now describe evidence to determine which pathway operates.

3 RESULTS

3.1 Substrate radicals

The most direct possible experiment would be to use EPR spectroscopy to detect either methyl (or other substrate) or hydroxyl radicals as they are formed. Unfortunately, the concentration of radical generated is too low for direct detection. However, the spin trapping technique has been specifically developed to overcome this problem [21–24]. Spin traps are molecules that react very rapidly with active radicals (generally $k > 10^9 \, M^{-1} \, s^{-1}$) converting them into stable radical adducts which accumulate and can be readily detected by EPR spectroscopy. These adducts are usually nitroxide radicals, because of their high stability. The best results in aqueous solution are obtained using nitrone precursors such as α-(4-pyridyl-1-oxide)-N-tert-butylnitrone (POBN) and 5,5-

dimethylpyrroline-1-oxide (DMPO). A typical reaction involves addition of the active radical across the double-bond of the nitrone, eqn. 1 (for DMPO).

$$(H_3C)_2 \underset{\underset{O^-}{N^+}}{\diagdown}\!\!\!-\!\!\!H \quad + R^\cdot \quad \longrightarrow \quad (H_3C)_2 \underset{\underset{O\cdot}{N}}{\diagdown}\!\!\!\overset{R}{\underset{H}{<}} \qquad (1)$$

A further advantage of this technique is that oxygen-based radicals such as ·OH or ·O$_2$H, which cannot be directly detected because of extremely broad EPR signals, are readily trapped and their adducts give distinctive spectra [25, 26]. Spin-trapping has been used to advantage in a large number of biological systems, especially by Mason and co-workers [21–24 and references therein].

Addition of POBN or DMPO to a conventional sMMO assay mixture with methane as substrate resulted in adduct formation with both traps [27–29]. Nitroxide radicals formed from nitrones like POBN and DMPO (Eqn. 1) normally show two hyperfine splitting constants in their EPR spectra, a doublet from the β proton and a triplet from the ^{14}N. The spectra of the POBN (a) and DMPO (b) methyl radical and POBN-methanol radical (c) spin-adducts are shown in Fig. 1. The solubility of methanol and its efficient conversion to formaldehyde by sMMO (2, 30) made it an ideal substrate for these experiments. A disadvantage of nitrone spin-traps is that the hyperfine splitting constants for all carbon-based adducts are similar (Table 1) so that identification of the specific trapped radical is difficult. However, the well defined extra doublet hyperfine splittings with ^{13}C-labelled methanol in the assay mixture showed unambiguously that the carbon-centered spin-adducts originate from substrate (Fig. 2). Carbon-based radicals were also trapped when many other known sMMO substrates (+ trap) were incubated with native enzyme under normal assay conditions, Table 1. Alkane, haloalkane, cycloalkane, alkene and aromatic substrates gave spin-adducts whose hyperfine splitting constants were all within the range of expected values [25,31]. Acetylene, a powerful mechanism-based inhibitor of sMMO [32,33] also gave a carbon-centered adduct with POBN. Substitution of *M.trichosporium* OB3b enzyme in the assay mixture also produced substrate radical adducts with POBN and DMPO.

No hydroxyl radicals (nor O based radicals of any kind) were trapped by POBN or DMPO in any of the native sMMO assay mixtures examined. Comparison of the reaction rate constants for ·OH with POBN or DMPO (k = 3.8 × 10^9 M^{-1} s^{-1} and 4.3 × 10^9 M^{-1} s^{-1} respectively) with that of ·OH with methane (k = 1.1 × 10^8 M^{-1} s^{-1}), and in these experiments the concentration of the trap was significantly higher than the concentration of methane, shows that hydroxyl radicals would certainly have been observed if present [34]. The EPR spectra of the ·OH spin adducts are quite different from those of carbon-based adducts (Table 1).

These spin-trapping results are the first *direct* experimental evidence for the P450-like mechanism proposed for sMMO [3,19,35]. Heterolytic cleavage of

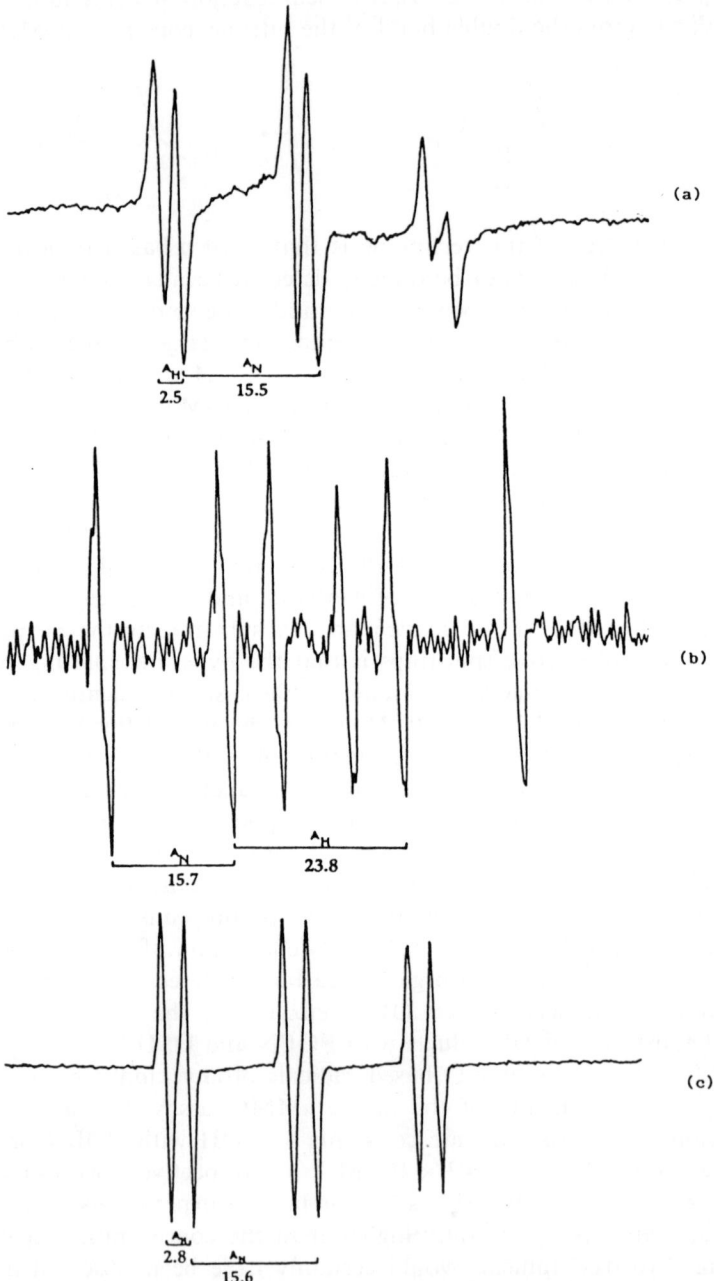

Figure 1 EPR spectra of the POBN · CH$_3$ (a) DMPO · CH$_3$ (b) and POBN · CH$_2$OH (c) spin-adducts. Instrument settings: (a) field center, 3350G; frequency, 9.446; range, 100G; modulation amplitude, 1G; time constant, 0.06 sec.; scan time, 7.8 min.; power, 1mW; ambient temperature.

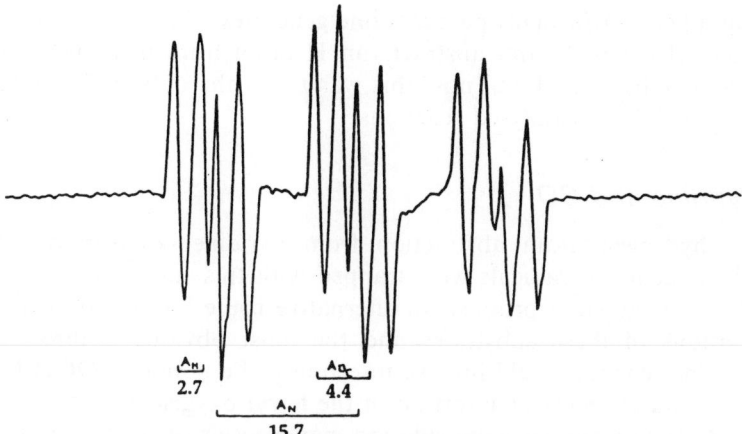

Figure 2 EPR spectrum of the $^{13}CH_2OH \cdot POBN$ adduct showing the additional splitting due to ^{13}C. Instrument settings as in Figure 1.

Table 1 Hyperfine splitting constants for sMMO substrate radical adducts with POBN and DMPO*

Substrate	POBN		DMPO	
	$A_N(G)$	$A_H(G)$	$A_N(G)$	$A_H(G)$
CH_4	15.5	2.5	15.7	23.8
CH_3OH	15.6	2.8	15.6	22.5
$^{13}CH_3OH$	15.7	2.7 4.4(^{13}C)	15.8	22.2 8.3(^{13}C)
CH_3CN	15.4	2.4	b	b
$HOC(CH_3)_3$	15.5	2.6	b	b
CH_2Cl_2	15.55	2.2	b	b
$CHCl_3$	a	a	15.6	22.2
CCl_4	b	b	b	b
cyclohexane	15.7	2.3	15.6	22.5
adamantane	15.5	2.5	15.5	22.2
$H_2C=CH_2$	15.8	2.3	15.6	22.3
allyl alcohol	15.6	2.7	15.6	22.7
acetylene	15.6	2.3	a	a
benzene	15.7	2.8	a	a
phenol	15.6	2.5	15.4	23.0
aniline	b	b	15.6	22.2
allyl alcohol (OB3b)	15.6	2.6	15.6	22.4
OH(Fenton)[(c)]	14.9	1.7	14.9	14.9
CO	b	b	b	b
pyridine	b	b	b	b

* Experimental conditions: [A] = 22-75 µM, saturating amounts of reductase and protein B; [Trap] = 0.1 M; water soluble substrates 0.1–0.5 M; sparingly soluble substrates 1–25 mM; 0.025 M MOPS, pH7; 45°, 3–10 min; (a) measurements not done; (b) no spin-adduct detected; (c) H_2O_2, UV irradiation.

the Fe bound O—O (formally peroxo) bond generates a ferryl species which acts as a strong electrophile and abstracts an H atom from methane and other hydrocarbon substrates. Until now this was really the only pathway believed to operate in sMMO catalysed reactions.

3.2 Pyridine and CO

Although hydrogen atom abstraction from pyridine could reasonably be expected to occur, no radicals were trapped with this substrate nor with CO [28]. We must therefore propose an alternative route for sMMO catalysis of the oxidations of these substrates and the most obvious is direct oxygen insertion. The reaction could involve insertion of Fe bound —OOH followed by O—O bond cleavage or insertion of the ferryl oxygen (Fe=O), i.e. oxene insertion. Catalysis of alkene epoxidation may also go, at least in part, by this route. The observation of relatively weak spin-adduct EPR signals for alkene radicals suggests this possibility [28].

3.3 A homolytic pathway?

Recent results with hydrogen peroxide-driven sMMO enzyme has led us now to propose the availability of a homolytic pathway (Fig. 3a), in addition to the heterolytic pathway (Fig. 3b), in the mechanistic cycle. Andersson et al. reported that H_2O_2 could supply the oxygen and reducing equivalents for *M. trichosporium* OB3b catalysed reactions [36]. We have also demonstrated this P450 like peroxide shunt for *M. capsulatus* (Bath) hydroxylase and have shown that substrate oxidation is proportional to H_2O_2 concentration (Fig. 4) [30]. Other surrogate O and electron donors such as iodosobenzene, periodate, chlorite and t-butylhydroperoxide, which can be used with P450, failed to activate the sMMO hydroxylase. The Km value for H_2O_2 in the propene to epoxypropene and methanol to formaldehyde reactions was approximately 66 mM (Fig. 4). This is orders of magnitude higher than the Km values for the reactants in the native sMMO complex [37], which may be a reflection of the fact that H_2O_2 acts as both oxygen and electron donor and that one of these functions is rate limiting. On the other hand the Km value for methanol in the H_2O_2/hydroxylase system is 1.2 mM which is very close to that with the complete sMMO complex (0.95 mM), suggesting that carbon substrate interactions are similar for both systems.

To show that H_2O_2 *per se* supplied the oxygen to substrate, rather than from peroxide decomposition to O_2, experiments were performed in the presence of labelled oxygen. In native sMMO catalysed reactions ^{18}O was readily incorporated into product when various amounts of $^{18}O_2$ were added to assay mixtures. In contrast, no incorporation of ^{18}O was observed in the H_2O_2-driven assays indicating that the oxygen atom is derived from H_2O_2 and not O_2.

The H_2O_2/hydroxylase system was shown to be catalytic for methane hydroxylation and for a variety of other known sMMO substrates (Table 2).

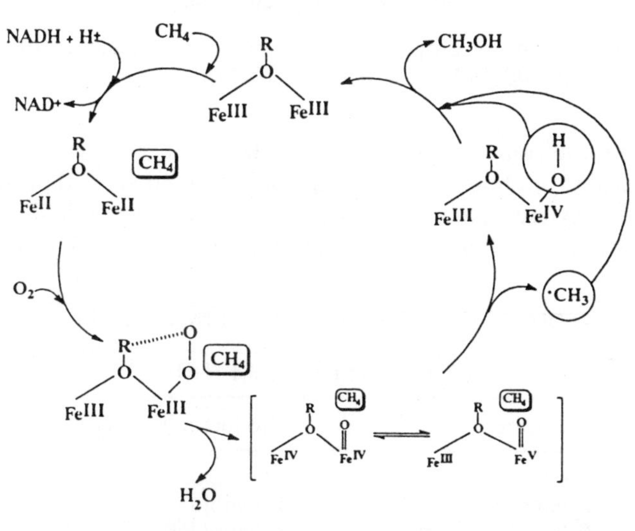

Figure 3 Mechanistic cycle for sMMO showing (a) the homolytic pathway and (b) the heterolytic pathway.

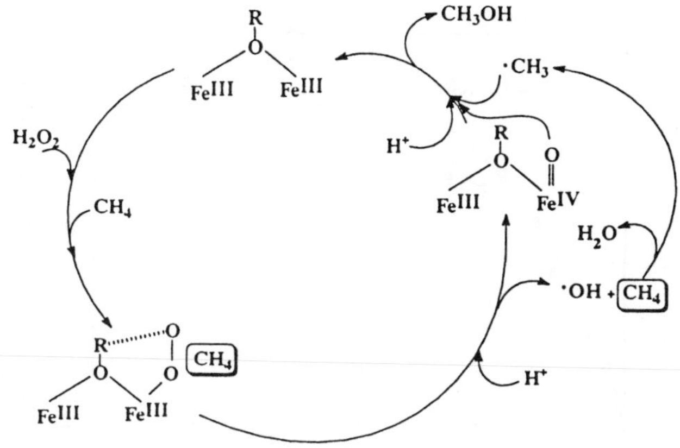

To eliminate purely chemical oxidation and in an attempt to identify possible reactants a number of experiments were carried out (Table 3). These show that there is no free iron (which could initiate a Fenton reaction) or superoxide present and that fully functional hydroxylase is required in the reactions.

The partial inhibition of activity by acetylene is a most intriguing result. Native sMMO is extremely susceptible to inhibition by acetylene where a concentration as low as 0.5% caused complete loss of activity by acting as a suicide substrate [32,33]. In the H_2O_2/hydroxylase system, 30% of the activity with propene and 45% with methanol remained in the presence of 10% acetylene (Table 3). Only when the amount of acetylene was increased to 40%

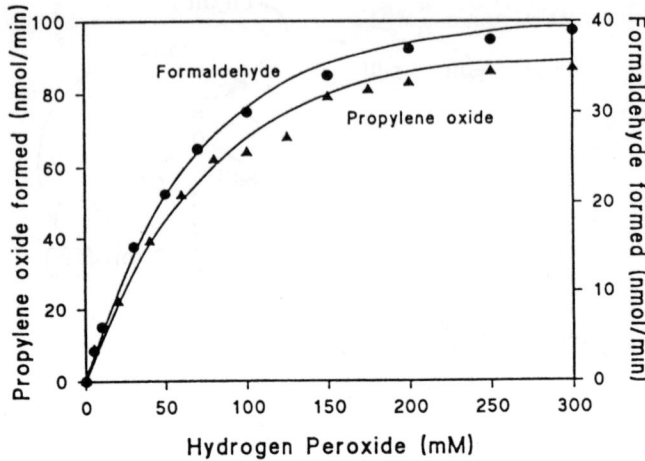

Figure 4 The oxidation of propene to epoxypropene and methanol to formaldehyde by the H_2O_2 (5-300 mM)/hydroxylase (24μM) system.

was complete inhibition observed. The results with acetylene suggest possible hydroxyl radical (from homolytic cleavage of Fe bound O—O) participation in the H_2O_2/hydroxylase system. The incomplete inhibition could be explained by assuming that acetylene is acting as a competitive inhibitor of propene

Table 2 Oxidation of substrates by the hydroxylase of sMMO from *Methylococcus capsulatus* (Bath) using 100 mM H_2O_2

Substrate	Product (nmol/min)
Methane[a]	Methanol (19)
Ethane[b]	Ethanol (55)
Ethane[b] (anaerobic)	Ethanol (50)
Propylene[c]	Propylene oxide (64)
Propylene[c] (anaerobic)	Propylene oxide (56)
Cyclohexane[d]	Cyclohexanol (11)
Cyclohexene[d]	Cyclohexene oxide (20)
Benzene[e]	Phenol (24)
Styrene[e]	Styrene oxide (17)
Methanol[f]	Formaldehyde (35)

[a] 120 μM hydroxylase; Ethyl acetate extracted; GC/BP-5 and GC/MS assay.
[b] 120 μM hydroxylase; GC and GC/MS assay.
[c] 24 μM hydroxylase; GC/MS assay.
[d] 80 μM hydroxylase; Ethyl acetate extracted; GC/BP-5 assay.
[e] 120 μM hydroxylase; Ethyl acetate extracted; GC/BP-1 assay.
[f] 24 μM hydroxylase; assayed by Nash method.

Table 3 Effect of various additions on substrate oxidation by the H_2O_2/hydroxylase system

Conditions	Products	
	Propylene oxide[a] (nmol/min)	Formaldehyde[b] (nmol/min)
None	64	35
−hydroxylase	0	0
Heated hydroxylase		
60°C, 5 mins	9.6	not done
100°C, 5 mins	0	0
+acetylene (10% (v/v) in air)	20	16
+catalase (5 units/ml)	0	0
+SOD (5 units/ml)	59	32
+mannitol (20–500 mM)	64	not done
+2,4,6-tri-*tert*-butylphenol* (4 mM–40 mM)	36–0	not done
Apohydroxylase	0	0

[a] 24 μM hydroxylase, 100 mM H_2O_2; GC assay.
[b] 24 μM hydroxylase, 100 mM H_2O_2; assayed by Nash method.
* 2,4,6-tri-*tert*-butylphenol was dissolved in methanol (0.4 M) as stock solution.

epoxidation. Reaction of acetylene with ·OH would produce vinyl radicals which in the presence of Fe^{2+} and H^+ would yield acetaldehyde in a manner similar to Fenton chemistry. Alternatively, ·OH and acetylene can react to form ·CH=CHOH radicals which rapidly combine with oxygen to form glyoxal.

Addition of the hydroxyl radical scavengers mannitol and 2,4,6-tri-t-butylphenol had little effect or were inconclusive. However if hydroxyl radicals are involved they may remain at the active site and not be accessible to scavengers.

A comparison of the product distribution in the H_2O_2/hydroxylase and complete sMMO complex suggests that different pathways predominate in the two systems. In the case of 2-methylpropane (Table 4) the product ratio was the same for the two systems, but for propane, 2-methylbutane and trans-2-butene the ratios were quite different. Native sMMO is seen to be non-selective, with perhaps a slight preference for attack at the secondary carbon in 2-methylbutane. This is completely consistent with the proposal of the strongly electrophilic ferryl species acting as H atom abstractor in catalyses by the complete enzyme complex. In contrast, in the H_2O_2/hydroxylase system the tertiary and secondary positions of 2-methylbutane were clearly favoured with only slight activation of one of the primary carbons. The selectivity in this reaction indicates a less electrophilic hydrogen abstracting species and is reminiscent of what would be expected for hydroxyl radicals. The two systems gave quite different product ratios in the oxidation of 2-butene where the

Table 4 Oxidation of alkanes and 2-butene by the H_2O_2/hydroxylase system and sMMO complex

Substrates	Products	sMMO + NADH[a]		Hydroxylase/H_2O_2[b]	
		Ratio (%)	Selectivity* (%)	Ratio (%)	Selectivity* (%)
propane	1-propanol	33	1°–, 14	5.5	1°–, 2
	2-propanol	67	2°–, 86	94.5	2°–, 98
2-methylpropane	2-methyl-1-propanol	51	1°–, 15	50	1°–, 14
	2-methyl-2-propanol	49	3°–, 85	50	3°–, 86
2-methylbutane	2-methyl-1-butanol	26	1°–, 16	0	1°–, 6
	3-methyl-1-butanol	26		27	
	3-methyl-2-butanol	36	2°–, 50	45	2°–, 42
	2-methyl-2-butanol	12	3°–, 34	28	3°–, 52
trans-2-butene	trans-2-epoxybutane	32[c]		97[d]	
	trans-2-butene-1-ol	68[c]		3[d]	

[a] Hydroxylase 8 μM, protein B (crude) 1.2 mg, protein C (crude) 1.5 mg, NADH 5 mM. Reaction 10 min. GC assay.
[b] Hydroxylase 80 μM, H_2O_2 100 mM. Reaction 30 min. GC and GC/MS assay.
[c] Reaction 5 min.
[d] Hydroxylase 24 μM, H_2O_2 100 mM. Reaction 15 min. GC assay.
* Selectivity was calculated as a percentage of the total amount of product divided by the number of available hydrogen atoms at that position. The data shown were average values from a number of experiments.

alcohol is almost completely absent in the H_2O_2-driven reaction. This is reasonable as a weaker electrophile ($\cdot$OH) would have more difficulty in forming the alcohol by C—H bond cleavage (—C=C—H, ~90 kcal/mol) than in forming the epoxide by breaking a π-bond (70 kcal/mol).

3.4 Substrate and (oxygen/electron) donor modulation

Our results have led us to propose three possible fates for the Fe-Fe-OOH species in the sMMO catalytic cycle, Eqn. 2, 3 and 4.

$$Fe^{III}\overset{R}{\underset{O}{\diagup\diagdown}}Fe^{III}\text{-OOH} \xrightarrow[H_2O]{+H^+} \left[Fe^{III}\overset{R}{\underset{O}{\diagup\diagdown}}Fe^{V}{=}O \longleftrightarrow Fe^{IV}\overset{R}{\underset{O}{\diagup\diagdown}}Fe^{IV}{=}O\right] +SH \rightarrow \left[Fe^{III}\overset{R}{\underset{O}{\diagup\diagdown}}Fe^{IV}\overset{OH}{\underset{}{|}} +\cdot S\right] \xrightarrow{k_2} Fe^{III}\overset{R}{\underset{O}{\diagup\diagdown}}Fe^{III} \quad (2)$$

$$SOH$$

$$Fe^{III}\overset{R}{\underset{O}{\diagup\diagdown}}Fe^{III}\text{-OOH} \longrightarrow Fe^{III}\overset{R}{\underset{O}{\diagup\diagdown}}Fe^{IV}{=}O + \cdot OH + SH + H^+ \xrightarrow{k_3} SOH + H_2O + Fe^{III}\overset{R}{\underset{O}{\diagup\diagdown}}Fe^{III} \quad (3)$$

$$Fe^{III}\overset{R}{\underset{O}{\diagup\diagdown}}Fe^{III}\text{-OOH} + S + H^+ \xrightarrow{k_4} SO + H_2O + Fe^{III}\overset{R}{\underset{O}{\diagup\diagdown}}Fe^{III} \quad (4)$$

The reactions appear to be kinetically controlled with the choice of pathway being dictated by the substrate which is presented to the active site of the enzyme and/or the O and electron donor used. Equation 2 describes heterolytic cleavage of the O—O bond which produces the non-selective, strongly electrophilic ferryl species. This pathway predominates with methane and other alkanes in native sMMO oxidations, i.e. $k_2 \gg k_3$ and k_4. In the H_2O_2/hydroxylase system the primary route may be via homolytic O—O bond breakage and $\cdot$OH formation, Eqn. 3, i.e. $k_3 \gg k_2$ and k_4. For pyridine and CO direct O insertion shown in Eqn. 4 must be favoured, i.e. $k_4 \gg k_2$ and k_3. Finally, in the case of alkenes and perhaps aromatics, product formation via all three pathways cannot be ruled out i.e. $k_2 \simeq k_3 \simeq k_4$.

3.5 Structural studies

The cloning and sequencing of the sMMO proteins from *M. capsulatus* (Bath) and *M. trichosporium* OB3b revealed a remarkably high degree (94%) of sequence identity in the α subunits of the hydroxylase from the two organisms [18, 38–40]. It was noted that there were two places in the α-subunit where the Glu-x-x-His sequence motif appeared which is known to coordinate the two irons in the R2 protein of ribonucleotide reductase [40]. The X-ray crystal structure of the R2 protein refined to 2.2 Å has recently been reported [41]. At this resolution the geometry at the diiron site can be reasonably well described.

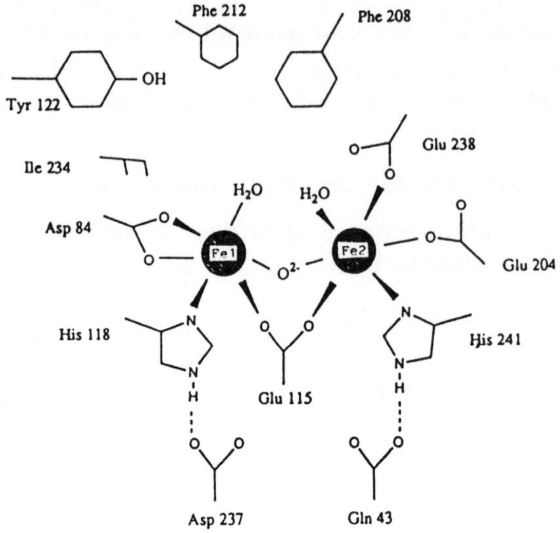

Figure 5 The diiron site in the R2 protein of ribonucleotide reductase.

Figure 5 shows the arrangement of the ligands around the irons in the R2 protein. Each iron has a water and a His (118, 241) bound to it and one other coordination site is taken up by the μ-oxo bridge. The other ligands are the bridging Glu 115, a bidentate Asp 84 on Fe1 and two monodentate Glu's (204, 238) on Fe2. Inspection of the amino acid sequences from ten different sources of R2 protein revealed that these six amino acids are completely conserved in each. Guided by the three dimensional structure of the *E. coli* protein and the two Glu-x-x-His sequences, Fig. 6, we have aligned the α-subunits of the *M.*

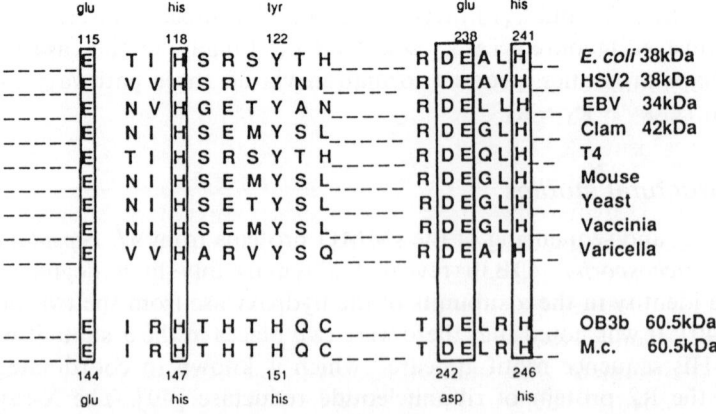

Figure 6 The sequence alignment of R2 proteins of ribonucleotide reductases with the α subunits of the sMMO hydroxylases.

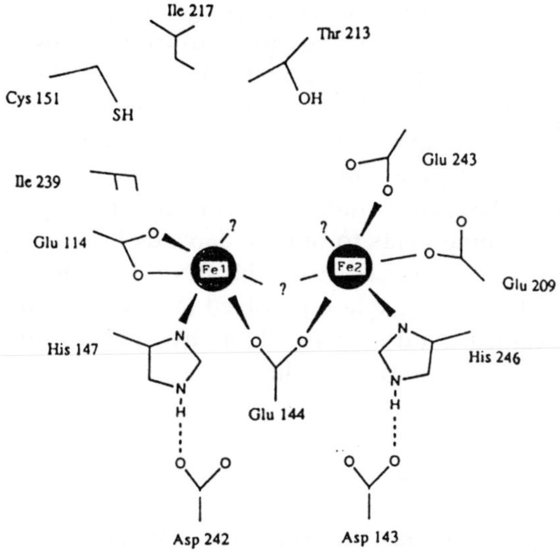

Figure 7 The active site model for the sMMO hydroxylase.

capsulatus (Bath) and *M. trichosporium* OB3b hydroxylases with the four helix iron coordination bundle of R2 [42]. When this is done, the corresponding iron ligands for the sMMO hydroxylase are His 147 and bidentate Glu 114 on Fe1, His 246 and monodentate Glu 209 and Glu 243 on Fe 2 and bridging Glu 144, Fig. 7. Cysteine 151 is found at the site of Tyr 122 on which a stable free radical resides during turnover of ribonucleotide reductase (RNR). The sMMO hydroxylase model could be built without any severe steric hindrance and most of the amino acids surrounding the oxygen binding site are smaller than the corresponding residues in the R2 protein. This is consistent with the fact that this pocket in the hydroxylase must also accommodate methane and larger substrates. Hydrogen bonding of the coordinated histidines to aspartates would given some anionic character to these ligands thereby helping to stabilize high valent iron species. The proposed model of the sMMO active site is in complete accord with recent EXAFS and ENDOR studies [43,44].

It also appears that a number of the non-iron bonding ligands at the active site of the MMO hydroxylase are effectively neutral in character (Thr, Ile). This prediction from the model building would also conform to what one might expect to observe from the MMO mechanism in which highly reactive radicals are produced. The presence of neutral amino acids would clearly minimize any damaging effects that these radicals could produce if they were to interact with the local environment. The spin-traps used in the experiments presented here need to be added in fairly high concentration ($\simeq 0.1$M) to be effective. Since they are larger and less hydrophobic than most of the radicals with which they interact it is not too surprising that they reduce activity by about 20% only at these concentrations. Thus one could envisage the cavity as being relatively hydrophilic at the periphery which possibly entraps the substrate methane once it has entered the hydrophobic core and retains the radical long enough to form the product methanol which then being much less hydrophobic can readily diffuse from the site. The spin trap can also enter the cavity and interact with the methyl radical and escape. This hypothesis is, of course, quite testable since one can now undertake specific mutagenesis on chosen amino acids at the active site. The choice of amino acids to mutate depends heavily upon the accuracy of the model we have produced from the RNR homology. Until the X-ray crystal structure of the hydroxylase is solved (already Lippard and colleagues at MIT have obtained good crystals of the hydroxylase from *M. capsulatus* (Bath)) our model will have to serve this role.

ACKNOWLEDGEMENTS

We are grateful to British Gas and to the Gas Research Institute, Chicago, for financial support.

REFERENCES

[1] Stanley, S. H., Prior, S. D., Leak, D. J. and Dalton, H. (1983) Copper stress underlies the fundamental change in intracellular location of methane monooxygenase in methane oxidizing organisms: studies in batch and continuous cultures. *Biotechnol. Lett.* **5**, 487–492.

[2] Colby, J., Stirling, D. I. and Dalton, H. (1977) The soluble methane monooxygenase of *Methylococcus capsulatus* (Bath): its ability to oxygenate n-alkanes, n-alkenes, ether, and alicyclic, aromatic and heterocyclic compounds. *Biochem. J.* **165**, 395–402.

[3] Green, J. and Dalton, H. (1989) Substrate specificity of soluble methane monooxygenase: mechanistic implications. *J. Biol. Chem.* **264**, 17698–17703.

[4] Colby, J. and Dalton, H. (1978) Purification and properties of component C, a flavoprotein. *Biochemical J.* **171**, 461–468.

[5] Fox, B. G., Froland, W. A., Dege, J. E. and Lipscomb, J. D. (1989) Methane monooxygenase from *Methylosinus trichosporium* OB3b. *J. Biol. Chem.* **264**, 10023–10033.

[6] Pilkington, S. J. and Dalton, H. (1991) Purification and characterization of the soluble methane monooxygenase from *Methylosinus sporium* 5 demonstrates the highly conserved nature of this enzyme in methanotrophs. *FEMS Microbiol. Lett.* **78**, 103–108.

[7] Patel, R. N. and Savas, J. C. (1987) Purification and properties of the hydroxylase component of methane monooxygenase. *J. Bacteriol.* **169**, 2313–2317.

[8] Patel, R. N. (1987) Methane monooxygenase: purification and properties of flavoprotein component. *Arch. Biochem. Biophys.* **252**, 229–236.

[9] Wilkins, P. C. and Wilkins, R. G. (1987) The coordination chemistry of the binuclear iron site in hemerythrin. *Coord. Chem. Revs.* **79**, 195–214.

[10] Le Gall, J., Pickril, B. C., Moura, I., Xavier, A. V., Moura, J. and Huynh, B–H. (1988) Isolation and characterization of rubrerythrin, a non-heme iron

protein from *Desulfovibrio vulgaris* that contains rubredoxin centers and a hemerythrin-like binuclear iron cluster. *Biochemistry* **27**, 1636–1642.
[11] Stubbe, J. (1990) Ribonucleotide reductases: amazing and confusing. *J. Biol. Chem.* **265**, 5329–5332.
[12] Doi, K., Antanaitis, B. C. and Aisen, P. (1988) The binuclear iron centers of uteroferrin and the purple acid phosphatases. *Struct. Bonding* **70**, 1–26.
[13] Hanna, P. M., Chen, Y. and Chasteen, N. D. (1991) Initial iron oxidation in horse spleen apoferritin. *J. Biol. Chem.* **266**, 886–893.
[14] Colby, J. and Dalton, H. (1979) Characterization of the second prosthetic group of the flavoenzyme NADH- acceptor reductase (component C) of the methane monooxygenase from *Methylococcus capsulatus* (Bath). *Biochem. J.* **177**, 903–908.
[15] Green, J. and Dalton, H. (1985) Protein B of soluble methane monooxygenase from *Methylococcus capsulatus* (Bath). *J. Biol. Chem.* **260**, 15795–15801.
[16] Fox, B. G., Liu, Y., Dege, J. E. and Lipscomb, J. D. (1991) Complex formation between the protein components of methane monooxygenase from *Methylosinus trichosporium* OB3b. *J. Biol. Chem.* **266**, 540–550.
[17] Liu, K. E. and Lippard, S. J. (1991) Redox properties of the hydroxylase component of methane monooxygenase from *Methylococcus capsulatus* (Bath). *J. Biol. Chem.* **266**, 12836–12839.
[18] Stainthorpe, A. C., Murrell, J. C., Salmond, G. P. C., Dalton, H. and Lees, V. (1989) Molecular analysis of methane monooxygenase from *Methylococcus capsulatus* (Bath). *Arch. Microbiol.* **152**, 154–159.
[19] Dalton, H., Smith, D. D. S. and Pilkington, S. J. (1990) Towards a unified mechanism of biological methane oxidation. *FEMS Microbiol. Revs.* **87**, 201–208.
[20] Ortiz de Montellano, P. R. (1986) Oxygen activation and transfer. In: Cytochrome P450, Structure, Function and Biochemistry (Ortiz de Montellano, P. R., Ed.), pp. 217–271, Plenum Press, New York.
[21] Perkins, M. J. (1980) Spin trapping. In: Advances in Physical Organic Chemistry, Vol. 17 (Gold, V. and Bethell, D., Eds), pp. 1–64, Academic Press, London.
[22] Janzen, E. G. (1980) A critical review of spin trapping in biological systems. In: Free Radicals in Biology, Vol. IV (Pryor, W. A., Ed.), pp. 115–154, Academic Press, New York.
[23] McCay, P. B., Noguchi, T., Fong, K.-L., Lai, E. K. and Poyer, J. L. (1980) Production of radicals from enzyme systems and the use of spin traps. In: Free Radicals in Biology, Vol. IV (Pryor, W. A., Ed.), pp. 155–186, Academic Press, New York.
[24] Mottley, C. and Mason, R. P. (1989) Nitroxide radical adducts in biology: chemistry, applications and pitfalls. In: Biological Magnetic Resonance, Vol. 8 (Berliner, L. J. and Reuben, J., Eds), pp. 489–546, Plenum Press, New York.
[25] Li, A. S. W., Cummings, K. B., Roethling, H. P., Buettner, G. R. and Chignell, C. F. (1988) A spin- trapping database implemented on the IBM PC/AT. *J. Magn. Res.* **79**, 140–142.
[26] Ozawa, T. and Hanaki, A. J. (1991) The first ESR spin-trapping evidence for the formation of hydroxyl radical from the reaction of copper (II) complex with hydrogen-peroxide in aqueous solution. *J. Chem. Soc., Chem. Commun*, 330–332.
[27] Deighton, N., Podmore, I. D., Symons, M. C. R., Wilkins, P. C. and Dalton, H. (1991) Substrate radical intermediates are involved in the soluble methane monooxygenase catalyzed oxidations of methane, methanol and acetonitrile. *J. Chem. Soc., Chem. Commun.* 1086–1088.
[28] Dalton, H., Wilkins, P. C., Deighton, N., Podmore, I. D. and Symons, M. C. R. (1992) Electron paramagnetic resonance studies of the mechanism of substrate oxidation by methane monooxygenase. *Faraday Discuss.* **93**, in press.

[29] Wilkins, P. C., Dalton, H., Podmore, I. D., Deighton, N. and Symons, M. C. R. (1992) Biological methane activation involves the intermediacy of carbon-centred radicals. *Eur. J. Biochem.*, in press.
[30] Jiang, Y., Wilkins, P. C. and Dalton, H. (1992) Activation of the hydroxylase of sMMO from *Methylococcus capsulatus* (Bath) by hydrogen peroxide. *Biochim. Biophys. Acta*, submitted for publication.
[31] Buettner, G. R. (1987) Spin trapping: ESR parameters of spin adducts. *Free Rad. Biol. Med.* **3**, 259–303.
[32] Prior, S. D. and Dalton, H. (1985) Acetylene as a suicide substrate and active site probe for methane monooxygenase from *Methylococcus capsulatus* (Bath). *FEMS Microbiol. Lett.* **29**, 105–109.
[33] Dalton, H. and Whittenbury, R. (1976) The acetylene reduction technique as an assay for nitrogenase activity in the methane oxidizing bacterium *Methylococcus capsulatus* strain Bath. *Arch. Microbiol.* **109**, 147–151.
[34] Buxton, G. V., Greenstock, C. L., Helman, W. P. and Ross, A. B. (1988) Critical review of rate constants for reactions of hydrated electrons, hydrogen atoms and hydroxyl radicals ($\cdot$OH/$\cdot$O$^-$) in aqueous solution. *J. Phys. Chem. Ref. Data* **17**, 513–886.
[35] Fox, B. G., Borneman, J. G., Wackett, L. P. and Lipscomb, J. D. (1990) Haloalkene oxidation by the soluble methane monooxygenase from *M. trichosporium* OB3b: mechanistic and environmental implications. *Biochemistry* **29**, 6419–6427.
[36] Andersson, K. K., Froland, W. A., Lee, S.-K. and Lipscomb, J. D. (1991) Dioxygen independent oxygenation of hydrocarbons by methane monooxygenase hydroxylase component. *New J. Chem.* **15**, 411–415.
[37] Green, J. and Dalton, H. (1986) Steady-state kinetic analysis of soluble methane monooxygenase from *Methylococcus capsulatus* (Bath). *Biochem. J.* **236**, 155–162.
[38] Stainthorpe, A. C., Lees, V., Salmond, G. P. C., Murrell, J. C. and Dalton, H. (1990) The methane monooxygenase gene cluster of *Methylococcus capsulatus* (Bath). *Gene* **91**, 27–34.
[39] Pilkington, S. J., Salmond, G. P. C., Murrell, J. C. and Dalton, H. (1990) Identification of the gene encoding the regulatory protein B of soluble methane monooxygenase. *FEMS Microbiol. Lett.* **72**, 345–348.
[40] Cardy, D. L. N., Laidler, V., Salmond, G. P. C. and Murrell, J. C. (1991) Molecular analysis of the methane monooxygenase (MMO) gene cluster of *Methylosinus trichosporium* OB3b. *Mol. Microbiol.* **5**, 335–342.
[41] Nordlund, P., Sjöberg, B.-M. and Eklund, H. (1990) The three-dimensional structure of the free radical protein of ribonucleotide reductase. *Nature* **345**, 593–598.
[42] Nordlund, P., Dalton, H. and Eklund, H. (1992) The active-site structure of methane monooxygenase is closely related to the binuclear iron center of ribonucleotide reductase. *FEBS Lett.* **307**, 257–262.
[43] De Witt, J. G., Bentsen, J. G., Rosenzweig, A. C., Hedman, B., Green, J., Pilkington, S. J., Papaefthymiou, G. C., Dalton, H., Hodgson, K. O. and Lippard, S. J. (1991) X-ray absorption, Mössbauer and EPR studies of the dinuclear iron center of the hydroxylase component of methane monooxygenase. *J. Amer. Chem. Soc.* **113**, 9219–9235.
[44] Hendrich, M. P., Fox, B. G., Andersson, K. K., Debrunner, P. G. and Lipscomb, J. D. (1992) Ligation of the diiron site of the hydroxylase component of methane monooxygenase. *J. Biol. Chem.* **267**, 261–269.

6

The Catalytic Cycle of Methane Monooxygenase and the Novel Roles Played by Protein Component Complexes During Turnover

Wayne A. Froland, Kristoffer K. Andersson, Sang-Kyu Lee, Yi Liu and John D. Lipscomb

Department of Biochemistry, Medical School, University of Minnesota, Minneapolis, MN 55455, USA

1 SUMMARY

The mechanism of soluble methane monooxygenase (MMO) is investigated by using spectroscopic, kinetic, and chemical techniques to provide evidence that: 1. the diiron cluster of the hydroxylase component is the reactive center, 2. in the natural cycle, the 2 electron reduced diiron cluster activates O_2, 3. H_2O_2 can supply both the reducing equivalents and the oxygen required for catalysis directly to the hydroxylase, and 4. a substrate radical is produced during catalysis. The studies strongly support our proposal that the mechanism resembles that of cytochrome P450. The roles of the reductase and B components are investigated through the use of catalytically functional subsystems of the reconstituted enzyme lacking one or both of these components. An unexpected role of these components is revealed by monitoring their marked effect on the product distribution resulting from turnover of complex substrates. It appears that structural changes in the hydroxylase active site occurring in response both to component binding and to the state of reduction of the diiron cluster change the manner in which substrates are presented to the reactive oxygen species. These conformational changes may be related to regulation in the catalytic cycle.

2 INTRODUCTION

The soluble methane monooxygenase from *Methylosinus trichosporium* OB3b [1] consists of three protein components: hydroxylase (($\alpha\beta\gamma)_2$) containing 2 oxygen bridged diiron clusters in which the bridging oxygen is protonated or otherwise substituted; reductase containing one mole each of a [2Fe-2S] center and FAD; and component B containing no recognized metals or cofactors. Reconstitution of all three components of MMO is required for efficient O_2 and NADH-coupled oxidation of methane to methanol.

Extensive spectroscopic studies of the hydroxylase [2–5] have shown that its diiron clusters can be stabilized at three different redox states: the diferric [Fe(III)·Fe(III)] resting state, a mixed valent [Fe(II)·Fe(III)] state, and the fully reduced diferrous [Fe(II)·Fe(II)] state. When nonenzymatically reduced to the diferrous state and then exposed to O_2, the hydroxylase in the absence of the other components was shown to be capable of catalyzing a single turnover to yield the same types of products as the reconstituted MMO [1]. This observation provided the first evidence based on catalysis that the diiron cluster of the hydroxylase is the site of activation of O_2 and product formation. The single turnover results demonstrated a new function for the diiron cluster in biology, and served to focus attention on the diferrous state of the cluster as an important precursor of oxygen activation. Recently, ENDOR studies [5] have shown that each iron of the diiron cluster is liganded by one or more histidines, and that it is largely sequestered from polar solvents. However, substrate-like molecules such as dimethylsulfoxide can approach within a few angstroms of the cluster, suggesting that substrates can bind nearby.

MMO is characterized by an exceptionally broad substrate range [6–8] reminiscent of that of the heme-based, liver microsomal cytochrome P450 monooxygenase superfamily [9]. In fact, several substrates are oxidized by both cyt. P450 and MMO to give the same products, suggesting that the mechanisms of the two enzymes may be similar. This is borne out by the observations made by several groups recently that mechanistically diagnostic substrates for P450 show the same reactivities when oxidized by MMO. For example, allylic migration [8,10], cyclopropane ring opening [11], isotope scrambling [10], and halide migration [12] have all been demonstrated for each system. Similarities such as these lead us to propose a mechanism for MMO [1,12] based on that of P450 [13,14] shown in Fig. 1 (outer circle).

The activated oxygen species of cytochrome P450 is believed to be formed after transfer of a second reducing equivalent to an O_2-complexed ferrous heme species; this precipitates heterolytic cleavage of bound oxygen which is formally at the oxidation state of peroxide [13,14]. The result is an oxygen atom (6 valence electrons) coordinated to the porphyrin iron and release of the first monooxygenase product, water. This oxygen atom is extremely electron deficient, and is apparently stabilized by withdrawal of one electron from the iron and a second from the porphyrin to give [O = Fe(IV)-porphyrin π-cation radical] in one resonance form. We have suggested that similar

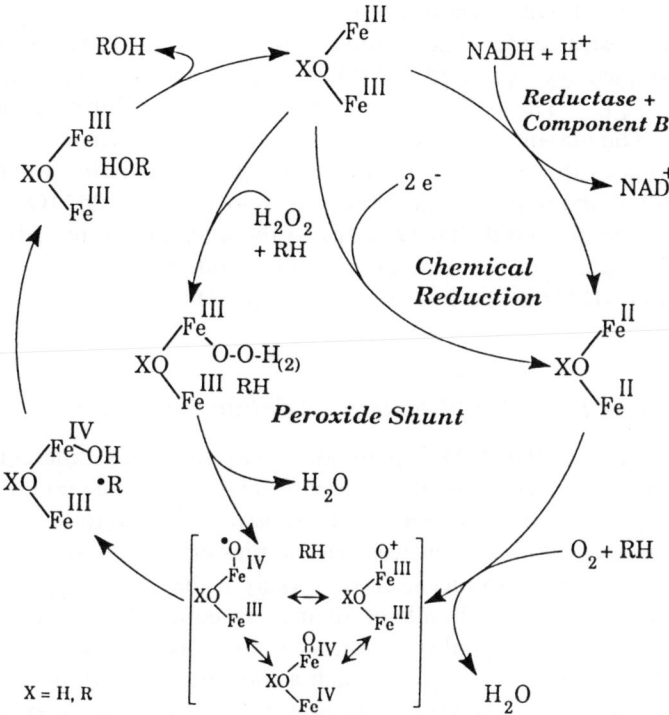

Figure 1 Proposed catalytic cycle of methane monooxygenase. The actual point of addition of substrate is not known. Formulated from the data in references [1,12,15,20].

electron delocalization can account for the ability of the diiron cluster of MMO to stabilize an analogous activated oxygen through formation of [O = Fe(IV)·Fe(IV)] cluster in one resonance form. Cytochrome P450 is proposed to utilize the activated oxygen species to abstract a hydrogen atom from substrate to leave a substrate radical and the redox equivalent of an iron-coordinated hydroxy radical. The substrate radical abstracts this iron-bound hydroxyl yielding product and completing the cycle. We propose a completely analogous reaction for MMO as depicted in Fig. 1.

3 RESULTS

3.1 The peroxide shunt of MMO hydroxylase component

Ferric resting cytochrome P450 has long been known to be able to utilize peroxides as the source of both the reducing equivalents and the oxygen required by the reaction stoichiometry [9,13,14]. Accordingly, when H_2O_2 is added to the oxidized hydroxylase component [Fe(III)·Fe(III)] in the presence of any MMO substrate, the same products are formed as are recovered from the NADH-coupled reconstituted MMO enzyme system and O_2 (see Fig. 1, peroxide shunt) [15]. The hydroxylase loses little activity during turnover in 10 mM H_2O_2. The maximum velocity of the H_2O_2 coupled reaction is approximately the same as that of the reconstituted enzyme system, but the high K_m for H_2O_2 (~250 mM) limits the observed velocity. The MMO peroxide shunt supports our proposal that O_2 adds to the diferrous state of the diiron cluster to form a transient peroxide intermediate, and confirms the location of the active site of MMO in the hydroxylase component.

3.2 The stereochemistry of alkane oxidation by MMO

It is conceivable that the MMO peroxide shunt functions differently than proposed in Fig. 1. For example, the iron coordinated peroxide could potentially react directly with the C—H bond of the substrate, or lead to the generation of diffusible hydroxy radicals. However, past studies of flavin oxygen activation indicate that the peroxide adduct itself is unlikely to be sufficiently reactive to attack the unactivated C—H bonds of alkanes. In contrast, hydroxyl radicals are capable of hydrogen atom abstraction from most alkanes (with the notable exception of methane). The reaction of such a radical outside of the active site as the principal mechanism of MMO can now be ruled out by the experiments described below.

The mechanism proposed in Fig. 1 implies the intermediacy of a non-iron bound, albeit sequestered, substrate radical. In contrast, other mechanisms have been advanced which invoke either an additional substrate carbocation intermediate [11] or concerted oxygen insertion into a substrate carbon-iron bond [16,17]. Recently, several studies have appeared that indicate the occurrence of a substrate intermediate not bound to the diiron cluster [10,11]. However, the substrates utilized used were rather unlike methane and were oxidized to several products. Thus these reactions may not have been typical of methane oxidation. We have approached this question by examining the stereochemistry of ethane turnover by MMO.

Ethane is the most similar molecule to methane that can be made chiral. In collaboration with Drs. H.G. Floss, N.D. Priestley, P.G. Williams, and H. Morimoto, carrier-free R- and S-$[1-^2H_1,^3H_1]$ethane were synthesized and used as substrates for both NADH/O_2-coupled reconstituted MMO and H_2O_2-coupled MMO hydroxylase [18]. It was shown by ^3H NMR that only the expected product, ethanol, was generated. From the ratios of the integrated areas of the peaks for CH_3CH^3HOH and $CH_3C^2H^3HOH$, the intramolecular primary kinetic deuterium isotope effect, $k_H/k_D = 4.2$, was determined. This value is similar to intramolecular kinetic isotope effects found by less precise techniques for other MMO substrates [10] as well as for some cytochrome P450 substrates [13,14]. The magnitude of the isotope effect clearly shows that

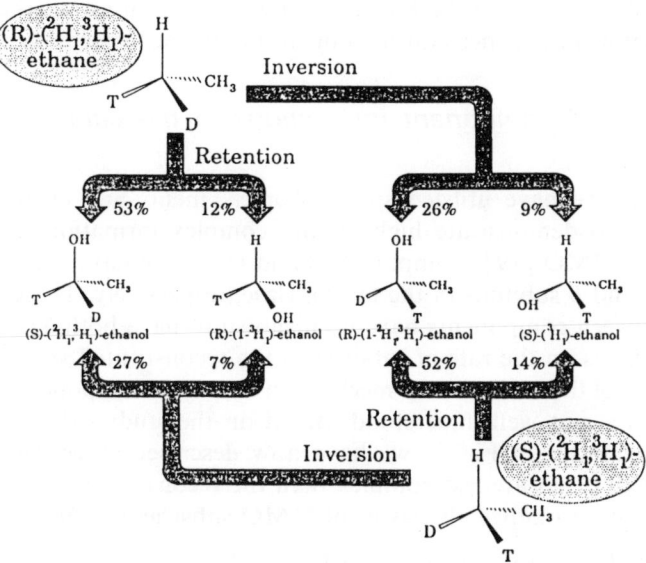

Figure 2 Distribution of products of chiral ethane oxidation catalyzed by methane monooxygenase. The data shown are for the NADH-coupled reconstituted MMO system. Essentially the same values were observed for the reaction catalyzed by the oxidized hydroxylase component without the reductase and component B in the presence of H_2O_2. Data from reference [18].

the C—H bond is completely broken at one stage of catalysis making a mechanism involving concerted oxygen insertion into this bond unlikely.

The ethanol obtained from the enzyme incubations of each enantiomer of ethane was converted to its (2R)-2-acetoxy-2-phenylethanoate derivative and subsequently analyzed by ^{3}H NMR. The results summarized in Fig. 2 show that MMO-catalyzed hydroxylation of ethane proceeds with predominant retention of configuration, accompanied by approximately 35% inversion. The fact that the product is not completely racemized shows that the reaction does not proceed by a mechanism such as the release of diffusible hydroxyl or substrate radicals into solution.

Partial inversion of the products implies the formation of a short-lived intermediate that can flip in the active site before the reaction is completed. This species is probably an ethyl radical because other possibilities, such as an ethyl cation would have too great an activation energy for direct formation and would be very unstable if formed. It is however, possible that there are additional intermediate species on the reaction path. This result strongly supports a radical-based mechanism for MMO catalyzed oxidation of alkanes, and makes improbable mechanisms not involving a free substrate intermediate sequestered in the enzyme active site. Finally, the ethane oxidation reaction has the same steric outcome regardless of whether the $NADH/O_2$-coupled MMO system or H_2O_2-coupled MMO hydroxylase alone is used as the catalyst, suggesting that a similar mechanism is operating in each case.

3.3 The role of component interactions in the catalytic cycle of MMO

In past studies, we have utilized spectroscopic, kinetic, and chemical cross-linking studies to demonstrate high affinity complex formation between the components of MMO [19]. Component B and reductase have specific binding sites on the α and β subunits of the hydroxylase, respectively. These and other complexes between components have been shown to have both activating and inactivating effects on the rate of catalysis in the reconstituted system. Despite the recognition of these effects, the mechanism by which component complexes affect catalysis is not well understood. Based on the studies that lead to the proposed mechanism of Fig. 1, we have now described three catalytic subsystems of MMO that are less complex than the reconstituted enzyme system but still function to oxidize all classes of MMO substrates [20]:

System I = Hydroxylase, reductase, NADH, and O_2
System II = Oxidized hydroxylase and H_2O_2
System III = Fully reduced hydroxylase and O_2

Since each system functions without component B, and in two cases also without reductase, the effects of adding these components can be readily ascertained.

The proposed mechanism (Fig. 1) emphasizes the predominant role of the hydroxylase diiron cluster in catalysis. If the only roles of the reductase and B components were to facilitate transfer of electrons to the hydroxylase, then we would expect to see effects of these components only in System I and only on the rate of reduction. No effects on the product forming aspect of the reaction in any of the systems is expected. Accordingly, we now know that all three subsystems produce the same types of products from any given substrate irrespective of the presence of the reductase and B components. These results, along with the identical stereochemical course of the NADH- and H_2O_2-coupled oxidation of chiral ethane described above, strongly suggests that the chemical oxidant and fundamental mechanism of the hydroxylase is the same regardless of the method used to bring it into the reactive state. However, two types of observations suggest that the reductase and component B may play other roles in catalysis. First, System II has no requirement for electron transfer, but component B has a dramatic (generally inhibitory) effect on rate as shown in Fig. 3. Second, dramatic, system specific differences in product distributions are observed for many classes of substrates that can be oxidized at more than one carbon [20]. These distributions are strongly affected by addition of component B.

The product distributions observed for isopentane in each system and the effect of component B on this distribution is shown in Table 1. In each case,

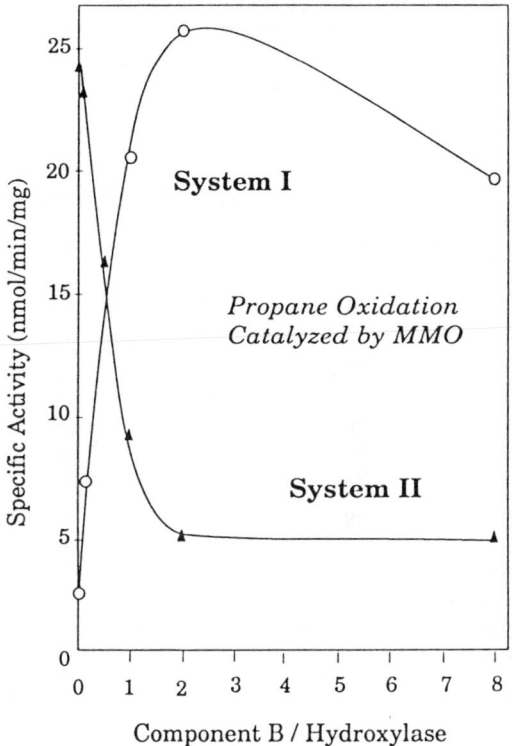

Figure 3 Effect of component B on the initial velocity of propane oxidation catalyzed by methane monooxygenase. System I is the reconstituted MMO system with reductase and hydroxylase at equal concentrations and the component B added in the ratio shown. System II contains only oxidized hydroxylase component plus hydrogen peroxide and the ratio of component B shown. An analogous plot of the effects of component B on isopentane turnover shows exactly the same trends. Data from references [15,19,20].

the distribution in the absence of component B favors hydroxylation at the least stable C—H bond as would be observed for hydroxylation of these substrates in solution by a catalyst utilizing Fenton chemistry. The distributions are all markedly different from those observed for the reconstituted MMO system (in effect, the results for System I plus component B in Table 1). The distributions also differ somewhat from each other. On addition of component B, the distributions of Systems I and III become identical to each other and to that observed for the reconstituted system. In each case, the hydroxylation at thermodynamically disfavored primary carbons is enhanced. Component B also affects the distribution from System II. Primary carbon hydroxylation is also enhanced, but the distribution remains different from those of the other systems. Since the underlying chemistries of Systems I, II, and III are unlikely to be different, the distribution changes must be related to some other difference in

Table 1 MMO Hydroxylase catalyzed oxidation of isopentane

System[a]	B:H[b] ratio	3° (OH)	2° (OH)	1° (HO)	1° (OH)
		Product Distribution (%)[c]			
I	0	27	30	30	13
NADH/O_2/	0.1	13	10	47	30
Reductase	1.0	11	8	51	30
	2.0	11	9	49	31
II	0	43	24	12	21
[H_2O_2]	0.1	42	23	13	22
	0.5	34	20	20	26
	1.0	26	19	29	26
	2.0	26	20	25	29
	8.0	26	19	28	27
III	0	38	26	22	14
Chemical	0.1	13	7	49	31
Reduction	0.5	12	8	53	27
+O_2	2.0	11	7	52	30

[a]All systems contain hydroxylase component
[b]Ratio of component B to hydroxylase (nmol/nmol)
[c]Data from reference [20]

the hydroxylase. A likely possibility is the state of the reduction of the hydroxylase diiron cluster during the time immediately before oxygen is activated. For Systems I and III, the diiron cluster is in the fully reduced state, while for System II the cluster is oxidized (see Fig. 1). If these changes in redox state are accompanied by structural changes in the active site, substrates may be presented differently to the active oxygen. Similarly the component B (and reductase) complexes with the hydroxylase may cause a change in the active site structure that affects substrate binding. Indeed, we have demonstrated that structural changes occur in the hydroxylase active site upon formation of a component B complex [19,20].

Another quite unexpected aspect of the effect of component interaction on product formation is noted by comparing the concentration dependence on component B of reaction rate (Fig. 3) and product distribution (Table 1). It is apparent that the change in product distribution of Systems I and III is complete for component B at ≤5% of the hydroxylase concentration. In contrast, the effects on rate in both Systems I and II and the effect on product distribution in System II maximize at stoichiometric amounts of component B. These effects might be explained by many schemes, most based on a large increase in the rate of catalysis when the hydroxylase-component B complex forms. However, these fail because the rate enhancement [19] caused by

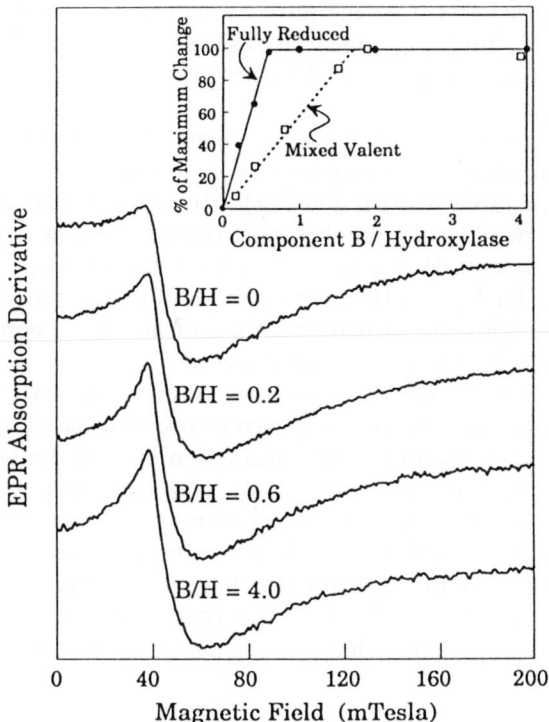

Figure 4 Effect of component B on the integer spin EPR spectrum of the fully reduced hydroxylase component. Component B was added in the ratio shown before freezing the sample for EPR measurements. *Inset*: Comparison of the component B concentration dependence of the changes in the EPR spectra of the fully reduced and mixed valent hydroxylase. Data from references [19,20].

component B is insufficient to account for the complete shift in product distribution at a 1:20 ratio of components. Moreover, any hypothesis must account for the complete shift in product distribution arising from System III in which only a single turnover occurs, and therefore, no amplification due to multiple turnover at a high rate accrues. One mechanism that does account for these results is that component B might induce a conformational change in the fully reduced hydroxylase that is stable for a time after component B dissociates. Thus a substoichiometric amount of component B could modify the entire population of reduced hydroxylase, giving rise to the complete shift in product distribution observed for Systems I and III. This is supported by EPR measurements (Fig. 4) indicating that the active site structure of the entire population of reduced hydroxylase is changed at a stoichiometry of approximately 0.3 B per hydroxylase active site [20]. Thus, an altered conformation persists after component B dissociates for a period at least comparable to the freezing time of the sample. In contrast, a stoichiometric concentration of component B is required to titrate the mixed valent hydroxylase [19] (Fig.4, *inset*), suggesting that rapid component B dissociation and/or hysteresis in relaxation of an altered conformation does not occur for the more oxidized states of the hydroxylase.

A similar mechanism would not apply to System II because the affinity of oxidized diferric hydroxylase for component B is too great ($K_d \approx 67$ nM) to postulate the required rapid dissociation [19]. Moreover, explanations based on rate enhancement would not apply because, as shown in Fig. 3, component B slows down the System II reaction. Therefore, the changes in product distribution observed for System II in the presence of component B would be expected to correlate with the formation of a stoichiometric component B-oxidized hydroxylase complex as we have observed.

Methane is the only physiologically relevant substrate of MMO. Thus, the structural changes in the hydroxylase proposed to occur upon reduction of the diiron cluster and/or component complex formation are not designed to shift product distribution *in vivo*. Nevertheless, we have demonstrated that both reduction and component-component interactions are integral parts of the natural catalytic cycle of MMO. Therefore, we propose that the putative conformational changes in the hydroxylase that occur in response to these events also occur during catalysis by the reconstituted enzyme system and are likely to play some role, perhaps related to regulation of the catalytic cycle.

4 CONCLUSION

The catalytic cycle shown in Fig. 1 for MMO based on that of cytochrome P450 appears to describe catalysis by the enzyme quite well. We have shown that: 1. the site of catalysis is the hydroxylase diiron cluster, 2. the diferrous cluster is the form which reacts with O_2, 3. H_2O_2 can provide both of the reducing equivalents and the oxygen required for catalysis, 4. a substrate intermediate free from the cluster, but still in the active site, is generated; the data strongly suggest this is a substrate radical, and 5. the component B must play at least two roles in catalysis since it affects rate and product distribution at very different concentrations relative to the hydroxylase.

The work described here indicates that although the reductase and component B are not required for catalysis, they do play multiple roles in directing the reaction. It seems likely that at least one of these roles is to regulate catalysis through cyclic formation of complexes which alter the structure of the hydroxylase. For example, the complexes might serve the essential function of coordinating oxygen activation with substrate binding, and/or facilitating both substrate binding and product release at the appropriate points in the catalytic cycle.

REFERENCES

[1] Fox, B.G., Froland W.A., Dege, J.E. and Lipscomb, J.D. (1989) Methane monooxygenase from *Methylosinus trichosporium* OB3b. Purification and properties of a three-component system with high specific activity from a type II methanotroph. *J. Biol. Chem.* **264**, 10023–10033.
[2] Fox, B.G., Surerus, K.K., Münck, E. and Lipscomb, J.D. (1988) Evidence for a μ-oxobridged binuclear iron cluster in the hydroxylase component of methane monooxygenase: Mösbauer and EPR studies. *J. Biol. Chem.* **263**, 10553–10556.
[3] Hendrich, M.P., Mück, E., Fox, B.G. and Lipscomb, J.D. (1990) Integer-spin EPR studies of the fully reduced methane monooxygenase hydroxylase component. *J. Am. Chem. Soc.* **112**, 5861–5865.
[4] DeWitt, J.G., Bentsen, J.G., Rosenzweig, A.C., Hedman, B., Green, J., Pilkington, S., Papaefthymiou, G.C., Dalton, H., Hodgson, K.O. and Lippard, S.J. (1991) X-ray absorption, Mössbauer, and EPR studies of the dinuclear iron center in the hydroxylase component of methane monooxygenase. *J. Am. Chem. Soc.* **113**, 9219–9235.
[5] Hendrich, M.P., Fox, B.G., Andersson, K.K., Debrunner, P.G. and Lipscomb, J.D. (1992) Ligation of the diiron site of the hydroxylase component of methane monooxygenase. An electron nuclear double resonance study. *J. Biol. Chem.* **267**, 261–269.
[6] Dalton, H. (1980) Oxidation of hydrocarbons by methane monooxygenases from a variety of microbes. *Adv. Appl. Microbiol.* **26**, 71–87.
[7] Higgins, I.J., Best, D.J. and Hammond, R.C. (1980) New findings in methane-utilizing bacteria highlight their importance in the biosphere and their commercial potential. *Nature* **286**, 561–564.
[8] Green, J. and Dalton, H. (1989) Substrate specificity of soluble methane monooxygenase: Mechanistic implications. *J. Biol. Chem.* **264**, 17698–17703.
[9] Porter, T.D. and Coon, M.J. (1991) Cytochrome P-450: Multiplicity of isoforms, substrates, and catalytic and regulatory mechanisms. *J. Biol. Chem.* **266**, 13469–13472.
[10] Rataj, M.J., Kauth, J.E. and Donnelly, M.I. (1991) Oxidation of deuterated compounds by high specific activity methane monooxygenase from *Methylosinus trichosporium* OB3b: Mechanistic implications. *J. Biol. Chem.* **266**, 18684–18690.
[11] Ruzicka, F., Huang, D.-S., Donnelly, M.I. and Frey, P.A. (1990) Methane monooxygenase catalyzed oxygenation of 1,1-dimethylcyclopropane. Evidence for radical and carbocationic intermediates. *Biochemistry* **29**, 1696–1700.
[12] Fox, B.G., Borneman, J.G., Wackett, L.P. and Lipscomb, J.D. (1990) Haloalkene oxidation by the soluble methane monooxygenase from *Methylosinus trichosporium* OB3b: mechanistic and environmental implications. *Biochemistry* **29**, 6419–6427.
[13] McMurry, T.J. and Groves, J.T. (1986) Metalloporphyrin models for cytochrome P-450. In: Cytochrome P-450 Structure, Mechanism, and Biochemistry (Ortiz de Montellano P.R., Ed.), pp. 1–28. Plenum, New York.
[14] Ortiz de Montellano, P.R. (1986) Oxygen activation and transfer. In: Cytochrome P-450 Structure, Mechanism, and Biochemistry (Ortiz de Montellano, P.R., Ed.), pp. 217–271. Plenum, New York.
[15] Andersson, K.K., Froland, W.A., Lee, S.-K. and Lipscomb, J.D. (1991) Dioxygen independent oxygenation of hydrocarbons by methane monooxygenase hydroxylase component. *New J. Chem.* **15**, 411–415.
[16] Barton, D.H.R., Csuhai, E., Doller, D., Ozbalik, N. and Balavoine, G. (1990) Mechanism of the selective functionalization of saturated hydrocarbons by Gif

systems: Relationship with methane monooxygenase. *Proc. Natl. Acad. Sci. USA* **87**, 3401–3404.

[17] Barton, D.H.R., Bévière, S.D., Chavasiri, W., Csuhai, E., Doller, D. and Liu, W.-G. (1992) The functionalization of saturated hydrocarbons. Part 20. Alkyl Hydroperoxides: Reaction intermediates in the oxidation of saturated hydrocarbons by Gif-type reactions and mechanistic studies on their formation. *J. Am. Chem. Soc.* **114**, 2147–2156.

[18] Priestley, N.D., Floss, H.G., Froland, W.A., Lipscomb, J.D., Williams, P. G. and Morimoto, H. (1992) Cryptic stereospecificity of methane monooxygenase. *J. Am. Chem. Soc.*, in press.

[19] Fox, B.G., Liu, Y., Dege, J.E. and Lipscomb, J.D. (1991) Complex formation between the protein components of methane monooxygenase from *Methylosinus trichosporium* OB3b: Identification of sites of component interactions. *J. Biol. Chem.* **266**, 540–550.

[20] Froland, W.A., Andersson, K.K., Lee, S.-K., Liu, Y. and Lipscomb, J.D. (1992) Methane monooxygenase component B and reductase alter the regioselectivity of the hydroxylase component catalyzed reactions: A novel role for protein-protein interactions in an oxygenase mechanism. *J. Biol. Chem.*, in press.

7

Biochemical and Biophysical Studies Toward Characterization of the Membrane-associated Methane Monooxygenase

Sunney I. Chan[1], Hiep-Hoa T. Nguyen[1], Andrew K. Shiemke[1] and Mary E. Lidstrom[2]

[1] A.A. Noyes Laboratory of Chemical Physics and [2] W.M. Keck Laboratory of Environmental Engineering, California Institute of Technology, Pasadena, CA 91125, USA

1 SUMMARY

The membrane-associated (particulate) methane monooxygenase (pMMO) found in methanotrophs converts methane to methanol in these bacteria. Recent results obtained in our laboratory confirm earlier findings that the pMMO has an obligatory requirement for copper ions. The pMMO activity is found to be directly proportional to the copper/total membrane protein ratio in membrane fractions obtained from *M. capsulatus* (Bath) grown at different copper concentrations. Electrophoretic analysis of these membrane fractions shows no substantial increase in the level of expression of the putative polypeptides that have been implicated with the pMMO. These results suggest that the pMMO is a copper-containing protein. The membrane-bound copper ions are EPR detectable and can be divided into two types on the basis of their EPR characteristics: (a) isolated type-2 copper ions; and (b) exchange-coupled trinuclear copper clusters. The analysis of magnetic susceptibility data clearly show that the bulk of the copper ions ($\sim 95\%$) in these membranes exists in the form of the trinuclear cluster. The trinuclear copper cluster is found to be ferromagnetically coupled with a quartet ground state. The exchange coupling constant (J) and the zero-field splitting parameters (D

Microbial Growth on C_1 Compounds
© Intercept Ltd, PO Box 716, Andover, Hampshire SP10 1YG, UK

& E) in the quartet state are determined from the temperature dependence of the magnetization ($J \sim 15$ cm^{-1}, $D < 0.05$ cm^{-1}, and $E \sim 0$). The high reactivity of the reduced cluster toward dioxygen implicates the trinuclear copper cluster as the site of monooxygenase activity of the pMMO system. A mechanism for the monooxygenase activity and methane activation based on the trinuclear copper cluster model proposed above is presented. Preliminary analysis of the copper/total membrane protein ratio suggests that the pMMO system contains several clusters per protein molecule.

2 INTRODUCTION

The enzyme methane monooxygenase (MMO) in methanotrophic bacteria catalyzes the conversion of methane to methanol using molecular oxygen as co-substrate [1]. Two forms of MMO differing in cellular location are known to exist [2]. The soluble methane monooxygenase (sMMO) found in the cytosolic fraction of the bacteria appears to be restricted to only *Methylococcus* and *Methylosinus* strains and is expressed only at limiting copper level in the growth medium [3–5]. The particulate methane monooxygenase (pMMO) found in the membrane fraction of the cells appears to be expressed in all types of methanotrophic bacteria [3–6]. Only the sMMO has been purified and extensively characterized [7–11].

Progress toward understanding the pMMO has been hampered by the difficulty in maintaining the enzyme activity *in vitro*. However, Dalton's pioneering work in this area has established the singly important role of copper ions to this system during protein synthesis and in MMO activity [3–5]. For strains capable of expressing both the sMMO and the particulate form of the enzyme, the pMMO will be expressed at high levels of copper ions, even though the copper concentration threshold to ensure pMMO expression is very low ($\ll 0.3$ μM). High copper concentrations also stabilize and enhance pMMO activity both *in vivo* and *in vitro* [4,5]. For instance, increased pMMO activity has been observed for membrane fractions obtained from *Methylococcus capsulatus* (Bath) when these cells are grown in copper-enriched media [5]. The pMMO activity is stimulated even further with the addition of copper ions during assay [5].

The pMMO is different from the sMMO in several aspects. The pMMO has a narrow substrate range, capable of oxidizing only C4 hydrocarbons or smaller [4]. The pMMO is sensitive to several inhibitors which do not affect sMMO. In addition, three new polypeptides of apparent molecular weight 46 kDa, 35 kDa, and 26 kDa appear on SDS-polyacryamide gels of these membranes when cells expressing sMMO are switched to express pMMO [3–5]. The pMMO expression in these organisms is associated with the formation of extensive intracytoplasmic membranes. A substantial increase in the carbon conversion efficiency of methane to biomass is also associated with the change from sMMO to pMMO expression [3].

Currently, the involvement of copper and iron in the pMMO has not been clearly established. Data to date consistently point to the importance of copper for the activity of the pMMO, thus suggesting that pMMO might be a copper-containing protein. Since the pMMO system has proven to be so refractory to purification, a different approach to study this enzyme may be necessary. Since pMMO activity is greatly affected by copper ions, this fact prompted us to carry out an extensive EPR and magnetic susceptibility studies on these membranes. Results from our laboratory strongly support the contention that pMMO is a copper-containing enzyme system.

3 THE EFFECTS OF COPPER IONS

Consistent with Prior and Dalton's earlier work [5], *M. capsulatus* (Bath) membrane fractions exhibit greater pMMO specific activity as the copper concentration of the growth medium is increased (Table 1). Addition of copper to the medium during the enzyme activity assay further stimulates the rate of propene oxidation as previously reported (Table 1) [5]. However, the addition of copper during activity assay appears to inhibit pMMO activity in membranes obtained from cells grown with 20 μM Cu^{2+} or higher. Also as expected, the soluble fractions from these cell cultures show no propene epoxidation activity, indicating that only the membrane-bound form of the enzyme is being expressed here. In cells grown under conditions allowing expression of the sMMO (low Cu concentration, high cell density), the sMMO activity was easily detected.

A connection between the propene epoxidation activity of pMMO and copper is suggested by metal analysis using direct current plasma emission (DCPE) spectroscopy. Copper analysis of whole cells of *M. capsulatus* (Bath) grown with varying copper concentration shows that most of the copper in the medium is eventually taken up by the bacteria (Table 1). The bulk of this intracellular copper is found in the membrane fractions, and the specific activity of pMMO is directly proportional to the membrane copper content (Table 1). When the specific activity of pMMO is plotted as a function of the copper content (copper/protein ratio) of the membranes, a linear relationship is observed (Fig. 1). This direct proportionality suggests that membrane copper is required to activate and/or stabilize the pMMO.

A strong correlation does not exist between the pMMO specific activity and the iron content (iron/protein ratio) of the membrane fractions. The iron content increases only by a factor of 4 over the total range of copper concentrations varied despite the fact that the initial iron concentration in the growth medium was constant (0.9 μM). The increase in iron content (iron/ protein ratio) which follows the increase in the copper content (copper/ protein ratio) of the membrane fractions appears to be fortuitous and represents a future line of inquiry. Moreover, the iron/protein ratio varies unpredictably in recent experiments. As a result, the present data do not permit any conclusion

Table 1 Effect of growth medium copper concentration on bacterial copper content and pMMO activity

Cu concentration[a] of media (μM)	Total intracellular[b] copper (μmol)	Cu content (μmol)[c]		Cu/protein (μg/mg)[d]		Specific activity(nmol/min/mg)[e]	
		Soluble	Membrane	Soluble	Membrane	As isolated	w/Copper added
<0.3	0.7 (~100)	0.17 (25)	0.53 (75)	0.41	1.32	0.68	19.5
2.0	4.4 (~100)	0.48 (11)	3.88 (89)	1.06	7.08	9.6	27.3
5.0	8.7 (87)	0.78 (9)	7.92 (91)	1.55	13.4	17.6	27.7
10.0	13.7 (69)	1.03 (8)	12.7 (92)	2.53	22.2	26.8	30.2
20.0	22.8 (57)	2.0 (9)	20.8 (91)	4.27	37.5	53.7	22.1

[a] Copper concentration of media after supplemental copper was added. The value of 0.3 μM is the upper limit of the copper concentration in the unsupplemented media.
[b] The total amount of copper found in the bacteria (whole cells), determined by DCPE. The number in parentheses is the percentage of the total copper available in the media that is ultimately found in the bacteria.
[c] The copper content of the soluble and membrane fractions. The number in parenthesis is the portion of total bacteria copper found in each fraction.
[d] The copper/protein ratio of the soluble and the membrane fractions. Protein concentration was determined by the Lowry method.
[e] The specific activity of the pMMO in the membrane fractions (per mg of total membrane protein) assayed as isolated or with copper added to the assay buffer (150 μM).

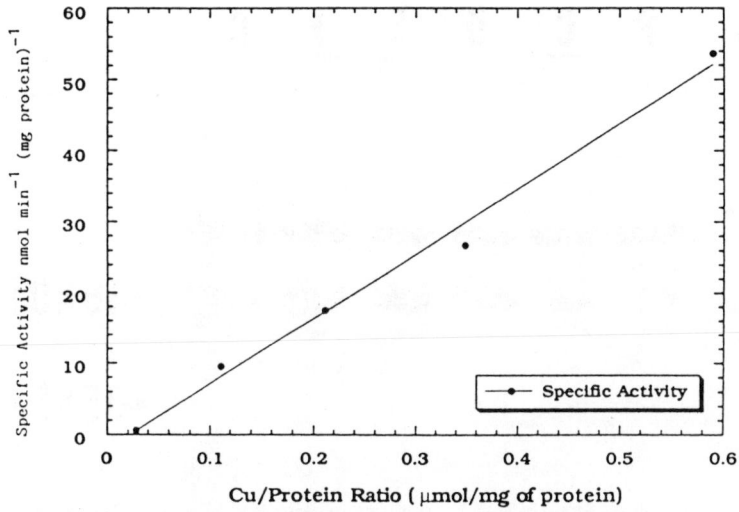

Figure 1 Effect of increasing Cu^{2+} concentration in growth medium on pMMO specific activity. Specific activity versus copper/protein ratio of membrane fractions obtained from *Methylococcus capsulatus* (Bath) grown at increasing Cu^{2+} concentrations. The cultures were grown in 500-ml batches in 2-l Erlenmeyer flasks using NMS medium, a 20% methane-to-air atmosphere, and continual shaking at 45°C. Cultures in which the copper concentration of the medium was varied were grown in parallel with identical medium and growth conditions. The membrane fractions were isolated as described previously [5]. The MMO activity of samples was measured by the standard propene epoxidation assay [5].

regarding the presence of iron in the pMMO. Our results also indicate that the iron detected appears to be associated with membrane-bound c-type cytochromes. Metal analysis also reveals no other metal ions (including Zn, Co, Mn, Ni, Mg), aside from Fe in the membrane preparations, at higher than trace levels.

The evidence presented above suggests the existence of a copper-deficient, inactive pMMO in cells grown at low Cu levels. This idea is supported by SDS-PAGE analysis of membrane fractions isolated from cells of *M. capsulatus* (Bath) grown with varying amounts of copper. Three bands corresponding to polypeptides of apparent molecular weights 46 kDa, 35 kDa, and 26 kDa have been implicated for the pMMO [3–5]. We have determined by SDS-PAGE the level of expression of these polypeptides as a function of the copper concentration in the growth medium. As illustrated in Fig. 2, the levels of expression of these polypeptides do not change significantly, whereas the specific activity of pMMO in these samples varies by nearly a factor of 80 (Table 1). The presence of the copper-deficient pMMO in cells grown at low Cu levels may explain the activity stimulation effect of copper ions, particularly during activity assay. The binding of copper ions to the apo-pMMO results in the activation of the

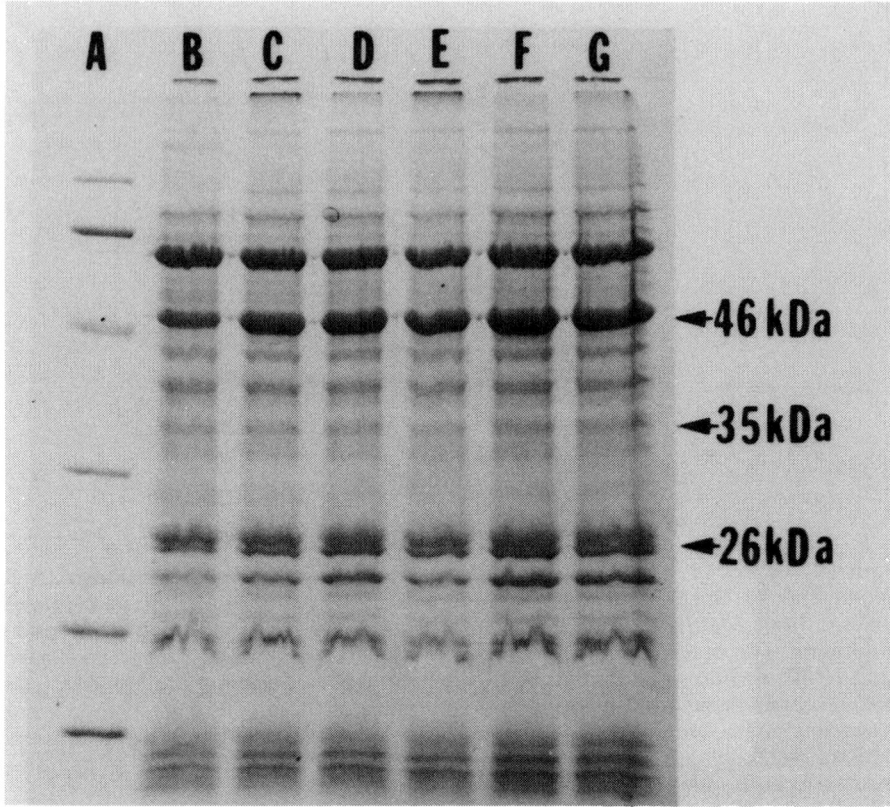

Figure 2 SDS-polyacrylamide gel electrophoresis of the membrane fractions from *Methylococcus capsulatus* (Bath) grown under increasing Cu^{2+} concentration. A, molecular weight standards (92, 66.2, 45, 31, 21.5 and 14.4 kDa); B, no added copper; C, 1 µM Cu^{2+}; D, 2.5 µM Cu^{2+}; E, 5 µM Cu^{2+}; F, 10 µM Cu^{2+}; G, 20 µM Cu^{2+}. Lanes B-G contain 75 µg total protein. The major band above the 46 kDa polypeptide is the large subunit of the methanol dehydrogenase, which is loosely bound to the membrane. Protein concentration was determined by the Lowry method. Electrophoresis was performed on a 15% slab gel according to the Laemmli method and the gel was stained with Coomassie Brilliant Blue.

enzyme; accordingly, an increase in the activity of the pMMO is observed. The specific activity of the enzyme appears to level off when the enzyme has acquired the optimum copper uptake as evidenced from activity assays carried out with an excess of copper.

4 EPR CHARACTERIZATION OF THE MEMBRANE-BOUND COPPER IONS

The membrane-associated copper is EPR detectable (Fig. 3a). The Cu EPR spectrum is observed in the membrane fractions only when the pMMO is

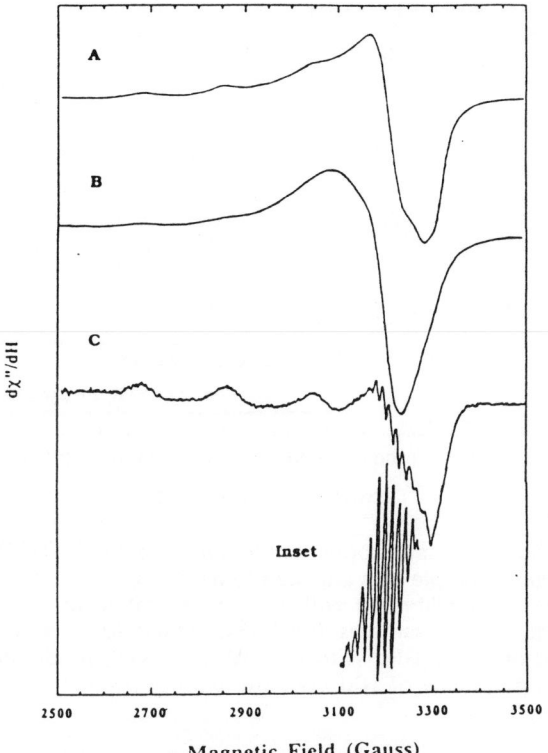

Figure 3 X-band EPR spectra of membrane fractions obtained from *Methylococcus capsulatus* (Bath) cells grown with 20 μM Cu^{2+}. (A) Spectrum obtained with microwave power of 0.2 mW; (B) Spectrum obtained with microwave power of 40 mW. EPR spectra (A) and (B) were recorded at 7 K with modulation amplitude of 16 G, modulation frequency of 100 kHz, and gain of 2.5×10^2 for (A) and 1.6×10^2 for (B); (C) Spectrum obtained after 3 hrs incubation with sodium dithionite and was recorded at 8.4 K with 1 mW of microwave power, modulation amplitude of 5 G, modulation frequency of 100 kHz, time constant of 0.25 sec., and gains of 3.2×10^3. Inset: Second derivative of the absorption at g = 2.06.

expressed. The intensity of the Cu EPR signal is found to correlate directly with the *in vitro* pMMO specific activity in isolated membrane fractions. A linear relationship is observed when the pMMO specific activity is plotted versus the EPR-detected copper concentration, i.e., the copper concentration inferred from the observed EPR intensity (Fig. 4). These results suggest that the pMMO activity is associated primarily with the membrane-bound copper ions. Interestingly, the intensity of the copper EPR observed for the membrane fractions isolated from organisms grown in high copper medium (>5 μM) accounts for only about one-third of the expected intensity based on the total copper content.

The analysis of the EPR spectrum of membrane fractions obtained from *M. capsulatus* (Bath) exhibiting high *in vitro* activity suggests the presence of two

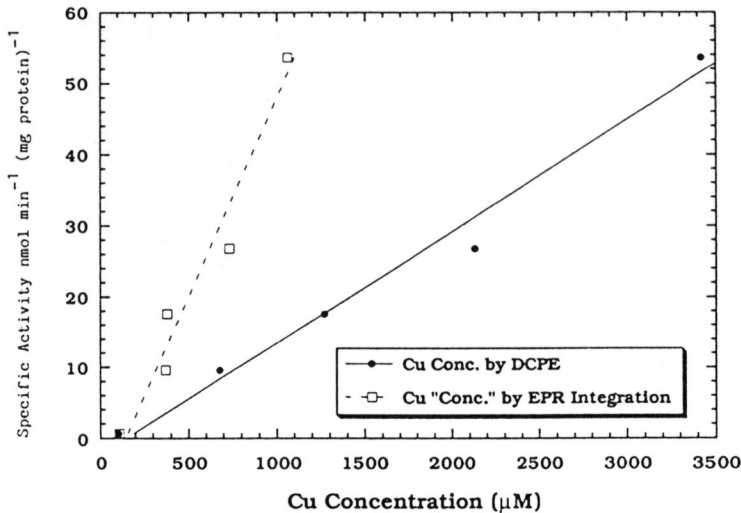

Figure 4 Specific activity versus copper content as detected by DCPE and by EPR double integration. Each sample was adjusted to a total protein concentration of 10 mg/ml. EPR intensity was calibrated with the concentration of a series of $CuSO_4$ standards. EPR spectra of the samples and $CuSO_4$ standards were obtained at 7 K at a microwave field of 9.126 GHz with 0.2 mW of power, modulation amplitude of 16 G, modulation frequency of 100 kHz, and gain of 5×10^3.

distinct copper signals. One signal (Fig. 3a) can be attributed to type 2 copper centers on the basis of A and g values ($g_{\parallel} = 2.25$, $A_{\parallel} \sim 18 \times 10^{-3}$ cm^{-1}, $g_{\perp} \sim 2.058$). The other signal occurs near $g \sim 2.06$. This signal is broad and nearly isotropic. The relative contribution of each type of Cu^{2+} ions signal to the EPR spectral intensity is very sensitive to temperature. The type 2 Cu^{2+} EPR signal can be easily saturated at low temperatures (<10 K) whereas the 'isotropic' signal is not, and hence can be isolated and directly observed by recording the spectrum at high microwave powers (Fig 3b). The appearance of this signal is not a result of lineshape distortion due to high microwave power but rather is a manifestation of dramatic difference in the relaxation characteristics of the two species which give rise to these two EPR signals. In addition, the intensity of the 'isotropic' signal is found to correlate directly with the *in vitro* pMMO activity.

These features of the 'isotropic' signal suggest that it may have origin in a multinuclear copper cluster. Consistent with this idea, the unusual signal decreases in intensity and gives way to a second isotropic signal at the same g value but with resolved hyperfine features ($|A| \sim 15$ Gauss) upon incubation and reduction with limiting, i.e., sub-stoichiometric amounts of dithionite (Fig. 3c). If the 'isotropic' signal arises from a multinuclear cluster, partial reduction might ultimately yield a mixed-valence system with an unpaired-electron spin delocalized over all the Cu ions of the cluster. The observed hyperfine interaction pattern (Fig. 3c and inset) is consistent with this expectation and suggests that the splitting arises from the coupling of an unpaired electron spin to three equivalent $I = 3/2$ nuclear spins. That is, the 'isotropic' signal with resolved ten nuclear hyperfine lines (Fig. 3, inset) arises from a mixed-valence trinuclear Cu cluster wherein the electron spin of a Cu(II) is delocalized over two additional Cu(I) ions. If this interpretation is correct, the 'isotropic' Cu EPR signal observed for the fully oxidized enzyme would arise from appropriate EPR transitions of an exchange-coupled trinuclear Cu cluster (*vide infra*).

5 MAGNETIC SUSCEPTIBILITY CHARACTERIZATION OF THE MEMBRANE-BOUND COPPER IONS

Metal analyses have shown that the membrane fractions obtained from *M. capsulatus* (Bath) grown at 20 μM Cu(II) in the growth medium contain a substantial amount of copper ions and the Cu/Fe ratio is high, ranging from 6–14. As shown by EPR spectra of these membrane fractions, the level of paramagnetic iron (probably low spin) is insignificant as compared with the amount of paramagnetic copper ions. The weak line observed at $g \sim 4.3$ in several cases obviously can be attributed to non-specific iron. As a result, only paramagnetic copper ions contribute significantly to the magnetization in magnetic susceptibility measurements.

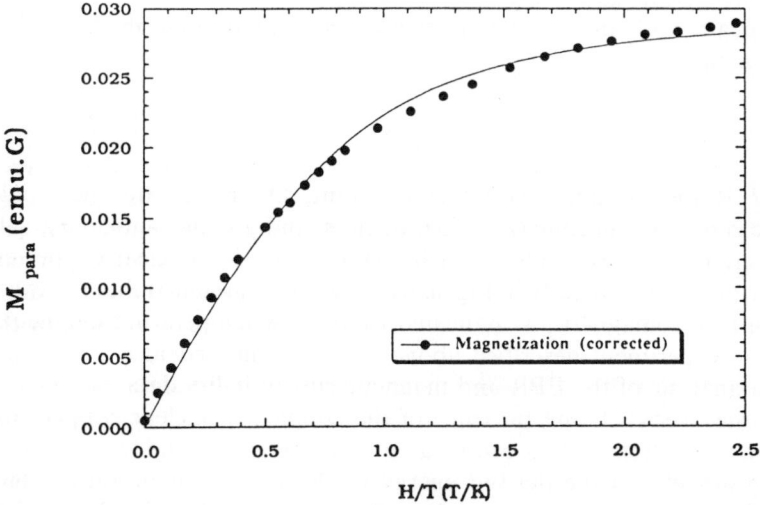

Figure 5 The Brillouin plot of the magnetization observed for membranes obtained from *Methylococcus capsulatus* (Bath) cells grown with 20 μM Cu^{2+}. The magnetization saturation data were obtained over a range of magnetic field (0–5.5 Tesla) and at 1.8 K.

Parallel magnetic susceptibility experiments on membranes obtained from *M. capsulatus* (Bath) confirm the existence of the above proposed exchange-coupled spin system. Coupling of the unpaired electron spins from three equivalent Cu(II) ions would give rise to a S = 3/2 state in addition to two S = 1/2 states. Since only the ground state contributes significantly to the magnetization at very low temperatures (<2 K), isothermal magnetization saturation experiments at these temperatures are expected to reveal the ground level, which can be either the quartet or the two degenerate doublets, depending on whether the exchange interactions among the Cu(II) ions are ferromagnetic or antiferromagnetic. Magnetization data measured at 1.8 K and varying field strengths fit the Brillouin function [12] with an effective spin (S_{av}) of 1.44 ± 0.05 (Fig. 5), indicating that the copper ions are ferromagnetically coupled and that the bulk of the copper ions in the membranes exists in the form of the trinuclear cluster. The value of the effective spin as deduced from the magnetic susceptibility experiment here provides strong support for our interpretation of the EPR results and establishes unequivocally the existence of a spin-coupled trinuclear copper cluster in the pMMO system.

6 ANALYSIS OF EPR AND MAGNETIC SUSCEPTIBILITY DATA: A SPIN-COUPLING MODEL

In the case of strong exchange-coupling, the spin-coupled system consists of multiplets of total spin S. Assuming that the three copper ions in the spin-coupled trinuclear cluster are equivalent, the energy levels of the system under the application of an external magnetic field can be described by the spin Hamiltonian:

$$\hat{H} = -2J \sum_{i>j}^{3} \hat{S}_i \cdot \hat{S}_j + D[\hat{S}_z^2 - \frac{1}{3} S(S+1)] + E[\hat{S}_y^2 - \hat{S}_y^2] + \beta g \hat{S} \cdot H$$

where J is the exchange coupling constant, D and E are the axial and rhombic zero-field splitting (ZFS) parameters, and g is the isotropic g value of the trimer. The energy levels for such a spin-coupled trinuclear copper cluster are illustrated schematically in Fig. 6, where we have assumed that the exchange coupling is ferromagnetic, as indicated by the low-temperature magnetization saturation experiment described above. This scheme provides a basis for the in-depth analysis of the EPR and magnetic susceptibility data.

With the overwhelming presence of the coupled trinuclear copper clusters, magnetic susceptibility data can be analyzed in greater detail according to the coupling model and the derived energy levels. The magnetic saturation data can be fitted to a quartet (S = 3/2) with about 5% of isolated type-2 Cu^{2+} ions (S = 1/2) if not less. The magnetization measured at constant field (1 Tesla) as a function of temperature (1.8 K–270 K) shows considerable departure from linearity (Fig. 7). The analysis of the temperature-dependent magnetization data

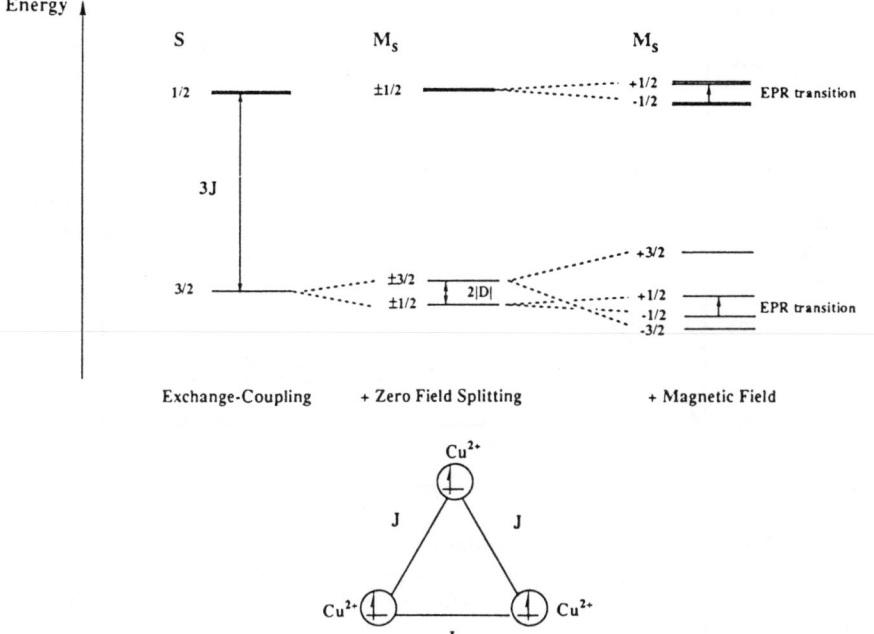

Figure 6 Energy level diagram for a ferromagnetically-coupled trinuclear copper cluster. J and D refer to the exchange-coupling constant and the axial zero-field splitting, respectively (see the equation in text). The spacing between the energy levels is not drawn to scale.

also shows that the ground state is indeed a quartet with $D < 0.05$ cm^{-1} and J is in the order of about 15–20 cm^{-1}. The inclusion of other species of $S = 1/2$ (presumably isolated type 2 Cu^{2+}) into the calculations also indicates that only $\sim 5\%$ of the total copper ions if not less are in the form of the type 2 Cu^{2+} ions.

For the D_{3h} model under consideration here, EPR signals at X-band frequency are expected from the $|-1/2\rangle \rightarrow |+1/2\rangle$ transitions of the two degenerate doublet manifolds and from the $|-1/2\rangle \rightarrow |+1/2\rangle$ transition within the quartet manifold (Fig. 6). At 7 K, the temperature at which the EPR measurements were made, the $|-1/2\rangle \rightarrow |+1/2\rangle$ transition of the quartet manifold should make the dominant contribution to the intensity of the EPR signal. However, for a system with $|D| \gg g\beta H$, this signal is highly anisotropic, with $g_{\text{eff}} \sim 2$ when the magnetic field is parallel to the normal of the equilateral triangle formed by the three copper ions and $g_{\text{eff}} \sim 4$ when the applied field is in the plane. These features at $g \sim 4$ are not observed, therefore $|D|$ must be rather small (<0.1 cm^{-1}). This is consistent with results from magnetic susceptibility data in which D is determined to be <0.05 cm^{-1}. For D of that magnitude, the $|-3/2\rangle \rightarrow |-1/2\rangle$ and $|+1/2\rangle \rightarrow |+3/2\rangle$ transitions of the

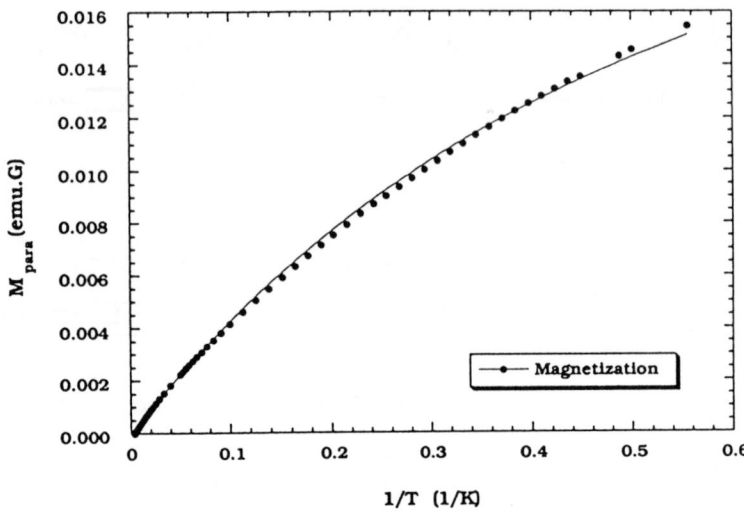

Figure 7 The Curie plot of the magnetization observed for membranes obtained from *Methylococcus capsulatus* (Bath) cells grown with 20 μM Cu^{2+}. Magnetization data were obtained at a fixed magnetic field of 1 Tesla as a function of temperature (1.8 K – 270 K).

quartet manifold may also contribute to the EPR spectra. Calculations of resonance fields show that these transitions are highly anisotropic. This would contribute to an apparent intensity anomaly when the intensity of the "isotropic" signal at g = 2.06 is used to quantify the copper concentration of the membrane fractions.

7 A MODEL FOR THE DIOXYGEN CHEMISTRY AND MECHANISM OF METHANE ACTIVATION

The correlation between the pMMO specific activity and the copper clusters in these membranes suggests that these clusters might be the active sites of the pMMO. An indication of this possible important catalytic role of the copper clusters is suggested by the reactivity of the partially reduced clusters toward dioxygen. Anaerobic titration of the membranes by dithionite leads to a substantial decrease in the intensity of the copper EPR signal. Upon re-exposure of the sample to air, we observed a full restoration of the copper EPR signal to its original intensity and appearance prior to electron reduction. Because the system reacts so readily with dioxygen, it seems likely that the copper clusters correspond to site(s) of oxygenase activity.

The exchange-coupled trinuclear copper clusters found in the pMMO system appear reminiscent of the interacting type 2 copper/type 3 binuclear copper cluster center in multiple blue copper oxidases in which the binuclear cluster

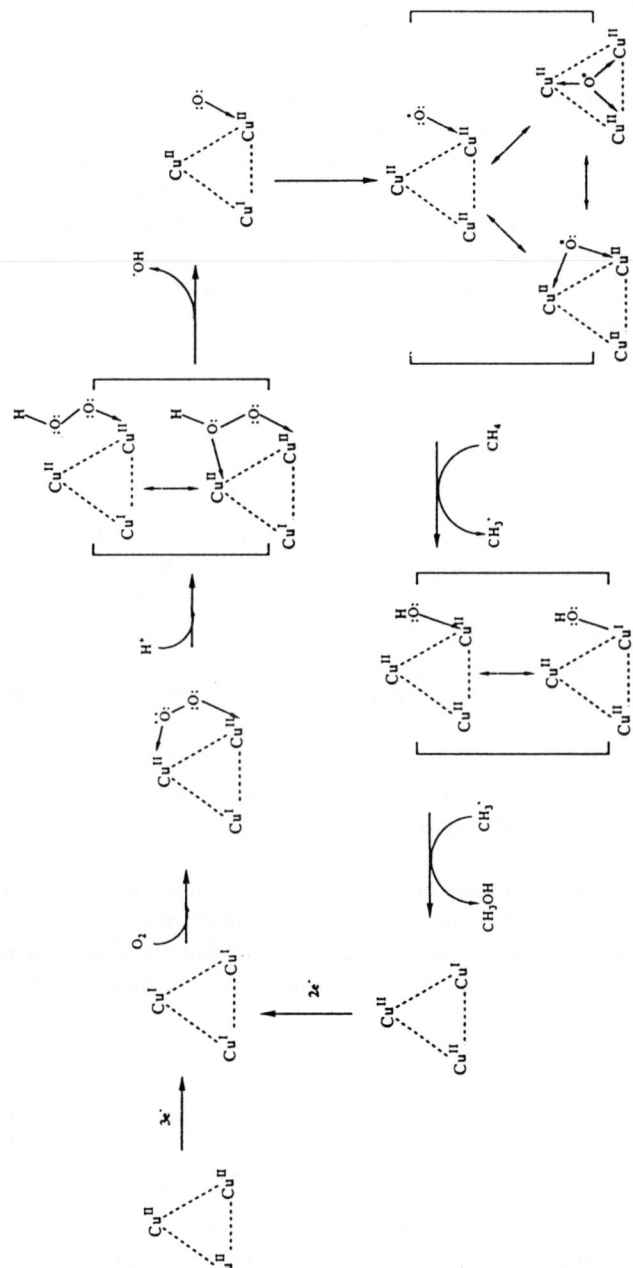

Figure 8 Mechanistic proposal for dioxygen and methane activation by a trinuclear copper cluster.

is the dioxygen binding and reduction site. The similarity between these two prosthetic groups appears to be a result of their similar function. Conceivably, the trinuclear clusters seen here have evolved specifically for the purpose of methane oxygenation. A mechanism for oxygen activation and methane oxidation involving these clusters can be proposed (Fig. 8).

According to this scheme, the trinuclear cluster will react most readily with dioxygen at the three-electron level of reduction. Subsequently, the O—O bond is cleaved heterolytically, resulting in the formation of an oxygen "radical" species. This highly reactive intermediate will abstract a hydrogen atom from methane, followed by the addition of the hydroxyl radical species into the methyl radical. The oxygenase reaction is obviously kinetically driven and the activation energy barrier is small. Several examples of this type of chemistry are known and once the oxygen intermediate is formed, the oxygenase reaction is expected to be facile.

8 CONCLUSION

Several intriguing aspects of these membranes emerge from this study. The finding that virtually all of the copper in the growth medium is incorporated into the membrane fraction suggests that these membranes, and/or the pMMO, act as a copper sponge. Our magnetic susceptibility studies indicate that the copper ions are arranged in trinuclear clusters and that there are many such clusters per protein molecule. Since the pMMO activity correlates with the copper content of the membranes in which the enzyme resides, and the copper ions appear to exist as trinuclear clusters with well-defined magnetic and redox properties, it seems reasonable to conclude that the copper clusters are associated primarily with the pMMO. If each trinuclear copper cluster can serve as a complete catalytic unit, capable of redox and dioxygen chemistry, then the pMMO system represents a new class of monooxygenase with virtually unknown properties. Efforts are now under way to obtain more direct evidence that the pMMO is indeed a copper protein, to establish the ligand structures of the copper clusters, and to elucidate the dioxygen and methane activation chemistry that might be at work here.

REFERENCES

[1] Anthony, C. (1982) The Biochemistry of Methylotrophs, Academic Press, London.
[2] Anthony, C. (1986) *Adv. Microb. Physiol.* **27**, 113–209.
[3] Stanley, S.H., Prior, S.D., Leak, D.J. and Dalton, H. (1983) Copper stress underlies the fundamental change in intracellular location of methane mono-oxygenase in methane-oxidising organisms: studies in batch and continuous cultures. *Biotechnol. Lett.* **5**, 487–492.

[4] Burrows, K.J., Cornish, A., Scott, D., and Higgins, I.J. (1984) Substrate specificities of the soluble and particulate methane mono-oxygenase of *Methylosinus trichosporium* OB3b. *J. Gen. Microbiol.* **130**, 3327–3333.
[5] Prior, S.D. and Dalton, H. (1985) The effect of copper ions on membrane content and methane monooxygenase activity in methanol-grown cells of *Methylococcus capsulatus* (Bath). *J. Gen. Microbiol.* **131**, 155–163.
[6] Bedard, C. and Knowles, R. (1989) Physiology, biochemistry, and specific inhibitors of CH_4, NH_4^+ and CO oxidation by methanotrophs and nitrifiers. *Microbiol. Rev.* **53**, 68–84.
[7] Fox, B.G. and Lipscomb, J.D. (1988) Purification of a high specific activity methane monooxygenase hydroxylase components from a Type II methanotroph. *Biochem. Biophys. Res. Commun.* **154**, 165–170.
[8] Fox, B.G., Suresus, K.K., Munck, E. and Lipscomb, J.D. (1988) Evidence for a μ-oxo-bridged binuclear iron cluster in the hydroxylase component of methane monooxygenase. *J. Biol. Chem.* **263**, 10553–10556.
[9] Woodland, M.P. and Dalton, H. (1984) Purification and characterization of component A of the methane monooxygenase from *Methylococcus capsulatus* (Bath). *J. Biol. Chem.* **259**, 53–60.
[10] Green, J. and Dalton, H. (1989) Substrate specificity of soluble methane monooxygenase. *J. Biol. Chem.* **264**, 17698–17703.
[11] Dewitt, J.G., Bentsen, J.G., Rosenzweig, A.C., Hedman, B., Green, J., Pilkington, S., Papaefthymiou, G.C., Dalton, H., Hodgson, K.O. and Lippard, S.J. (1991) X-ray absorption, Mössbauer, and EPR studies of the binuclear iron center in the hydroxylase component of methane mono-oxygenase. *J. Am. Chem. Soc.* **113**, 9219–9235.
[12] Carlin, R.L. (1986) Magneto Chemistry, Springer–Verlag, Berlin.
[13] Allendorf, M.D., Spira, D.J. and Solomon, E.I. (1985) Low temperature magnetic circular dichroism studies of native laccase: spectroscopic evidence for exogenous ligand bridging at a trinuclear copper active site. *Proc. Natl. Acad. Sci. U.S.A.* **82**, 3063–3067.
[14] Messerschmidt, A., Rossi, A., Ladenstein, R., Huber, R., Bolognesi, M., Gatti, G., Marchesini, A., Petruzzelli, R., Finazzi-Agro, A. (1989) X-ray crystal structures of the blue oxidase Ascorbate oxidase from zucchini. Analysis of the polypeptide fold and a model of the copper sites and ligands. (1989) *J. Mol. Biol.* **206**, 513–529.

8
Molecular Biology of Methane Oxidation

J. C. Murrell

Department of Biological Sciences, University of Warwick, Coventry CV4 7AL, UK

1 SUMMARY

The genes encoding the soluble methane monooxygenase enzyme systems from the methanotrophs *Methylococcus capsulatus* (Bath) and *Methylosinus trichosporium* OB3b were cloned into *Escherichia coli* and DNA sequenced. In both organisms the genes lie on the chromosome in the order *mmo*X, *mmo*Y, *mmo*B, *mmo*Z, orfY, *mmo*C. These genes encode the α and β subunits of Protein A, Protein B, the γ subunit of Protein A, a protein of unknown function and Protein C respectively, of the soluble methane monooxygenase enzyme complex. *mmo*B and *mmo*C have been expressed in *E. coli* and the proteins obtained were functionally active. Studies on the transcriptional organization of the soluble methane monooxygenase gene cluster from *Methylosinus* indicated that transcriptional regulation occurs from an RpoN-like promoter, 5' of *mmo*X and that there are promoters internal to the gene cluster which are functional during expression of the gene cluster. Promoter-probe analysis confirmed that a similar regulatory system occurs in *Methylococcus*. Mutants defective in methane oxidation were isolated by marker-exchange mutagenesis and were found to be more stable than those isolated by mutagenesis procedures using the suicide-substrate dichloromethane. Knowledge of the soluble methane monooxygenase gene sequences were exploited to detect methane monooxygenase gene-specific DNA in natural environmental samples using the polymerase chain reaction.

Microbial Growth on C_1 Compounds
© Intercept Ltd, PO Box 716, Andover, Hampshire SP10 1YG, UK

2 INTRODUCTION

Methanotrophs have attracted considerable interest over the last twenty years due to their potential for the production of bulk chemicals, single cell protein and for use in biotransformations. More recently, their ability to cooxidize chlorinated compounds such as trichloroethylene has also been examined in detail. These studies have resulted in a great deal of information on the physiology and biochemistry of methanotrophs [1,2], particularly concerning the methane monooxygenase (MMO) enzyme complexes from *Methylococcus capsulatus* (Bath) and *Methylosinus trichosporium* OB3b (see Chapters 5 and 6). However, the molecular biology and genetics of methanotrophs have only relatively recently been examined in detail. This is in part due to the refractory nature of these organisms to conventional genetic analysis (reviewed in [3–5]) but also to the recent availability of protein sequence information concerning the soluble (cytoplasmic) methane monooxygenase (sMMO).

Both *M. trichosporium* OB3b and *M. capsulatus* (Bath) synthesize a sMMO with similar biochemical characteristics during copper-deficient growth [7–13]. sMMO consists of three components A, B and C. Protein A, the hydroxylase component, comprises of three subunits α, β and γ of molecular masses 60, 45 and 20 kDa, $(\alpha, \beta$ and $\gamma)_2$ and contains a binuclear iron-oxo centre believed to be the site of the monooxygenation reaction. Protein B, 16 kDa, acts as an effector of electron transfer in the catalytic mechanism. Protein C is a 39 kDa NADH reductase containing 1 mol each of FAD and a (2Fe-2S) cluster and acts as a moderator between hydride transfer from NADH and electron transfer to the hydroxylase (reviewed in [14]).

A membrane-bound (particulate) methane monooxygenase (pMMO) is expressed during copper-excess growth of *M. capsulatus* and *M. trichosporium* (conditions in which the sMMO is not expressed) [8,14,15]. Although most methanotrophs appear to possess the pMMO, difficulties in purifying this enzyme complex have hindered its characterization [16,17] until recently (see Chapter 7).

3 CLONING AND SEQUENCING OF METHANE MONOOXYGENASE GENES

Biochemical characterization of the sMMO from *M. capsulatus* (Bath) by Dalton and co-workers has generated N-terminal amino acid sequence information on the β and γ subunit polypeptides of Protein A [14]. This has been used to design degenerate nucleotide probes which were subsequently used to clone the soluble MMO gene cluster from *M. capsulatus* (Bath) [18–20]. DNA sequencing of all the genes in the sMMO gene cluster of *Methylococcus* showed that the genes encoding the α, β and γ subunits of Protein A (*mmo*X, Y and Z), Protein B (*mmo*B) and Protein C (*mmo*C) are all linked on the chromosome.

The sMMO gene cluster of *Methylococcus* has been used as a hybridization probe for the detection of sMMO genes in a number of representative strains of methanotrophs in the University of Warwick culture collection. Only *Methylococcus* and *Methylosinus* species were found to possess sMMO gene homologues. No sMMO genes were detected in *Methylocystis parvus* OBBP or type I methanotrophs, including representatives of *Methylomonas albus*, *Methylomonas methanica*, *Methylomonas agile* and *Methylobacter capsulatus* [21]. The homology between sMMO genes of *Methylococcus* and *M. trichosporium* OB3b homologues was used to clone the entire sMMO gene cluster from OB3b [22,23]. This gene cluster has also been DNA sequenced and the genes are arranged in the same order as in *Methylococcus*. The location of these genes is shown in Fig. 1. Sequence comparisons have been made between the two sMMO gene clusters. The deduced polypeptide sequences of *M. trichosporium* OB3b and *M. capsulatus* sMMO components showed a high degree of similarity (Table 1) and codon usage in OB3b was similar to that found for *Methylococcus* [4].

Searches of current protein sequence banks showed that the α subunit of Protein A from both methanotrophs exhibited significant homologies with two conserved domains of the B2 subunit of *E. coli* ribonucleotide reductase. It has been predicted that the active site of sMMO resides on the α subunit(s) and that a binuclear iron-oxo centre, involved in catalysis, is associated with this polypeptide [14]. This type of iron centre occurs in ribonucleotide reductase

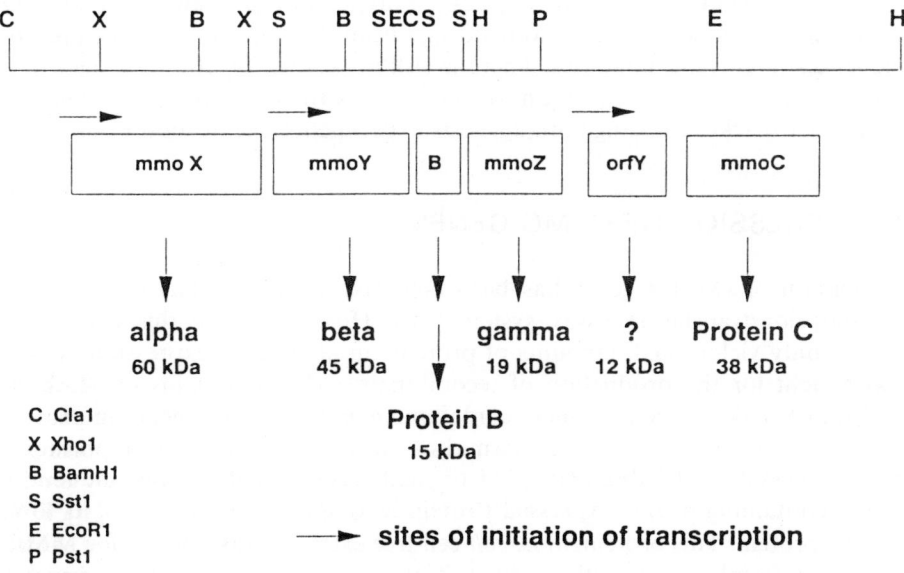

Figure 1 The sMMO gene cluster of *Methylosinus trichosporium* OB3b.

Table 1 Comparative analysis of the deduced polypeptide sequences of *M. trichosporium* OB3b and *M. capsulatus* sMMO components

M. capsulatus (Bath) sMMO			*M. trichosporium* OB3b sMMO			Percentage similarity
Component	No. of amino acids	Mwt (kDa)	Component	No. of amino acids	Mwt (kDa)	
Protein A:			Protein A:			
α	527	60.6	α	525	60	94
β	387	44.7	β	394	45	83.5
γ	170	19.8	γ	169	19.3	85
Protein B	141	16	Protein B	138	14.9	89.4
Protein C	348	38.5	Protein C	340	38	78

and X-ray crystal structure experiments with this protein have identified the ligands associated with the binuclear iron-oxo centre of this protein. Alignment of amino acids of the α subunit of Protein A and the B2 protein revealed two regions of the α subunit where perfect complementarity can be made with the sequences for the iron-binding site of the B2 protein [4,14,20]. Application of site-directed mutagenesis to these putative Fe binding sites in the α-subunit polypeptide may reveal further information on the structure of the active site of sMMO.

The N-terminal amino acids of Protein C exhibited strong homologies with ferredoxins of plant and bacterial origin [20] confirming biochemical evidence for a 2Fe-2S centre in Protein C [6,24]. Polypeptides β, γ and B of sMMO appear to be unique since there were no significant homologies with any protein sequences in current databases. There is still no known function for the *orf* Y gene product from either methanotroph. *orf* Y is therefore an ideal candidate for further marker-exchange mutagenesis experiments (see Section 6).

4 EXPRESSION OF sMMO GENES

Expression of sMMO genes has been achieved using a Zubay-type *in vitro* transcription/translation assay system [22]. However, since this expression system only yields small amounts of protein, an alternative expression system was sought for the production of recombinant sMMO proteins of *Methylococcus* in *E. coli*. A DNA fragment containing *mmo*B, the gene encoding Protein B, was subcloned into pT7-5, a plasmid of the T7 RNA polymerase promoter expression system of Tabor *et al.* [25], to yield plasmid pEB51. Upon induction, *E. coli* containing pEB51 expressed Protein B to approximately 2% of its total soluble protein. This protein, in *E. coli* cell-free extracts, was assayed for MMO activity in combination with purified Proteins A and C from *M. capsulatus* (Bath). It was found to be functionally active. When purified, the recombinant

Protein B, assayed in reconstitution experiments, had a specific activity of 8200 nmol min^{-1} (mg protein)$^{-1}$ which is comparable to Protein B purified from *M. capsulatus* (Bath) [26]. Similar expression experiments using the T7 expression plasmid pT7-7, which has an integral ribosome-binding site and an ATG start codon with the correct spacing and base composition for efficient translation, have also been done to produce Protein C from *Methylococcus* in *E. coli*. Recombinant Protein C in cell-free extracts of *E. coli* cross-reacted with antibody raised against *Methylococcus* Protein C and was functionally active with a specific activity of around 90 nmol min^{-1} (mg protein)$^{-1}$. Production of Protein C by this expression system was approximately 2% of the *E. coli* soluble cell-free extract [26].

A recombinant plasmid containing the *mmo*X, Y and Z genes of *Methylococcus* has been constructed in pT7-7. However, the expression of the α polypeptide of Protein A was poor (as determined by Western blotting experiments using antibody raised against Protein A) and sMMO activity could not be reconstituted by the addition of purified Proteins B and C to recombinant Protein A (C. West and J.C. Murrell, unpublished). The reasons for this are unclear at present but better expression of sMMO may be achieved in another heterologous host. To date, the entire sMMO gene cluster from *Methylococcus* has been cloned into the broad host-range plasmid pVK100 [27] and subsequently transfered into *Methylomonas albus* BG8 and *Methylocystis parvus* OBBP. This recombinant is stably maintained in these methanotrophs, both of which possess only a pMMO. Preliminary evidence suggests that at least Protein C is expressed in these organisms and efforts are now directed at manipulating the growth conditions of both organisms to allow functional expression of sMMO (Murrell *et al.*, unpublished).

5 TRANSCRIPTIONAL REGULATION OF sMMO

Although the first information on the primary sequence of sMMO genes arose from studies on *Methylococcus capsulatus* (Bath), regulatory studies of sMMO are more easily achieved using *M. trichosporium* OB3b since this organism has the ability to grow well on both methane and methanol, facilitating the study of induction/derepression of methane oxidation systems when the organism is switched from growth with methanol to growth on methane. OB3b also possesses both a pMMO and sMMO enzyme system, the expression and activity of which respond to the Cu^{2+}-to-biomass ratio achieved under different growth regimes [14]. OB3b is also the more amenable of the two organisms with regard to genetic manipulations [5].

Biochemical analysis has shown that sMMO of OB3b is only expressed under growth conditions when the Cu^{2+}-to-biomass ratio is low. In order to determine the basis for differential expression of sMMO in the absence or presence of Cu^{2+}, mRNA was extracted from *M. trichosporium* OB3b grown in 'high Cu^{2+}' or 'Cu^{2+}-free' medium. This mRNA was then used to generate cDNAs by reverse transcriptase. mRNA from OB3b grown in the presence of Cu^{2+} (i.e. containing pMMO) did not yield any cDNA products. However, the reverse was true when Cu^{2+} was absent from the growth medium (sMMO containing cells). Furthermore, Northern blotting experiments using these mRNAs and sMMO gene probes confirmed that sMMO was expressed only during growth of OB3b in the absence of Cu^{2+} (Murrell et al., unpublished observation). This supports previous findings [8,28] which showed that sMMO is only observed when OB3b is grown in Cu^{2+}-depleted medium.

5.1 Primer extension analysis

The major start points of mRNA synthesis in the sMMO gene cluster of OB3b, as determined by primer extension analysis, are shown in Fig. 2. Upstream of mmoX, the major start of mRNA synthesis is a C-residue at position 423. Preceding this are two hexamers (indicated by underlining), which exhibit a high degree of homology to -12, -24 conserved regions of promoters recognised by a minor form of E. coli RNA polymerase containing σ^{54} (NtrA, RpoN). Another set of promoter elements exhibiting a high degree of homology to the consensus -10, -35 sequences of E. coli exist 9 bp 'upstream' of the -24 hexamer (indicated by overlining). However, there is no evidence to suggest this putative promoter is functional during any growth conditions tested.

Another major transcript initiating at G residue 2161, which is preceded by a putative -10, -35 E. coli consensus sequence, was observed upstream of mmoY. Preliminary studies have also identified a major start point of transcription upstream of orfY (Murrell et al., unpublished). Again putative promoter elements including an RpoN-like -12, -24 promoter at nucleotide positions 4443-4458 (indicated by underlining) and a -10, -35 promoter at nucleotides 4340-4366 (double underlined) are shown in Fig. 2. Northern analysis indicates a single transcript encoding orfY and mmoC and the determination of the exact mRNA start point for these genes is currently being carried out in the author's laboratory.

5.2 Promoter probe analysis

Several regions of the Methylococcus sMMO gene cluster have been subcloned into the broad host-range promoter probe vector pAA182 [34]. Relative activities of β-galactosidase obtained with the five fragments of the sMMO gene cluster are shown in Fig. 3. Results obtained indicate that fragments 2 and 4 give significantly higher β- galactosidase activities than controls i.e. that there is a promoter between mmoX and mmoY and either upstream of orfY or mmoC. The second promoter is likely to reside upstream of orfY due to the larger intergenic region between mmoZ and orfY. Fragments 1 and 5 were expected to exhibit promoter activity since promoter elements have been observed upstream of mmoX [20]. The reason for lack of β-galactosidase activity is unclear and warrants further investigation.

```
        10        20        30        40        50        60        70        80        90       100       110       120
ATCGATCCGGCGCGCGTCACGCGTCTCGCATTGCGCAACGCCGTCGCCACCGCATCGTCATTGATGACCGTCGAATGCGCGGTCACTCATATTCCGCCGAGCGATCCGACATACGGCTTC
       130       140       150       160       170       180       190       200       210       220       230       240
GATCCGCATCTCGCCGCTGCGACGCGTGAAGATCCGCGCAGCTGAAGCAGCGAATGGCGAAAGAATGAAAGCAGCTGCGGCGACTGTTCGCCGCAGCGTTTTTTATGCGCGGATGCGCC
       250       260       270       280       290       300       310       320       330       340       350       360
CGTGATTTCATCGAATGCAGTCATCATCACGATACGTATGGTTCTTTGACATGTGTATCGCGCATCGCCATAGCACGATCATTTTTACGCGTCGCACGTAGTTGCGAATGCGACGAATCC
__     370       380______ 390       400       410       420 +1--> 430       440       450       460       470       480
CAACAAATACCGATGAAATCTATTATCCGCAGCGAGTGGCACAGGCCTTGCCAAATAAGAAGCGTCGACGCTTCCGCGCAGATGCGCAAGGACGAGAGACGTTTCGAAAACCCGAGGTTT
       490       500       510       520       530       540       550       560       570       580       590       600
GGAAAGCGGCGACGCGGTTCGGATGATCGACGCGCGCAATGAGCGCGAGCACGACGGTCCGAAACAAAAGAAAACTCTGTCAGGAGGAACAAGCATGGCGATCAGTCTCGCTACGAAAGC
                                                                                               M  A  I  S  L  A  T  K  A
                                                                                               Alpha ---->
       610       620       630       640       650       660       670       680       690       700       710       720
GGCGACCGATGCTCTGAAGGTCAACCGCGCTCCGGTCGGCGTGGAGCCTCAGGAGGTCCACAAATGGCTGCAGAGCTTCAACAGGGACTTCAAAGAGAACCGGACGAAGTATCCGACCAA
 A  T  D  A  L  K  V  N  R  A  P  V  G  V  E  P  Q  E  V  H  K  W  L  Q  S  F  N  R  D  F  K  E  N  R  T  K  Y  P  T  K


      2050      2060      2070      2080      2090      2100      2110      2120      2130      2140      2150      2160
CGACGGCAAGACGCTGATCGCTCAGCCGCACACGCGCGGCGACAATCTGTGGACGCTCGAGGACATCAAGCGGGCCGGCTGCGTGTTCCCCGATCCGCTCGCCAAATTCTGATCTCACGA
 D  G  K  T  L  I  A  Q  P  H  T  R  G  D  N  L  W  T  L  E  D  I  K  R  A  G  C  V  F  P  D  P  L  A  K  F  *
+1--> 2170      2180      2190      2200      2210      2220      2230      2240      2250      2260      2270      2280
GATTCCGGCGCGCGTCGCTCGCAACGCGCGTCGGATGAACCGCCGGGCCTCCCGCGTCCGGCGTCTCCTTCGACGCGAGCACGAAGCACCAAAAATCCTTCTGTGGAGGAAGAACTCACC
      2290      2300      2310      2320      2330      2340      2350      2360      2370      2380      2390      2400
ATGTCACAGCCTCAGAGCTCCCAAGTCACCAAGCGGGGTCTCACCGATCCCGAGCGTGCGGCCATCATCGCCGCCGCTGTGCCGGATCATGCGCTCGACACGCAGCGCAAATATCACTAC
 M  S  Q  P  Q  S  S  Q  V  T  K  R  G  L  T  D  P  E  R  A  A  I  I  A  A  A  V  P  D  H  A  L  D  T  Q  R  K  Y  H  Y
Beta ---->

      4330      4340      4350      4360      4370      4380      4390      4400      4410      4420      4430      4440
TGCGGCGCTCGGCACGGTGCTGATGGAGATCCGCAACACCGATTATTACGGCACTCCGCTCGAAGGGCTGCGCAAGGAGCCGGGCGTCAAGGTCCTGCATCTGCAGGCGTGACGGTACGC
 A  A  L  G  T  V  L  M  E  I  R  N  T  D  Y  Y  G  T  P  L  E  G  L  R  K  E  P  G  V  K  V  L  H  L  Q  A  *
      4450      4460      4470      4480      4490      4500      4510      4520      4530      4540      4550      4560
ATTGGCGAAAGACATGCGTTCGGGCGGCCGCGTCGCCCGAACGACAAAGAAAAGCGGGAGGACGAAAATGGACCAACAGACGGCGCACGAAGTCCGGCAAACCTTGATTCACGCCGACGA
                                                                    M  D  Q  Q  T  A  H  E  V  R  Q  T  L  I  H  A  D  E
```

Figure 2 Sequences of the sMMO gene cluster of *Methylosinus trichosporium* OB3b containing the major start points of mRNA synthesis and putative promoter elements. Transcriptional start sites as determined by primer extension mapping are indicated by +1. Putative promoter elements are underlined or double underlined. Putative Shine-Dalgarno sequences are double underlined and in bold lettering. Sequence data are taken from [22,23]. The N-terminal amino acid sequences of the α and β polypeptides of Protein A are indicated (derived from nucleotides 575-721 and 2281-2400 respectively). Nucleotides 4321-4560 encode the C-terminal of the γ polypeptide of Protein A and the N-terminal of *OrfY*.

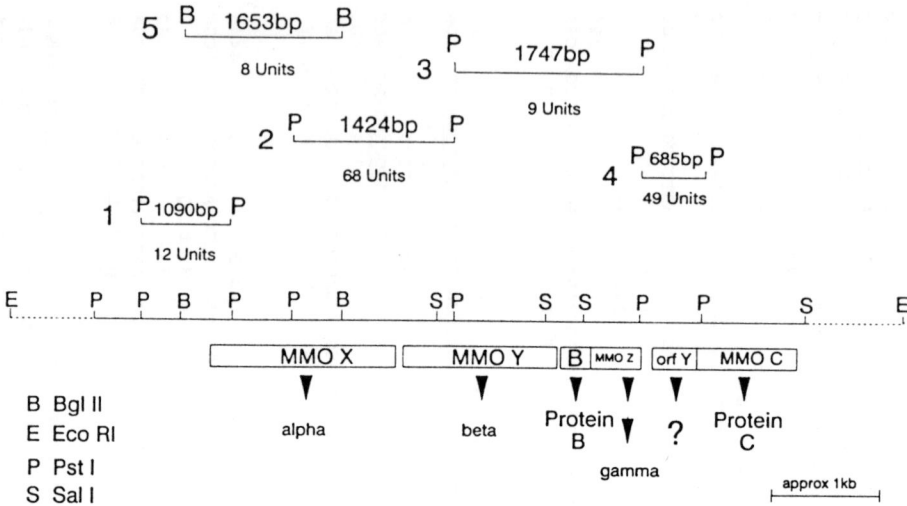

Figure 3 The sMMO gene cluster of *Methylococcus capsulatus* (Bath) indicating the five DNA fragments subcloned into the promoter-probe vector pAA182 [34]. The relative β-galactosidase activities of subsequent recombinants in *Escherichia coli* is indicated below the relevant DNA fragment.

Examination of sMMO gene sequences from *Methylococcus* and *Methylosinus* has revealed DNA sequences, which would result in the formation of potential stem-loop 'terminator' sequences upon transcription, between *mmo*X and *mmo*Y and immediately downstream of *mmo*Z, which together with the primer extension analysis detailed above suggests that the sMMO gene cluster is expressed as several transcriptional units rather than one long polycistronic mRNA. The precise regulation of transcription by Cu^{2+} remains to be elucidated.

6 MUTAGENESIS STUDIES

Several workers have isolated methanotrophs that are defective in growth on methane. This has been achieved using the 'suicide-substrate' dichloromethane (DCM) which is cooxidized by MMO to carbon monoxide, a potentially lethal product. Methanol-adapted *Methylosinus* [29] and *Methylomonas albus* [30] were incubated on methanol plates in the presence of DCM. This gave rise to DCM-resistant colonies at frequencies of 10^{-4} to 10^{-5}. On analysis, these DCM-resistant mutants were unable to grow on methane. Since *M. albus* only contains pMMO it was assumed that the mutations lay in the gene(s) for pMMO [30]. The DCM-resistant *M. trichosporium* OB3b mutants were not analysed in detail, although it is likely that the growth conditions employed would have resulted in the isolation of pMMO mutants only [29].

We have succeeded in isolating numerous DCM-resistant mutants of *M. trichosporium* OB3b by similar methods. However, many of these mutants are unstable and do not survive continued subculture. Twenty-one stable DCM-resistant mutants were analysed under carefully defined growth conditions. They were maintained on Cu^{2+}-free methanol plates and subsequently tested for growth with methane on agar medium in the presence and absence of Cu^{2+}. They all exhibited a pMMO$^-$, sMMO$^-$ phenotype. However, in liquid culture they exhibited a pMMO$^+$, sMMO$^-$ phenotype. Due to the unstable nature of these DCM-resistant mutants, an alternative strategy, marker-exchange mutagenesis, was employed which utilized the cloned sMMO genes. A DNA fragment containing *mmo*X of OB3b was cloned into pBR325. An *Xho*I fragment internal to the *mmo*X gene was replaced with a kanamycin resistance cassette and the *mob* fragment was cloned into pBR325 to facilitate transfer by conjugation of the resulting recombinant plasmid pHM32 from *E. coli* S17-1 to OB3b (Fig. 4). Homologous recombination of the mutated *mmo*X gene with the wild-type *mmo*X gene in the chromosome and subsequent loss of the unstable pBR325-based plasmid allowed the isolation of six kanamycin resistant OB3b colonies. These were DCM-resistant and could be maintained on Cu^{2+}-free NMS/methanol plates. Careful growth experiments in liquid culture

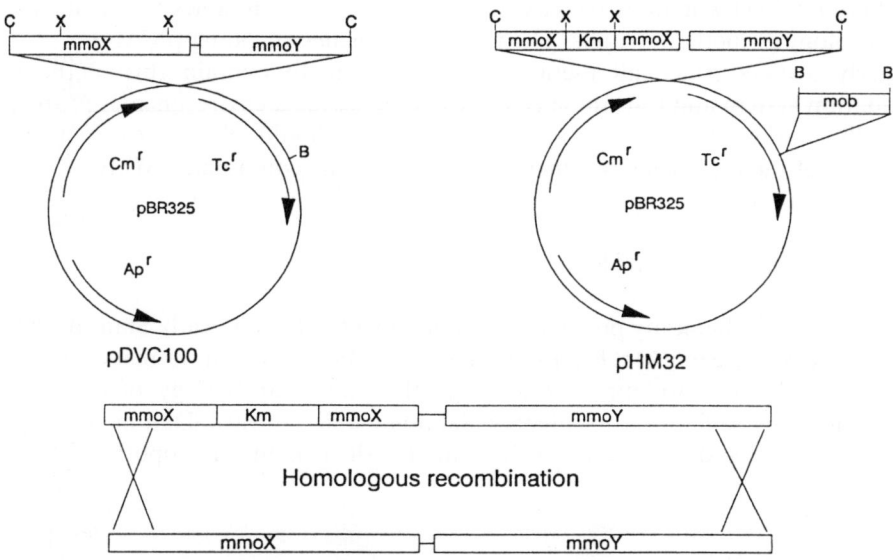

Figure 4 Marker exchange mutagenesis using the *mmo*X gene of *Methylosinus trichosporium* OB3b. A *Cla*I fragment containing *mmo*X and *mmo*Y of OB3b was subcloned into pBR325. A *Xho*I fragment internal to *mmo*X was replaced with a kanamycin resistance cassette and a mob fragment was cloned into the *Bam*H1 site of pBR325 to construct pHM32. Homologous recombination of the mutated *mmo*X gene on pHM32 with the 'wild-type' sMMO gene in OB3b yielded a sMMO$^-$ OB3b strain.

showed them to be sMMO$^-$ but pMMO$^+$ i.e they grew on methane only in the presence of excess Cu^{2+} (J.C. Murrell and H. Martin, unpublished). This method of marker-exchange mutagenesis will be invaluable for the future isolation of stable MMO mutants.

7 CONCLUDING REMARKS

The cloning and expression in *E. coli* of methane oxidation genes from methanotrophs should now allow site-directed mutagenesis studies to elucidate the mechanism of action and active site of MMO. Further analysis of the organization and expression of sMMO genes is clearly necessary, since a novel mechanism involving Cu^{2+} and possibly methane or its oxidation products is likely. The involvement of RpoN-like promoter elements also warrants further investigation. Knowledge of the primary sequences of methane monooxygenase genes has allowed the detection of methane-oxidizing bacteria by conventional colony hybridization/gene probing technology [21,31] and also by the use of the polymerase chain reaction where MMO-specific oligonucleotide primers have been used to amplify sMMO genes from total bacterial DNA isolated from the natural environment [32]. This, together with 16S rRNA probing technology [33] will allow increased use of molecular techniques to investigate the ecology of methanotrophs. The isolation of genes encoding pMMO is also eagerly awaited since all methanotrophs appear to contain this methane oxidation system and thus these genes would make ideal environmental probes. Finally, marker-exchange mutagenesis techniques should allow us to elucidate the role of the *orf*Y gene product in methanotrophs containing sMMO.

ACKNOWLEDGEMENTS

I would like to thank my postdocs and students Don Cardy, Andy Stainthorpe, Charlotte West, Roni McGowan and Howard Martin, who have made significant contributions to our knowledge of the molecular biology of methane oxidation, my collaborators George Salmond and Howard Dalton and the SERC, NERC and Gas Research Institute for their financial support.

REFERENCES

[1] Anthony, C. (1986) Bacterial oxidation of methane and methanol. *Adv. Microbiol. Physiol.* **27**, 113–210.
[2] Anthony, C. (1982) The Biochemistry of Methylotrophs. Academic Press, London.
[3] Lidstrom, M.E. and Stirling, D.I. (1990) Methylotrophs: genetics and commercial applications. *Ann. Rev. Microbiol.* **44**, 27–58.

[4] Murrell, J.C. (1992) The genetics and molecular biology of obligate methane-oxidizing bacteria. In: Methane and Methanol Utilizers (Murrell, J.C. and Dalton, H., Eds), pp. 115–148, Plenum Press, New York.
[5] Murrell, J.C. (1992) Genetics and molecular biology of methanotrophs. *FEMS Microbiol. Rev.* **88**, 233–248.
[6] Fox, B.G., Froland, W.A., Dege, J.E. and Lipscomb, J.D. (1989) Methane monooxygenase from *Methylosinus trichosporium* OB3b. Purification and properties of a three-component system with high specific activity from a type II methanotroph. *J. Biol. Chem.* **264**, 10023–10033.
[7] Fox, B.G., Lin, Y., Dege, J.E. and Lipscomb, J.D. (1991) Complex formation between the protein components of methane monooxygenase from *Methylosinus trichosporium* OB3b, identification of sites of component interaction. *J. Biol. Chem.* **266**, 540–550.
[8] Burrows, K.J., Cornish, A., Scott, D. and Higgins, I.J. (1984) Substrate specificities of the soluble and particulate methane monooxygenase of *Methylosinus trichosporium* OB3b. *J. Gen. Microbiol.* **130**, 3327–3333.
[9] Colby, J. and Dalton, H. (1979) Characterization of the second prosthetic group of the flavoenzyme NADH-acceptor reductase (Component C) of the methane monooxygenase from *Methylococcus capsulatus* (Bath). *Biochem. J.* **177**, 903–908.
[10] Green, J. and Dalton, H. (1985) Protein B of soluble methane monooxygenase from *Methylococcus capsulatus* (Bath). *J. Biol. Chem.* **260**, 15795–15801.
[11] Woodland, M.P. and Dalton, H. (1984) Purification and characterization of component A of the methane monooxygenase from *Methylococcus capsulatus* (Bath). *J. Biol. Chem.* **259**, 53–60.
[12] Fox, B.G., Surerus, K.K., Münck, E. and Lipscomb, J.D. (1988) Evidence for a μ-oxo-bridged binuclear iron cluster in the hydroxylase component of methane monooxygenase. *J. Biol. Chem.* **263**, 10553–10556.
[13] Ericson, A., Hedman, B., Hodgson, K.O., Green, J., Dalton, H., Bentsen, J.G., Beer, R.H. and Lippard, S.J. (1988) Structural characterization by EXAF's spectroscopy of the binuclear iron centre in component A of methane monooxygenase from *Methylococcus capsulatus* (Bath). *J. Am. Chem. Soc.* **110**, 2330–2332.
[14] Dalton, H. (1992) Methane oxidation by methanotrophs, physiological and mechanistic implications. In: Methane and Methanol Utilizers (Murrell, J.C. and Dalton, H., Eds), pp. 85–114, Plenum Press, New York.
[15] Stanley, S.H., Prior, S.D., Leak, D.J. and Dalton, H. (1983) Copper stress underlies the fundamental change in intracellular location of methane monooxygenase in methane oxidising organisms: studies in batch and continuous culture. *Biotechnol. Lett.* **5**, 487–492.
[16] Akent'eva, N.F. and Gvozdev, R.I. (1988) Purification and physiochemical properties of methane monooxygenase from membrane structures of *Methylococcus capsulatus*. *Biokhimiya*, **53**, 91–96.
[17] Smith, D.D.S. and Dalton, H. (1989) Solubilization of methane monooxygenase from *Methylococcus capsulatus* (Bath). *Eur. J. Biochem.* **182**, 667–671.
[18] Mullens, I.A. and Dalton, H. (1987) Cloning of the gamma-subunit methane monooxygenase from *Methylococcus capsulatus*. *Biotechnology* **5**, 490–493.
[19] Stainthorpe, A.C., Murrell, J.C., Salmond, G.P.C., Dalton, H. and Lees, V. (1989) Molecular analysis of methane monooxygenase from *Methylococcus capsulatus* (Bath). *Arch. Microbiol.* **152**, 154–159.
[20] Stainthorpe, A.C., Lees, V., Salmond, G.P.C., Dalton, H. and Murrell, J.C. (1990) The methane monooxygenase gene cluster of *Methylococcus capsulatus* (Bath). *Gene* **91**, 27–34.
[21] Stainthorpe, A.C., Salmond, G.P.C., Dalton, H. and Murrell, J.C. (1990) Screening

of obligate methanotrophs for soluble methane monooxygenase genes. *FEMS Microbiol. Lett.* **72**, 345–342.
[22] Cardy, D.L.N., Laidler, V., Salmond, G.P.C. and Murrell, J.C. (1991) Molecular analysis of the methane monooxygenase (MMO) gene cluster of *Methylosinus trichosporium* OB3b. *Mol. Microbiol.* **5**, 335–342.
[23] Cardy, D.L.N., Laidler, V., Salmond, G.P.C. and Murrell, J.C. (1991) The methane monooxygenase gene cluster of *Methylosinus trichosporium*: cloning and sequencing of the *mmo*C gene. *Arch. Microbiol.* **156**, 477–483.
[24] Lund, J. and Dalton, H. (1985) Further investigations of the FAD and Fe_2S_2 redox centres of component C NADH: acceptor reductase of the soluble methane monooxygenase from *Methylococcus capsulatus* (Bath). *Eur. J. Biochem.* **147**, 291–296.
[25] Tabor, S. and Richardson, C.C. (1985) A bacteriophage T7 RNA polymerase/promoter system for controlled exclusive expression of specific genes. *Proc. Natl. Acad. Sci. USA* **82**, 1074–1078.
[26] West, C.A, Salmond, G.P.C., Dalton, H. and Murrell, J.C. (1992) Functional expression in *Escherichia coli* of proteins B and C from soluble methane monooxygenase of *Methylococcus capsulatus* (Bath). *J. Gen. Microbiol.* **138**, 1301–1307.
[27] Knauf, V.C. and Nester, E.W. (1982) Wide host range cloning vectors: a cosmid clone bank of *Agrobacterium* Ti plasmids. *Plasmid* **8**, 45–54.
[28] Brusseau, G.A., Tsien, H.-G., Hanson, R.S. and Wackett, L.P. (1990) Optimization of trichloroethylene oxidation by methanotrophs and the use of a colorimetric assay to detect soluble methane monooxygenase activity. *Biodegradation* **1**, 19–29.
[29] Nicolaidis, A.A. and Sargent, A.W. (1987) Isolation of methane monooxygenase-deficient mutants from *Methylosinus trichosporium* OB3b using dichloromethane. *FEMS Microbiol. Lett.* **41**, 47–52.
[30] McPheat, W.L., Mann, N.H. and Dalton, H. (1987) Isolation of mutants of the obligate methanotroph *Methylomonas albus* defective in growth on methane. *Arch. Microbiol.* **148**, 40–43.
[31] Alvarez-Cohen, L., McCarty, P.L., Boulygina, E., Hanson, R.S., Brusseau, G.A. and Tsien, H.C. (1992) Characterization of a methane-utilizing bacterium from a bacterial consortium that rapidly degrades trichloroethylene and chloroform. *Appl. Environ. Microbiol.* **58**, 1886–1893.
[32] Murrell, J.C., McGowan, V. and Cardy, D.L.N. (1992) Detection of methylotrophic bacteria in natural samples by molecular probing techniques. *Chemosphere*, in press.
[33] Tsien, H.-C., Bratina, B.J., Tsuji, K. and Hanson, R.S. (1990) Use of oligonucleotide signature probes for identification of physiological groups of methylotrophic bacteria. *Appl. Environ. Microbiol.* **56**, 2858–2865.
[34] Lodge, J., Williams, R., Bell, A., Chan, B. and Busby, S. (1990). Comparison of promoter activities in *Escherichia coli* and *Pseudomonas aeruginosa* : use of a new broad-host-range promoter-probe plasmid. *FEMS Microbiol. Lett.* **67**, 221–226.

9

Degradation of Trichloroethylene by Methanotrophic Bacteria

Roelof Oldenhuis and Dick B. Janssen

Department of Biochemistry, University of Groningen, Nijenborgh 4, 9747 AG Groningen, The Netherlands

1 SUMMARY

The degradation of trichloroethylene and other chlorinated aliphatic hydrocarbons by *Methylosinus trichosporium* OB3b and *Methylomonas* GJ6 was determined. *M. trichosporium*, expressing soluble methane monooxygenase (MMO), showed the highest rates of chlorinated hydrocarbon transformation with values exceeding 100 nmol min^{-1} mg of cells^{-1} for compounds such as chloroform, the 1,2-dichloroethylenes and trichloroethylene (TCE). These rates are higher than those found with other cultures capable of cometabolic transformation. With trichloroethylene, however, inactivation of the organisms occurred, and the resulting toxic effect limited the amount of trichloroethylene that could be converted in a continuous culture to 0.03 nmol min^{-1} mg of cells^{-1}. At higher loadings, washout occurred. *Methylomonas* strain GJ6 was less sensitive to trichloroethylene but did not significantly degrade this compound. The low rate of degradation was not caused by suicide enzyme inactivation, as shown by the lack of [^{14}C]-trichloroethylene incorporation. Instead, trichloroethylene was poorly recognized as a substrate by the monooxygenase.

2 INTRODUCTION

One of the promising applications of methanotrophic bacteria is the cleanup of waste streams or polluted aquifers that are contaminated with low molecular

---

Microbial Growth on C_1 Compounds
© Intercept Ltd, PO Box 716, Andover, Hampshire SP10 1YG, UK

weight halogenated hydrocarbons, such as chloroform, trichloroethylene (TCE), dichloroethylenes, 1,1,1-trichloroethane, and chloropropanes. These compounds have never been found to serve as a growth substrate for bacterial cultures, leaving cometabolic conversion as the only option for biotransformation under aerobic conditions.

The capacity of methanotrophs to convert halogenated compounds was recognized by Colby et al. [1], who showed that methylbromide could be used as a model substrate for the methane monooxygenase (MMO) reaction. Further work was stimulated by the observation of Wilson and Wilson [2], who demonstrated that methane could stimulate biological TCE degradation in soil. Since then, various pure cultures of methanotrophic bacteria were found to be capable of cometabolic transformation of halogenated hydrocarbons [3–8]. Large differences between various cultures were observed, and high TCE degradation rates were found with *M. trichosporium* OB3b cultivated under conditions of copper limitation and incubated with formate instead of methane as the source of reducing equivalents [6]. Cells grown under copper limitation express exclusively the soluble type MMO [9,10], and the conclusion that it is this form of the enzyme that is responsible for TCE oxidation was confirmed by *in vitro* incubations with the purified enzyme [11].

The activities found with these cultures are attractive, but difficulties in applying cultures with high TCE degradation rates are expected because of the reactivity and toxicity of cometabolic transformation products [11–13]. From studies with cytochrome P_{450} catalyzed oxidation of TCE, it is known that a variety of unstable products can be formed, such as trichloroethylene epoxide, dichloroacetylchloride and trichloroacetaldehyde [14]. Other problems with practical application are the sensitivity of viable cultures to substrate starvation [12,15] and the fact that some methanotrophs show very poor conversion of TCE (see below). The low degradation rates found with some cultures could theoretically be caused by extreme sensitivity to enzyme inactivation or by a low intrinsic activity of the MMO of these organisms. In both cases, it will be necessary for successful application to develop ways to specifically stimulate methanotrophs that show high conversion rates. Reducing the availability of copper could be a way to do this, but will be difficult to manipulate if it is present in the material to be treated.

Here, we present an overview on the range of compounds that can be converted by cells of *M. trichosporium* OB3b and *Methylomonas* strain GJ6. The former organism can produce either a soluble or a membrane-bound MMO [9], while *Methylomonas* produces a membrane-bound enzyme only [16]. Strain GJ6 is a pink pigmented methanotroph isolated from freshwater sediment [17]. Similar cultures are frequently obtained when enrichments are carried out in the presence of copper in the medium. We also present data on the maximum rates of TCE removal that could be achieved with *M. trichosporium* OB3b in continuous culture, and compare the results with the activities of other cultures known to degrade chlorinated aliphatic hydrocarbons.

3 DEGRADATION OF CHLORINATED HYDROCARBONS BY M. Trichosporium OB3b AND Methylomonas MGJ6

Degradation of 13 chlorinated aliphatic hydrocarbons was tested with high density suspensions of cells of *Methylomonas* strain GJ6 [17] grown in continuous culture, and the results were compared to the data obtained with cells of copper-limited and copper-excess continuous cultures of *M. trichosporium* OB3b [6,12]. Incubations were carried out in serum flasks closed with Viton and butylrubber septa, and substrates were added at 0.2 mM. In all cases, formate (20 mM) was used as the electron donor, since we found this compound gave significantly higher degradation rates than methane or methanol [6]. Cells were added directly from the fermenter to a density of 0.3–1.0 mg of cells ml^{-1}. After 22 h at 30°C, incubations were analyzed by gas chromatography [4].

In all incubations, dichloromethane and 1,2-dichloroethane were degraded to below the detection limit, with release of stoichiometric amounts of inorganic chloride (Table 1). *Trans*-1,2-dichloroethylene was also completely converted. Conversion rates of these compounds were at least 0.2 nmol min^{-1} mg of cells^{-1}. Carbon tetrachloride and tetrachloroethylene were not degraded by cells expressing either of the monooxygenases. Thus, the presence of a carbon-hydrogen bond seems to be required for conversion by all MMOs tested here.

Compounds exclusively converted by the soluble MMO of *M. trichosporium* were TCE, 1,1,1-trichloroethane, and 1,1-dichloroethane. The rate of TCE conversion by *M. trichosporium* expressing the soluble MMO is also much higher than the rates found with other organisms, especially at high (>10 μM) substrate concentrations (Table 2). Cells with soluble MMO converted the following substrates much faster than either of the two particulate enzymes (Table 1): chloroform, 1,1-dichloroethylene, *cis*-1,2-dichloroethylene, and 1,2-dichloropropane. The observation that the soluble MMO of *M. trichosporium* is capable of much faster TCE degradation than its particulate enzyme or the MMO of *Methylomonas* strains, does, of course, not exclude the possibility that organisms producing a particulate MMO with high activity for TCE do not exist. One such culture was described by Henry and Grbić-Galić [8], who found metal speciation as a very important factor for the control of MMO activity.

Chlorinated intermediates were detected by GC analyses with 1,1,1-trichloroethane, *cis*-1,2-dichloroethylene, *trans*-1,2-dichloroethylene, and TCE. The identity of the intermediates was determined by GC-MS analysis or by retention time comparison with authentic standards (Table 3). Quantitative chloride production was found with the chloromethanes, as expected from the chemical instability of chloromethanols. Chloride production was also found with 1,1-dichloroethane and 1,2-dichloropropane, indicating oxygen incorporation at a carbon atom to which a chlorine atom is bound. The epoxides of 1,2-dichloroethylenes are known to decompose spontaneously, and with the

Table 1 Biodegradation of chlorinated aliphatics by *Methylosinus trichosporium* OB3b and *Methylomonas* GJ6[a]

Chlorinated hydrocarbon	Degradation (%)			Dechlorination (%)		
	t-sMMO	t-pMMO	M-pMMO	t-sMMO	t-pMMO	M-pMMO
Dichloromethane	>98	86	>98	>95	68	>95
Chloroform	>99	96	91	>95	73	92
Carbon tetrachloride	10	<10	<10	<5	<5	<5
1,1-Dichloroethane	<10	12	97	<5	<5	10
1,2-Dichloroethane	>98	>98	>98	85	65	>95
1,1,1-Trichloroethane	62	<10	<10	<10	<10	<10
1,1-Dichloroethylene	40	18	<10	35	<5	<5
trans-1,2-Dichloroethylene	>98	>98	>98	67	0	55
cis-1,2-Dichloroethylene	>98	56	24	33	37	12
Trichloroethylene	>99	<10	<10	>95	<5	<5
Tetrachloroethylene	<5	<5	<5	<2	<2	<2
1,2-Dichloropropane	>99	51	13	27	5	10
trans-1,3-Dichloropropylene	85	48	45	52	5	37

[a] Degradation experiments were carried out at 30°C with suspensions of 0.3–0.5 mg ml^{-1} *M. trichosporium* containing soluble methane monooxygenase (t-sMMO) or particulate MMO (t-pMMO). Incubations with *Methylomonas* (M-pMMO) contained 0.9 mg cells ml^{-1}. All incubations contained 20 mM formate, and 0.2 mM halogenated substrate was added. Concentrations of remaining compounds and inorganic chloride were measure after 22–24 h.

Table 2 Degradation of trichloroethylene by pure bacterial cultures that produce monooxygenases

Culture	Substrate[a]	Rate (nmol mg^{-1} min^{-1})	Concentration (μM)	Reference
Alcaligenes eutrophus JMP134	phenol	0.2	25	Harker and Kim [26]
Escherichia coli with toluene oxidation genes	LB + IPTG	1–2	—	Winter and Ensley [27]
Methylosinus trichosporium OB3b	methane	1	100	Zylstra et al. [28]
		145	145 (K_s)	Oldenhuis et al. [4]; Brusseau et al. [7]
Methylomonas GJ6	methane	0.0005	100	This paper
Methylomonas	methane			Henry and Grbic-Galic [8]
Mycobacterium convolutum, other *Mycobacterium* strains	propane	0.5	10	Wackett et al. [29]
Nitrosomonas europaea	ammonia	0.4	12	Arciero et al. [30]
Pseudomonas cepacia G4	phenol	4	3 (K_s)	Folsom et al. [31]
Pseudomonas putida F1	toluene	2	80	Wackett and Gibson [32]
Rhodiccus erythropolis JE77	isoprene	1	1000	Ewers et al. [33]
Strain 46–1	methane	0.2	1–2	Little et al. [3]

[a] Substrate for growth of the culture and induction of the catabolic enzyme.

Table 3 Products identified after incubation of chlorinated hydrocarbons with *Methylosinus trichosporium* OB3b or *Methylomonas* GJ6

Substrate	Product
Trichloroethylene	chloride, trichloroethanol, trichloroacetaldehyde dichloroacetic acid
cis-1,2-Dichloroethylene	chloride, *cis*-1,2-dichloroethylene epoxide
trans-1,2-Dichloroethylene	chloride, *trans*-1,2-dichloroethylene epoxide
1,2-Dichloropropane	chloride, 1,2-dichloro-3-propanol
1,1,1-Trichloropropane	trichloroethanol

epoxides produced in medium with *M. trichosporium* we found half lives of 18 h and 22 h for the *cis* and *trans* epoxide, respectively.

After incubations of cells of *Methylomonas* strain GJ6 with 1,1,1-trichloroethane or TCE, no significant decrease in concentrations of these substrates occurred. One of the expected transformation products is trichloroethanol, which can be determined very sensitively with an electron capture detector [12]. Very low concentrations of trichloroethanol were indeed detected (4.4 μM and 0.02 μM from 1,1,1-trichloroethane and TCE, respectively), showing that some conversion occurred. Assuming that also with strain GJ6 only 3% of the TCE degraded is converted to trichloroethanol, this would mean a degradation rate of about 0.0005 nmol min^{-1} mg of protein^{-1}. Chloride liberation from 1,1,1-trichloroethane or its oxidation products trichoroethanol, trichloroactaldehyde or trichlororacetic acid by methanotrophs has never been observed.

4 TOXICITY OF TRICHLOROETHYLENE

Table 1 shows that also at high concentrations of substrate and cells (0.2 mM TCE and 0.9 mg cells ml^{-1}), TCE was not significantly degraded by *Methylomonas* strain GJ6. It has been found that TCE causes suicide inactivation of MMO activity [11,12]. TCE conversion products were covalently bound to cellular components. Thus, the lack of significant TCE degradation by *Methylomonas* strain GJ6 could be due either to a low activity of the particulate MMO of this strain with TCE or to extremely high sensitivity to inactivation by transformation products. The latter should allow a higher percentage of the TCE added to be converted if the substrate concentration is lowered. Therefore degradation experiments (0.2 mg cells ml^{-1}) were carried out with TCE at 0.2, 0.4, 1.1 and 4.8 μM, but no significant degradation was found and no trichloroethanol could be detected with an electron capture detector (detection limit 0.4 nM). Furthermore, it was found that cells preincubated for 10 min with TCE at concentrations varying from 2 to 47 μM still degraded *trans*-1,2-dichloroethylene, add at 60 μM. The rate of propylene oxide production was also measured to determine the activity of MMO, and found to remain constant during short term preincubation with TCE. Thus, TCE at low concentrations did not inactivate MMO of the *Methylomonas* cells under these conditions and TCE was not a strong competitive inhibitor of the enzyme.

In agreement with these observations, incorporation of TCE degradation products in cellular components of *Methylomonas* GJ6 or *M. trichosporium* OB3b producing only particulate MMO, did not occur. Experiments with [^{14}C]-TCE were carried out as described elsewhere [12], using cells of a culture of *Methylomonas* GJ6 grown in batch in the presence of copper. As shown in Fig. 1, the only case with significant incorporation of ^{14}C is with cells of *M.*

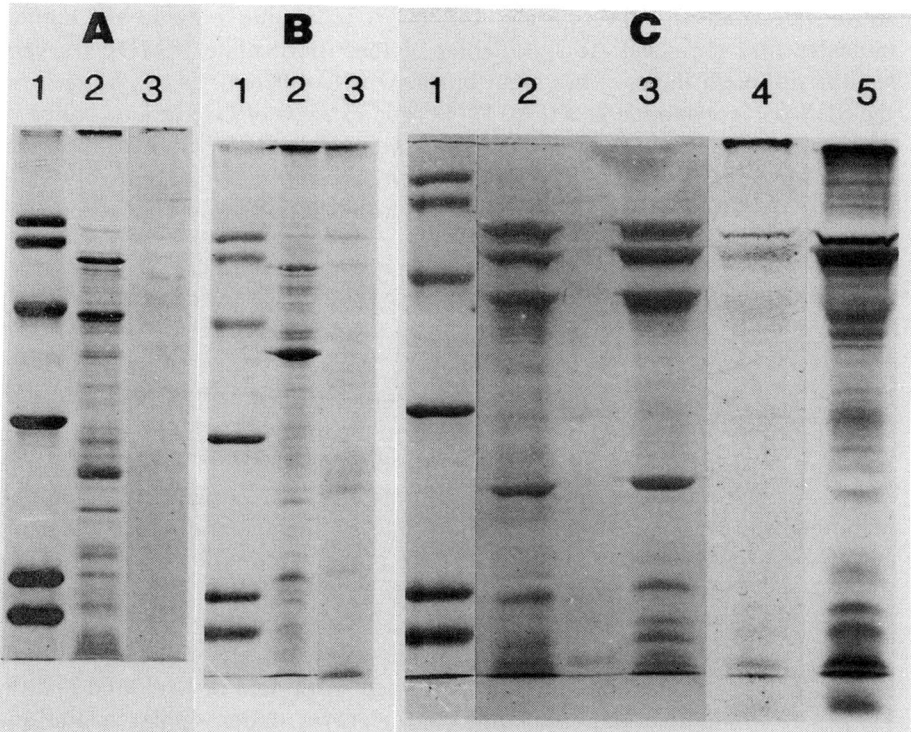

Figure 1. Binding of [^{14}C]-TCE transformation products to cellular proteins. Whole cells were incubated with [^{14}C]-TCE and crude extracts were separated on SDS-polyacrylamide gels. The coomassie stained gels and the corresponding fluorographs are shown. Molecular weight markers (slots with number 1) were ovotransferrin (78 kDa), bovine serum albumin (67 kDa), ovalbumin (45 kDa), carbonic anhydrase (30 kDa), myoglobin (17 kDa), and cytochrome *c* (12 kDa). (A) *Methylomonas* GJ6 grown in batch with 0.07 μM copper exposed to TCE. Slot 2, protein stain; slot 3, fluorograph. (B) *Methylosinus trichosporium* OB3b [4] grown with copper in the medium. Slot 2, protein stain; slot 3, fluorograph. (C) Strain OB3b grown without copper. Slot 2, protein stain of extract of cells not exposed to TCE; slot 3, protein stain cells exposed to TCE; slots 4 and 5, fluorograph of slot 3 [2].

trichosporium cultivated in the absence of copper. Up to 10% of the label added (2.6 μCi; 460 μM TCE; 2 mg cells ml^{-1}) was incorporated in the α-subunit of soluble MMO [12]. A large amount of radioactivity did not enter the gel, and thus was present in non-protein material. After incubation of cells of OB3b or GJ6 expressing particulate MMO with [^{14}C]-TCE (0.64 μCi; 160 μM TCE; 2.5 mg cells ml^{-1}), less than 0.5% of the label added was recovered in the crude extracts. The fluorographs showed that no ^{14}C was bound to specific proteins (Fig. 1). [^{14}C]-TCE or its derivatives were not significantly incorporated, and the lack of TCE degradation was not due to covalent binding of TCE conversion products to particulate MMO.

The similarity of the results of degradation experiments with *Methylomonas* strain GJ6 and *M. trichosporium* OB3b grown in the presence of copper indicates that the substrate specificities of their particulate MMOs are very similar, although the enzymes seem biochemically different, as can be seen on the SDS-polyacrylamide gels (Fig. 1) of crude extracts of GJ6 (major bands of 43 kDa, 27 kDa, and 24 kDa) and *M. trichosporium* OB3b (39 kDa, 26 kDa, and 19 kDa). The particulate MMO of *Methyloccus capsulatus* (Bath) (49 kDa, 23 kDa, and 22 kDa) [18] is also different from that of strain GJ6.

5 DEGRADATION OF TRICHLOROETHYLENE IN CONTINUOUS CULTURES OF *M. trichosporium*

Continuous degradation of TCE by growing cells of *M. trichosporium* OB3b was studied by adding air containing TCE to a continuous culture growing on methane in a copper-free medium. For all these degradation experiments, the bioreactor settings were kept constant, only the TCE concentration in the incoming gas was varied. The dilution rate of the reactor was 0.028 h^{-1}. Initially, *M. trichosporium* OB3b cells were grown continuously without TCE for 3 weeks. This resulted in a steady-state with a cell density of 2.5 mg ml^{-1}. The growth yield on methane (Y_{CH_4}) was 0.51 g cells per g CH_4 and the ratio between oxygen and methane consumption (O_2/CH_4) was 1.5 (mol/mol) (Table 4). These values were in good agreement with data of Leak and Dalton [19], who found that *M. capsulatus* (Bath) grown under similar conditions (nitrate as nitrogen source, methane limited, copper stress), had a Y_{CH_4} of 0.50 and an O_2/CH_4 of 1.50. They also showed that *M. capsulatus* (Bath) grown with copper, thus producing pMMO, had a Y_{CH_4} of 0.79, [19]. which was also found by us with cultures of *M. trichosporium* OB3b [4].

The degradation of TCE in the continuous culture was subsequently estimated at 4 different concentrations, varying from 0.02 μmol to 3.1 μmol TCE per litre of air in the incoming gas (Table 4). Because of the low dilution rate, the conditions were held constant for at least 3 weeks before determining a number of steady states at time intervals of 3–4 days. The differences in TCE concentrations and cell densities measured at a certain steady state were rather low (<5%) and the mean values are given in Table 1.

Table 4 Degradation of trichloroethylene in a continuous culture of *Methylosinus trichosporium* OB3b with increasing influent concentrations of trichloroethylene

Steady state number[a]	1	3	2	4	5[b]
Cell density (mg l^{-1})	2.5	2.3	2.0	1.6	$1.4 \to 0.13$
Total gas flow (ml min^{-1})	44	44.5	45	46	46
Concentrations[c]					
TCE (g, in) (μmol l^{-1})	$<10^{-5}$	0.02	0.105	0.55	3.1
TCE (g, out) (μmol l^{-1})	$<10^{-5}$	10^{-4}	1.4×10^{-3}	1.9×10^{-2}	$2.4 \times 10^{-1} \to 2.15$
TCE (l, out) (μM)	10^{-5}	4×10^{-4}	2×10^{-3}	1.4×10^{-2}	$(1.8 \times 10^{-1} \to 3.3)$
Trichloroethanol (l, out) (μM)	$<10^{-5}$	6.3×10^{-2}	3.6×10^{-1}	1.33	$(14.5 \to 0.53)$
Chloride (l, out) (mM)	0.02	0.03	0.07	0.26	$(1.4 \to 0.53)$
CH$_4$ (g, in) (%)	12.7	12.5	12.4	12.2	12.1
CH$_4$ (g, out) (%)	8.6	8.7	9.1	9.5	$9.6 \to 11.9$
O$_2$ (g, in) (%)	18.0	18.0	18.1	18.3	18.2
O$_2$ (g, out) (%)	11.9	12.3	13.1	14.3	$14.5 \to 17.9$
Performance					
Elimination (%)	—	99.5	98.7	96.7	$99.2 \to 29.9$
Volumetric rate (nmol l^{-1} reactor min^{-1})	—	0.9	4.7	25	$130 \to 43$
Degradation rate (nmol mg cells^{-1} min^{-1})	—	8×10^{-4}	6×10^{-3}	2.4×10^{-2}	$1.6 \times 10^{-1} \to 5.1 \times 10^{-1}$
Michaelis-Menten rate[d]	—	8×10^{-4}	4.2×10^{-2}	2.9×10^{-2}	$3.6 \times 10^{-1} \to 6.45$
Growth yield on methane (g/g)	0.51	0.50	0.49	0.47	$(0.45 \to 0.52)$

[a] Bioreactor conditions were as described elsewhere [4].
[b] No steady-state was reached, because washout occurred. The values given were measured on days 1 and 9. Values based on the liquid phase are inaccurate, because of the low dilution rate of the reactor, and are given in parenthesis.
[c] Compounds were measured in gas and liquid phase as indicated with g or l, respectively.
[d] Predicted degradation rate of TCE by Michaelis Menten equation with K_m of 145 μM and V_{max} of 290 nmol min^{-1} mg of cells^{-1} [12].

The lowest TCE concentration tested was 0.02 μmol l^{-1} of air in the inlet gas stream. At this TCE feed, complete degradation occurred (99.5%). At higher TCE concentrations (0.105 and 0.55 μmol l^{-1}), the percentages of TCE degraded were 98.7% and 96.7%, respectively. If TCE would be completely dechlorinated, the chloride concentration in the medium should have increased to 0.008, 0.045, and 0.234 mM for the three steady states, respectively, which is close to the observed values of 0.01, 0.05, and 0.24 mM ($\pm$0.005 mM) (Table 4). In all cases it was found that 2–3% of the TCE degraded was converted to trichloroethanol, which was only found in the liquid phase. Previous work showed that a similar amount of trichloroethanol was formed in batch incubations [4]. Trichloroacetaldehyde was detected neither in the liquid phase nor in the gas phase.

The highest TCE degradation rate observed was 2.4×10^{-2} nmol min^{-1} mg of cells^{-1}, at a TCE concentration in the liquid phase of 1.33 μM (steady state 4). This value is in close agreement with the rates predicted by Michaelis Menten kinetics, using the previously determined values for V_{max} and K_m ([12], Table 4). The degradation rates found with the first two steady states were also in good agreement with the calculated values (Table 4). The presence of methane as a growth substrate thus did not result in significant competitive inhibition.

At the higher TCE loadings, the obtained steady-states showed lower cell densities. When the TCE influent concentration was increased to 3.1 μmol TCE per l of air, no steady-state was reached, and the culture washed out within 9 days. The cell density decreased to 10% of its initial value, resulting in a drop in the degree of TCE elimination from 92% to 30% (Table 4). At this TCE feed, cell inactivation was apparently too high compared to production of new cells. At the timepoint when the TCE feed was increased, the ratio between TCE feed and cell production was 0.29 μmol TCE mg of cells^{-1}. This value was apparently too high and became even higher when the cell mass decreased, causing increased toxicity and washout of the culture.

The toxicity of TCE, as found by various investigators [4,8,11–13] thus implies that only a small amount of TCE added can be degraded per amount of active cells formed. It is not known whether cells can recover from TCE inactivation. We have previously found that about 2 μmol of TCE converted resulted in activation of 1 mg of cells [12]. Alvarez-Cohen and McCarty [13,20] using mixed cultures of methanotrophs, observed a TCE transformation capacity of 0.27 μmol of TCE mg of cells^{-1}, which is in good agreement with our data. The Y_{CH4} also decreased when TCE was increased (Table 4). The ratio between methane consumption compared to TCE consumption varied from 10^5 mol CH$_4$ per mol of TCE at the lowest TCE loading to 2.6×10^3 mol CH$_4$ per mol of TCE at the highest TCE concentrations giving good degradation.

The highest stable TCE degradation rate observed in our continuous culture experiments was 2.4×10^{-2} nmol min^{-1} mg of cells^{-1}. At this loading rate, the amount of CH$_4$ consumed compared to TCE removed, was reduced to 2.7×10^3 mol of CH$_4$ per mol TCE. The highest volumetric elimination rate was thus 0.2 g TCE h^{-1} m^{-3} reactor volume. Strandberg and coworkers [21] used a fixed-film packed-bed bioreactor with a consortium of methanotrophs to treat a synthetic groundwater containing TCE and *trans*-1,2-dichloroethylene. They found a TCE degradation rate of 7 mg h^{-1}, with a degradation percentage of only 40%. The volumetric elimination capacity of this type of reactor is then 0.6 g TCE h^{-1} m^{-3} reactor, which is in the same order of magnitude as found in our work. The amount of CH_4 required for degrading TCE was five-fold lower than in our system. It may well be that a mixed culture or biofilm is more stable. Ewers *et al.* [22] found that TCE and several other chlorinated ethylenes were cooxidized by 2-methyl-1,3-butadiene (isoprene) grown bacteria. A packed bed bioreactor was inoculated with the isoprene-utilizer *Rhodococcus erythropolis* strain JE77, which was proposed to be capable of active TCE epoxide metabolism. The highest elimination rate given was 0.06 g TCE h^{-1} m^{-3} reactor volume, which is, however, significantly lower than the level at which toxicity is observed with methanotrophs.

All these degradation rates are low compared to what can be achieved with microorganisms that utilize chlorinated solvents as a carbon source. Hartmans *et al.* [23], for example, found with *Mycobacterium* L1 in a conventional fermentor an elimination percentage of 93% and a volumetric elimination rate of 28 g vinyl chloride h^{-1} m^{-3} reactor. From the results of Reineke and Knackmuss [24] a volumetric elimination capacity of 0.2 kg chlorobenzene h^{-1} m^{-3} reactor by strain WR1306 could be calculated. A very high value was found by Gälli [25], with *Pseudomonas* strain DM5 degrading dichloromethane at a rate of 1.6 kg dichloromethane h^{-1} m^{-3} reactor volume in a fluidized bed. It is evident that the isolation or genetic construction of a bacterium capable of growth on trichloroethylene would allow the development of more efficient degradation processes than those based on cometabolism.

6 CONCLUSIONS

The available data on the degradation of chlorinated hydrocarbons by methanotrophs indicate large differences between transformation capacity of various cultures. Expression of the soluble methane monooxygenase in *M. trichosporium* allows rapid degradation of components with at least one carbon-hydrogen bond. The substrate range is wider and the degradation rates are higher than those found with cultures that produce other monooxygenases. Because trichloroethylene can be transformed at a high rate, some methanotrophs are sensitive to toxic effects of TCE, which limits the transformation capacity of a culture in a continuous system. Application possibilities are best if dilute contaminations have to be treated, such as with groundwater cleanup. The significance of methanotrophs in the conversion of chlorinated hydrocarbons in natural environments remains to be established. Furthermore, the development of methods to monitor and specifically stimulate methanotrophs expressing monooxygenases that tackle TCE are desirable.

REFERENCES

[1] Colby, J., Dalton, H. and Whittenbury, R. (1975) An improved assay for bacterial mono-oxygenase: some properties of the enzyme from *Methylomonas methanica*. *Biochem. J.* **151**, 459–462.
[2] Wilson, J.T. and Wilson B.H. (1985) Biotransformation of trichloroethylene in soil. *Appl. Environ. Microbiol.* **49** 242–243.
[3] Little, C.D., Palumbo, A.V., Herbes, S.E., Lidstrom, M.E., Tyndall, R.L. and Gilmer, P.J. (1988) Trichloroethylene biodegradation by a methane-oxidizing bacterium. *Appl. Environ. Microbiol.* **54**, 951–956.
[4] Oldenhuis, R., Vink, R.L.J.M., Janssen, D.B. and Witholt, B. (1989) Degradation of chlorinated aliphatic hydrocarbons by *Methylosinus trichosporium* OB3b expressing soluble methane monooxygenase. *Appl. Environ. Microbiol.* **55**, 2819–2826.
[5] Tsien, H.-C., Brusseau, G.A., Hanson, R.S. and Wackett, L.P. (1989) Biodegradation of trichloroethylene by *Methylosinus trichosporium* OB3b. *Appl. Environ. Microbiol.* **55**, 3155–3161.
[6] Uchiyama, H., Nakajima, T., Yagi, O. and Tabuchi, T. (1989) Aerobic degradation of trichloroethylene by a new type II methane-utilizing bacterium, strain M. *Agric. Biol. Chem.* **53**, 2903–2907.
[7] Brusseau, G.A., Tsien, H.-C., Hanson, R.S. and Wackett, L.P. (1990) Optimization of trichloroethylene oxidation by methanotrophs and the use of a colorimetric assay to detect soluble methane monooxygenase activity. *Biodegration*, **1**, 19–29.
[8] Henry, S.M., and Grbić-Galić, D. (1991) Influence of endogenous and exogenous electron donors and trichloroethylene oxidation toxicity on trichloroethylene oxidation by methanotrophic cultures from a groundwater aquifer. *Appl. Environ. Microbiol.* **57**, 236–244.
[9] Burrows, K.J., Cornish, A., Scott, D. and Higgins, I.J. (1984) Substrate specificities of the soluble and particulate methane monooxygenases of *Methylosinus trichosporium* OB3b. *J. Gen. Microbiol.* **130**, 3327–3333.
[10] Stanley, S.H., Prior, S.D., Leak, D.J. and Dalton, H. (1983) Copper stress underlies the fundamental change in intracellular location of methane monooxygenase in methane-oxidizing organisms: studies in batch and continuous cultures. *Biotechnol. Lett.* **5**, 487–492.
[11] Fox, B.G., Borneman, J.G., Wackett, L.P. and Lipscomb, J.D. (1990) Haloalkene oxidation by the soluble methane monooxygenase from *Methylosinus trichosporium* OB3b: mechanistic and environmental implications. *Biochemistry* **29**, 6419–6427.
[12] Oldenhuis, R., Oedzes, J.Y., van der Waarde, J.J. and Janssen, D.B. (1991) Kinetics of chlorinated hydrocarbon degradation by *Methylosinus trichosporium* OB3b and toxicity of trichloroethylene. *Appl. Environ. Microbiol.* **57**, 7–14.
[13] Alvarez-Cohen, L. and McCarty, P.L. (1991) Effects of toxicity, aeration, and reductant supply on trichloroethylene transformation by a mixed methanotrophic culture. *Appl. Environ. Microbiol.* **57**, 228–235.
[14] Anders, M.W. and Pohl, L.R. (1985) Halogenated alkanes. In: Bioactivation of Foreign Compounds (Anders, M.W., Ed.), pp. 283–315, Academic Press, New York.
[15] Janssen, D.B., Oldenhuis, R., van den Wijngaard, A.J. and van der Waarde, J.J. (1991) Biochemistry and kinetics of aerobic degradation of chlorinated aliphatic hydrocarbons. In: On Site Bioreclamation Processes for Xenobiotic and Hydrocarbon Treatment (Hinchee, R.E. and Olfenbuttel, R.F. Eds), Butterworth-Heineman Publishing, Boston.

[16] Dalton, H., Prior, S.D., Leak, D.J. and Stanley, S.H. (1984) Regulation and control of methane monooxygenase. In: Microbial Growth on C_1 Compounds (Crawford, R.L. and Hanson, R.S., Eds), pp.75–82. American Society for Microbiology, Washington, DC.
[17] Janssen, D.B., Grobben, G., Hoekstra, R., Oldenhuis, R. and Witholt, B. (1988) Degradation of *trans*-1,2-dichloroethene by mixed and pure cultures of methanotrophic bacteria. *Appl. Microbiol. Biotechnol.* 29, 392–399.
[18] Smith, D.D.S. and Dalton, H. (1989) Solubilisation of methane monooxygenase from *Methylococcus capsulatus* (Bath). *Eur. J. Biochem.* 182, 667–671.
[19] Leak, D.J. and Dalton, H. (1986) Growth yields of methanotrophs. 1. Effect of copper on the energetics of methane oxidation. *Appl. Microbiol. Biotechnol.* 23, 470–476.
[20] Alvarez-Cohen, L. and McCarty, P.L. (1991) A cometabolic biotransformation model for halogenated aliphatic compounds exhibiting product toxicity. *Environ. Sci. Technol.* 25, 1381–1387.
[21] Strandberg, G.W., Donaldson, T.L. and Farr, L.L. (1989) Degradation of trichloroethylene and *trans*-1,2-dichloroethylene by a methanotrophic consortium in a fixed-film, packed-bed bioreactor. *Environ. Sci. Technol.* 23, 1422–1425.
[22] Ewers, J., Clemens, W. and Knackmuss, H.J. (1991) Biodegradation of chloroethenes using isoprene as co-substrate. In: Proc. International Symposium Environmental Biotechnology Oostende (Verachtert, H. and Verstraete, W., Eds), pp. 77-83, Royal Flemish Society of Engineers, Belgium.
[23] Hartmans, S., de Bont, J.A.M., Tramper, J. and Luyben, K.Ch.A.M. (1985) Bacterial degradation of vinylchloride. *Biotechnol. Lett.* 7, 383–388.
[24] Reineke, W. and Knackmuss, H.-J. (1984) Microbial metabolism of haloaromatics: isolation and properties of a chlorobenzene-degrading bacterium. *Appl. Environ. Microbiol.* 47, 395–402.
fluidized bed bioreactor. *Appl. Microbiol. Biotechnol.* 27, 206–213.
[26] Harker, A.R. and Kim, Y. (1990) Trichloroethylene degradation by two independent aromatic degrading pathways in *Alcaligenes eutrophus* JMP134. *Appl. Environ. Microbiol.* 56, 1179–1181.
[27] Winter, R.B., Yen, K.M. and Ensley, B.D. (1989) Efficient degradation of trichloroethylene by a recombinant *Escherichia coli*. *Bio/Technol.* 7, 282–285.
[28] Zylstra, G.J., Wackett, L.P. and Gibson, D.T. (1989) Trichloroethylene degradation by *Escherichia coli* containing the cloned *Pseudomonas putida* F1 toluene dioxygenase genes. *Appl. Environ. Microbiol.* 55, 3162-3166.
[29] Wackett, L.P., Brusseau, G.A., Householder, S.R. and Hanson, R.S. (1989) Survey of microbial oxygenases: trichloroethylene degradation by propane-oxidizing bacteria. *Appl. Environ. Microbiol.* 55, 2960–2964.
[30] Arciero, D., Vannelli, T., Logan, M. and Hooper, A.B. (1989) Degradation of trichloroethylene by the ammonia-oxidizing bacterium *Nitrosomonas europaea*. *Biochem. Biophys. Res. Commun.* 159, 640–643.
[31] Folsom, B.R., Chapman, P.J. and Pritchard, P.H. (1990) Phenol and trichloroethylene degradation by *Pseudomonas cepacia* G4: kinetics and interactions between substrates. *Appl. Environ. Microbiol.* 56, 1279–1285.
[32] Wackett, L.P. and Gibson, D.T. (1988) Degradation of trichloroethylene by toluene dioxygenase in whole-cell studies with *Pseudomonas putida* F1. *Appl. Environ. Microbiol.* 54, 1703–1708.
[33] Ewers, J., Freier-Schröder, D. and Knackmuss, H.-J. (1990) Selection of trichloroethylene (TCE) degrading bacteria that resist inactivation by TCE. *Arch. Microbiol.* 154, 410–413.

10

Methanopterin, its Structural Diversity and Functional Uniqueness

Jan T. Keltjens, Petronella C. Raemakers-Franken and Godfried D. Vogels

Department of Microbiology, University of Nijmegen, Nijmegen, The Netherlands

1 SUMMARY

Methanopterins constitute a group of closely related compounds found in methanogenic bacteria. With the pterin group present in the 5,6,7,8-tetrahydro state of reduction (H_4MPT) the compounds function as C_1 carriers in the same series of reactions that are also known for 5,6,7,8-tetrahydrofolate (H_4folate). However, by some subtle modifications near the catalytic centre the thermodynamics of individual reactions are markedly changed with respect to the analogous H_4folate-dependent steps. These changes make methanopterins particularly suited to act as a catalyst in the methanogenic reactions. In addition, methanopterins are involved in a number of novel and unique reactions in formyl and methyl group transfer and formyl (methenyl) group reduction.

2 INTRODUCTION

Methanogenic bacteria derive the energy for growth from the conversion of hydrogen and CO_2, formate, methylated C_1 compounds or acetate. An important contribution to our understanding of these processes was the discovery and subsequent identification of methanopterin (MPT) as a structural analogon of folic acid [1–3]. Escalante-Semerena *et al.* [4,5] established that 5,6,7,8-

tetrahydromethanopterin (H_4MPT) served as a C_1 carrier in an identical series of reactions as known for 5,6,7,8-tetrahydrofolate (H_4folate).

MPT is only one representative of a group of structurally related compounds that can be isolated from the various methanogenic species. The derivatives have some structural elements in common that distinguish methanopterins from folic acids. The structural differences with respect to folic acid lend the methanogenic cofactors some unique chemical and biochemical properties. Previous reviews [6–9] focused on the analogy between methanopterins and folates or dealt with the enzymology of H_4MPT-dependent reactions. Here, we will investigate how the specific structural features of methanopterins effect their catalytic properties and incorporate some novel advances in the study of their biochemistry.

3 STRUCTURAL DIVERSITY OF THE METHANOPTERINS

With a single exception, methanopterins have only been found in methanogenic bacteria. The exception concerns the thermophilic sulfate-reducing archeon *Archaeoglobus* where H_4MPT participates in the same series of reactions as in methanogens [10]. Among the methanogenic bacteria a number of methanopterin derivatives is found (Fig. 1). All derivatives have as a common structure N-[1'-(2''-amino-4''-hydroxy-6''-pteridinyl)ethyl-4-[2',3',4',5'-tetrahydroxy-D-ribit-1'-yl(5' → 1'')-O-α-ribofuranosyl-5''-phosphoric acid] aniline. The last common element is α-hydroxyglutarate which is esterified with the phosphate group. The unusual long aniline side-chain has been denoted as methaniline [2,3]. Parent methanopterin (MPT), which is usually found in hydrogen- or formate-utilizing organisms [11], contains an additional methyl group bound at the C-7 position of the pterin. Methylotrophic bacteria are characterized by the presence of a derivative called sarcinapterin [3,11], which has an additional L-glutamate bound to α carboxylic group of hydroxyglutarate. Sarcinapterin was named after *Methanosarcina barkeri* from which it originally was purified. The pterins isolated from *Methanogenium tationis* and *Methanoculleus thermophilicum* harbour L-aspartate (tatiopterin-0, thermopterin) and aspartylglutamate (tatiopterin-1) (Fig. 1) [12–14]. Moreover, thermopterin and the tatiopterins do not contain the C-13 methyl group, whereas in thermopterin two additional hydroxy functions are present at the 2- and 5-positions of the aniline moiety. Future investigations may reveal still other derivatives. In *Methanoplanus endosymbiosus* and in *Methanospirillum hungatei*, for instance, the above compounds are absent, but they contain substantial amounts of chromatographically distinct pterins [11].

4 BIOCHEMICAL SPECIFICITY OF THE METHANOPTERINS

Inspection of the structures of the methanopterins immediately reveals the close structural relationship with folic acid. Most importantly the catalytic centre

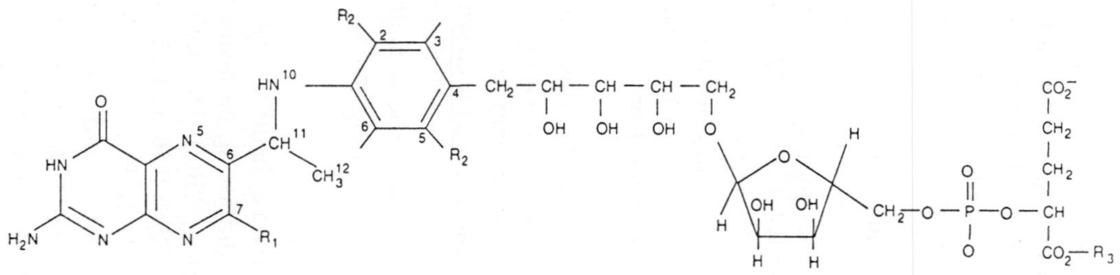

Methanopterin: diversity and uniqueness

	R_1	R_2	R_3
methanopterin	$-CH_3$ (C-13)	$-H$	$-H$
sarcinapterin	$-CH_3$ (C-13)	$-H$	$-$ glutamate
tatiopterin-0	$-H$	$-H$	$-$ aspartate
tatiopterin-1	$-H$	$-H$	$-$ aspartylglutamate
thermopterin	$-H$	$-OH$	$-$ aspartate

Figure 1 Structures of methanopterins isolated from methanogenic bacteria.

around the N-5 and N-10 nitrogen atoms is conserved. As a consequence, H_4MPT catalyzes essentially the same reactions as H_4folate [6–9]. Yet, methanopterins are quite specifically used by the methanogenic enzymes and H_4folates are generally completely inactive as substrates [4, 16–18]. There is one notable exception. Very recently it was shown that 5-methyl-H_4folate serves, albeit with reduced activity, as a substrate in the 5-methyl-H_4MPT: coenzyme (HS-CoM) methyltransferase reaction [18]. This substantiates early reports in which methane formation from methyl-H_4folate was described [19,20]. Conversely, methanopterins may be inactive in folate-dependent systems. The former neither acted as a substrate nor as an inhibitor of dihydrofolate reductase [21].

Despite their structural diversity, methanopterins derived from one organism are recognized and interconverted by the enzymes from another species. H_4MPT obtained from *Methanobacterium thermoautotrophicum*, for instance, is used with high catalytic efficiency by enzymes isolated from *M. barkeri*, which contains tetrahydrosarcinapterin (H_4SPT) as the natural cofactor [22–24]. H_4MPT also has a strong cross reactivity with the enzymes from *M. tationis* and *M. thermophilicum*, whereas the cofactors present in cell extracts of the latter organisms serve as substrates for the enzymes from *M. thermoautotrophicum* [17].

The biochemical specificities, of course, may be the result of the gross differences in the overall structures. Methanopterins with their huge side-chains may simply not fit in the active sites of folate-dependent enzymes, and the folates may show an insufficient affinity for the methanogenic enzymes. Specificity may also result from structural differences near the catalytic centre of the cofactors. Comparison between the structures of the various methanopterins and folic acid shows that the former share two distinct elements:

1. The presence of the electron-donating para-substituted ribityl unit in methaniline as opposed to the electron-withdrawing para-substituted carboxy function in p-aminobenzoylglutamate (pABG). The amine of pABG has a $pK_a = 3.03$, which is decreased to $pK_a = 0.36$ and $pK_a = -1.25$ in folic acid and H_4folate, respectively (Table 1). In H_4folate $\Delta pK_a \sim 6$ exists between N-5 and N-10. Studies with H_4folate and model compounds thereof have shown that the relative basicities of the nitrogen atoms and the ΔpK_a between these are of critical importance in the cofactor function, catalytic properties and situation of chemical equilibria [see for reviews: 25,26]. The amine group of methaniline is more basic and shows pK_a values of 5.1 and 4.0 in the free and methanopterin-bound forms, respectively (Table 1). In H_4MPT the latter value may be lowered by one or two units. In 6-methyl-5,6,7,8-tetrahydropterin and in its 6,7-dimethyl derivative the $pK_a(N-5)$ values are about equal, which is expected also to be the case in H_4MPT and H_4folate. Thus, a ΔpK_a of only 2–3 units is estimated for the N-5 and N-10 of H_4MPT.

Table 1 Dissociation constants of methanopterin, folic acid and related compounds

Compound	pK$_a$		Reference
	N-5	N-10	
Methanopterin	N.D.	4.0	[3]
Methaniline	—	5.1	[3]
6,7-dimethyl-5,6,7,8-H$_4$pterin	5.4	—	[27]
Folic acid	< −1.5	0.36	[28]
H$_4$folate	4.82	−1.25	[27]
p-Aminobenzoylglutamate	—	3.03	[27]
6-methyl-5,6,7,8-H$_4$pterin	5.6	—	[29]

N.D., not determined

2. All methanopterins isolated thus far contain a methyl group (C-12) bound to the C-11 bridge function. This methyl group will influence the flexibility of the catalytic centre, the targeting in the addition reactions and group eliminations, which will be reflected in shifts in the chemical equilibria with respect to the analogous reactions involving H$_4$folates.

The introduction of the C-12 methyl group makes the C-11 become an asymmetric carbon. Thus, together with C-6 and C-7, three asymmetric carbon atoms are present in the proximity of the catalytic centre of H$_4$MPT and H$_4$SPT, which give rise to 8 possible diastereomers. The absolute configuration of biochemically active H$_4$MPT is not known, but part of the question can be solved by high-resolution H^1-NMR. In 5,10-methenyl-H$_4$folate, whose absolute configuration has been established by X-ray analysis [30], the benzene, the pyrimidine and the imidazolium rings are held as planar structures interconnected by an extended conjugated system. Hereby is the tetrahydropyrazine part of the pterin forced into a half-chair conformation with the C-6 proton in an axial position. In view of the structural equivalences, this will also hold for 5,10-methenyl-H$_4$MPT. When the coupling constants of the relevant protons of both 5,10-methenyl species are compared (Table 2) it can be concluded that in the H$_4$MPT derivative the proton bound to C-11 takes an axial position, whereas the proton at C-7 is (pseudo)equatorial. This results in a conformation as depicted in Fig. 2. In Fig. 2, the *R* configuration for C-6 is arbitrarily assumed like in biologically active 5,10-methenyl-H$_4$folate [30]. The structure of the compound with C-6 in the *S* configuration is simply the mirror image with respect to the pterin plane: both C-6 diastereomers will show identical NMR spectra.

We will now discuss how the specific changes in the H$_4$MPT molecule with respect to H$_4$folate influence the (bio)chemistry and thermodynamics of the individual reactions. For the sake of comparison, the reactions have been compiled in Table 3.

Table 2 ^{1}H-NMR properties of 5,10-methenyl-H_4MPT and 5,10-methenyl-H_4folate.

Compound	Coupling constant (J, Hz) of proton bound to[a]					
	C-6		C-7		C-11	
5,10-methenyl-H_4MPT[b]	—[c]		J_{H-H6} ~	0	J_{H-H6} =	6.4
			J_{H-H13} =	6.4	J_{H-H12} =	6.4
5,10-methenyl-H_4folate[d]	J_{H-H7eq} =	4.2	J_{Heq-H6} =	4.2	J_{Hax-H6} =	9.2
	J_{H-H7ax} =	10.6	J_{Hax-H6} =	10.6	J_{Heq-H6} =	11.2
	$J_{H-H11ax}$ =	9.2	J_{vic} =	−12.7	J_{vic} =	−11.8
	$J_{H-H11eq}$ =	11.2				

[a] A numbering system is used as shown in Fig. 1.
[b] Coupling constants were obtained by J. Keltjens (unpublished results).
[c] The signal was buried under a complex series of lines derived from protons attached to the ribityl and furanosyl units.
[d] Coupling constants were taken from [31].
ax, axial; eq, equatorial; vic, vicinal.

5 H_4MPT- AND H_4FOLATE-DEPENDENT REACTIONS

5.1 *Formyl transfer reactions*

H_4MPT enters the pathway of CO_2 reduction to methane by a formyl group transfer from formylmethanofuran (reaction 1a). The reaction is catalyzed by the structurally unique formylmethanofuran: H_4MPT formyltransferase and proceeds in the direction of the thermodynamically more stable product,

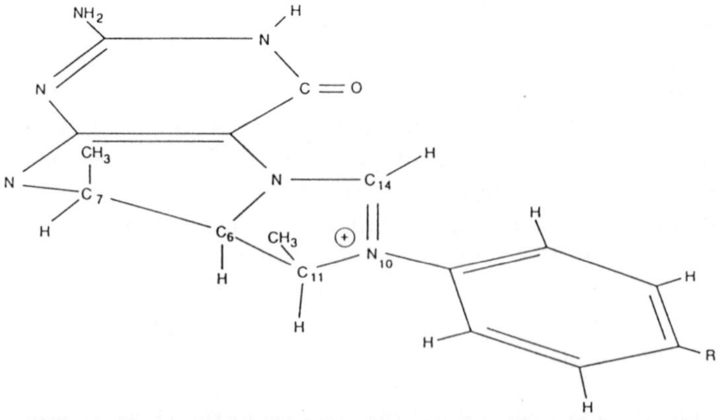

Figure 2 Schematic representation of the proposed conformation of 5,10-methenyl-H_4MPT. In the figure the configuration of C-6 as *R* is arbitrarily chosen; *R* represents the side-chain starting from the D-ribityl unit.

Table 3 Equilibrium constants, free energy changes and redox potentials of H_4MPT- and $H_4folate$-dependent reactions at pH7

Reaction number	Reaction	K_{eq} ($\Delta G^{\circ\prime}$, kJ/reaction; $E_0^\prime$, mV)	Reference
	H_4-dependent reaction		
1a	Formylmethanofuran + H_4MPT = methanofuran + 5-formyl-H_4MPT	4 (-3.4 kJ/mol)	[24]
2a	Formylmethanofuran + H_4MPT = methanofuran + 10-formyl-H_4MPT	0.11 ($+5.5$ kJ/mol)	—[a]
3a	5-Formyl-H_4MPT + H^+ = 5,10-methenyl-H_4MPT + H_2O	2.21/M (-2.0 kJ/mol)	[22][b]
4a	10-Formyl-H_4MPT + H^+ = 5,10-methenyl-H_4MPT + H_2O	78.4/M (-10.8 kJ/mol)	[32]
5a	5,10-Methenyl-H_4MPT + $F_{420}H_2$ = 5,10-methylene-H_4MPT + F_{420} + H^+	0.44 M ($+2.0$ kJ/mol; -362 mV)	[33]
6a	H_4MPT + HCHO = 5,10-methylene-H_4MPT + H_2O	9.1×10^3 (-22.6 kJ/mol)	[34]
7a	5,10-Methylene-H_4MPT + $F_{420}H_2$ = 5-methyl-H_4MPT + F_{420}	2–5 (-3 kJ/mol; -330 mV)	[23, 35]
8a	5-methyl-H_4MPT + HS-CoM = CH_3CoM + H_4MPT	—	
	H_4folate-dependent reaction		
3b	5-Formyl-H_4folate + H^+ = 5,10-methenyl-H_4folate + H_2O	6.5×10^{-5}/M ($+23.8$ kJ/mol)	[36]
4b	10-Formyl-H_4folate + H^+ = 5,10-methenyl-H_4folate + H_2O	0.09/M ($+5.9$ kJ/mol)	[36]
5b	5,10-Methenyl-H_4folate + NADH = 5,10-methylene-H_4folate + NAD^+	2.0 (-1.7 kJ/mol; -311 mV)	[37]
6b	H_4folate + HCHO = 5,10-methylene-H_4folate + H_2O	32×10^3 (-25.7 kJ/mol)	[38]
7b	5,10-Methylene-H_4folate + NADH + H^+ = 5-methyl-H_4folate + NAD^+	6×10^3/M (-22 kJ/mol; -200 mV)	[39]
8b	5-methyl-H_4folate + homocysteine = methionine + H_4folate	1.4×10^5 (-29.4 kJ/mol)	[40]

[a] The equilibrium constant of this hypothetical reaction can be derived from the K_{eq} values of the reactions 1a, 3a and 4a.
[b] In the cited paper the reaction was improperly attributed to reaction 4a (cfr. [24]).

5-formyl-H_4MPT [24,41–43]. The reaction has no apparent equivalent in folate biochemistry. Contrary to earlier suggestions by our group [7–9,22; see also: 24,43] 10-formyl-H_4MPT does not appear to be an intermediate in methanogenesis. From data collected in [44] and in Table 3, one may estimate that the free energies of hydrolysis of 5- and 10-formyl-H_4MPT, 5- and 10-formyl-H_4folate to formic acid and the corresponding tetrahydropterins amount to -9, -18, -4 and -22 kJ/mol, respectively. Thus, at this stage the binding energies of the C_1 unit to the N-5 or N-10 positions of H_4MPT are about equal to those of H_4folate.

The next step in the methanogenic process involves the dehydration of 5-formyl-H_4MPT to the cyclic 5,10-methenyl derivative (reaction 3a). Here a most striking difference with the analogous H_4folate-dependent reaction (3b) comes to the fore. The reaction equilibrium (3a) exceeds the one of (3b) by a factor of 3×10^4; even at neutral pH, 5,10-methenyl-H_4MPT formation is somewhat favored. At physiological pH, the enzymic conversion of 5-formyl- into 10-formyl- or 5,10-methenyl-H_4folate is only brought about at the expense of stoichiometric ATP hydrolysis [36]. The favorable change in the thermodynamics can largely be attributed to the presence of the methaniline side-chain. Studies with tetrahydroquinoxalines that closely mimic the folate cofactor function demonstrated that a mere substitution in the aniline ring of a carboxylic ester for a methyl group increased the K_{eq} about 1000-fold [45]. The additional increase of the K_{eq} may be due to the methyl group at C-11 which supports the reaction by correct targeting the ring closure and by its thermodynamic attractive equatorial positioning in the imidazolium ring (Fig. 2).

As in the formyl-H_4folates and model compounds the reversible cyclohydrolysis is likely to proceed *via* the tetrahydral intermediate **2** in Fig. 3 [45–47]. During the enzymic reaction performed in D_2O the methenyl proton is exchanged for deuterium (J. Keltjens, unpublished results), which indicates that the ylide **4** (Fig. 3) is to some extent an intermediate; a similar solvent exchange has been observed with 5,10-methenyl-H_4folate [47]. Under alkaline conditions, the non-enzymic hydrolysis of 5,10-methenyl-H_4MPT proceeds in the thermodynamically unfavorable direction of the 10-formyl derivative **5** [32]. This implies the reaction has now come under kinetic control. Upon neutralization, a rapid isomerization occurs to the more stable 5-formyl species [24]: with 10-formyl-H_4folate the analogous isomerization is only observed after prolonged heating.

5.2 *Methenyl reduction*

5,10-Methenyl-H_4MPT is enzymatically reduced to 5,10-methylene-H_4MPT according to reaction 5a. In the reaction reduced F_{420} serves as the reductant. Enzyme kinetics are in agreement with a direct electron (hydride) transfer between the reactants [22,33]. In eubacterial and eukaryotic organisms the analogous 5,10-methylene-H_4folate reduction is catalyzed by NAD(P)(H)-

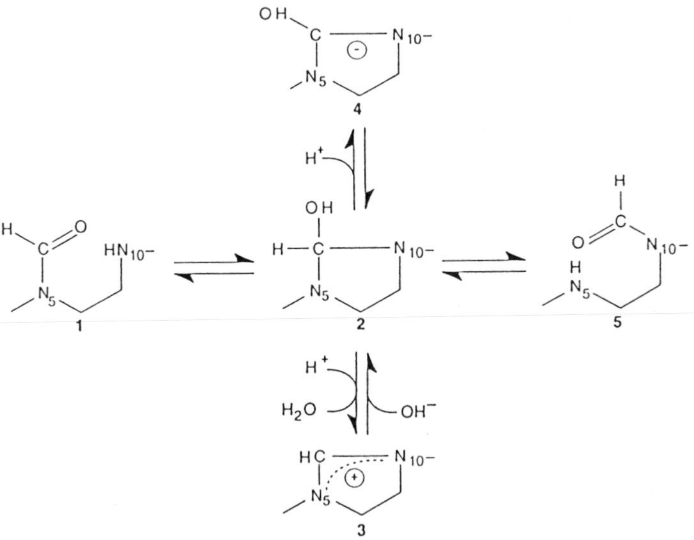

Figure 3 Scheme of the interconversion of formyl-H$_4$MPT and formyl-H$_4$folate derivatives. The compounds are represented by the partial structures around the N-5 and N-10 atoms of the catalytic centre.

dependent enzymes (reaction 5b). The midpoint potential ($E'_0 = -311$ mV) is 50 mV higher than in the methenyl-H$_4$MPT reduction ($E'_0 = -362$ mV). Bearing in mind the differences between the pK$_a$ values of the N-10 atoms, the electron density of the methenyl group will be relatively lower in 5,10-methenyl-H$_4$folate, which facilitates the reduction.

Hydrogenotrophic methanogens contain an unique methylene dehydrogenase, which is genetically distinct from the F$_{420}$-dependent enzyme (both enzymes may be encountered in a same organism) and which catalyzes the following reversible reaction [48–50]:

$$5,10\text{-methenyl-H}_4\text{MPT} + H_2 \rightleftharpoons 5,10\text{-methylene-H}_4\text{MPT} + H^+$$

The enzyme apparently functions as a hydrogenase. Yet, prosthetic groups that are characteristic for this class of enzymes, like [4Fe-4S] clusters or nickel, seem to be absent [47]. The way in which bond breaking and bond formation of hydrogen gas occurs is unknown and it would be quite interesting to see if the H$_4$MPT part plays a direct role herein.

5.3 Reactions at the formaldehyde state of reduction

Analogously to H$_4$folate (reaction 6b), formaldehyde binds in a non-enzymic way to H$_4$MPT under the formation of 5,10-methylene-H$_4$MPT (reaction 6a). This property led to the discovery of H$_4$MPT as a central C$_1$ carrier [4,5] and

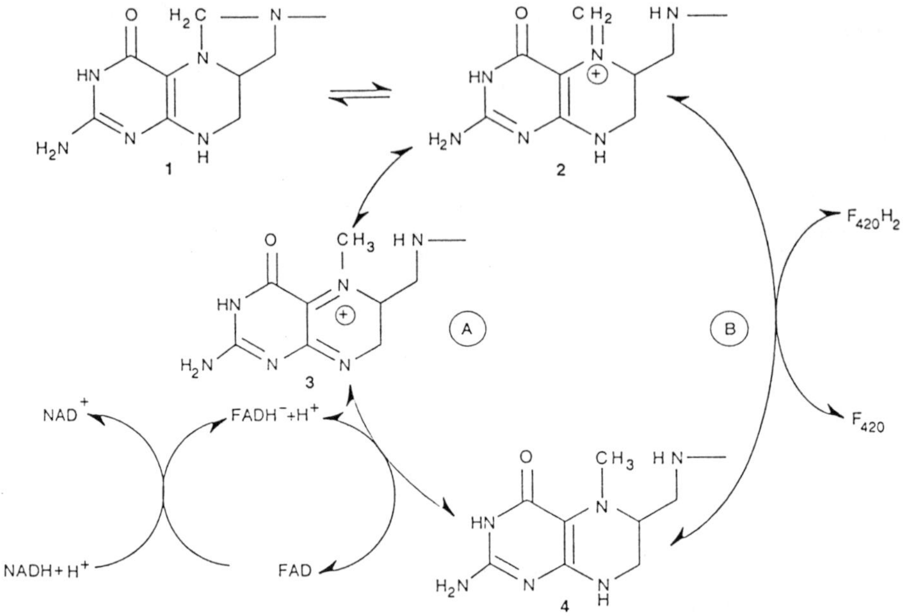

Figure 4 Mechanism of the NAD(P)H-dependent 5,10-methylene-H$_4$folate (A) and of the proposed F$_{420}$H$_2$-dependent 5,10-methylene-H$_4$MPT (B) reductase reactions. H$_4$folate and H$_4$MPT derivatives are represented by their pterin moieties.

it is still frequently used as an easy means to introduce the C$_1$ unit at the formaldehyde state of reduction into the methanogenic pathway. The equilibrium constant of reaction 6a is about three-fold lower than that of 6b, but methylene formation is still highly favored in both reactions.

5,10-Methylene-H$_4$MPT can be chemically [4,5] and enzymically reduced to 5-methyl-H$_4$MPT. In the enzymic reaction (7a) F$_{420}$H$_2$ is the specific reductant [23,35,51]. 5,10-Methylene-H$_4$folate reductases either use NAD(P)H (reaction 7b), ferredoxin and artificial dyes as cosubstrates. Comparison between the methylene reduction reactions (7a) and (7b) shows the former to be less exergonic by about 19 kJ/mol, whereas the midpoint potential is 130 mV lower. The first step in methylene reduction (Fig. 4) must involve the imidazole ring opening to produce the N-5- bound imine cation **2**, followed by its reduction to the methyl group [26,52]. Since the reduction potentials of the cationic imines bound to H$_4$MPT and H$_4$folate are not expected to differ much, the less favourable energetics of the methylene-H$_4$MPT reduction must be associated with the ring opening reaction. Apparently, the presence of the ribityl side-chain and the methyl group at C-11 make N-10 a relative poor leaving group, while it favors addition of N-10 and accompanying ring closure.

Methylene-H$_4$folate reductases isolated thus far all contain a flavin (FAD) prosthetic group; the enzymes catalyze the reaction according to route A in

Fig. 4 by a ping-pong bi-bi mechanism [26,52]. In this route the imine cation 2 undergoes an intramolecular oxidation-reduction to quinonoid 5-methyldihydrofolate 3. Subsequent reduction of the dihydropterin by $FADH_2$ completes the sequence. In the NAD(P)H-dependent enzymes $FADH_2$ is regenerated by NADH oxidation. Methyl- to methylene-H_4folate oxidation is feasible with artificial electron acceptors of sufficient high redox potential. With $NAD(P)^+$ the reaction in this direction does not proceed, which must be due the large $\Delta E'_0$ between the flavin and the nicotinamide. In contrast, methylene-H_4MPT reductases catalyze the reaction (7a) in either direction, though methyl oxidation only occurs in the presence of methylene-H_4MPT dehydrogenase [23,35,51]. The methanogenic enzymes lack a flavin prosthetic group and enzyme kinetics are in agreement with a ternary complex mechanism. These findings might indicate that methylene (imine cation) reduction and methyl oxidation here occur by a direct hydride transfer with $F_{420}(H_2)$ as proposed in Fig. 4 (route B).

5.4 Methyl transfer reactions

Though H_4MPT also plays a role as a methyl group carrier in acetyl-CoA synthesis and degradation [7,8,53] we will restrict our discussion to the 5-methyl-H_4MPT: HS-CoM methyltransferase reaction (reaction 8a). This reaction is unique for methanogens. With respect to the folates an equivalent, however, may be found in the B_{12}-dependent 5-methyl-H_4folate: homocysteine methyltransferase (methionine synthetase) reaction (reaction 8b) [54,55]. Methionine synthesis occurs by a two-step process (Fig. 5). First, enzyme-bound Cob(I)alamin (B_{12s}), nature's most powerful nucleophile [56], accepts the methyl group from methyl-H_4folate and methyl-Cob(III)alamin is formed. In the second half of the catalytic cycle the methyl group is transferred to homocysteine. Methionine synthesis is quite exergonic [40]. Since the methyl-H_4folate: B_{12s} methyl transfer step is reversible ($K_{eq} \sim 1$; [55]), energy release is associated with the second partial reaction. Methylcoenzyme M synthesis according to reaction (8a) is catalyzed by a membrane protein which contains 5-hydroxybenzimidazolyl cobamide (B_{12}HBI) as a prosthetic group [18,57,58]. The reaction proceeds with the intermediary formation of methyl-B_{12}HBI by a similar two-step mechanism as the methyl-H_4folate: homocysteine methyl group transfer (Fig. 5). The $\Delta G^{0'}$ of methylcoenzyme M synthesis may compare with the one of methionine synthesis in which the methyl group transfer from methyl-B_{12}HBI to HS-CoM then constitutes the more exergonic part. Methanogenic bacteria are able to conserve the energy by pumping sodium across the cell membrane [18]. The way sodium pumping is achieved is not known.

Methyl-H_4MPT: HS-CoM methyltransferase and methionine synthetase are only catalytically active when the central cobalt is present in the reduced Co(I) state. Systems are available for both enzymes to reactivate oxidized species. Reactivation of methionine synthetase is brought about by a coupled reduction and trapping traces of B_{12s} by methylation with S-adenosylmethionine [55]

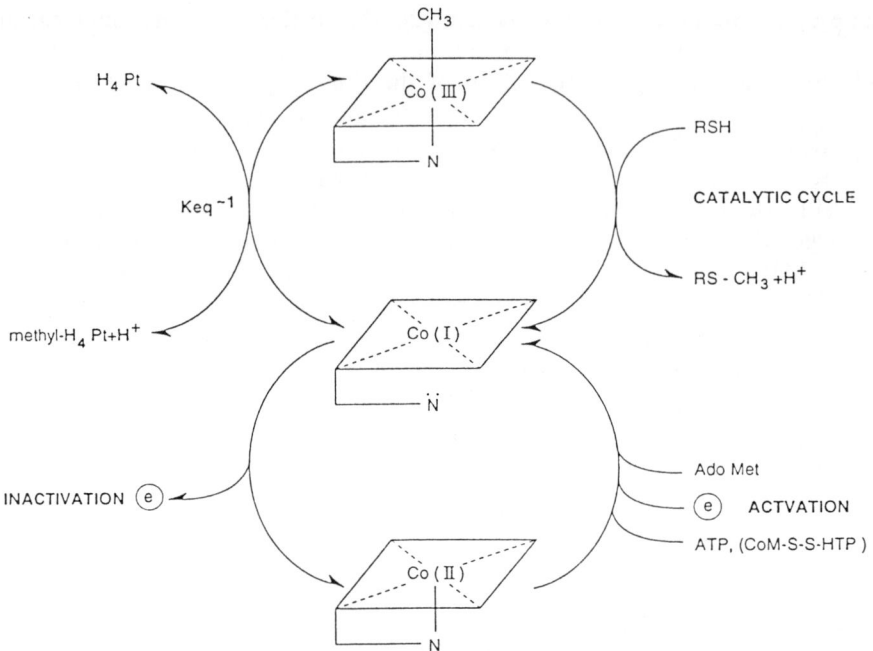

Figure 5 Scheme of the cycles of catalysis, inactivation and reductive activation of the corrinoid-dependent methyl-H$_4$folate: homocysteine (methionine synthetase) and methyl-H$_4$MPT:coenzyme M (methylcoenzyme M synthetase) methyltransferase reactions. The proteins are represented by the schematically drawn corrinoid prosthetic groups. Methionine synthesis: H$_4$Pt, H$_4$folate; methyl-H$_4$PT, 5-methyl-H$_4$folate; RSH, homocysteine; RS-CH$_3$, methionine; N,5,6-dimethylbenzimidazolyl nucleotide. Reactivation involves the action of S-adenosylmethionine (AdoMet). Methylcoenzyme M synthesis: H$_4$Pt, H$_4$MPT; methyl-H$_4$Pt, methyl-H$_4$MPT; RSH. coenzyme M; RS-CH$_3$, methylcoenzyme M; N,5-hydroxybenzimidazolyl nucleotide. Reactivation requires ATP and the contingent action of CoM-S-S-HTP.

(Fig. 5). Reactivation of the methyltransferase from *M. barkeri* requires the presence of a reducing system and the action of ATP alone (P. Daas, unpublished results), whereas the enzyme from *M. thermoautotrophicum* is reactivated by the concerted action of ATP and the heterodisulfide (CoM-S-S-HTP) of HS-CoM and 7-mercaptoheptanoylthreonine phosphate (HS-HTP) [59]. The mechanism underlying the reductive activation of the methanogenic enzymes is not clear.

ACKNOWLEDGEMENT

The work of J.T. Keltjens was supported by a senior fellowship of the Royal Netherlands Society of Arts and Sciences (KNAW).

REFERENCES

[1] Keltjens, J.T., Huberts, M.J., Laarhoven, W.H. and Vogels, G.D. (1983) Structural elements of methanopterin, a novel pterin present in *Methanobacterium thermoautotrophicum*. *Eur. J. Biochem.* **130**, 537–544.
[2] Van Beelen, P., Stassen, A.P.M., Bosch, J.W.G., Vogels, G.D., Guijt, W., and Haasnoot, C.A.G. (1984) Elucidation of the structure of methanopterin, a coenzyme from *Methanobacterium thermoautotrophicum*, using two-dimensional nuclear-magnetic-resonance techniques. *Eur. J. Biochem.* **138**, 563–571.
[3] Van Beelen, P., Labro, J.F.A., Keltjens, J.T., Geerts, W.J., Vogels, G.D., Laarhoven, W.H., Guijt, W. and Haasnoot, C.A.G (1984) Derivatives of methanopterin, a coenzyme involved in methanogenesis. *Eur. J. Biochem.* **139**, 359–365.
[4] Escalante-Semerena, J.C., Leigh, J.A., Rinehart, K.L., Jr. and Wolfe, R.S. (1984) Formaldehyde activation factor, tetrahydromethanopterin, a coenzyme of methanogenesis. *Proc. Natl. Acad. Sci. USA* **81**, 1976–1980.
[5] Escalante-Semerena, J.C., Rinehart, K.L., Jr. and Wolfe, R.S. (1984) Tetrahydromethanopterin, a carbon carrier in methanogenesis. *J. Biol. Chem.* **259**, 9447–9455.
[6] DiMarco, A.A., Bobik, T.A. and Wolfe, R.S. (1990) Unusual coenzymes of methanogenesis. *Annu. Rev. Biochem.* **59**, 355-394.
[7] Keltjens, J.T. and Vogels, G.D. (1988) Methanopterin and methanogenic bacteria. *BioFactors* **1**, 95–103.
[8] Keltjens, J.T., te Brömmelstroet, B.W., Kengen, S.W.M., van der Drift, C. and Vogels, G.D. (1990) 5,6,7,8-tetrahydromethanopterin-dependent enzymes involved in methanogenesis. *FEMS Microbiol. Rev.* **87**, 327–332.
[9] Keltjens, J.T., te Brömmelstroet, B.W., Kengen, S.W.M., van der Drift, C. and Vogels, G.D. (1990) Methanopterin: a structural and functional folic acid analogon in the biochemistry of methanogenic bacteria. In: Chemistry and Biology of Pteridines 1989. Pteridines and Folic Acid Derivatives (Curtius, H.-Ch., Ghisla, S. and Blau, N. Eds), pp. 285–293. Walter de Gruyter, Berlin.
[10] Möller-Zinkhan, D., Börner, G. and Thauer, R.K. (1989) Function of methanofuran, tetrahydromethanopterin and coenzyme F_{420} in *Archaeoglobus fulgidus*. *Arch. Microbiol.* **152**, 362–368.
[11] Gorris, L.G.M. and van der Drift, C. (1986) Methanogenic cofactors in pure cultures of methanogens in relation to substrate utilization. In: Biology of Anaerobic Bacteria (Dubourguier, H.C., Albagnac, G., Montreuil, H.J., Romond C., Sautière, P. and Guillaume J., Eds), pp. 144–150. Elsevier Science Publishers, Amsterdam.
[12] Raemakers-Franken, P.C., Voncken, F.G.J., Korteland, J., Keltjens, J.T., van der Drift, C. and Vogels, G.D. (1989) Structural characterization of tatiopterin, a novel pterin isolated from *Methanogenium tationis*. *BioFactors* **2**, 117–122.
[13] Raemakers-Franken, P.C., van Elderen, C.H.M., van der Drift, C. and Vogels, G.D. (1991) Identification of a novel tatiopterin derivative in *Methanogenium tationis*. *BioFactors* **3**, 127–130.
[14] Raemakers-Franken, P.C., Bongaerts, R., Fokkens, R., van der Drift, C. and Vogels, G.D. (1991) Characterization of two pterin derivatives isolated from *Methanoculleus thermophilicum*. *Eur. J. Biochem.* **200**, 783–787
[15] Ferry, J.G., Sherod, D.W., Peck, H.D., Jr. and Ljungdahl, L.G. (1976) Levels of formyltetrahydrofolate synthetase and methylenetetrahydrofolate dehydrogenase in methanogenic bacteria. In: Microbial Production and Utilization of Gases (H_2, CH_4, CO) (Schlegel, H.G., Gottschalk, G. and Pfennig, N., Eds) pp. 151–155. E. Goltze KG, Göttingen, Germany.
[16] Hartzell, P.L., Zvilius, G., Escalante-Semerena, J.C. and Donnelly, M.I. (1985) Coenzyme F_{420} dependence of the methylenetetrahydromethanopterin dehydro-

genase of *Methanobacterium thermoautotrophicum*. *Biochem. Biophys. Res. Commun.* **133**, 884–890.
[17] Raemakers-Franken, P.C., Kortstee, A.J., van der Drift, C. and Vogels, G.D. (1990) Methanogenesis involving a novel carrier of C_1 compounds in *Methanogenium tationis*. *J. Bacteriol.* **172**, 1157–1159.
[18] Becher, B., Müller, V. and Gottschalk, G. (1992) The methyl-tetrahydro methanopterin: coenzyme M methyltransferase of *Methanosarcina* strain Göl is a primary sodium pump. *FEMS Microbiol. Lett.* **91**, 239–244.
[19] Wood, J.M. and Wolfe, R.S. (1965) The formation of CH_4 from N^5-methyltetrahydrofolate monoglutamate by cell-free extracts of *Methanobacillus omelianskii*. *Biochem. Biophys. Res. Commun.* **19**, 306–311.
[20] Blaylock, B.A. (1968) Cobamide-dependent methanol-cyanocob(I)alamin methyltransferase from *Methanosarcina barkeri*. *Arch. Biochem. Biophys.* **124**, 314–324.
[21] Raemakers-Franken, P.C., De Abreu, R.A., Willems, J.G., van der Drift, C. and Vogels, G.D. (1991) In vitro inhibition of cell growth of MOLT-4 malignant human T-lymphoblasts by coenzyme F_{420}. *Biochem. Pharmacol.* **4**, 561–566.
[22] te Brömmelstroet, B.W., Hensgens, C.M.H., Geerts, W.J., Keltjens, J.T., van der Drift, C. and Vogels, G.D. (1990) Purification and properties of the 5,10-methenyltetrahydromethanopterin cyclohydrolase from *Methanosarcina barkeri*. *J. Bacteriol.* **172**, 564–571.
[23] te Brömmelstroet, B.W., Geerts, W.G., Keltjens, J.T., van der Drift, C. and Vogels, G.D. (1991) Purification and properties of 5,10-methylenetetrahydromethanopterin dehydrogenase and 5,10-methylenetetrahydromethan opterin reductase, two coenzyme F_{420}- dependent enzymes, from *Methanosarcina barkeri*. *Biochim. Biophys. Acta* **1079**, 293–302.
[24] Keltjens, J.T., Brugman, A.J.A.M., Kesseleer, J.M.A., te Brömmelstroet, B.W.J., van der Drift, C. and Vogels G.D. (1992) 5-Formyl-5,6,7,8-tetrahydromethanopterin is the intermediate in the process of methanogenesis in *Methanosarcina barkeri*. *BioFactors* **3**, 249–255.
[25] Benkovic, S.J. (1978) On the mechanism of folate cofactors. *Acc. Chem. Res.* **11**, 314–320.
[26] Benkovic, S.J. (1980) On the mechanism of action of folate- and biopterin-requiring enzymes. *Annu. Rev. Biochem.* **49**, 227–251.
[27] Kallen, R.G. and Jencks, W.P. (1966) The dissociation constants of tetrahydrofolic acid. *J. Biol. Chem.* **241**, 5845–5850.
[28] Poe, M. (1977) Acidic dissociation constants of folic acid, dihydrofolic acid, and methotrexate. *J. Biol. Chem.* **252**, 3724–3728.
[29] Whiteley, J.M. and Huennekens, F.M. (1967) 2-Amino-4-hydroxy-7,8-dihydropteridine as a model for dihydrofolate. *Biochemistry* **6**, 2620–2625.
[30] Fontecilla-Camps, J.C., Bugg, C.E., Temple, C., Jr., Rose, J.D., Montgomery, J.A. and Kisliuk, R. (1979) Absolute configuration of biological tetrahydrofolates. A crystallographic determination. *J. Am. Chem. Soc.* **101**, 6114–6115.
[31] Khalifa, E., Bieri, J. and Viscontini, M. Konformationsanalyse von 5,10-methenyl-(6RS)-5,6,7,8-tetrahydro-L-folsäure. *Helv. Chim. Acta* **62**, 1340–1344.
[32] DiMarco, A.A., Donnelly, M.I. and Wolfe, R.S. (1986) Purification and properties of the 5,10-methenyltetrahydromethanopterin cyclohydrolase from *Methanobacterium thermoautotrophicum*. *J. Bacteriol.* **168**, 1372–1377.
[33] te Brömmelstroet, B.W., Hensgens, C.M.H., Keltjens, J.T., van der Drift, C. and Vogels, G.D. (1991) Purification and characterization of coenzyme F_{420}-dependent 5,10-methylenetetrahydromethanopterin dehydrogenase from *Methanobacterium thermoautotrophicum* strain ΔH. *Biochim. Biophys. Acta* **1073**, 77–84.
[34] Keltjens, J.T., Caerteling, C.G., van der Drift, C. and Vogels G.D. (1986) Methanopterin and the intermediary steps of methanogenesis. *System. Appl. Microbiol.* **7**, 370–375.

[35] te Brömmelstroet, B.W., Hensgens, C.M.H., Keltjens, J.T., van der Drift, C. and Vogels, G.D. (1990) Purification and properties of 5,10-methylenetetrahydromethanopterin reductase, a coenzyme F_{420}-dependent enzyme, from *Methanobacterium thermoautotrophicum* strain ΔH. *J. Biol. Chem.* **265**, 1852–1857.
[36] Kay, L.D., Osborn, M.J., Hatefi, Y. and Huennekens, F.M. (1960) The enzymatic conversion of N^5-formyl tetrahydrofolic acid (folinic acid) to N^{10}-formyl tetrahydrofolic acid. *J. Biol. Chem.* **235**, 195–201.
[37] Ragsdale, S.W. and Ljungdahl, L.G. (1984) Purification and properties of NAD-dependent 5,10-methylenetetrahydrofolate dehydrogenase from *Acetobacterium woodii*. *J. Biol. Chem.* **259**, 3499–3503.
[38] Kallen, R.G. and Jencks, W.P. (1966) The mechanism of the condensation of formaldehyde with tetrahydrofolic acid. *J. Biol. Chem.* **241**, 5851–5863.
[39] Wohlfarth, G. and Diekert, G. (1991) Thermodynamics of methylenetetrahydrofolate reduction to methyltetrahydrofolate and its implications for the energy metabolism of homoacetogenic bacteria. *Arch. Microbiol.* **155**, 378–381.
[40] Rüdiger, H. and Jaenicke, L. (1969) Methionine synthesis: demonstration of the reversibility of the reaction. *FEBS Lett.* **4**, 316–318.
[41] Donnelly, M.I. and Wolfe, R.S. (1986) The role of the formylmethanofuran: tetrahydromethanopterin formyltransferase in methanogenesis from carbon dioxide. *J. Biol. Chem.* **261**, 16653–16659.
[42] DiMarco A.A., Sment K.A., Konisky, J and Wolfe, R.S. (1990) The formylmethanofuran: tetrahydromethanopterin formyltransferase from *Methanobacterium thermoautotrophicum*: nucleotide sequence and functional expression of the cloned gene. *J. Biol. Chem.* **265**, 472–476.
[43] Breitung, J. and Thauer, R.K. (1991) Formylmethanofuran: tetrahydromethanopterin formyltransferase from *Methanosarcina barkeri*. Identification of N^5-formyltetrahydromethanopterin as the product. *FEBS Lett.* **275**, 226–230.
[44] Keltjens, J.T. and van der Drift, C. (1986) Electron transfer reactions in methanogens. *FEMS Microbiol. Rev.* **39**, 259–303.
[45] Benkovic, S.J., Bullard, W.P. and Benkovic, P.A. (1972) Studies on models for tetrahydrofolic acid. III. Hydrolytic interconversion of the tetrahydroquinoxaline analogs at the formate level of oxidation. *J. Am. Chem. Soc.* **94**, 7542–7549.
[46] Robinson, D.R. and Jencks, W.P. (1967) Mechanism and catalysis of the hydrolysis of methenyltetrahydrofolic acid. *J. Am. Chem. Soc.* **89**, 7098–7103.
[47] Stover, P. and Schirch, V. (1992) Evidence for the accumulation of a stable intermediate in the nonenzymatic hydrolysis of 5,10-methenyltetrahydropteroylglutamate to 5-formyltetrahydropteroylglutamate. *Biochemistry* **31**, 2148–2155.
[48] Zirngibl, C., Hedderich, R. and Thauer, R.K. (1990) N^5,N^{10}-methenyltetrahydromethanopterin dehydrogenase from *Methanobacterium thermoautotrophicum* has hydrogenase activity. *FEBS Lett.* **261**, 112–116.
[49] Schwörer, B. and Thauer, R.K. (1991) Activities of formylmethanofuran dehydrogenase, methylenetetrahydromethanopterin dehydrogenase, methylenetetrahydromethanopterin reductase, and heterodisulfide reductase in methanogenic bacteria. *Arch. Microbiol.* **155**, 459–465.
[50] Von Bünau, R., Zirngibl, C. and Thauer, R.K. (1991) Hydrogen-forming and coenzyme-F_{420}-reducing methylene tetrahydromethanopterin dehydrogenase are genetically distinct enzymes in *Methanobacterium thermoautotrophicum*. *Eur. J. Biochem.* **202**, 1205–1208.
[51] Ma, K. and Thauer, R.K. (1990) Purification and properties of N^5,N^{10}-methenyltetrahydromethanopterin reductase from *Methanobacterium thermoautotrophicum* (strain Marburg) *Eur. J. Biochem.* **191**, 187–193.
[52] Matthews. R.G. and Haywood, B. (1979) Inhibition of pig liver methylene tetrahydrofolate reductase by dihydrofolate: some mechanistic and regulatory implications. *Biochemistry* **18**, 4845–4881.

[53] Grahame, D.A. (1991) Catalysis of acetyl-CoA cleavage and tetrahydrosarcinapterin methylation by a carbon monoxide dehydrogenase-corrinoid enzyme complex. *J. Biol. Chem.* **266**, 22227–22233.
[54] Matthews, R.G., Banerjee, R.V. and Ragsdale, S.W. (1990) Cobamide-dependent methyl transferases. *BioFactors* **3**, 147–152.
[55] Banerjee, R.V., Harder, S.R., Ragsdale, S.W. and Matthews, R.G. (1990) Mechanism of reductive activation of cobalamin-dependent methionine synthetase: an electron paramagnetic resonance spectroelectrochemical study. *Biochemistry* **29**, 1129–1135.
[56] Kräutler, B. (1990) Chemistry of methylcorrinoids related to their roles in bacterial C_1 metabolism. *FEMS Microbiol. Rev.* **87**, 349–354.
[57] Kengen, S.W.M., Daas, P.J.H., Duits, E.F.G., Keltjens, J.T., van der Drift, C. and Vogels, G.D. (1992) Isolation of a 5-hydroxybenzimidazolyl-containing enzyme involved in the methyltetrahydromethanopterin: coenzyme M methyltransferase reaction in *Methanobacterium thermoautotrophicum*. *Biochim. Biophys. Acta* **1118**, 249–260.
[58] Stupperich, E., Juza, A., Eckerskorn, C. and Edelmann, L. (1990) An immunological study of corrinoid proteins from bacteria revealed homologous antigenic determinants of a soluble corrinoid-dependent methyl transferase and corrinoid-containing membrane proteins from *Methanobacterium* species. *Arch. Microbiol.* **155**, 28–34.
[59] Kengen, S.W.M., Daas, P.J.H., Keltjens, J.T. and Vogels, G.D. (1990) Stimulation of the methyltetrahydromethanopterin: coenzyme M methyltransferase reaction in cell-extracts of *Methanobacterium thermoautotrophicum* by the heterodisulfide of coenzyme M and 7- mercaptoheptanoylthreonine phosphate. *Arch. Microbiol.* **154**, 156–161.

11
Enzymes Involved in Methanogenesis from CO_2

Rudolf K. Thauer, Beatrix Schwörer and Carmen Zirngibl

Max-Planck-Institut für Terrestrische Mikrobiologie and Laboratorium für Mikrobiologie des Fachbereichs Biologie der Philipps-Universität, D-3550 Marburg/L., Germany

1 SUMMARY

The reduction of CO_2 to CH_4 in methanogenic archaea proceeds via carrier bound C_1-units. Formylmethanofuran, N^5-formyltetrahydromethanopterin (N^5-formyl-H_4MPT), N^5, N^{10}-methenyl-H_4MPT, N^5, N^{10}-methylene-H_4MPT, N^5-methyl-H_4MPT, Co-methyl-5-hydroxybenzimidazolyl-cobamide, methyl-coenzyme M, and most likely Ni-methyl-coenzyme F_{430} are the intermediates identified. All the enzymes involved in this C_1-metabolism have been purified and characterized. This review concentrates on the H_2-forming methylene-H_4MPT dehydrogenase, which is a novel hydrogenase containing neither nickel nor iron-sulfur clusters and which catalyzes a two-electron hydride transfer and thus differs from all other hydrogenases known to date.

2 INTRODUCTION

Most methanogenic archaea can grow on H_2 and CO_2 as sole energy sources (reaction a).

(a) $CO_2 + 4H_2 \longrightarrow CH_4 + 2H_2O$ $\Delta G^{\circ\prime} = -131$ kJ/mol

The pathway of CO_2 reduction has been elucidated mainly by the work of Ralph Wolfe and collaborators [1]. It involves methanofuran (MFR), tetrahydromethanopterin (H_4MPT), and coenzyme M (H-S-CoM) as C_1-unit car-

---
Microbial Growth on C_1 Compounds
© Intercept Ltd, PO Box 716, Andover, Hampshire SP10 1YG, UK

riers. Formyl-MFR, N^5-formyl-H_4MPT, N^5, N^{10}-methenyl-H_4MPT, N^5, N^{10}-methylene-H_4MPT, N^5-methyl-H_4MPT, and methyl-coenzyme M have been identified as intermediates [2]. The conversion of CO_2 to CH_4 via these C_1-intermediates is catalyzed by formylmethanofuran dehydrogenase, formylmethanofuran: H_4MPT formyltransferase, N^5, N^{10}-methenyl- H_4MPT cyclohydrolase, N^5, N^{10}-methylene-H_4MPT dehydrogenase, N^5, N^{10}-methylene-H_4MPT reductase, N^5-methyl-H_4MPT: coenzyme M methyltransferase, methyl-coenzyme M reductase, and heterodisulfide reductase. The properties of these eight enzymes, which are found in all methanogens investigated to date [3,4], will be reviewed with special emphasis on the H_2-forming methylene-H_4MPT dehydrogenase, which turned out to be a novel type of hydrogenase. Five of the eight enzymes have also been found in the sulfate-reducing archaeon *Archaeoglobus fulgidus* [5,6]. Their properties are also included.

Most of the enzymes involved in methane formation from CO_2 have been isolated from several methanogens belonging to phylogenetically distinct orders [7]: *Methanobacterium thermoautotrophicum* to the order of *Methanobacteriales*; *Methanosarcina barkeri* to the order of *Methanomicrobiales*; *Methanococcus voltae* to the order of *Methanococcales*; and *Methanopyrus kandleri* to an unnamed novel order [8]. *A. fulgidus* is phylogenetically most closely related to the order of *Methanomicrobiales* [9].

Most work referred to here was performed with purified or at least partially purified enzyme preparations following the famous dictum of Ephraim Racker: 'Don't waste clean thinking on dirty enzymes' [10].

3 FORMYLMETHANOFURAN DEHYDROGENASE

The dehydrogenase catalyzes the reduction of CO_2 plus methanofuran to formylmethanofuran and H_2O (reaction b). The physiological electron donor for the enzyme is not known. For activity determination, the enzyme is tested in the direction of formylmethanofuran oxidation with viologen dyes as artificial electron acceptors [11].

(b) $2[H] + CO_2 +$ [structure] NH_3^+ ⇌ [structure] $+ H_2O + H^+$

$E^{o'} = -497$ mV

The enzyme, which is a peripheral membrane protein, has been purified from *M. thermoautotrophicum* (strain Marburg) [12], from *M. barkeri* [13], from *Methanobacterium wolfei* [14,15], and from *A. fulgidus* [6]. The very oxygen labile purified enzyme exhibits specific activities of up to 175 U/mg protein (1U = 1μmol/min) [13]. It is a multi-subunit molybdenum iron sulfur protein with molybdenum bound to molybdopterin guanine dinucleotide [16]. The enzyme from *M. thermoautotrophicum* additionally contains molybdopterin adenine dinucleotide and molybdopterin hypoxanthine dinucleotide [12]. In *M. wolfei* two formylmethanofuran dehydrogenases are found: one is a molybdenum iron-sulfur protein [14,15] and the other is a tungsten iron-sulfur protein [17]. Both enzymes contain molybdopterin guanine dinucleotide. The tungsten enzyme is 'induced' when the growth medium is deprived of molybdate and supplemented with tungstate [17].

4 FORMYLMETHANOFURAN:H₄MPT FORMYLTRANSFERASE

The formyltransferase catalyzes the formation of N^5-formyl-H$_4$MPT from formylmethanofuran and tetrahydromethanopterin (reaction c).

(c) CHO-MFR + [structure] ⇌ [structure] + MFR

$\Delta G^{o\prime} = -4.4$ kJ/mol

The enzyme, which is a cytoplasmic protein, has been purified from *M. thermoautotrophicum* (strain ΔH) [18], from *M. barkeri* [19,20], from *M. kandleri* [21], and from *A. fulgidus* [21]. The purified enzyme exhibits specific activities between 2500 U/mg and 4500 U/mg and is relatively stable in air. It is composed of only one type of subunit of apparent molecular mass 30 kDa to 41 kDa and lacks a chromophoric prosthetic group. The catalytic mechanism has been determined to be of the ternary complex type [21]. The gene for the enzyme of *M. thermoautotrophicum* has been cloned, sequenced, and expressed in *Escherichia coli* yielding an active enzyme [22]. The N-terminal amino acid sequences of the four enzymes investigated are almost identical [21].

5 N^5,N^{10}-METHENYL-H₄MPT CYCLOHYDROLASE

The cyclohydrolase catalyzes the reversible hydrolysis of N^5, N^{10}-methenyl-H$_4$MPT to N^5-formyl-H$_4$MPT (reaction d).

(d) [structure] + H⁺ ⇌ [structure] + H₂O

$\Delta G^{o\prime} = -4.6$ kJ/mol

The enzyme, which is a cytoplasmic protein, has been purified from *M. thermoautotrophicum* [23], from *M. barkeri* [20], from *M. kandleri* [24], and from *A. fulgidus* [25]. Under optimal conditions the purified enzyme exhibits specific activities of between 2000 U/mg and 8500 U/mg [25]. The enzyme is relatively stable in air. It is composed of only one type of subunit of apparent molecular mass 39 kDa to 41 kDa and lacks a chromophoric prosthetic group. A comparison of N-terminal amino acid sequences reveals a high degree of similarity between the four enzymes isolated [25].

6 N^5,N^{10}-METHYLENE-H_4MPT DEHYDROGENASE

Most methanogens contain two N^5, N^{10}-methylene-H_4MPT dehydrogenases, an H_2-forming dehydrogenase, and a coenzyme F_{420}-reducing dehydrogenase. The methanogens belonging to the order of *Methanomicrobiales* appear to contain only the F_{420}-reducing enzyme [3].

6.1 H_2-forming N^5,N^{10}-methylene-H_4MPT dehydrogenase

This enzyme catalyzes the reversible reduction of N^5, N^{10}-methenyltetrahydromethanopterin with H_2 to N^5, N^{10}-methylene-H_4MPT (reaction e). Since H_2 is either a substrate or a product of the enzyme it has to be classified as a hydrogenase.

(e) [chemical reaction scheme with $\Delta G^{\circ\prime} = -5.5$ kJ/mol]

The novel hydrogenase, which is a cytoplasmic protein, has been purified from *M. thermoautotrophicum* [26], from *M. wolfei* [27] and from *M. kandleri* [28]. The purified enzyme exhibits specific activities (V_{max}) of 2000 U/mg [27]. Its activity is rapidly lost when in contact with dioxygen. The enzyme is composed of only one type of subunit of apparent molecular mass 43 kDa and lacks a chromophoric prosthetic group. The UV/visible spectrum is almost identical to that of albumin. Nickel and iron, components of classical hydrogenases, were looked for but were not found. The only transition metal present in significant amounts appears to be zinc.

The gene encoding for the polypeptide has been cloned and sequenced for the enzyme from *M. thermoautotrophicum* [29] and *M. kandleri* [27]. The two sequences show a high degree of similarity and are devoid of characteristic binding motives for nickel or for iron-sulfur clusters [27].

The purified enzyme does not catalyze the reduction of viologen dyes with either H_2 or N^5, N^{10}-methylene-H_4MPT nor of any other dyes tested. It also does not catalyze an isotopic exchange between H_2 and H^+ in the absence of the electron acceptor N^5, N^{10}-methenyl-H_4MPT [27]. It is not inhibited by carbon monoxide, acetylene, cyanide or azide.

The purified enzyme catalyzes the formation of ^{3}H-labelled N^5, N^{10}-methylene-H$_4$MPT from ^{3}H$_2$ and N^5, N^{10}-methenyl-H$_4$MPT. The specific radioactivity of the labelled product was 25% of that of the ^{3}H$_2$. Evidence is available that ^{3}H was stereospecifically incorporated into the methylene group [27].

The formation of H$_2$, HD, and of D$_2$ was studied in experiments in which either methylene-H$_4$MPT or water were labelled with deuterium. In the case of CD$_2$=H$_4$MPT and H$_2$O, the dihydrogen formed was composed of approximately 50% HD and 50% H$_2$. In the case of CH$_2$=H$_4$MPT and D$_2$O, the dihydrogen generated was composed of approximately 60% HD and 40% D$_2$. The kinetics indicate that H$_2$ and D$_2$ were formed in parallel to HD. The possibility was excluded that the H$_2$ and D$_2$ were formed by consecutive isotope exchange between HD and H$_2$O or between HD and D$_2$O, respectively [30]. The results indicate that the novel hydrogenase catalyzes a hydride transfer but that during hydride transfer there is significant exchange with H$_2$O.

6.2 Coenzyme F_{420}-reducing N^5, N^{10}-methylene-H$_4$MPT dehydrogenase

The dehydrogenase catalyzes the reversible reduction of N^5, N^{10}-methenyl-H$_4$MPT with reduced coenzyme F$_{420}$ to N^5, N^{10}-methylene-H$_4$MPT (reaction f). Reduction proceeds via hydride transfer and is probably stereospecific.

(f) $F_{420}H_2$ + [structure] $\rightleftharpoons$ F_{420} + [structure] + H$^+$

$\Delta G^{\circ\prime} = +5.5$ kJ/mol

The enzyme, which is a cytoplasmic protein, has been purified from *M. thermoautotrophicum* strain Marburg [31] and strain ΔH [32], from *M. barkeri* strain Fusaro [33] and strain MS [34], and from *A. fulgidus* [35]. The purified enzyme has a specific activity of up to 4000 U/mg protein and is relatively stable in air. It is composed of only one type of subunit of apparent molecular mass 31 kDa to 36 kDa and lacks a chromophoric prosthetic group. The catalytic mechanism is of the ternary complex type. The N-terminal amino acid sequence of the dehydrogenases isolated show a high degree of similarity [33].

7 N^5, N^{10}-METHYLENE-H$_4$MPT REDUCTASE

The reductase catalyzes the reduction of N^5, N^{10}-methylene-H$_4$MPT to N^5-methyl-H$_4$MPT with reduced coenzyme F$_{420}$ as electron donor (reaction g).

Reduction is via hydride transfer and probably proceeds stereospecifically.

(g) $F_{420}H_2$ + [structure] ⇌ F_{420} + [structure]

$\Delta G^{o'} = -6.2$ kJ/mol

The enzyme has been purified from *M. thermoautotrophicum* strains ΔH [36] and Marburg [37,38], from *M. barkeri* strain MS [34] and strain Fusaro [39], from *M. kandleri* [40], and from *A. fulgidus* [6]. The purified enzyme exhibits specific activities of between 450 U/mg and 6000 U/mg protein and is relatively stable in air. It is composed of only one type of subunit of apparent molecular mass 35 kDa to 38 kDa. The catalytic mechanism is of the ternary complex type. The N-terminal amino acid sequences of the isolated enzymes show a high degree of similarity [6].

8 N^5-METHYL-H_4MPT:COENZYME M METHYLTRANSFERASE

The methyltransferase catalyzes the formation of methyl-coenzyme M and H_4MPT from N^5-methyl-H_4MPT and coenzyme M (reaction h).

(h) [structure] + H-S-CH₂CH₂-SO₃⁻ ⟶ H_4MPT + CH₃-S-CH₂CH₂-SO₃⁻

$\Delta G^{o'} = -29.7$ kJ/mol

The enzyme, which is an integral membrane protein [41], has been purified from *M. thermoautotrophicum* [42]. The purified enzyme exhibits a specific activity of 3 U/mg and is extremely sensitive towards inactivation by dioxygen. It is composed of seven different subunits of apparent molecular mass of 12.5 kDa, 13.5 kDa, 21 kDa, 23 kDa, 24 kDa, 28 kDa, and 34 kDa. The 23 kDa polypeptide is the corrinoid-binding polypeptide. The complex (156 kDa) contains per mol: 1.6 mol 5-hydroxybenzimidazolyl cobamide, 8 mol non-heme iron, and 8 mol acid-labile sulfur. Evidence is available that the enzyme is only active in the Co(I) form and that Co-methyl-5-hydroxybenzimidazolyl cobamide is an inter-mediate in the catalytic mechanism.

A report has recently appeared, that the methyltransferase from *M. thermoautotrophicum* (strain ΔH) is composed of only three subunits of apparent molecular mass 35 kDa, 33 kDa, and 31 kDa [43]. The preparation contained only 0.2 mol 5-hydroxybenzimidazolyl cobamide and is therefore not considered to be pure.

9 METHYL-COENZYME M REDUCTASE

The reductase catalyzes the irreversible reduction of methyl-coenzyme M with 7-mercaptoheptanoylthreonine phosphate to methane and the heterodisulfide of coenzyme M and 7-mercaptoheptanoylthreonine phosphate (reaction i).

(i)

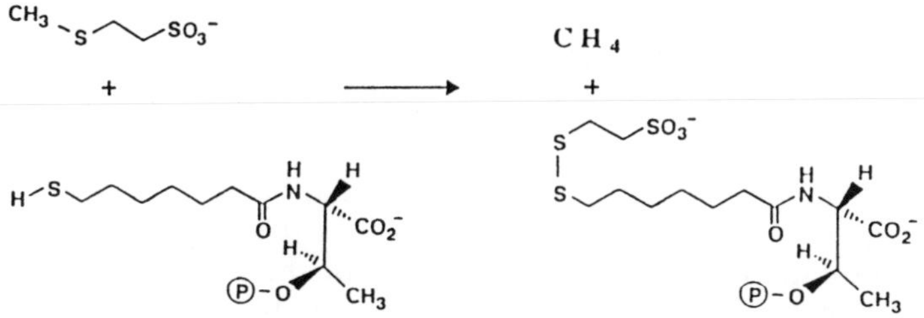

$\Delta G^{o'} = -45$ kJ/mol

The enzyme has been purified from *M. thermoautotrophicum* strain ΔH [44] and strain Marburg [45,46], from *M. barkeri* [47,48], from *Methanosarcina thermophila* [49], from *Methanothrix soehngenii* [50], from *Methanopyrus kandleri* [4], and from *Methanococcus voltae* [51]. Most of the preparations were virtually inactive. Only the purified enzyme from *M. thermoautotrophicum* showed significant activity (≈ 1 U/mg) [45,46]. Recently a preparation with as much as 20 U/mg was described [52]. The soluble enzyme with an apparent molecular mass of 300 kDa is composed of three different subunits in an $\alpha_2\beta_2\gamma_2$ arrangement and contains 2 mol coenzyme F_{430} tightly but not covalently bound. Coenzyme F_{430} is a nickel porphinoid of unique structure [53,54]. EPR studies indicate that in the active enzyme, coenzyme F_{430} is in the reduced Ni(I) state [55]. Ni-Methyl-coenzyme F_{430} has been proposed as intermediate [56,57].

The genes encoding the three subunits of methyl-coenzyme M reductase have been cloned and sequenced from *M. thermoautotrophicum* (strain Marburg), from *Methanococcus vannielii*, from *M. voltae*, from *Methanothermus fervidus*, and from *M. barkeri* [58,59]. In all these organisms the three genes were found to be organized in one transcription unit, which additionally contains two open reading frames, encoding for two small polypeptides of molecular masses below 20 kDa. The two polypeptides, which can be detected in cell extracts [60], appear to be absent in purified active methyl-coenzyme M reductase preparations [46,52]. This indicates that they have no apparent function in the catalytic cycle.

Recently it was found that *M. thermoautotrophicum* harbors two genetically distinct methyl-coenzyme M reductases [61] the expressions of which are differentially regulated by the growth conditions [62].

10 HETERODISULFIDE REDUCTASE

This enzyme catalyzes the reduction of CoM-S-S-HTP with reduced viologen dyes as electron donor (reaction j) and the oxidation of H-S-HTP and H-S-CoM to CoM-S-S-HTP with methylene blue as electron acceptor [63]. The physiological electron donor is not known.

(j)

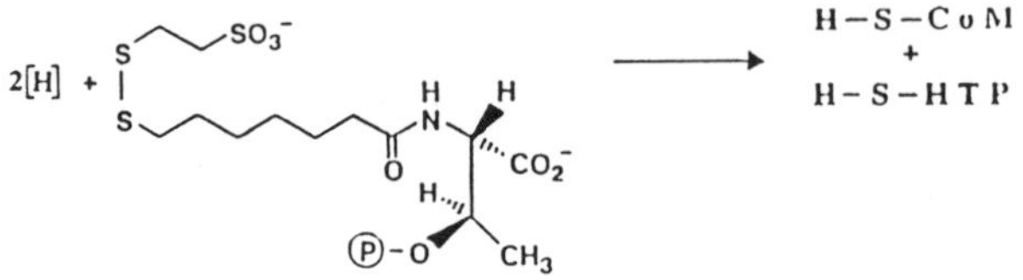

$E^{o'} = -210$ mV

Heterodisulfide reductase is associated with the membrane fraction. In *M. thermoautotrophicum* the enzyme is less hydrophobic than in *M. barkeri*. It has been purified from these two organisms.

Heterodisulfide reductase from *M. thermoautotrophicum* in the native form has an apparent molecular mass of 550 kDa and is composed of three different subunits of apparent molecular masses of 80 kDa, 36 kDa, and 21 kDa, respectively. Per $\alpha_4\beta_4\gamma_4$ polymer the enzyme contains 4 mol FAD and 72 mol non-heme iron and the same amount of acid-labile sulfur [64].

The enzyme from methanol grown *M. barkeri* as isolated is composed of three subunits of apparent molecular mass of 45 kDa, 26 kDa, and 24 kDa, respectively. The preparation also contains cytochrome b in addition to FAD and iron sulfur clusters [unpublished results]. It thus appears that the electron transport chain from H_2 to CoM-S-S-HTP in *M. barkeri* differs from that in *M. thermoautotrophicum* which, like all methanogens growing solely on H_2 and CO_2, lacks cytochromes [65].

ACKNOWLEDGEMENT

This work was supported by grants from the Deutsche Forschungsgemeinschaft and by the Fonds der Chemischen Industrie.

REFERENCES

[1] Wolfe, R.S. (1991) My kind of biology. *Ann. Rev. Microbiol.* **45**, 1–35.
[2] DiMarco, A.A., Bobik, T.A. and Wolfe, R.S. (1990) Unusual coenzymes of methanogenesis. *Ann. Rev. Biochem.* **59**, 355–394.
[3] Schwörer, B. and Thauer, R.K. (1991) Activities of formylmethanofuran dehydrogenase, methylenetetrahydromethanopterin dehydrogenase, methylenetetrahydromethanopterin reductase, and heterodisulfide reductase in methanogenic bacteria. *Arch. Microbiol.* **155**, 459–465.

[4] Rospert, S., Breitung, J., Ma., K., Schwörer, B., Zirngibl, C., Thauer, R.K., Linder, D., Huber, R. and Stetter, K.O. (1991) Methyl-coenzyme M reductase and other enzymes involved in methanogenesis from CO_2 and H_2 in the extreme thermophile *Methanopyrus kandleri*. *Arch. Microbiol.* **156**, 49–55.
[5] Möller-Zinkhan, D., Börner, G. and Thauer, R.K. (1989) Function of methanofuran, tetrahydromethanopterin, and coenzyme F_{420} in *Archaeoglobus fulgidus*. *Arch. Microbiol.* **152**, 362–368.
[6] Schmitz, R.A., Linder, D., Stetter, K.O. and Thauer, R.K. (1991) N^5, N^{10}-methylenetetrahydromethanopterin reductase (coenzyme F_{420}-dependent) and formylmethanofuran dehydrogenase from the hyperthermophile *Archaeoglobus fulgidus*. *Arch. Microbiol.* **156**, 427–434.
[7] Woese, C.R. (1987) Bacterial evolution. *Microbiol. Rev.* **51**, 221–271.
[8] Burggraf, S., Stetter, K.O., Rouviére, P. and Woese, C.R. (1991) *Methanopyrus kandleri*: An archaeal methanogen unrelated to all other known methanogens. *System. Appl. Microbiol.* **14**, 346–351.
[9] Woese, C.R., Achenbach, L., Rouviére, P. and Mandelco, L. (1991) Archaeal phylogeny: Reexamination of the phylogenetic position of *Archaeoglobus fulgidus* in light of certain composition-induced artifacts. *System. Appl. Microbiol.* **14**, 364–371.
[10] Racker, E. (1992) Remembering Ef. *Trends in Biol. Sci.* **17**, 6.
[11] Börner, G., Karrasch, M. and Thauer, R.K. (1989) Formylmethanofuran dehydrogenase activity in cell extracts of *Methanobacterium thermoautotrophicum* and of *Methanosarcina barkeri*. *FEBS Lett.* **244**, 21–25.
[12] Börner, G., Karrasch, M. and Thauer, R.K. (1991) Molybdopterin adenine dinucleotide and molybdopterin hypoxanthine dinucleotide in formylmethanofuran dehydrogenase from *Methanobacterium thermoautotrophicum* (Marburg). *FEBS Lett.* **290**, 31–34.
[13] Karrasch, M., Börner, G., Enßle, M. and Thauer, R.K. (1990) The molybdoenzyme formylmethanofuran dehydrogenase from *Methanosarcina barkeri* contains a pterin cofactor. *Eur. J. Biochem.* **194**, 367–372.
[14] Schmitz, R.A., Albracht, S.P.J. and Thauer, R.K. (1992) A molybdenum and a tungsten isoenzyme of formylmethanofuran dehydrogenase in the thermophilic archaeon *Methanobacterium wolfei*. *Eur. J. Biochem.* **209**, 1013–1018.
[15] Schmitz, R.A., Albracht, S.P.J. and Thauer, R.K. (1992) Properties of the tungsten substituted molybdenum formylmethanofuran dehydrogenase from *Methanobacterium wolfei*. *FEBS Lett.*, in press.
[16] Karrasch, M., Börner, G. and Thauer, R.K. (1990) The molybdenum cofactor of formylmethanofuran dehydrogenase from *Methanosarcina barkeri* is a molybdopterin guanine dinucleotide. *FEBS Lett.* **274**, 48–52.
[17] Schmitz, R.A., Richter, M., Linder, D. and Thauer, R.K. (1992) A tungsten-containing active formylmethanofuran dehydrogenase in the thermophilic archaeon *Methanobacterium wolfei*. *Eur. J. Biochem.* **207**, 559–565.
[18] Donnelly, M.I. and Wolfe, R.S. (1986) The role of formylmethanofuran:tetrahydromethanopterin formyltransferase in methanogenesis from carbon dioxide. *J. Biol. Chem.* **261**, 16653–16659.
[19] Breitung, J. and Thauer, R.K. (1990) Formylmethanofuran:tetrahydromethanopterin formyltransferase from *Methanosarcina barkeri*. Identification of N^5-formyltetrahydromethanopterin as the product. *FEBS Lett.* **275**, 226–230.
[20] Keltjens, J.T., Brugman, A.J.A.M., Kesseleer, J.M.A., te Brömmelstroet, B.W.J., van der Drift, C. and Vogels, G.D. (1992) 5-Formyl-5,6,7,8-tetrahydromethanopterin is the intermediate in the process of methanogenesis in *Methanosarcina barkeri*. *BioFactors* **4**, 249–255.
[21] Breitung, J., Börner, G., Scholz, S., Linder, D., Stetter, K.O. and Thauer, R.K. (1992) Formylmethanofuran:tetrahydromethanopterin formyltransferase from

the extreme thermophile *Methanopyrus kandleri*: salt dependence, kinetic properties, and catalytic mechanism. *Eur. J. Biochem.*, submitted.

[22] DiMarco, A.A., Sment, K.A., Konisky, J. and Wolfe, R.S. (1990) The formylmethanofuran:tetrahydromethanopterin formyltransferase from *Methanobacterium thermoautotrophicum* ΔH. Nucleotide sequence and functional expression of the cloned gene. *J. Biol. Chem.* **265**, 472–476.

[23] DiMarco, A.A., Donnelly, M.I. and Wolfe, R.S. (1986) Purification and properties of the 5,10-methenyltetrahydromethanopterin cyclohydrolase from *Methanobacterium thermoautotrophicum*. *J. Bacteriol.* **168**, 1372–1377.

[24] Breitung, J., Schmitz, R.A., Stetter, K.O. and Thauer, R.K. (1991) N^5, N^{10}-methenyltetrahydromethanopterin cyclohydrolase from the extreme thermophile *Methanopyrus kandleri*: increase of catalytic efficiency (k_{cat}/K_M) and thermostability in the presence of salts. *Arch. Microbiol.* **156**, 517–524.

[25] Klein, A., Breitung, J., Linder, D., Stetter, K.O. and Thauer, R.K. (1993) N^5, N^{10}-Methenyltetrahydromethanopterin cyclohydrolase from the extremely thermophilic sulfate reducing *Archaeoglobus fulgidus*: Comparison of its properties with those of the cyclohydrolase from the extremely thermophilic *Methanopyrus kandleri*. *Arch. Microbiol.*, in press.

[26] Zirngibl, C., Hedderich, R. and Thauer, R.K. (1990) N^5, N^{10}-methylenetetrahydromethanopterin dehydrogenase from *Methanobacterium thermoautotrophicum* has hydrogenase activity. *FEBS Lett.* **261**, 112–116.

[27] Zirngibl, C., van Dongen, W., Schwörer, B., von Bünau, R., Richter, M., Klein, A. and Thauer, R.K. (1992) H_2-forming methylenetetrahydromethanopterin dehydrogenase, a novel type of hydrogenase without iron-sulfur clusters in methanogenic archaea. *Eur. J. Biochem.* **208**, 511–520.

[28] Ma, K., Zirngibl, C., Linder, D., Stetter, K.O. and Thauer, R.K. (1991) N^5, N^{10}-methylenetetrahydromethanopterin dehydrogenase (H_2-forming) from the extreme thermophile *Methanopyrus kandleri*. *Arch. Microbiol.* **156**, 43–48.

[29] Von Bünau, R., Zirngibl, C., Thauer, R.K. and Klein, A. (1991) Hydrogen-forming and coenzyme-F_{420}-reducing methylene tetrahydromethanopterin dehydrogenase are genetically distinct enzymes in *Methanobacterium thermoautotrophicum* Marburg). *Eur. J. Biochem.* **202**, 1205–1208.

[30] Schwörer, B., Fernandez, V.M., Zirngibl, C. and Thauer, R.K. (1993) H_2-forming N^5, N^{10}-methylenetetrahydromethanopterin dehydrogenase from *Methanobacterium thermoautotrophicum*: Studies of the catalytic mechanism with hydrogen isotopes. *Eur. J. Biochem.*, submitted.

[31] Mukhopadhyay, B. and Daniels, L. (1989) Aerobic purification of N^5, N^{10}-methylenetetrahydromethanopterin dehydrogenase, separated from N^5, N^{10}-methenyltetrahydromethanopterin cyclohydrolase, from *Methanobacterium thermoautotrophicum* strain Marburg. *Can. J. Microbiol.* **35**, 499–507.

[32] Te Brömmelstroet, B.W., Hensgens, C.M.H., Keltjens, J.T., van der Drift, C. and Vogels, G.D. (1991) Purification and characterization of coenzyme F_{420}-dependent 5,10-methylenetetrahydromethanopterin dehydrogenase from *Methanobacterium thermoautotrophicum* strain ΔH. *Biochim. Biophys. Acta* **1073**, 77–84.

[33] Enßle, M., Zirngibl, C., Linder, D. and Thauer, R.K. (1991) Coenzyme F_{420} dependent N^5, N^{10}- methylenetetrahydromethanopterin dehydrogenase in methanol grown *Methanosarcina barkeri*. *Arch. Microbiol.* **155**, 483–490.

[34] Te Brömmelstroet, B.W.J., Geerts, W.J., Keltjens, J.T., van der Drift, C. and Vogels, G.D. (1991) Purification and properties of 5,10-methylenetetrahydromethanopterin dehydrogenase and 5,10-methylenetetrahydromethanopterin reductase, two coenzyme F_{420}-dependent enzymes, from *Methanosarcina barkeri*. *Biochim. Biophys. Acta* **1079**, 293–302.

[35] Schwörer, B., Breitung, J., Klein, A.J., Stetter, K.O. and Thauer, R.K. (1993) Formylmethanofuran:tetrahydromethanopterin formyltransferase and N^5, N^{10}-methylene tetrahydro methanopterin dehydrogenase from the sulfate-reducing *Archaeoglobus fulgidus*: similarities with the enzymes from methanogenic archaea. *Arch. Microbiol*, in press.

[36] Te Brömmelstroet, B.W., Hensgens, C.M.H., Keltjens, J.T., van der Drift, C. and Vogels, G.D. (1990) Purification and properties of 5,10-methylenetetrahydromethanopterin reductase, a coenzyme F_{420}-dependent enzyme, from *Methanobacterium thermoautotrophicum* strain ΔH. *J. Biol. Chem.* **265**, 1852–1857.

[37] Ma, K. and Thauer, R.K. (1990) Purification and properties of N^5, N^{10}-methylenetetrahydromethanopterin reductase from *Methanobacterium thermoautotrophicum* (strain Marburg). *Eur. J. Biochem.* **191**, 187–193.

[38] Ma, K. and Thauer, R.K. (1990) Single step purification of methylenetetrahydromethanopterin reductase from *Methanobacterium thermoautotrophicum* by specific binding to Blue Sepharose CL-6B. *FEBS Lett.* **268**, 59–62.

[39] Ma, K. and Thauer, R.K. (1990) N^5, N^{10}-Methylenetetrahydromethanopterin reductase from *Methanosarcina barkeri*. *FEMS Microbiol. Lett.* **70**, 119–124.

[40] Ma, K., Linder, D., Stetter, K.O. and Thauer, R.K. (1991) Purification and properties of N^5, N^{10}-methylenetetrahydromethanopterin reductase (coenzyme F_{420}-dependent) from the extreme thermophile *Methanopyrus kandleri*. *Arch. Microbiol.* **155**, 593–600.

[41] Fischer, R., Gärtner, P., Yeliseev, A. and Thauer, R.K. (1992) N^5-Methyltetrahydromethanopterin:coenzyme M methyltransferase in methanogenic archaebacteria is a membrane protein. *Arch. Microbiol.* **158**, 208–217.

[42] Gärtner, P., Ecker, A., Fischer, R., Fuchs, G. and Thauer, R.K. (1993) Purification and properties of N^5-methyltetrahydromethanopterin:coenzyme M methyltransferase from *Methanobacterium thermoautotrophicum*. *Eur. J. Biochem.*, submitted.

[43] Kengen, S.W.M., Daas, P.J.H., Duits, E.F.G., Keltjens, J.T., van der Drift, C. and Vogels, G.D. (1992) Isolation of a 5-hydroxybenzimidazolyl cobamide-containing enzyme involved in the methyltetrahydromethanopterin:coenzyme M methyltransferase reaction in *Methanobacterium thermoautotrophicum*. *Biochim. Biophys. Acta* **1118**, 249–260.

[44] Ellefson, W.L. and Wolfe, R.S. (1981) Component C of the methylreductase system of *Methanobacterium*. *J. Biol. Chem.* **256**, 4259–4262.

[45] Ellermann J., Hedderich, R., Böcher, R. and Thauer, R.K. (1988) The final step in methane formation. Investigations with highly purified methyl-CoM reductase (component C) from *Methanobacterium thermoautotrophicum* (strain Marburg). *Eur. J. Biochem.* **172**, 669–677.

[46] Ellermann, J., Rospert, S., Thauer, R.K., Bokranz, M., Klein, A., Voges, M. and Berkessel, A. (1989) Methyl-coenzyme-M reductase from *Methanobacterium thermoautotrophicum* (strain Marburg). Purity, activity and novel inhibitors. *Eur. J. Biochem.* **184**, 63–68.

[47] Moura, I., Moura, J.J.G., Santos, H., Xavier, A.V., Burch, G., Peck Jr, H.D. and LeGall, J. (1983) Proteins containing the factor F_{430} from *Methanosarcina barkeri* and *Methanobacterium thermoautotrophicum*. Isolation and properties. *Biochim. Biophys. Acta* **742**, 84–90.

[48] Hoppert, M. and Mayer, F. (1990) Electron microscopy of native and artificial methylreductase high-molecular-weight complexes in strain Göl and *Methanococcus voltae*. *FEBS Lett.* **267**, 33–37.

[49] Jablonski, P.E. and Ferry, J.G. (1991) Purification and properties of methyl coenzyme M methylreductase from acetate-grown *Methanosarcina thermophila*. *J. Bacteriol.* **173**, 2481–2487.

[50] Jetten, M.S.M., Stams, A.J.M. and Zehnder, A.J.B. (1990) Purification and some properties of the methyl-CoM reductase of *Methanothrix soehngenii*. *FEMS Microbiol. Lett.* **66**, 183–186.

[51] Konheiser, U., Pasti, G., Bollschweiler, C. and Klein, A. (1984) Physical mapping of genes coding for two subunits of methyl-CoM reductase component C of *Methanococcus voltae*. *Mol. Gen. Genet.* **198**, 146–152.

[52] Rospert, S., Böcher, R., Albracht, S.P.J. and Thauer, R.K. (1991) Methyl-coenzyme M reductase preparations with high specific activity from H_2-preincubated cells of *Methanobacterium thermoautotrophicum*. *FEBS Lett.* **291**, 371–375.

[53] Färber, G., Keller, W., Kratky, C., Jaun, B., Pfaltz, A., Spinner, C., Kobelt, A. and Eschenmoser, A. (1991) Coenzyme F_{430} from methanogenic bacteria: complete assignment of configuration based on an X-ray analysis of 12,13-Diepi-F_{430} pentamethyl ester and on NMR spectroscopy. *Helv. Chim. Acta* **74**, 697–716.

[54] Friedmann, H.C., Klein, A. and Thauer, R.K. (1991) Biochemistry of coenzyme F_{430}, a nickel porphinoid involved in methanogenesis. In: Biosynthesis of Tetrapyrroles (Jordan, P.M., Ed.), pp. 139–154, Elsevier Science Publishers, Amsterdam.

[55] Rospert, S., Voges, M., Berkessel, A., Albracht, S.P.J. and Thauer, R.K. (1992) Substrate analogue induced changes in the nickel EPR spectrum of active methyl-coenzyme M reductase from *Methanobacterium thermoautotrophicum*. *Eur. J. Biochem.*, **210**, 101–107.

[56] Lin, S.-K. and Jaun, B. (1991) Coenzyme F_{430} from methanogenic bacteria: Detection of a paramagnetic methylnickel(II) derivative of the pentamethyl ester by ^{2}H-NMR spectroscopy. *Helv. Chim. Acta* **74**, 1725–1738.

[57] Berkessel, A. (1991) Methyl-coenzyme M reductase: model studies on pentadentate nickel complexes and a hypothetical mechanism. *Bioorg. Chem.* **19**, 101–115.

[58] Allmansberger, R., Bokranz, M., Kröckel, L., Schallenberg, J. and Klein, A. (1989) Conserved gene structures and expression signals in methanogenic archaebacteria. *Can. J. Microbiol.* **35**, 52–57.

[59] Weil, C.F., Sherf, B.A. and Reeve, J.N. (1989) A comparison of the methyl reductase genes and gene products. *Can. J. Microbiol.* **35**, 101–108.

[60] Sherf, B.A. and Reeve, J.N. (1990) Identification of the *mcrD* gene product and its association with component C of methyl coenzyme M reductase in *Methanococcus vannielii*. *J. Bacteriol.* **172**, 1828–1833.

[61] Rospert, S., Linder, D., Ellermann, J. and Thauer, R.K. (1990) Two genetically distinct methyl-coenzyme M reductases in *Methanobacterium thermoautotrophicum* strain Marburg and ΔH. *Eur. J. Biochem.* **194**, 871–877.

[62] Bonacker, L.G., Baudner, S. and Thauer, R.K. (1992) Differential expression of the two methyl-coenzyme M reductases in *Methanobacterium thermoautotrophicum* as determined immunochemically via isoenzyme-specific antisera. *Eur. J. Biochem.* **206**, 87–92.

[63] Hedderich, R., Berkessel, A. and Thauer, R.K. (1989) Catalytic properties of the heterodisulfide reductase involved in the final step of methanogenesis. *FEBS Lett.* **255**, 67–71.

[64] Hedderich, R., Berkessel, A. and Thauer, R.K. (1990) Purification and properties of heterodisulfide reductase from *Methanobacterium thermoautotrophicum* (strain Marburg). *Eur. J. Biochem.* **193**, 255–261.

[65] Köhn, W. and Gottschalk, G. (1983) Characterization of the cytochromes occurring in *Methanosarcina* species. *Eur. J. Biochem.* **135**, 89–94.

12

Enzymology of the Methanogenic Fermentation of Acetate by *Methanosarcina thermophila*

Birgit E. Alber, Andrew P. Clements, Peter E. Jablonski, Matthew T. Latimer, and James G. Ferry

Department of Anaerobic Microbiology, Virginia Polytechnic Institute and State University, Blacksburg, VA 24061, USA

1 SUMMARY

The initial step in the pathway of acetate fermentation to methane and carbon dioxide by *Methanosarcina thermophila* involves an activation to acetyl-CoA catalyzed by acetate kinase and phosphotransacetylase. A CO dehydrogenase (CODH) complex, comprised of a CO-oxidizing Ni/Fe-S enzyme component and a Co/Fe-S enzyme component, cleaves the C-C and C-S bonds of acetyl-CoA. The Ni/Fe-S component is the proposed site for the cleavage reactions and oxidation of the carbonyl group to CO_2; a ferredoxin is the electron acceptor. Carbonic anhydrase catalyzes the hydration of CO_2 to carbonic acid. The Co/Fe-S component is thought to accept the methyl group from the Ni/Fe-S component. The methyl group is ultimately transferred to HS-CoM and CH_3-S-CoM is reductively demethylated to methane catalyzed by the CH_3-S-CoM methylreductase.

2 INTRODUCTION

Approximately two-thirds of the methane produced in anaerobic freshwater environments originates from the methyl group of acetate, and about one-third from the reduction of CO_2 with electrons derived from the oxidation of H_2 or formate. The pathways for conversion of the methyl group of acetate to

---

Microbial Growth on C_1 Compounds
© Intercept Ltd, PO Box 716, Andover, Hampshire SP10 1YG, UK

methane and reduction of CO_2 to methane are fundamentally different but merge at the final step in which CH_3-S-CoM is reductively demethylated to methane.

Most acetate-utilizing respiratory anaerobes in the *Bacteria* cleave acetyl-CoA and oxidize the methyl and carbonyl groups completely to CO_2 [1]. However, the methanogenic *Archaea* ferment acetate by a pathway in which the molecule is cleaved and the methyl group reduced to methane with electrons derived from oxidation of the carbonyl group to CO_2 (reaction 1). Acetate-utilizing anaerobes from

$$CH_3COO^- + H^+ \longrightarrow CH_4 + CO_2 \quad (1)$$

$$\Delta G^{\circ\prime} = -36 \text{ kJ/mol } [2]$$

both the *Archaea* and *Bacteria* domains contain CODH which catalyzes the acetyl-CoA cleavage reaction [1]. Although most of the methane produced in nature derives from the methyl group of acetate, only species in the genera *Methanosarcina* and *Methanothrix* can utilize acetate for growth.

3 ACETATE KINASE AND PHOSPHOTRANSACETYLASE

Prior to cleavage by the CODH complex, *Methanosarcina* spp activate acetate to acetyl-CoA [3,4] by the combined activities of acetate kinase (reaction 2) and phosphotransacetylase (reaction 3). These enzymes are absent in acetate-grown

$$CH_3COO^- + ATP \longrightarrow CH_3CO_2PO_3^{2-} + ADP \quad (2)$$

$$CH_3CO_2PO_3^{2-} + CoA \longrightarrow CH_3COSCoA + Pi \quad (3)$$

$$CH_3COO^- + CoA + ATP \longrightarrow CH_3COSCoA + AMP + PPi \quad (4)$$

Methanothrix soehngenii, but this organism contains high levels of acetyl-CoA synthetase [5] which catalyzes the activation to acetyl-CoA (reaction 4). The energy of one high-energy phosphate bond is apparently lost for each acetate metabolized and represents a considerable investment when the small amount of energy available for ATP synthesis (reaction 1) is considered. The acetate kinase from *M. thermophila* is air stable and is purified from the soluble fraction as an α_2 homodimer with a subunit M_r of 53 kDa [6]. The K_m for ATP is 2.8 mM and the K_m for acetate is 22 mM which reflects the relatively high K_s for acetate uptake [7]. The air-stable monomeric ($M_r = 42$ kDa) phosphotransacetylase is also purified from the soluble fraction of *M. thermophila* [8]. The amounts of both enzymes are several-fold higher in acetate-grown cells when compared to methanol-grown cells [7,8]; thus, it follows that the genes encoding both enzymes appear to be organized in an operon [9].

4 CO DEHYDROGENASE

The five-subunit CODH enzyme complex from *M. thermophila* contains two enzyme components: a 200 kDa CO-oxidizing Ni/Fe-S component which contains 89 kDa and 19 kDa subunits, and a 100 kDa Co/Fe-S component which contains 60 kDa and 58 kDa subunits [10]. The fifth subunit (71 kDa) has not been characterized. The CO-oxidizing Ni/Fe-S component has several properties in common with the CODH (acetyl-CoA synthase) from the homoacetogenic organism *Clostridium thermoaceticum*. Both enzymes, when reduced with CO, exhibit a nearly identical EPR spectrum of a spin-coupled Ni-Fe-C center which is the proposed site for synthesis or cleavage of the acetyl moiety of acetyl-CoA [11–13]. In addition, the acetyl-CoA synthase and the Ni/Fe-S component associate with a corrinoid-containing iron-sulfur protein and donate electrons to the cobalt atom reducing it to the strongly nucleophilic Co^{1+} which can then accept a methyl group [10]. The Co/Fe-S component from the *M. thermophila* CODH enzyme complex contains cobalt in factor III (Coα-[α-(5-hydroxybenzimidazolyl)]-cobamide) which is in the base-off configuration [10,14]. Redox titration of the $Co^{2+/1+}$ couple reveals a midpoint potential of approximately -500 mV [14] similar to that reported for the analogous *C. thermoaceticum* corrinoid/Fe-S protein which is also in the base-off configuration [15]. The base-off configuration of corrinoids changes the potential of the $Co^{2+/1+}$ couple to a less negative value which is within range of physiological electron donors [10,15]. Both the Ni/Fe-S and Co/Fe-S components of the CODH complex from *M. thermophila* contain Fe-S centers [14] with properties similar to those reported for the analogous *C. thermoaceticum* proteins [16,17]. Two of the three Fe-S centers in the Ni/Fe-S component have EPR spectra typical of bacterial-like 4Fe-4S centers; however, the third Fe-S center has an atypical spectrum. It is not known which, if any, of the Fe atoms from the Fe-S centers participate in formation of the Ni-Fe center. The Co/Fe-S component contains a 4Fe-4S center with a midpoint potential similar to the 4Fe-4S center present in the analogous *C. thermoaceticum* corrinoid/Fe-S protein (P.J. Jablonski, W.-P. Lu, S.W. Ragsdale and J.G. Ferry unpublished).

In summary, the composition and properties of component enzymes in the CODH complex from *M. thermophila* are consistent with a proposed acetyl-CoA cleavage mechanism which is analogous to a reversal of the mechanism proposed for acetyl-CoA synthesis in *C. thermoaceticum* [12,13]. In the proposed mechanism, the Ni/Fe-S component cleaves the C-C and C-S bonds of acetyl-CoA at a Ni-Fe site [10] and transfers the methyl group to the Co^{1+} atom of the Co/Fe-S component. It is further proposed that the Ni-Fe center binds the carbonyl group and oxidizes it to CO_2. The proposed function is supported by the ability of the Ni/Fe-S component to oxidize CO and reduce a ferredoxin purified from *M. thermophila* [10].

The CODH purified from *M. barkeri* has properties similar to the CO-oxidizing Ni/Fe-S component of the *M. thermophila* complex [18,19] and can be purified in a complex which also contains a corrinoid protein [20]. The complex from *M. barkeri* catalyzes cleavage of acetyl-CoA and transfer of the methyl group to tetrahydrosarcinapterin [20]; however, the properties and function of the corrinoid protein were not investigated. The *M. barkeri* CODH contains Ni and Fe. No EPR signals attributable to Ni have been reported; however, 4Fe-4S centers are indicated by core extrusion experiments and low temperature EPR spectroscopy [19,21]. Similar to the CO-oxidizing Ni/Fe-S component from *M. thermophila*, a second low temperature EPR signal is obtained from the reduced *M. barkeri* CODH which is atypical of known Fe-S centers. The *M. soehngenii* CODH has properties similar to the enzymes from *M. barkeri* except that it has not been purified in association with a corrinoid-containing protein [22,23].

5 METHYL-COENZYME M METHYLREDUCTASE

After cleavage of acetyl-CoA, the methyl group is ultimately transferred to HS-CoM (2-mercaptoethanesulfonic acid) [24] which implies a requirement for one or more methyltransferases. Several corrinoid-containing proteins, with the potential to function as methyltransferases, have been identified in *M. barkeri* [25,26].

The reductive demethylation of CH_3-S-CoM to CH_4 catalyzed by CH_3-S-CoM methylreductase is the final step in the various pathways for the utilization of all methanogenic substrates. The electron donor to the methylreductase is HS-HTP (7-mercaptoheptanoylthreonine phosphate) [24] and the heterodisulfide CoM-S-S-HTP is a product of the reaction (reaction 5).

$$CH_3-S-CoM + HS-HTP \longrightarrow CH_4 + CoM-S-S-HTP \qquad (5)$$

$$\Delta G^{\circ\prime} = -45 \text{ kJ/mol [29]}$$

In the pathway for methanogenesis from acetate, the mixed disulfide is reduced to the corresponding sulfhydryl forms of the cofactors with electrons originating from oxidation of the carbonyl group of acetyl-CoA [27,28].

The native methylreductase purified from *M. thermophila* contains one mol of F_{430} and has a subunit composition of $\alpha_1\beta_1\gamma_1$ with M_rs of 69, 42, and 33 kDa [30]. The enzymes from *Methanosarcina mazei* and *M. soehngenii* have similar subunit sizes, but in a $\alpha_2\beta_2\gamma_2$ configuration [31,32]. The enzyme from *M. thermophila* requires a reductive reactivation which can be accomplished with a ferredoxin purified from the same organism [33]; however, the physiological significance of this reactivation is unknown. Immunogold labelling of several acetate-grown *Methanosarcina* species and *M. soehngenii* indicates that the methylreductase of these acetotrophic organisms is primarily located in the cytoplasm [32]. However, the cells were grown with abundant nickel in the growth medium, conditions which may have influenced the amount of cytoplasmic methylreductase relative to membrane-associated enzyme [34].

6 FERREDOXIN

Ferredoxin is required for methanogenesis from acetate in extracts of *M. barkeri* [35] and is a direct acceptor of electrons from the Ni/Fe-S component of the CODH enzyme complex of *M. thermophila* [10], results which strongly implicate an involvement in electron transport. In addition, a ferredoxin-dependent evolution of CO_2 and H_2 from acetyl-CoA has been demonstrated in extracts of *M. barkeri* [35] and a CO-oxidizing: H_2-evolving system in *M. thermophila* (reaction 6) has been reconstituted with the CODH

$$CO + H_2O \longrightarrow CO_2 + H_2 \qquad (6)$$
$$\Delta G^{\circ\prime} = -20 \text{ kJ/mol [36]}$$

complex, ferredoxin and purified membranes which contain a hydrogenase linked to cytochrome *b* [37].

Ferredoxins from three different methane-producing organisms have been purified and characterized [33,38,39]. EPR spectroscopy of the ferredoxin purified from *M. thermophila* indicates two 4Fe-4S centers with similar midpoint potentials of -407 mV (A.P. Clements, W.-P. Lu, S.W. Ragsdale, and J.G. Ferry, unpublished). The ferredoxin gene has a deduced amino acid sequence which contains eight cysteines in a spacing highly characteristic of 2[4Fe-4S] ferredoxins from the *Bacteria* [40]. Northern analyses indicate that the gene is expressed in *M. thermophila* cells grown on acetate, trimethylamine or methanol, a result which suggests that the ferredoxin is required for growth on all three substrates [40].

7 CARBONIC ANHYDRASE

Growth of *M. barkeri* and *M. thermophila* on acetate induces carbonic anhydrase activity (reaction 7) [41,42] which suggests an involvement of this enzyme in the

$$CO_2 + H_2O \longrightarrow HCO_3^- + H^+ \qquad (7)$$

pathway. It is proposed that CO_2, a product of the oxidation of the carbonyl group of acetyl-CoA, is converted to carbonic acid which is stoichiometrically required in an antiport mechanism for transport of the acetate anion into the cell [41,42]. The enzyme, purified from *M. thermophila*, is a homodimer with a subunit molecular mass of 40 kDa and contains an N-terminal sequence with significant identity to the zinc binding site of the human carbonic anhydrases (B.E. Alber and J.G. Ferry, unpublished).

ACKNOWLEDGEMENTS

Work in the authors' laboratory was supported by: the Office of Naval Research (ONR N0014-91-J-1900); the Department of Energy, Basic Energy Sciences (DE-FG05-85ER13730); and the National Institutes of Health (1 R01 GM44661-01A1).

REFERENCES

[1] Thauer, R.K., Moller-Zinkhan, D. and Spormann, A.M. (1989) Biochemistry of acetate catabolism in anaerobic chemotrophic bacteria. *Ann. Rev. Microbiol.* **43**, 43–67.
[2] Vogels, G.D., Keltjens, J.T. and van der Drift, C. (1988) Biochemistry of methane production. In: Biology of Anaerobic Microorganisms (Zehnder, A.J.B., Ed.), pp. 707–770, Wiley, New York.
[3] Fischer, R. and Thauer, R.K. (1988) Methane formation from acetyl phosphate in cell extracts of *Methanosarcina barkeri*. Dependence of the reaction on coenzyme A. *FEBS Lett.* **228**, 249–253.
[4] Grahame, D.A. and Stadtman, T.C. (1987) *In vitro* methane and methyl coenzyme M formation from acetate; evidence that acetyl-CoA is the required intermediate activated form of acetate. *Biochem. Biophys. Res. Commun.* **147**, 254–258.
[5] Jetten, M.S.M., Stams, A.J.M. and Zehnder, A.J.B. (1989) Isolation and characterization of acetyl-coenzyme A synthetase from *Methanothrix soehngenii*. *J. Bacteriol.* **171**, 5430–5435.
[6] Aceti, D.J. and Ferry, J.G. (1988) Purification and characterization of acetate kinase from acetate-grown *Methanosarcina thermophila*. *J. Biol. Chem.* **263**, 15444–15448.
[7] Jetten, M.S.M., Stams, A.J.M., and Zehnder, A.J.B. (1990) Acetate threshold values and acetate activating enzymes in methanogenic bacteria. *FEMS Microb. Ecol.* **73**, 339–344.
[8] Lundie, L.L. and Ferry, J.G. (1989) Activation of acetate by *Methanosarcina thermophila*. Purification and characterization of phosphotransacetylase. *J. Biol. Chem.* **264**, 18392–18396.
[9] Latimer, M.T. and Ferry, J.G. (1991) Cloning of the genes encoding the acetate activating enzymes in *Methanosarcina thermophila*, Abstr. I-113, p. 209, Abstr. 91st Ann. Meet. Am. Soc. Microbiol.
[10] Abbanat, D.R. and Ferry, J.G. (1991) Resolution of component proteins in an enzyme complex from *Methanosarcina thermophila* catalyzing the synthesis or cleavage of acetyl-CoA. *Proc. Natl. Acad. Sci. USA*. **88**, 3272–3276.
[11] Terlesky, K.C., Barber, M.J., Aceti, D.J. and Ferry, J.G. (1987) EPR properties of the Ni-Fe-C center in an enzyme complex with carbon monoxide dehydrogenase activity from acetate-grown *Methanosarcina thermophila*. Evidence that acetyl-CoA is a physiological substrate. *J. Biol. Chem.* **262**, 15392–15395.
[12] Lu, W.-P., Harder, S.R. and Ragsdale, S.W. (1990) Controlled potential enzymology of methyl transfer reactions involved in acetyl-CoA synthesis by CO dehydrogenase and the corrinoid/iron-sulfur protein from *Clostridium thermoaceticum*. *J. Biol. Chem.* **265**, 3124–3133.
[13] Lu, W.-P. and Ragsdale, S.W. (1991) Reductive activation of the coenzyme A/acetyl-CoA isotopic exchange reaction catalyzed by carbon monoxide dehydrogenase from *Clostridium thermoaceticum* and its inhibition by nitrous oxide and carbon monoxide. *J. Biol. Chem.* **266**, 3554–3564.

[14] Jablonski, P.E., Lu, W.-P., Ragsdale, S.W. and Ferry, J.G. (1991) EPR spectroelectrochemical studies of the carbon monoxide dehydrogenase enzyme complex, nickel/iron-sulfur protein, and corrinoid/iron-sulfur protein of acetate-grown *Methanosarcina thermophila*, abstr. K-136, p. 237. Abstr. 91st Ann. Meet. Am. Soc. Microbiol.
[15] Harder, S.R., Lu, W.-P., Feinberg, B.A. and Ragsdale, S.W. (1989) Spectroelectrochemical studies of the corrinoid iron-sulfur protein involved in acetyl coenzyme-A synthesis by *Clostridium thermoaceticum*. Biochemistry. 28, 9080–9087.
[16] Lindahl, P.A., Ragsdale, S.W. and Münck, E. (1990) Mössbauer study of CO dehydrogenase from *Clostridium thermoaceticum*. J. Biol. Chem. 265, 3880–3888.
[17] Lindahl, P.A., Münck, E. and Ragsdale, S.W. (1990) CO dehydrogenase from *Clostridium thermoaceticum*. EPR and electrochemical studies in CO_2 and argon atmospheres. J. Biol. Chem. 265, 3873–3879.
[18] Grahame, D.A. and Stadtman, T.C. (1987) Carbon monoxide dehydrogenase from *Methanosarcina barkeri*. Disaggregation, purification, and physiochemical properties of the enzyme. J. Biol. Chem. 262, 3706–3712.
[19] Krzycki, J.A., Mortenson, L.E. and Prince, R.C. (1989) Paramagnetic centers of carbon monoxide dehydrogenase from aceticlastic *Methanosarcina barkeri*. J. Biol. Chem. 264, 7217–7221.
[20] Grahame, D.A. (1991) Catalysis of acetyl-CoA cleavage and tetrahydrosarcinapterin methylation by a carbon monoxide dehydrogenase-corrinoid enzyme complex. J. Biol. Chem. 266, 22227–22233.
[21] Krzycki, J.A. and Prince, R.C. (1990) EPR observation of carbon monoxide dehydrogenase, methylreductase and corrinoid in intact *Methanosarcina barkeri* during methanogenesis from acetate. Biochim. Biophys. Acta. 1015, 53–60.
[22] Jetten, M.S.M., Hagen, W.R., Pierik, A.J., Stams, A.J.M. and Zehnder, A.J.B. (1991) Paramagnetic centers and acetyl-coenzyme A/CO exchange activity of carbon monoxide dehydrogenase from *Methanothrix soehngenii*. Eur. J. Biochem. 195, 385–391.
[23] Jetten, M.S.M., Pierik, A.J. and Hagen, W.R. (1991) EPR characterization of a high-spin system in carbon monoxide dehydrogenase from *Methanothrix soehngenii*. Eur. J. Biochem. 202, 1291–1297.
[24] DiMarco, A.A., Bobik, T.A. and Wolfe, R.S. (1990) Unusual coenzymes of methanogenesis. Ann. Rev. Biochem. 59, 355–394.
[25] Cao, X. and Krzycki, J.A. (1991) Acetate-dependent methylation of two corrinoid proteins in extracts of *Methanosarcina barkeri*. J. Bacteriol. 173, 5439–5448.
[26] Grahame, D.A. (1989) Different isozymes of methylcobalamin-2-mercaptoethanesulfonate methyltransferase predominate in methanol-grown versus acetate-grown *Methanosarcina barkeri*. J. Biol. Chem. 264, 12890–12894.
[27] Hedderich, R., Berkessel, A. and Thauer, R.K. (1989) Catalytic properties of the heterodisulfide reductase involved in the final step of methanogenesis. FEBS Lett. 255, 67–71.
[28] Schwörer, B. and Thauer, R.K. (1991) Activities of formylmethanofuran dehydrogenase, methylenetetrahydromethanopterin dehydrogenase, methylenetetrahydromethanopterin reductase, and heterodisulfide reductase in methanogenic bacteria. Arch. Microbiol. 155, 459–465.
[29] Thauer, R.K. (1990) Energy metabolism of methanogenic bacteria. Biochim. Biophys. Acta. 1018, 256–259.
[30] Jablonski, P.E. and Ferry, J.G. (1991) Purification and properties of methyl coenzyme M methylreductase from acetate-grown *Methanosarcina thermophila*. J.Bacteriol. 173, 2481–2487.
[31] Jetten, M.S.M., Stams, A.J.M. and Zehnder, A.J.B. (1990) Purification and some properties of the methyl-CoM reductase of *Methanothrix soehngenii*. FEMS Microbiol. Lett. 66, 183–186.

[32] Thomas, I., Dubourguier, H.-C., Presier, G., Debeire, P. and Albagnac, G. (1987) Purification of component C from *Methanosarcina mazei* and immunolocalization in *Methanosarcinaceae*. *Arch. Microbiol.* **148**, 193–201.

[33] Terlesky, K.C. and Ferry, J.G. (1988) Purification and characterization of a ferredoxin from acetate-grown *Methanosarcina thermophila*. *J. Biol. Chem.* **263**, 4080–4082.

[34] Aldrich, H.C., Beimborn, D.B., Bokranz, M. and Schönheit, P. (1987) Immunocytochemical localization of methyl-coenzyme M reductase in *Methanobacterium thermoautotrophicum*. *Arch. Microbiol.* **147**, 190–194.

[35] Fischer, R. and Thauer, R.K. (1990) Ferredoxin-dependent methane formation from acetate in cell extracts of *Methanosarcina barkeri* (strain MS). *FEBS Lett.*, **269**, 368–372.

[36] Bott, M., and Thauer, R.K. (1989) Proton translocation coupled to the oxidation of carbon monoxide to CO_2 and H_2 in *Methanosarcina barkeri*. *Eur. J. Biochem.* **179**, 469–472.

[37] Terlesky, K.C. and Ferry, J.G. (1988) Ferredoxin requirement for electron transport from the carbon monoxide dehydrogenase complex to a membrane-bound hydrogenase in acetate-grown *Methanosarcina thermophila*. *J. Biol. Chem.*, **263**, 4075–4079.

[38] Hatchikian, E.C., Fardeau, M.L., Bruschi, M., Belaich, J.P., Chapman, A. and Cammack, R. (1989) Isolation, characterization, and biological activity of the *Methanococcus thermolithotrophicus* ferredoxin. *J. Bacteriol.* **171**, 2384–2390.

[39] Hausinger, R.P., Moura, I., Moura, J.J.G., Xavier, A.V., Santos, M.H., LeGall, J. and Howard, J.B. (1982) Amino acid sequence of a 3Fe:3S ferredoxin from the 'Archaebacterium' *Methanosarcina barkeri* (DSM 800). *J. Biol. Chem.* **257**, 14192–14197.

[40] Clements, A.P. and Ferry, J.G. (1992) Cloning, nucleotide sequence, and transcriptional analyses of the gene encoding a ferredoxin from *Methanosarcina thermophila*. *J. Bacteriol.*, **174**, 5244–5250.

[41] Jablonski, P.E., DiMarco, A.A., Bobik, T.A., Cabell, M.C. and Ferry, J.G. (1990) Protein content and enzyme activities in methanol- and acetate-grown *Methanosarcina thermophila*. *J. Bacteriol.* **172**, 1271–1275.

[42] Karrasch, M., Bott, M. and Thauer, R. K. (1989) Carbonic anhydrase activity in acetate grown *Methanosarcina barkeri*. *Arch. Microbiol.* **151**, 137–142.

13
Mechanisms of Energy Conservation in Methanogenic Bacteria

Michael Blaut, U. Deppenmeier, B. Kamlage,
D. Westenberg, B. Becher, V. Müller and G. Gottschalk

*Institut für Mikrobiologie der Georg-August-Universität Göttingen,
W-3400 Göttingen, Germany*

1 SUMMARY

Methanogenesis from methanol or $H_2 + CO_2$ involves the exergonic reduction of the mixed disulfide of coenzyme M and of 7-mercaptoheptanoylthreonine phosphate (CoM-S-S-HTP). The reaction occurs at the membranes with either H_2 or reduced coenzyme F_{420} as electron donor. The electron transport from either reductant to CoM-S-S-HTP was shown in inverted vesicles of *Methanosarcina* strain Gö1 to generate a transmembrane electrochemical gradient of protons which drives the synthesis of ATP from ADP and P_i. ATP synthesis is catalyzed by an archaebacterial type of ATP synthase. Other unique reactions of the methanogenic pathways have been identified that are involved in energy conservation. Their role and importance for the methanogenic metabolism are described.

2 INTRODUCTION

Methanogenesis is restricted to a specialized group of strictly anaerobic archaebacteria. The biochemical routes leading from a limited number of substrates ($H_2 + CO_2$, formate, methanol, and acetate) to methane have been largely elucidated in the past two decades [1]. The conversion of all of these substrates

Microbial Growth on C_1 Compounds
© Intercept Ltd, PO Box 716, Andover, Hampshire SP10 1YG, UK

leads to the formation of the central intermediate methyl-CoM (CH_3-S-CoM) which is reductively cleaved to methane and coenzyme M (HS-CoM). The latter reaction is associated with a free energy change of -85 kJ/mol and was shown in *Methanosarcina barkeri* and *Methanosarcina* strain Göl to be coupled with the generation of ATP by electron transport phosphorylation [2,3]. After the discovery that methyl-CoM reduction consists of two steps [4,5], the question arose: which of the two is associated with the generation of a transmembrane electrochemical gradient of protons ($\Delta\tilde{\mu}_{H+}$)? In the first step, methyl-CoM is reductively cleaved with 7-mercaptoheptanoylthreonine phosphate (HS-HTP) as the reductant to methane and the heterodisulfide of HS-CoM and HS-HTP (CoM-S-S-HTP) (Eqn. 1). In the second step, the latter is reduced to HS-CoM and HS-HTP (Eqn. 2).

$$CH_3\text{-S-CoM} + \text{HS-HTP} \longrightarrow CH_4 + \text{CoM-S-S-HTP} \quad (1)$$

$$\text{CoM-S-S-HTP} + 2H^+ + 2e_- \longrightarrow \text{HS-CoM} + \text{HS-HTP} \quad (2)$$

Evidence was presented that only the reaction underlying Eqn. 2 is associated with electron transport-driven H^+ translocation [6]. Here we report on the role and function of two heterodisulfide reductase systems which differ in the electron donor utilized.

3 ENERGY CONSERVATION BY HETERODISULFIDE REDUCTION

3.1 *The H_2-dependent heterodisulfide reductase system*

After it became evident that only the heterodisulfide reduction is coupled to energy conservation, a vesicle system free of cytoplasmic proteins was developed [7]. Washed inside-out vesicles of *Methanosarcina* strain Göl catalyze a H_2-dependent reduction of CoM-S-S-HTP ($\Delta G^{\circ\prime} = -42$ kJ/mol). This electron transport along membrane-bound electron carriers is accompanied by H^+ translocation into the lumen of the vesicles leading to the generation of a $\Delta\tilde{\mu}_{H+}$ which in turn drives the synthesis of ATP from ADP and P_i. The vesicle system exhibits a stringent coupling between ATP synthesis and H_2-dependent CoM-S-S-HTP reduction. This is evident from a decrease in the electron transport rate upon omission of ADP from the assay or upon the addition of the ATP synthase inhibitor N,N'-dicyclohexylcarbodiimide (DCCD) as well as a reversal of this inhibition by protonophores. Protonophores inhibit ATP synthesis but stimulate the heterodisulfide reduction. The system exhibits maximal stoichiometries of 1 H^+ translocated/e^- and 1 ATP sythesized/$4e^-$.

3.2 *Components involved in H_2-dependent heterodisulfide reduction*

The H_2-dependent heterodisulfide reductase depends on a hydrogenase for H_2 oxidation. Two types of hydrogenase are found in methanogenic bacteria: one type uses F_{420} and viologen dyes as electron acceptor (F_{420}-reactive hydrogenase), whereas the other does not react with F_{420} but only with viologen dyes (F_{420}-nonreactive hydrogenase, often referred to as viologen-dependent hydrogenase). In the washing procedure used for the preparation of inverted vesicles from *Methanosarcina* strain Göl, the F_{420}-dependent hydrogenase activity is almost completely washed off the membranes, whereas F_{420}-nonreactive hydrogenase remains largely bound to the membranes, suggesting that only the latter type of hydrogenase plays a role in H_2-dependent heterodisulfide reduction. It was therefore proposed that the F_{420}-dependent hydrogenase plays only a role in the $F_{420}H_2$-dependent reactions of the carbon dioxide reduction pathway during growth on $H_2 + CO_2$ [7]. The F_{420}-nonreactive hydrogenase was recently purified from membranes of *Methanosarcina* strain Göl after solubilization with the detergent CHAPS [8].

The other part of the H_2-dependent heterodisulfide reductase system is the actual heterodisulfide reductase which is assayed by following the CoM-S-S-HTP-dependent oxidation of reduced benzylviologen or the oxidation of HS-CoM and HS-HTP with methylene blue as an oxidant. The heterodisulfide reductase was purified from the soluble fraction of *Mb. thermoautotrophicum* after cell breakage [9]. In contrast, in *Methanosarcina* strains the enzyme is membrane-bound and detergents have to be employed for their solubilization (U. Deppenmeier, unpublished results). The physiological electron donor for CoM-S-S-HTP reduction is not yet known. However, it is reasonable to assume that an additional electron carrier mediates the electron transfer from the hydrogenase to the heterodisulfide reductase, since otherwise it is difficult to see how the electron transport can be coupled to proton translocation. So far, there is no hint as to the nature of such as postulated electron carrier. Cytochromes have been suggested to be involved in electron transport from H_2 to methyl-CoM in *Ms. barkeri* [10], but more recent experiments indicate that the H_2-dependent cytochrome reduction observed may be unphysiological [11]. It is important to note that the majority of the methanogens must rely on other electron carriers since cytochromes have only been detected in the Methanosarcinaceae.

3.3 The $F_{420}H_2$-dependent heterodisulfide reductase system

Washed inverted vesicles of *Methanosarcina* strain Göl and membranes from *Methanolobus tindarius* contain another energy-transducing heterosulfide reductase system which does not depend on H_2, but on $F_{420}H_2$ as the reductant ($\Delta G^{o\prime} = -29$ kJ/mol) [12,13]. This $F_{420}H_2$-dependent heterosulfide reduction is also associated with $\Delta\tilde{\mu}_{H^+}$-driven synthesis of ATP from ADP and P_i. Protonophores stimulate the heterodisulfide reduction, but prevent the establishment of $\Delta\tilde{\mu}_{H^+}$ and ATP synthesis. The ATP synthase inhibitor DCCD decreases the rate of $F_{420}H_2$-dependent heterodisulfide reduction. The reversal of this DCCD-caused inhibition by protonophores and the stimulation of the $F_{420}H_2$-dependent heterodisulfide reduction by ADP indicate a stringent coupling between electron transport and ATP synthesis. The $F_{420}H_2$-dependent heterosulfide reductase system displays stoichiometries of $1 H^+$ translocated/e^- and 0.8 ATP synthesized/4 e^-.

3.4 Components involved in $F_{420}H_2$-dependent heterodisulfide reduction

The oxidation of $F_{420}H_2$ at the membrane is catalyzed by $F_{420}H_2$ dehydrogenase. The physiological electron acceptor is not known, but the enzyme can be assayed using metronidazole as electron acceptor and methylviologen as mediator. The $F_{420}H_2$ dehydrogenase was recently isolated from membranes of *Methanolobus tindarius* after solubilization with the detergent CHAPS [14]. It is composed of five different subunits with molecular masses of 45, 40, 22, 18, and 17 kDa, and contains Fe and S, but flavin could not be detected. Recent experiments with membranes of *Methanosarcina* strain Göl strongly suggest the participation of one or several cytochromes in electron transport from $F_{420}H_2$ to CoM-S-S-HTP [11]. The addition of CoM-S-S-HTP to reduced membranes of *Methanosarcina* strain Göl leads to a very rapid cytochrome oxidation, whereas the addition of $F_{420}H_2$ causes a very fast cytochrome reduction. It is important to note that the cytochromes appear to be restricted to the Methanosarcinaceae [15,16].

$F_{420}H_2$ dehydrogenase and heterodisulfide reductase most likely face the cytoplasmic side of the cytoplasmic membrane, since the substrates $F_{420}H_2$ and CoM-S-S-HTP are not converted by whole cells but only by inverted vesicles. Furthermore, the solubilization behavior of the two enzymes does not support the idea of an intrinsic membrane location. Due to this fact, it is difficult to see how proton translocation can be achieved unless an additional membrane-intrinsic electron carrier is postulated. A direct participation of cytochromes in proton translocation coupled to $F_{420}H_2$-dependent heterodisulfide reduction is conceivable, but it is also possible that other additional components are involved.

3.5 The physiological role of the two heterodisulfide reductase systems

The primary function of the two heterodisulfide reductase systems is clearly the generation of $\Delta\tilde{\mu}_{H+}$ for ATP synthesis via an ATP synthase. It has been proposed that the H_2-dependent system is needed for hydrogenotrophic growth, and the $F_{420}H_2$-dependent system for methylotrophic growth [7]. This suggestion is based on the fact that *Methanosarcina* strain Göl, which may either grow on methanol or on $H_2 + CO_2$, harbors both systems, whereas *Ml. tindarius*, which grows exclusively on methyl-containing substrates, contains only the $F_{420}H_2$-dependent heterodisulfide reductase. From the physiological point of view it makes sense that the $F_{420}H_2$-dependent dehydrogenase plays a prominent role during methylotrophic growth because two redox reactions involved

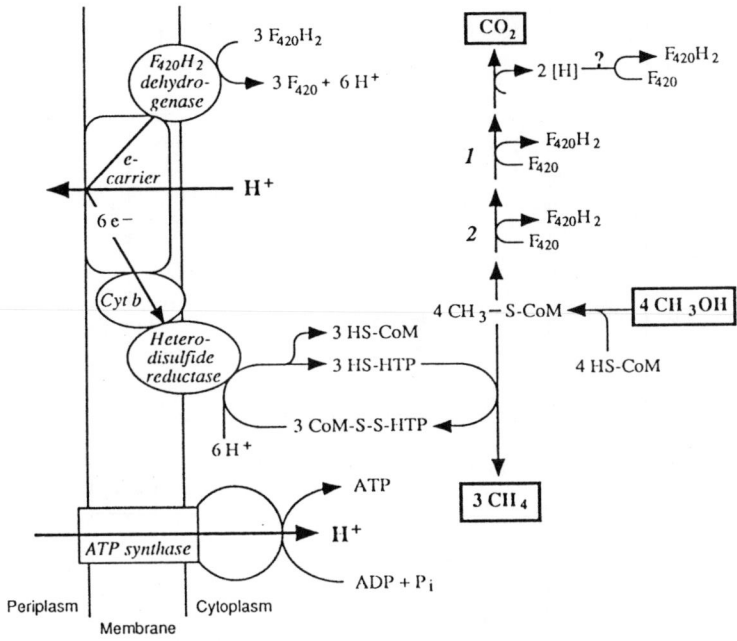

Figure 1 Model depicting the role of the $F_{420}H_2$-dependent heterodisulfide reductase system in energy transduction in *Methanosarcina* strain Gö1 during methanogenesis from methanol. The membrane location of the methyl-H_4MPT: coenzyme M methyltransferase and of the formyl-MF dehydrogenase is not taken into consideration. The numbers represent the following enzymes: 1. Methylene-H_4MPT dehydrogenase. 2. Methylene-H_4MPT reductase. The question mark indicates that the physiological electron donor of the reaction has not yet been identified. In order to show the stoichiometry of methanol conversion, it is assumed that the electrons derived from formyl-MF oxidation are transferred F_{420}.

in methyl group oxidation lead to the formation of $F_{420}H_2$. The $F_{420}H_2$-dependent heterodisulfide reductase serves then to regenerate F_{420} and to channel the electrons to the heterodisulfide (Fig. 1). During hydrogenotrophic growth the situation is different. Carbon dioxide reduction to the level of methyl-H_4MPT involves at least two redox reactions in which $F_{420}H_2$ is used as electron donor. This is supplied by the F_{420}-dependent oxidation of H_2 as catalyzed by the F_{420}-reactive hydrogenase (Fig. 2).

4 ATP SYNTHASE

$\Delta \tilde{\mu}_{H^+}$-driven ATP synthesis from ADP and P_i in inverted vesicles of *Methanosarcina* strain Gö1 requires the presence of an ATP synthase. Inatomi [17] isolated the catalytic part of an ATPase from membranes of *Ms. barkeri* and

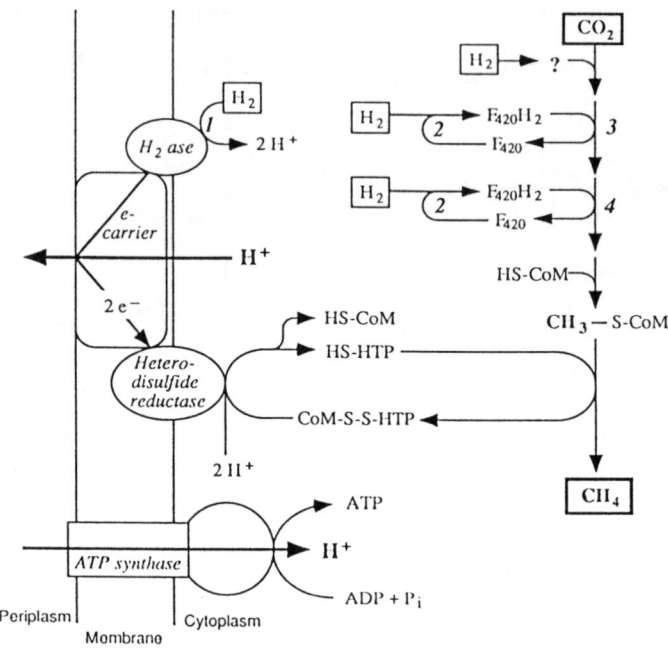

Figure 2 Model depicting the role of the H_2-dependent heterodisulfide reductase system in energy transduction in *Methanosarcina* strain Gö1 during methanogenesis from $H_2 + CO_2$. The membrane location of the methyl-H_4/MPT:coenzyme M methyltransferase and of the formyl-MF dehydrogenase is not taken into consideration. The numbers represent the following enzymes: 1. F_{420}-nonreactive hydrogenase. 2. F_{420}-reactive hydrogenase. 3. Methylene-H_4MPT dehydrogenase. 4. Methylene-H_4MPT reductase. The question mark indicates that the physiological electron donor of the reaction has not yet been identified.

sequenced the corresponding genes. Interestingly, the deduced amino acid sequence revealed over 50% identity with the corresponding polypeptides of the vacuolar (V-type) ATPase from *Neurospora crassa* or the ATPase from *Sulfolobus acidocaldarius*, but only 20% identity with the corresponding subunits of the F_1F_0-type ATPase from *Escherichia coli* [18]. It was suggested that all three types of enzymes (the archaebacterial, the V-type, and the F_1F_0-type ATPase) evolved from a common ancestral protein. Although the catalytic subunits of the archaebacterial ATPase show a higher degree of similarity to the corresponding subunits of the V-type than to those of those of the F_1F_0-type ATPase, the available data suggest that, from the functional point of view, the archaebacterial ATPase is more closely related to the F_1F_0-type ATPase than to the V-type ATPase [19, 20]. The physiological role of the V-type ATPases is ATP-driven H^+ translocation, whereas the archaebacterial and the F_1F_0-type ATPases serve primarily as ATP synthases.

A completely different type of ATPase was purified from *Methanococcus voltae* [21]. The amino acid sequence as deduced from the corresponding genes shows no homology to other ATPases [22]. At present, a role in ATP synthesis appears unlikely.

5 OTHER REACTIONS OF METHANOGENIC PATHWAYS ASSOCIATED WITH ENERGY TRANSDUCTION

5.1 *Methanogenesis from H_2 + CO_2*

H_2 + CO_2 is the most common substrate of methanogenic bacteria. The first reaction of the CO_2 reduction pathway, the formation of formyl-methanofuran (HCO-MF) from methanofuran (MF), CO_2, and H_2 is endergonic under standard conditions (Eqn. 3).

$$CO_2 + H_2 + MF \longrightarrow HCO-MF + H_2O \quad \Delta G^{o\prime} = +16 \text{ kJ/mol} \quad (3)$$

At a hydrogen partial pressure of about 10^{-4} to 10^{-5} atm, which is typical of natural habitats, $\Delta G'$ increases to $+40$ kJ/mol. The necessity for an energetic coupling of HCO—MF formation with an exergonic process is indicated by a number of observations. (1) Methyl-CoM greatly stimulates methanogenesis from H_2 + CO_2 [23]. (2) CoM-S-S-HTP activates an unknown low potential electron carrier which activates HCO-MF synthesis [24]. (3) The low potential electron donor Ti(III)citrate activates HCO-MF synthesis in the absence of added CoM-S-S-HTP [24]. It is not yet understood how CoM-S-S-HTP brings about this activation, but it is conceivable that the H_2-dependent CoM-S-S-HTP reduction, which generates a transmembrane electrochemical gradient of protons ($\Delta\tilde{\mu}_{H^+}$), is used to drive the H_2-dependent reduction of a low potential electron carrier with a midpoint potential ($E_{m,7}$) in the range of the CO_2/CHO-MF couple (-500 mV). Such a reverse electron transport could also explain the inhibition of methanogenesis from H_2 + CO_2 by agents that affect the transmembrane electrical potential ($\Delta\Psi$) [25,26,27]. The involvement of membrane components in CHO-MF formation is furthermore indicated by the finding that 60% of the enzyme is found in the membrane fraction of *Methanobacterium thermoautotrophicum* [28]. Other experiments suggest that CHO-MF formation is driven by a transmembrane gradient of sodium ions ($\Delta\tilde{\mu}_{Na^+}$) [29] which is generated by a novel type of transferase reaction.

The methyl-H_4MPT:HS-CoM methyltransferase is the second reaction in the CO_2 reduction pathway that is coupled to energy transduction. The activity of this membrane-bound corrinoid protein depends on the presence of sodium ions, and the methyl group transfer was recently shown in inverted vesicles of *Methanosarcina* strain Göl to be coupled to the translocation of Na^+ and the generation of a primary transmembrane electrochemical gradient of Na^+ [30].

5.2 *Methanogenesis from methanol*

Methanogenesis from methanol involves a reductive branch which includes the energy-conserving heterodisulfide reductase and an oxidative branch in which methyl-CoM is oxidized to CO_2. Methyl-CoM reduction to CO_2 involves the same reactions that are used for CO_2 reduction to methyl-CoM, but these operate in the reverse directions [31,32]. As a consequence it can be concluded that the endergonic methyl group transfer from methyl-CoM to H_4MPT has to be driven by $\Delta\tilde{\mu}_{Na^+}$. This is in accordance with the observation that methanol utilization depends on an energized membrane and on the presence of Na^+ [33, 34]. During methanol utilization $\Delta\tilde{\mu}_{Na^+}$ is generated secondarily by means of a Na^+/H^+ antiporter which takes advantage of the proton gradient formed during heterodisulfide reduction [35].

It is not yet clear how the electrons derived from the oxidation of HCO-MF are channeled to the heterodisulfide during growth on methanol. There are no experimental indications that H_2 or $F_{420}H_2$ are intermediates in this electron transfer. However, since the reverse reaction needs the input of energy, it has to be concluded that the oxidation of HCO-MF generates energy. In accordance with this notion Kaesler and Schönheit described a primary Na^+ translocation in response to the oxidation of formaldehyde (by way of methylene-H_4MPT and HCO-MF) to CO_2 and H_2 by intact cells of *Ms. barkeri*, which can be observed when methanogenesis is inhibited [36]. Other experiments favor the idea that this reaction is coupled to the generation of a $\Delta\tilde{\mu}_{H^+}$ [37].

6 CONCLUSIONS

The central role of the heterodisulfide reductase in energy transduction is now well established. The mode of energy generation coupled to this reaction is undoubtedly electron transport phosphorylation. Although many details concerning the electron carriers involved in electron transport from H_2 or $F_{420}H_2$ to CoM-S-S-HTP are still unknown, the structures of the two systems have been worked out and their functions have been experimentally demonstrated. Considerable progress has been made in the identification of other reactions involved in energy transduction. Notably the demonstration of a sodium ion-translocating methyltransferase is intriguing and represents a novel type of energy transduction. Only the mechanism of the energetic coupling of HCO-methanofuran formation is still obscure. It is remarkable that methanogenic bacteria use both H^+ and Na^+ as coupling ions for energy transduction.

REFERENCES

[1] Rouvière, P. and Wolfe, R.S. (1988) Novel biochemistry of methanogenesis. *J. Biol. Chem.* **263**, 7913–7916.

[2] Blaut, M. and Gottschalk, G. (1984) Coupling of ATP synthesis and methane formation from methanol and molecular hydrogen in *Methanosarcina barkeri*. *Eur. J. Biochem.* **141**, 217-222.

[3] Peinemann, S., Blaut, M. and Gottschalk, G. (1989) ATP synthesis coupled to methane formation from methyl-CoM and H_2 catalyzed by vesicles of the methanogenic bacterial strain Göl. *Eur. J. Biochem.* **186**, 175–180.

[4] Bobik, T.A., Olson, K.D., Noll, K.M. and Wolfe, R.S. (1987) Evidence that the heterodisulfide of coenzyme M and 7-mercaptoheptanoylthreoninephosphate is a product of the methylreductase reaction in *Methanobacterium. Biochem. Biophys. Res. Commun* 149, 455–460.
[5] Hedderich, R. and Thauer, R.K. (1988) *Methanobacterium thermoautotrophicum* contains a soluble enzyme system that specifically catalyzes the reduction of the heterodisulfide of coenzyme M and 7-mercaptoheptanoylthreonine phosphate with H_2. *FEBS Lett.* 234, 223–227.
[6] Peinemann, S., Hedderich, R., Blaut, M., Thauer, R.K. and Gottschalk, G. (1990) ATP synthesis coupled to electron transfer from H_2 to the heterodisulfide of 2-mercaptoethane sulfonate and 7-mercaptoheptanoylthreonine phosphate in vesicle preparations of the methanogenic bacterium strain Göl. *FEBS Lett.* 263, 57–60.
[7] Deppenmeier, U., Blaut, M. and Gottschalk, G. (1991) H_2:heterodisulfide oxidoreductase, a second energy conserving system in the methanogenic strain Göl. *Arch. Microbiol.* 155, 272–277.
[8] Deppenmeier, U., Blaut, M., Schmidt, B. and Gottschalk, G. (1992) Purification and properties of a F_{420}-nonreactive, membrane-bound hydrogenase from *Methanosarcina* strain Göl. *Arch. Microbiol.* 157, 505–511.
[9] Hedderich, R., Berkessel, A and Thauer, R.K. (1990) Purification and properties of heterodisulfide reductase from *Methanobacterium thermoautotrophicum* (strain Marburg). *Eur. J. Biochem.* 193, 255–261.
[10] Kemner, J.M., Krzycki, J.A., Prince, R.C. and Zeikus, J.G. (1987) Spectroscopic and enzymatic evidence for membrane-bound electron transport carriers and hydrogenase and their relation to cytochrome *b* function in *Methanosarcina barkeri. FEMS Microbiol. Lett.* 48, 267–272.
[11] Kamlage, B. and Blaut, M. (1992) Characterization of cytochromes from *Methanosarcina* strain Göl and their involvement in electron transport during growth on methanol. *J. Bacteriol.* 174, 3921–3927.
[12] Deppenmeier, U., Blaut, M., Mahlmann, A. and Gottschalk, G. (1990) Reduced coenzyme F_{420} heterodisulfide oxidoreductase, a proton translocating redox system in methanogenic bacteria. *Proc. Natl. Acad. Sci. USA.* 87, 9449–9453.
[13] Deppenmeier, U., Blaut, M., Mahlmann, A. and Gottschalk, G. (1990) Membrane-bound $F_{420}H_2$-dependent heterodisulfide reductase in methanogenic bacterium strain Göl and *Methanolobus tindaris. FEBS Lett.* 261, 199–203.
[14] Haase, P., Deppenmeier, U., Blaut, M. and Gottschalk, G. (1992) Purification and characterization of $F_{420}H_2$ dehydrogenase from *Methanolobus tindarius. Eur. J. Biochem.* 203, 527–531.
[15] Kühn, W., Fiebig, K., Hippe, H., Mah, R.A., Huser, B.A. and Gottschalk, G. (1983) Distribution of cytochromes in methanogenic bacteria. *FEMS Microbiol. Lett.* 20, 407–410.
[16] Jussofie, A. and Gottschalk, G. (1986) Further studies on the distribution of cytochromes in methanogenic bacterium. *FEMS Microbiol. Lett.* 22, 15–18.
[17] Inatomi, K.I. (1986) Characterization and purification of the membrane-bound ATPase of the archaebacterium *Methanosarcina barkeri. J. Bacteriol.* 167, 837–841.
[18] Inatomi, K.I., Eya, S., Maeda, M. and Futai, M. (1989) Amino acid sequence of the alpha and beta subunits of *Methanosarcina barkeri* ATPase deduced from cloned genes. *J. Biol. Chem.* 264, 10954–10959.
[19] Inatomi, K., Maeda, M. and Futai, M. (1989) Dicyclohexylcarbodiimide-binding protein is a subunit of the *Methanosarcina barkeri* ATPase complex. *Biochem. Biophys. Res. Commun.* 162, 1585–1590.
[20] Scheel, E. and Schäfer, G. (1990) Chemiosmotic energy conversion and the membrane ATPase of *Methanolobus tindarius. Eur. J. Biochem.* 187, 727–735.

[21] Dharmavaram, R.M. and Konisky, J. (1987) Identification of a vanadate-sensitive, membrane-bound ATPase in the archebacterium *Methanococcus voltae*. *J. Bacteriol.* **169**, 3921–3925.

[22] Dharmavaram, R., Gillevet, P. and Konisky, J. (1990) Nucleotide seqence of the gene encoding the vanadate sensitive membrane-associated ATPase of *Methanococcus voltae*. *J. Bacteriol.* **173**, 2131–2133.

[23] Gunsalus, R.P. and Wolfe, R.S. (1977) Stimulation of CO_2 reduction to methane by methyl-coenzyme M in extracts of *Methanobacterium*. *Biochem. Biophys. Res. Commun.* **76**, 790–795.

[24] Bobik, T.A. and Wolfe, R.S. (1989) Activation of formylmethanofuran synthesis in cell extracts of *Methanobacterium thermoautotrophicum*. *J. Bacteriol.* **171**, 1423–1427.

[25] Roberton, A.M. and Wolfe, R.S. (1970) Adenosine triphosphate pools in *Methanobacterium*. *J. Bacteriol.* **102**, 43–51.

[26] Jarrell, K.F. and Sprott, G.D. (1983) Measurement and significance of the membrane potential in *Methanobacterium bryantii*. *Biochem. Biophys. Acta* **725**, 280–288.

[27] Butsch, B.M. and Bachofen, R. (1984) The membrane potential in whole cells of *Methanobacterium thermoautotrophicum*. *Arch. Microbiol.* **138**, 293–298.

[28] Börner, G., Karrasch, M. and Thauer, R.K. (1989) Formylmethanofuran dehydrogenase activity in cells extracts of *Methanobacterium thermoautotrophicum* and of *Methanosarcina barkeri*. *FEBS. Lett.* **244**, 21–55.

[29] Kaesler, B. and Schönheit, P. (1989) The sodium cycle in methanogenesis-CO_2 reduction to the formaldehyde level in methanogenic bacteria is driven by a primary electrochemical potential of Na^+ generated by formaldehyde reduction to CH_4. *Eur. J. Biochem.* **186**, 309–316.

[30] Becher, B., Müller, V. and Gottschalk, G. (1992) The methyl-tetrahydromethanopterin:coenzyme M methyltransferase of *Methanosarcina* strain Göl is a primary sodium pump. *FEMS Microbiol. Lett.* **91**, 239–244.

[31] Fischer, R. and Thauer, R.K. (1989) Methyltetrahydromethanopterin as an intermediate in methanogenesis from acetate in *Methanosarcina barkeri*. *Arch. Microbiol.* **151**, 459–465.

[32] Mahlmann, A., Deppenmeier, U. and Gottschalk, G. (1989) Methanofuran b is required for CO_2 formation from formaldehyde by *Methanosarcina barkeri*. *FEMS Microbiol. Lett.* **61**, 115–120.

[33] Blaut, M., Müller, V., Fiebig, K. and Gottschalk, G. (1985) Sodium ions and an energized membrane required by *Methanosarcina barkeri* for the oxidation of methanol to the level of formaldehyde. *J. Bacteriol.* **164**, 95–101.

[34] Müller, V., Blaut, M. and Gottschalk, G. (1988) The transmembrane electrochemical gradient of Na^+ as driving force for methanol oxidation in *Methanosarcina barkeri*. *Eur. J. Biochem.* **172**, 601–606.

[35] Müller, V., Blaut, M. and Gottschalk, G. (1987) Generation of a transmembrane gradient of Na^+ in *Methanosarcina barkeri*. *Eur. J. Biochem.* **162**, 461–466.

[36] Kaesler, B. and Schönheit, P. (1989) The role of sodium ions in methanogenesis-formaldehyde oxidation to CO_2 and $2H_2$ in methanogenic bacteria is coupled with primary electrogenic Na^+ translocation at a stoichiometry of 2-3 Na^+/CO_2. *Eur. J. Biochem.* **184**, 223–232.

[37] Winner, C. and Gottschalk, G. (1989) H_2 and CO_2 production from methanol or formaldehyde by the methanogenic bacterium strain Göl treated with 2-bromoethanesulfonic acid. *FEMS Microbiol. Lett.* **65**, 259–264.

14

Genes Encoding the Methyl Viologen-reducing Hydrogenase, Polyferredoxin and Methyl Coenzyme M Reductase II are Adjacent in the Genomes of *Methanobacterium thermoautotrophicum* and *Methanothermus fervidus*

Vanessa J. Steigerwald, Aidan N. Hennigan, Todd D. Pihl and John N. Reeve

Department of Microbiology, The Ohio State University, 483 W. 12th Avenue, Columbus, OH 43210-1292, USA

1 SUMMARY

Methanobacterium thermoautotrophicum strains ΔH and Marburg and *Methanothermus fervidus* have been shown to contain two methyl coenzyme M reductase (MR) encoding regions. The genes (*mrt*) encoding *m*ethyl coenzyme M *r*eductase II (*t*wo), the isoenzyme synthesized predominantly in exponentially growing cells, have been located immediately downstream from the genes encoding the *m*ethyl *v*iologen-reducing *h*ydrogenase (*mvh*DGA) and the polyferredoxin (*mvh*B) in these *Methanobacteriales*. As a very similar pattern, synthesis in growing cells and decrease on entry into stationary phase, is observed for both the polyferredoxin and MRII in *M. thermoautotrophicum* strain ΔH, the *mvh* and *mrt* regions which encode these proteins may not only be physically adjacent but also coordinately expressed. Comparisons of the amino acid sequences of the β-subunits of the MRs indicate that the MRI and MRII isoenzymes in different *Methanobacteriales* are more similar to each other than are the MRI and MRII isoenzymes in the same methanogen.

2 INTRODUCTION

Methanogenesis appears to be obligatory for methanogens. There are no known 'facultative' methanogens that can gain sufficient energy for growth by methanogenesis or by an alternative metabolic pathway such as fermentation, respiration or photosynthesis. This essential nature of methanogenesis, coupled with the substantial technical difficulties inherent in growing and manipulating these fastidiously anaerobic microorganisms, has limited genetic studies of methanogenesis and of the regulation of this process. Despite these problems, evidence for regulation of expression of some genes, directly involved in methanogenesis, has been accumulated [1]. The *Methanosarcinaceae* respond to the availability of either methanol, mono-, di- or tri-methylamines or acetate, as the methanogenic substrate, by synthesizing different substrate-specific methyl-transferases [2,3] or carbon monoxide dehydrogenase [4,5]. Transcription of the genes that encode the subunits of formate dehydrogenase, the enzyme that initiates the catabolism of formate to methane, is regulated by molybdenum availability in *Methanobacterium formicicum* [6] and, recently, it has been shown that some *Methanobacteriales* contain two methyl coenzyme M reductases (MRI and MRII) whose syntheses are dependent on the growth conditions [7,8]. MRII is synthesized predominantly in substrate-sufficient, exponentially growing cells whereas MRI is synthesized in substrate-limited cells entering the stationary phase of growth. These *Methanobacteriales* must therefore contain two MR encoding regions that are expressed differently in response to a growth-rate determined parameter. If both MR activities are not always essential, it may be possible to identify conditions [8] under which mutations can be introduced into genes encoding one of the two MRs without loss of viability. Here we document experimentally that there are two MR encoding regions in the genomes of *Methanobacterium thermoautotrophicum* strains ΔH and Marburg and in *Methanothermus fervidus*, and compare amino-acid sequences encoded by the *m*ethyl *c*oenzyme M *r*eductase I (one) β-subunit encoding genes (*mcr*B) with the sequences encoded by the methyl coenzyme M reductase II (*t*wo) β-subunit encoding genes (*mrt*B). The *mrt*B genes have been located immediately downstream from operons that encode the methyl viologen-reducing hydrogenase (*mvh*DGA) and polyferredoxin (*mvh*B) in the genomes of *M. thermoautotrophicum* strain ΔH [9] and *M. fervidus* [10]. These methanogens therefore contain physically adjacent clusters of genes, involved in methanogenesis, that may be coordinately expressed in response to substrate availability.

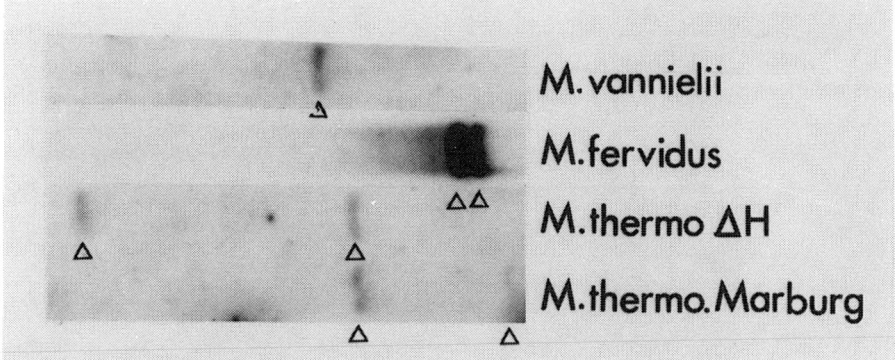

Figure 1 Southern blot analyses of methanogen genomic DNAs for MR-encoding sequences. Genomic DNAs, isolated from *Methanococcus vannielii, Methanothermus fervidus, Methanobacterium thermoautotrophicum* strains ΔH and Marburg were digested with *Eco*RI, separated by agarose gel electrophoresis, denatured, transferred to nitrocellulose and hybridized to a [^{32}P]-labeled oligonucleotide probe with the sequence, 5'CAGCTGTAGGTTTCTCATTCTTCAGTCACTCAATC-TATGGTG GTGGAGGACC-3'.. The sequence of this 52-mer was based on a region, highly conserved in all *mcr*B genes, that encodes amino-acid residues 357 through 374 of the β-subunit of MR [11,12,20,21]. The arrow-heads indicate *Eco*RI fragments that hybridized specifically to this probe. The single strongly hybridizing fragment in the *M. vannielii* digest is 1268 bp in length [21].

3 DEMONSTRATION OF TWO METHYL COENZYME M REDUCTASE ENCODING REGIONS IN *M. thermoautotrophicum* AND *M. fervidus*

Comparisons of the five sequenced *mcr* operons identified several highly conserved regions [11,12] that can be synthesized *in vitro* and used, as hybridization probes, either to identify and quantitate methanogens in environmental samples or to isolate additional MR-encoding sequences. A 52-mer, with a sequence based on the conserved nucleotide sequence that encodes amino-acid residues 357 through 374 of the β-subunits of MR [11] was ^{32}P-end-labeled and used in Southern hybridizations to probe restriction digests of several methanogen genomic DNAs. As shown in Fig. 1, this oligonucleotide hybridized strongly to two *Eco*RI restriction fragments in the genomes of *Methanothermus fervidus* and *Methanobacterium thermoautotrophicum* strains Marburg and ΔH but to only one *Eco*RI fragment from the genome of *Methanococcus vannielii*. These *Methanobacteriales* and *Methanobacterium wolfei* ([8]; results not shown) therefore do contain two, physically separate, MR encoding regions whereas the *Methanococcales* appear to contain only one region with a sequence closely related to that of the MR encoding genes so far sequenced [13,14].

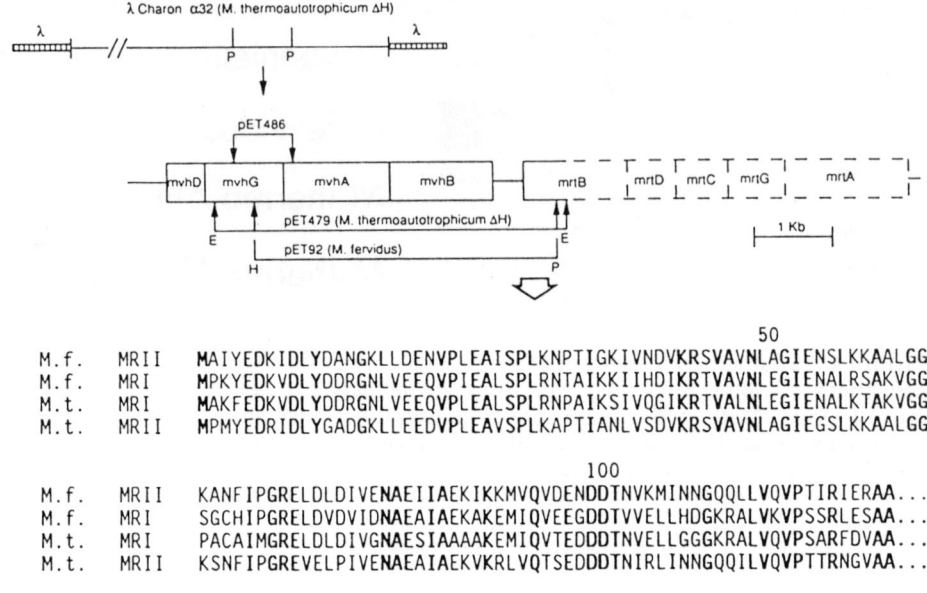

Figure 2 Cloning and genomic linkage of the genes encoding the methyl viologen-reducing hydrogenase and methyl coenzyme M reductase II in *M. thermoautotrophicum* strain ΔH and *M. fervidus*. The recombinant phage, λCharon-α32, contains ~20 Kb of *M. thermoautotrophicum* strain ΔH genomic DNA cloned from a Sau3A partial digest [9]. The 714 bp PstI (P) fragment indicated, subcloned into pUC8 to obtain pET486, was used as the hybridization probe to isolate pET479 [16] and pET92 [10]. Sequencing the EcoRI (E) fragment of *M. thermoautotrophicum* strain ΔH DNA in pET479, and the HindIII (H)-PstI (P) fragment of *M. fervidus* DNA in pET92, revealed the mvhG, mvhA, mvhB and mrtB genes organized as indicated. The sequences of the intergenic regions separating the polyferredoxin-encoding mvhB genes and the MRII β-subunit encoding mrtB genes are given in Fig. 3.

A mrtBDCGA operon is shown with the same organization of five genes that has been found in all mcr operons [11,12,20,21] but this has not yet been proven directly. Two additional recombinant phages, λCharon-A6 and λCharon-H1, that were isolated by the same procedure and at the same time as λCharon-α32 [9], do contain both the *M. thermoautotrophicum* strain ΔH mvh operon and sequences that hybridize to probes derived from the *M. vannielii* mcrB, and mcrGA genes. These phages appear therefore to contain mrtB, mrtG and mrtA sequences although their arrangement, as genes, has yet to be determined.

The sequences of the amino-terminal 126 amino-acid residues of the β-subunits of the methyl coenzyme M reductase Is (MRI) encoded by the mcrB genes in *Methanothermus fervidus* (M.f.) [11] and *Methanobacterium thermoautotrophicum* strain Marburg (M.t.) [20] and the β-subunits of the methyl coenzyme M reductase IIs (MRII) encoded by the mrtB genes in *Methanobacterium thermoautotrophicum* strain ΔH [9] and *M. fervidus* are shown aligned, using the single-letter amino-acid code, with amino-acid residues found in all four sequences indicated by bold type.

4 CLONING OF THE MRII ENCODING REGIONS BY LINKAGE TO THE MVH OPERONS

The genes (*mvhDGA*) that encode the subunits of the methyl viologen-reducing hydrogenase (MVH) were cloned from a *Sau3A* partial digest of *M. thermoautotrophicum* strain ΔH genomic DNA, as a ~20 Kb DNA fragment, in the λCharon35 vector [9]. This generated a recombinant phage that was designated λCharon-α32 (Fig. 2). Plaques containing λCharon-α32 were identified by the presence of antigens that bound antibodies raised against the α-subunit of the factor F_{420}-reducing hydrogenase (FRH) from *M. thermoautotrophicum* strain ΔH. Subsequent cloning and sequencing of the FRH encoding genes [15] demonstrated that MVH and FRH, from *M. thermoautotrophicum* strain ΔH, do contain several common amino acid sequences so that some antibodies raised against the α-subunit of FRH would be expected to bind to the α-subunit of MVH synthesized by λCharon-α32 infected *E. coli* cells. A 714 bp *Pst*1 fragment was subcloned from λCharon-α32 into pUC8 generating pET486 (Fig. 2) which also directed the synthesis of antigens in *E. coli* that bound anti-FRH antibodies. Using this *Pst*1 fragment as the hybridization probe, *E. coli* colonies were identified which contained the recombinant plasmids pET479 [16] and pET92 [10] that were subsequently shown to contain the *M. thermoautotrophicum* strain ΔH and *M. fervidus* genomic DNA fragments indicated in Fig. 2. Sequencing these DNA fragments revealed the presence of the polyferredoxin encoding *mvh*B genes very tightly linked, and probably cotranscribed, with the hydrogenase encoding *mvhDGA* genes [9,10,16]. Immediately downstream from the *mvh*B genes are several oligo-T sequences that conform to a transcription terminating motif found frequently in thermophilic *Archaea* [17,18] (Fig. 3). A truncated open reading frame (ORF146) in pET479 [9,16], which begins 445 bp downstream of *mvh*B in the genome of *M. thermoautotrophicum* strain ΔH, and a related truncated ORF containing 126 codons (ORF126) in pET92 [10], but separated by only 220 bp from the *mvh*B gene in the genome of *M. fervidus*, have also been sequenced. The amino-acid sequences encoded by ORF146 and ORF126 have been identified as the amino-terminal sequences of the β-subunits of the *M. thermoautotrophicum* strain ΔH and *M. fervidus* MRII isoenzymes, respectively. These ORFs have therefore been re-designated *mrt*B as they encode *m*ethyl coenzyme M *r*eductase II (*t*wo) *b*eta subunits (Fig. 2). The longer intergenic region that separates the *mvh*B and *mrt*B genes in *M. thermoautotrophicum* strain ΔH contains a relatively short ORF (ORF56), located on the opposite DNA strand, that is not present in *M. fervidus* (Fig. 3). Whether this is a *bona fide* gene remains uncertain; it does not encode an amino acid sequence clearly related to that of any known protein.

Immediately upstream of the two *mrt*B genes are conserved sequences that appear likely to be promoter elements and ribosome binding sites [17–19]. There are two copies of an A/T-rich 12 bp inverted repeat in the *M. thermoautotrophicum* strain ΔH intergenic sequence and a related 21 bp A/T-rich inverted repeat, at approximately the same location, in *M. fervidus* (Fig. 3).

186 V.J. Steigerwald *et al.*

```
M.t.     mvhB-TAA CCCCCTCCTTTTTTGTTTTCAGATGCAGGATTCCAGTTATCACAGACCATGTTATTATCACTGTTAATATTT
M.f.     mvhB-TAA TTTTATTTTTTCTTTTTCTCCATGATTATTACTAATTATTTGTTATTTATGATTTATTTAACATTTTCTTTTT

M.t.  73 TTATGAAAAATGATTTTTCATATTTAAAATTTTTCTTTTTTTCCTTTTGAAAACATTTATATTATCCTTTTTTCTT
M.f.  73 TATTTTTTTAGTAAAGGATATAAACAGTTTTTAACATAGCTTCAAAAAACATCTGGATGCTGATGAATATATTTGT

M.t. 149 ATTCAATAGAGTCATAGATGTCAAAAAACTGAAGCATAAGCGGTTATATTCAGATTAATACTGGATAATAACGAGA
                                                                 o  y  q  i  i  v  l
M.f. 149 GA-----------------------------------------------------------------------

M.t. 225 TGTACAGTATGGCTTGTTCTATAATGAGCTGTGCTCTGGAGGTTGGAGGATGAACTTCGGTATAACCTGAAGGTCC
            h  v  t  h  s  t  r  y  h  a  t  s  q  l  n  s  s  s  s  r  y  l  r  f  t
M.f.     --------------------------------------------------------------------------
                                                                               ⌐ ORF56
M.t. 301 ATTTCGAGACATTGGTGATGGATACGTCCTTCATGAGGTGACCATTTCCATGGATTATCGCTGGCAATCCCATAAC
            w  k  s  v  n  t  i  s  v  d  k  m  l  h  g  n  g  h  i  i  a  p  l  g  m
M.f.     ----------------------------------------------------------------------- TT

M.t. 377 CCCATCAGTTTTATTAATAAAATAGTAAATTTTATTAATAAATAAATAAAACAAGAGGTGTGAATACC ATG-mrtB
M.f. 153 TTAAATTATGATGTTGGCCAAAAATTTTATTATTAATAAAAAATTACATAAAAATAT GAGGTGACAAA ATG-mrtB
                                                                    RBS
```

Figure 3 Sequences of the intergenic regions separating the *mvh*B and *mrt*B genes in *Methanobacterium thermoautotrophicum* strain ΔH (M.t.) and *Methanothermus fervidus* (M.f.). The translation termination codons of the polyferredoxin encoding *mvh*B genes and translation initiation codons of the MRII β-subunit encoding *mrt*B genes are boxed. The amino-acid sequence encoded by ORF56, located on the opposite DNA strand from *mvh*B and *mrt*B, that is present in *M. thermoautotrophicum* but not in *M. fervidus*, is indicated. Sequences likely to be ribosome binding sites (RBS) are identified and the locations of conserved A/T-rich inverted-repeat sequences are indicated by two-headed arrows. These A/T rich regions contain sequences that are similar, but not identical, to the consensus TATA-box element determined for methanogen promoters [19]. The M.t. sequence was published previously [9] in which the mrtB gene was designated ORF146.

These A/T rich regions have sequences resembling the consensus for the methanogen TATA-box promoter element [19] but, being inverted repeats, they are also structurally similar to the binding sites for many dimeric regulatory proteins.

5 COMPARISON OF THE MRI AND MRII β-SUBUNIT SEQUENCES

The amino-terminal sequences of the β-subunits of MRI and MRII from *M. thermoautotrophicum* strains Marburg [20] and ΔH [9], respectively, and from *M. fervidus* [11] are aligned in Fig. 2 and quantitative pair-wise comparisons, based on this alignment, are listed in Table 1. Fifty-three of the 126 amino-acid residues (42%), are present in all four sequences. The two MRI sequences are 75% identical and the two MRII sequences are 74% identical, but the MRI and MRII sequences within *M. thermoautotrophicum* strains Marburg and ΔH and within *M. fervidus* are only 52% and 57% identical to each other, respectively (Table 1). Comparing MRI from *M. fervidus* with MRII from *M. thermoautotrophicum* strain ΔH, or MRI from *M. thermoautotrophicum* strain

Table 1 Quantitative pairwise comparisons of the amino acid sequences of the MRI and MRII β-subunits

Sequences Compared[a]	% Identical Amino Acid Residues[c]
MRI[b] (M.t.) v. MRII[b] (M.t.)	52
MRI (M.f.) v. MRII (M.f.)	57
MRI (M.t.) v. MRI (M.f.)	75
MRII (M.t.) v. MRII (M.f.)	74
MRI (M.t.) v. MRII (M.f.)	60
MRI (M.f.) v. MRII (M.t.)	51

[a] The amino-terminal 126 amino-acid residues of the β-subunits of the MRIs and MRIIs, aligned as shown in Fig. 2, are compared.
[b] The MRI sequence for *M. thermoautotrophicum* (M.t.) was obtained from strain Marburg [20] whereas the MRII sequence is from strain ΔH [9]; M.f. = *Methanothermus fervidus* [11; Fig. 2]).
[c] 43% of the amino acid residues are identical in all four sequences.

Marburg with MRII from *M. fervidus*, reveals similar levels of conservation, these sequences contain 51 and 60% identical amino acid residues, respectively. The two MRI and two MRII encoding sequences appear therefore more similar to each other than do the MRI and MRII sequences in the same methanogen. It would seem that the *mcr* and *mrt* regions must have been evolving, separated from each other, for an extended period of time. The DNA duplication event that presumably gave rise to the ancestors of the current *mcr* and *mrt* operons most probably occurred before the separation of *Methanobacterium* species from *Methanothermus* species. The amino-terminal sequence of the β-subunit of MR from *Methanococcus vannielii* [21] appears to be equally similar to the MRI and MRII sequences of the *Methanobacteriales*.

6 GROWTH-DEPENDENT SYNTHESIS OF THE POLYFERREDOXIN AND OF MRII

Synthesis of MRII has been shown to be growth rate dependent in *M. thermoautotrophicum* strains ΔH and Marburg, occurring at the highest levels in cells growing exponentially in a substrate-sufficient environment at temperatures below 65° and pH $>$ 7.0 [7,8]. When cell growth was limited by substrate supply, and on entry into stationary phase, MRII synthesis was very much reduced and MRII replaced by MRI [7,8]. A similar pattern of synthesis during exponential growth and decrease on entry into stationary phase has also been demonstrated for the polyferredoxin in *M. thermoautotrophicum* strain ΔH [22]. Antibodies raised against the product of a *lacZ-mvh*B gene fusion detected substantial amounts of the polyferredoxin in extracts of growing cells but very little polyferredoxin in extracts from cells entering the stationary phase of

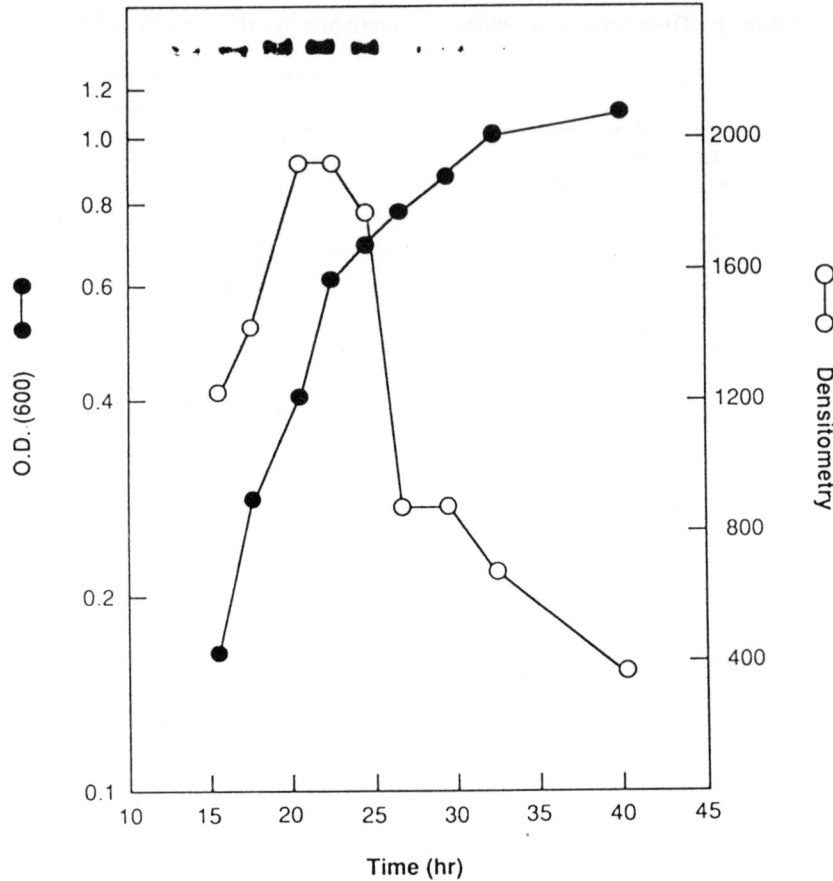

Figure 4 Growth phase dependent synthesis and decay of the polyferredoxin in *M. thermoautotrophicum* strain ΔH. Samples were removed from a culture, growing on CO_2 plus H_2 at 65°, at the times indicated and the optical density at 600 nm (●) and amounts of protein in each sample were determined. The Western blot shown was generated by using affinity-purified rabbit antibodies which were raised against the product of a *lacZ-mvh*B gene fusion synthesized in, and purified from, *E. coli* [22]. Aliquots of the *M. thermoautotrophicum* proteins (10 μg) were separated by SDS/PAGE and a soft-laser-scanning densitometer was used to quantitate the amount of polyferredoxin in each sample as demonstrated by the Western blots. The values obtained, indicating the relative amounts of polyferredoxin at each time point, are plotted in arbitrary units (○).

growth (Fig. 4). It is intriguing therefore that, as shown in Fig. 2, the polyferredoxin and MRII encoding genes, whose expression seems to be similarly regulated, are physically adjacent in the genomes of *M. thermoautotrophicum* strain ΔH and *M. fervidus*.

7 CONCLUSION

The functional significance, if any, of the close physical association of the *mvh* and *mrt* regions in the genomes of *M. thermoautotrophicum* and *M. fervidus* is currently unclear. A polyferredoxin-encoding *mvh*B gene, linked to *mvh*DGA genes, has also been cloned and sequenced from *Methanococcus voltae* but this does not appear to be adjacent to a MR encoding operon [14,23]. The similarity in the patterns of synthesis and decay of the polyferredoxin and MRII in *M. thermoautotrophicum* does hint at a coordination in the expression of their encoding genes but a comparison of the intergenic regions upstream of the *mvh* and *mrt* regions in *M. thermoautotrophicum* strain ΔH [9,16] does not reveal any obviously conserved regulatory 'box'. Two additional recombinant phages (λCharon-A6, and λCharon-H1) were isolated at the same time and by the same procedure as λCharon-α32 and these phages do contain both *mvh* and *mrt* sequences (results not shown). Using the *M. thermoautotrophicum* strain ΔH DNAs cloned in these phages, as hybridization probes in Northern blots, it should now be possible to determine if transcription of the *mvh* and *mrt* regions is similarly regulated. By using the promoter regions from the *mcr* [11] and *mrt* operons (Fig. 3) from *M. fervidus* as DNA templates with a cell-free transcription system that functions accurately with *M. fervidus* promoters [24], it should also be possible to determine if the growth rate dependent expression of these operons occurs *in vitro* at the level of transcription initiation.

ACKNOWLEDGEMENTS

We thank B. Baum for bringing the identity of ORF146 to our attention. This research was supported by grant DE-FG02-87ER13731-A003 from the U.S. Department of Energy. VJS was the recipient of a predoctoral fellowship from the Office of Naval Research. The 52-mer used in Fig. 1 was generously provided by J. Tiedje.

REFERENCES

[1] Reeve, J.N. (1992) Molecular biology of methanogens. *Ann. Rev. Microbiol.* **46**, 165–191.
[2] van der Meijden, P., Heythuysen, H.J., Pouwels, A., Houwen, F., van der Drift, C. and Vogels, G.D. (1983) Methyl transferases involved in methanol conversion by *Methanosarcina barkeri*. *Arch. Microbiol.* **134**, 238–242.
[3] Nauman, E., Fahlbusch, K. and Gottschalk, G. (1984) Presence of a trimethylamine: HS-coenzyme M methyl transferase in *Methanosarcina barkeri*. *Arch. Microbiol.* **138**, 79–83.
[4] Terlesky, K.C., Nelson, J.K. and Ferry, J.G. (1986) Isolation of an enzyme complex with carbon monoxide dehydrogenase activity containing corrinoid and nickel from acetate-grown *Methanosarcina thermophila*. *J. Bacteriol.* **168**, 1053–1058.

[5] Bhatnagar, L., Krzycki, J.A. and Zeikus, J.G. (1987) Analysis of hydrogen metabolism in *Methanosarcina barkeri*, regulation of hydrogenase and role of CO-dehydrogenase in H_2 production. *FEMS Microbiol. Lett.* **41**, 337–343.

[6] Shuber, A.P., Orr, E.C., Recny, M.A., Schendel, P.F., May, H.D., Shauer, N.L. and Ferry, J.G. (1986) Cloning, expression, and nucleotide sequence of the formate dehydrogenase genes from *Methanobacterium formicicum*. *J. Biol. Chem.* **261**, 12942–12947.

[7] Rospert, S., Linder, D., Ellermann, J. and Thauer, R.K. (1990) Two genetically distinct methyl-coenzyme M reductases in *Methanobacterium thermoautotrophicum* strains Marburg and ΔH. *Eur. J. Biochem.* **194**, 871–877.

[8] Bonacker, L.G., Baudner, S. and Thauer, R.K. (1992) Differential expression of the two methyl-coenzyme M reductases in *Methanobacterium thermoautotrophicum* as determined immunochemically via isoenzyme-specific antisera. *Eur. J. Biochem.* **206**, 87–92.

[9] Reeve, J.N., Beckler, G.S., Cram, D.S., Hamilton, P.T., Brown, J.W., Krzycki, J.A., Kolodziej, A.F., Alex, L.,Orme-Johnson, W.H. and Walsh, C.T. (1989) A hydrogenase-linked gene in *Methanobacterium thermoautotrophicum* strain ΔH encodes a polyferredoxin. *Proc. Natl. Acad. Sci. USA* **86**, 3031–3035.

[10] Steigerwald, V.J., Beckler, G.S. and Reeve, J.N. (1990) Conservation of hydrogenase and polyferredoxin structures in the hyperthermophilic archaebacterium *Methanothermus fervidus*. *J. Bacteriol.* **172**, 4715–4718.

[11] Weil, C.F., Cram, D.S., Sherf, B.A. and Reeve, J.N. (1988) Structure and comparative analysis of the genes encoding component C of methyl coenzyme M reductase in the extremely thermophilic archaebacterium *Methanothermus fervidus*. *J. Bacteriol.* **170**, 4718–4726.

[12] Weil, C.F., Sherf, B.A. and Reeve, J.N. (1989) A comparison of the methyl reductase genes and gene products. *Can. J. Microbiol.* **35**, 101–108.

[13] Hennigan, A.N., Stroup, D., Palmer, J.P. and Reeve, J.N. (1991) Identification and quantitation of the transcript and products of the methyl coenzyme M reductase operon in *Methanococcus vannielii*. Abs. Ann. Meeting Amer. Soc. Microbiol. I118.

[14] Sitzmann, J. and Klein, A. (1991) Physical and genetic map of the *Methanococcus voltae* genome. *Mol. Microbiol.* **5**, 505–513.

[15] Alex, L.A., Reeve, J.N., Orme-Jonson, W.H. and Walsh, C.T. (1990) Cloning, sequence determination, and expression of the genes encoding the subunits of the nickel-containing 8-hydroxy-5-deazaflavin reducing hydrogenase from *M. thermoautotrophicum* strain ΔH. *Biochemistry* **24**, 7237–7244

[16] Reeve, J.N. and Beckler, G.S. (1990) Conservation of primary structure in prokaryotic hydrogenases. *FEMS Microbiol. Rev.* **87**, 419–424.

[17] Zillig, W., Palm, P., Reiter, W.-D., Gropp, F., Puhler, G. and Klenk, H.P. (1988) Comparative evaluation of gene expression in archaebacteria. *Eur. J. Biochem.* **173**, 473–482.

[18] Brown, J.W., Daniels, C.J. and Reeve, J.N. (1989) Gene structure, organization, and expression in archaebacteria. *C.R.C. Crit. Rev. Microbiol.* **16**, 287–338.

[19] Hausner, W., Frey, G. and Thomm, M. (1991) Control regions of an archaeal gene: a TATA box and an initiator element promote cell-free transcription of the tRNA^{Val} gene of *Methanococcus vannielii*. *J. Mol. Biol.* **222**, 495–508.

[20] Bokranz, M, Bäumner, G., Allmansberger, R., Ankel-Fuchs, D. and Klein, A. (1988) Cloning and characterization of the methyl coenzyme M reductase genes from *Methanobacterium thermoautotrophicum*. *J. Bacteriol.* **170**, 568–577.

[21] Cram, D.S., Sherf, B.A., Libby, R.T., Mattaliano, R.J., Ramachandran, K.L. and Reeve, J.N. (1987) Structure and expression of the genes *mcrBDCGA*, which encode the subunits of component C of methyl coenzyme M reductase in *Methanococcus vannielii*. *Proc. Natl. Acad. Sci. USA* **84**, 3992–3996.

[22] Steigerwald, V.J., Pihl, T.D. and Reeve, J.N. (1992) Identification and isolation of the polyferredoxin from *Methanobacterium thermoautotrophicum* strain ΔH. *Proc. Natl. Acad. Sci. USA* **89**, 6929–6933.

[23] Halboth, S. and Klein, A. (1992) *Methanococcus voltae* harbors four gene clusters potentially encoding two [NiFe] and two [NiFeSe] hydrogenases, each of the cofactor F_{420}-reducing or F_{420}-non-reducing types. *Mol. Gen. Genet.* **233**, 217–224.

[24] Thomm, M., Sandman, K., Frey, G., Koller, G. and Reeve, J.N. (1992) Transcription *in vivo* and *in vitro* of the histone-encoding gene *hmf*B from the hyperthermophilic archaeon *Methanothermus fervidus*. *J. Bacteriol.* **174**, 3508–3513.

15

A Molecular Analysis of Peroxisome Biogenesis and Function in the Methylotrophic Yeast *Hansenula polymorpha*

G. Sulter, M. Veenhuis and W. Harder

Biological Center, University of Groningen, Kerklaan 30, 9751 NN Haren, The Netherlands

1 SUMMARY

Microbodies are ubiquitous in eukaryotic cells. They may be involved in a variety of metabolic processes; their number, volume fraction and physiological function changes with the conditions of growth. Especially in methylotrophic yeasts the accumulated knowledge and experience now enables us to adjust precisely both the level of microbody induction and their protein composition by selecting appropriate growth conditions. Also the role of the organelles in methanol metabolism is currently well understood. These aspects are reviewed, taking *Hansenula polymorpha* as an example.

During the last few years a series of peroxisome-deficient (*per*) mutants of *H. polymorpha* have been isolated. Phenotypically these mutants are characterized by the fact that they are not able to grow on methanol. Three mutant phenotypes were defined on the basis of morphological criteria, including a) mutants completely lacking *per*oxisomes (Per$^-$; 13 complementation groups), b) mutants containing few small peroxisomes which are partly impaired in the *per*oxisomal *im*port of matrix proteins (Pim$^-$; 5 complementation groups) and c) mutants with aberrations in the *per*oxisomal *subs*tructure (Pss$^-$; 2 complementation groups). In addition several conditional Per$^-$, Pim$^-$ and Pss$^-$ mutants have been obtained. In all cases the mutant phenotype was shown to be caused by a recessive mutation in one gene. A detailed genetic analysis revealed that several *PER* genes, essential for peroxisome biogenesis, are tightly linked and organized in a hierarchical fashion.

The potential of both constitutive and conditional *per* mutants for current and future studies of the molecular mechanisms controlling peroxisome biogenesis and function is discussed.

2 INTRODUCTION

Microbodies (peroxisomes) are ubiquitous single-membrane bounded organelles, present in all eukaryotes examined so far except archaezoa [1]. Despite their simple morphology, the organelles show a great diversity in enzyme repertoires and, as a consequence, may be involved in a variety of metabolic processes [2–4]. These functions are partly inducible by specific substrates present in the growth medium but may also vary with the developmental stage and/or other growth conditions of the organism harboring the organelles. This multi-functionality of microbodies is unique among subcellular organelles in eukaryotes. Their importance in cellular metabolism is exemplified by the fact that peroxisomal dysfunctioning may lead to metabolic abnormalities and occasionally can be lethal, particularly in man (*Zellweger* syndrome; [5]).

In yeasts, microbodies were first described by Avers [6]. In the intervening years much progress has been made towards our understanding of the mechanisms involved in the biogenesis and metabolic function of these organelles which, based on their function, are called peroxisomes [4, 7–10]. This paper summarizes major recent developments in the studies on peroxisome biogenesis/function in the methylotrophic yeast *Hansenula polymorpha* and emphasizes the potential of this organism as a model system in future studies on this comparatively neglected organelle.

3 PEROXISOMES IN WILD TYPE CELLS OF *H. polymorpha*

3.1 *Peroxisome induction and morphology*

In yeasts the proliferation and metabolic significance of peroxisomes is largely prescribed by cultivation conditions [7–10]. In *H. polymorpha* optimal induction (where their volume fraction may reach up to 80% of that of the cytoplasm) is achieved during growth of cells in a methanol-limited chemostat [7]. Other substrates known to induce microbodies in this organism are ethanol, primary amines, D-amino acids and purines [7,10] but not oleic acid [11]. Peroxisomes play an crucial role during growth of cells under these conditions since they contain key enzymes of the metabolism of the respective carbon or nitrogen sources. The current view on the compartmentalization of methanol metabolism is depicted in Fig. 1.

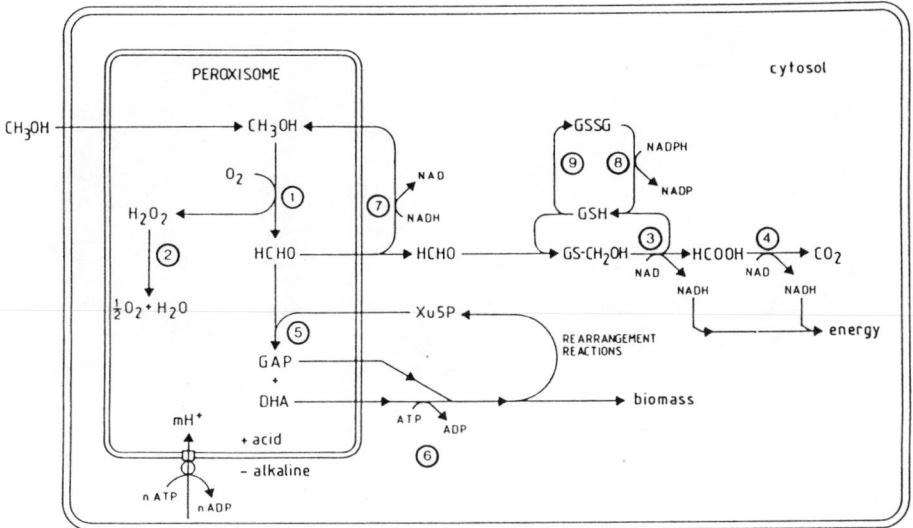

Figure 1 Schematic representation of the compartmentalization and function of peroxisomes in methanol metabolism in *H. polymorpha* wild type cells. 1. alcohol oxidase, 2. catalase, 3. formaldehyde dehydrogenase, 4. formate dehydrogenase, 5. dihydroxyacetone synthase, 6. dihydroxyacetone kinase, 7. formaldehyde reductase, 8. glutathione reductase, 9. oxidation of glutathione. From [43].

Also the mature size and substructure of peroxisomes contained in *H. polymorpha* is a reflection of the cultivation conditions and more in particular is determined by i) the nature of the carbon/nitrogen source used for growth [7,10] and ii) the mode of cultivation and/or the phase of growth of the culture [10]. Large cuboid, completely crystalline organelles (Fig. 2) have only been observed in cells grown in methanol-limited chemostat cultures at low dilution rates and are due to the synthesis (and crystallization) of excessive amounts of alcohol oxidase protein. In cells from the exponential phase in batch cultures, rounded organelles are predominant; this morphology is largely independent of the growth substrate. Furthermore, differences in the shape of individual organelles present in one cell may reflect differences in their developmental stage [12].

3.2 Development of peroxisomes

There is now general consensus that peroxisomes in yeasts develop from mature, pre-existing organelles [4,7,8]. In *H. polymorpha*, peroxisome proliferation has been studied by ultrastructural methods following a shift of cells from glucose to methanol-containing media. In glucose-grown cells (Fig. 2) generally a single, small peroxisome is present in each cell [10]. Upon a shift of cells to methanol this organelle rapidly increased in size due to the import of newly synthesized

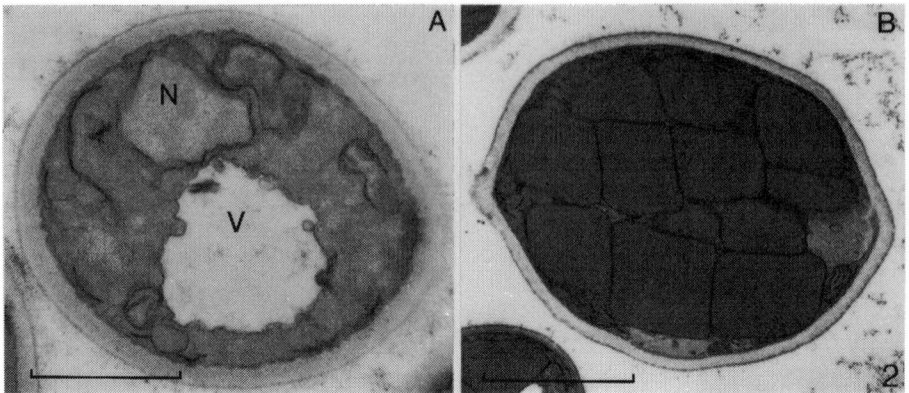

Figure 2 Survey of cells of *H. polymorpha* from a batch culture in the mid-exponential growth phase on glucose (Fig. 2A) and from a methanol-limited chemostat culture (Fig. 2B) to show the overall cell morphology and the shape of peroxisomes.
Electron micrographs. Cells are fixed with $KMnO_4$ unless otherwise stated. Abbreviations: A – cytosolic alcohol oxidase crystalloid; N – nucleus; V – vacuole. The marker represents 1.0 μm unless otherwise indicated.

matrix proteins (e.g. alcohol oxidase and dihydroxyacetone synthase). Growth of the organelles continued until a mature peroxisome was formed from which small peroxisomes were separated, and which in turn developed. The results obtained indicated that the capacity to import proteins was strictly associated with the capacity of the organelle to divide. As a consequence, the microbody population in one cell is heterogeneous [12]; only small immature organelles are capable of incorporating newly synthesized proteins and of subsequent fission. This mechanism of microbody multiplication is probably general in yeasts [13].

Proliferation of peroxisomes is not simply triggered by the synthesis of their matrix proteins. In different species (over)expression of matrix proteins only resulted in a considerable increase in size but not in the number of organelles; this implies that upon saturation of the import capacity of the organelles, additionally expressed matrix proteins will remain in the cytosol where they generally accumulate into large proteinaceous aggregates [14,15]. Fission of peroxisomes is also not dependent on the size of the organelles as was evident in mutants of *H. polymorpha* which are partly blocked in matrix protein import [16].

3.3 The peroxisomal membrane

The current level of understanding of the peroxisomal membrane has hardly surpassed the descriptional level. In yeast, the membrane of peroxisomes can be morphologically distinguished from all other cellular membranes by its relatively small width (ca. 8 nm in ultrathin sections) and unique, rather smooth, architecture in freeze etch replicas [17]. Analysis of variously grown *H. polymorpha* revealed that the polypeptide composition of the peroxisomal membrane, in contrast to its phospholipid composition, greatly differed from other membranes in the cell. Most of these proteins appeared to represent constitutive components of the membrane; however, others were shown to be inducible by growth substrates and therefore probably are related to specific peroxisomal functions [17]. Participation of peroxisomes in different metabolic pathways often requires extensive metabolic interactions with other cell compartments (for instance mitochondria, ER and the cytosol) and thus an intensive traffic of various low and high molecular weight components must take place. The molecular mechanisms involved in these transport processes are still unknown. *In vivo* yeast peroxisomal membranes are probably not permeable for small molecules; this view is mainly based on the observed pH gradient across the peroxisomal membrane [18,19] and the physiological function of the organelles which necessitates selective permeability (see Fig. 1). On the other hand, isolated yeast peroxisomal membranes are invariably leaky [20] and show permeability properties comparable to those of isolated peroxisomal membranes from rat liver cells [21]. Recent studies indicated that the permeability of peroxisomes of *H. polymorpha* may be attributed to the presence of a pore-forming protein. This protein, which is a constitutive 31 kDa integral peroxisomal membrane protein (PMP), has been purified and biochemically characterized [22]. The 31 kDa PMP was weakly cation-selective and showed permeability properties comparable to its mitochondrial counterpart. The main difference between the two porins appeared to be their mode of regulation; whereas the mitochondrial porin was voltage-dependent, the peroxisomal 31 kDa PMP was regulated by calcium ions [23]. These results have now opened the way for future studies on the possible energy dependency of various transport processes across the peroxisomal membrane.

3.4 Peroxisomal protein targeting and assembly

The activity of peroxisomal matrix enzymes in wild type cells is strictly confined to the peroxisomal matrix [7]. Peroxisomes lack nucleic acid and ribosomes and therefore all peroxisomal proteins are encoded by nuclear genes. Precursors of these proteins are synthesized in the cytosol on free polysomes and are imported posttranslationally into the target organelle where assemblage and activation takes place. Precursors of matrix proteins do not contain cleavable targeting sequences; consequently, topogenic information has to be contained in the amino acid sequence of these proteins. Recently, a highly conserved general targeting signal (PTS), located at the extreme C-terminus of various matrix proteins has been identified (SKL-COOH [24]). However, this SKL-motif is certainly not the only peroxisomal targeting signal. This is indicated by the finding that not all peroxisomal proteins contain an SKL or SKL-like motif which is particularly true for yeast enzymes [25]. Indeed, other sorting signals have been identified [26]. These signals are also located predominantly at the extreme C-terminus, although exceptions have been encountered [27]. In baker's yeast this is probably true for thiolase [9] whereas in *H. polymorpha* amine oxidase does not contain a C-terminal PTS [28].

Additional components essential for protein import/assembly probably are constitutively present in *H. polymorpha* since both alcohol oxidase and dihydroxyacetone synthase, artificially expressed under non-methylotrophic growth conditions, were imported and correctly assembled into the active protein inside the target organelle [29,30]. On the other hand amine oxidase, when artificially expressed under ammonium excess conditions (the homologous gene is then fully repressed), was only partially imported into *H. polymorpha* peroxisomes. However, this partial inhibition of amine oxidase import was fully abolished in the presence of an amine substrate (K.N. Faber and M. Veenhuis, unpublished results). The above phenomena might be explained by the presence of different receptor/translocator proteins for distinct classes of precursors (e.g. based on similarities in targeting sequences). This is not a hypothetical possibility as examples exist, for instance in mitochondrial protein import (recently reviewed by Wienhues and Neupert [31]). So far, two receptor proteins have been identified on the outer membrane of mitochondria of *Neurospora crassa*. MOM19 serves as 'master receptor' and functions in the specific recognition of most mitochondrial precursors, including those that carry an N-terminal targeting sequence. Another receptor protein, MOM72, specifically recognizes the inner membrane-bound ATP-ADP carrier (AAC), which contains an internal targeting signal. However, MOM72 mutants still display a residual AAC import which now occurs via MOM19. Evidently, MOM19 can substitute for MOM72, although at low efficiency. By analogy, import of AO and DHAS in *H. polymorpha*, which both contain a C-terminal PTS [26], may be facilitated by a general, constitutive import mechanism [16] which may partly substitute for a specific, substrate inducible, pathway involving a receptor for amine oxidase. A most intriguing question remaining, however, is whether different receptors interact with a common protein translocating machinery? Also, the question whether protein import/assembly is dependent on the energy status of the target organelle is still a matter of debate (see above).

3.5 Degradation of peroxisomes

As a rule, peroxisomal enzymes are not inactivated after a shift of cells to a new growth environment in which these enzymes have become redundant. Generally, the observed decrease in their specific activities can be accounted for by dilution of existing enzyme protein among newly formed cells following repression of their synthesis in the new medium. However, two main exceptions have been encountered. These include selective inactivation (e.g. by excess glucose) of i) alcohol oxidase in *H. polymorpha* and ii) amine oxidase in *Trichosporon cutaneum* [10]. In both organisms a rapid degradative turnover of the peroxisomal population was observed under these conditions. This process was energy-dependent but independent of protein synthesis. Peroxisomes were degraded individually by means of an autophagic process; hydrolytic enzymes required for the degradation of the peroxisomal contents were supplied by the vacuole [32].

The mechanisms triggering peroxisome turnover are still unknown. However, cytosolic peroxisomal enzymes in peroxisome-deficient mutants of *H. polymorpha* (for details see below) are no longer susceptible to carbon catabolite inactivation [33]. These results indicate that the mechanisms initiating proteolytic turnover of peroxisomal enzymes are not directed against the enzyme protein but instead to the peroxisomal membrane.

4 PEROXISOME DEFICIENT MUTANTS OF *H. polymorpha*

4.1 Mutant isolation and phenotypical characterization

All mutants described were identified within a collection of 260 methanol-utilization-defective (Mut$^-$) mutants of *Hansenula polymorpha*, previously described [33]. After incubation of the mutants in methanol-containing media, 85 strains were identified by electron microscopy as having one of the following peroxisomal defects: 1) complete absence of peroxisomes (Per$^-$ phenotype; 58 strains), 2) presence of only a few small peroxisomes along with the presence of the bulk of the peroxisomal matrix proteins in the cytosol (Pim$^-$ phenotype; 20 strains), and 3) aberrations in the peroxisomal sub-structure, i.e. presence of electron dense inclusions in the crystalline peroxisomal matrix (Pss$^-$ phenotype; 7 strains). In addition, a series of conditional mutants (6 complementation groups) have been isolated. The mutant phenotypes were genetically analysed and shown to be determined by single recessive genomic mutations. The different mutant phenotypes are described in more detail below.

(a) Per$^-$ phenotype. The Per$^-$ mutants were subdivided into 13 complementation groups. Representatives of the different groups showed comparable morphological phenotypes in that i) they are impaired to grow on methanol, but grow well on a range of other compounds including those that require the activity of peroxisomal enzymes in wild type cells; ii) all peroxisomal matrix and membrane proteins tested are normally induced and assembled; iii) all peroxisomal proteins, both inducible and constitutive, are located in the cytosol. Peroxisomal membrane proteins were located in small proteinaceous/phospholipid aggregates [35]. In fully derepressed cells cytosolic alcohol oxidase forms large crystalloids (Fig. 3A,D) in which the bulk of the other matrix proteins, except catalase, is also incorporated [36].

(b) Pim$^-$ phenotype. The Pim$^-$ mutants were organized into five different complementation groups which, as was the case with the Per$^-$ mutants, showed comparable phenotypes. The growth properties and peroxisomal enzyme ex-

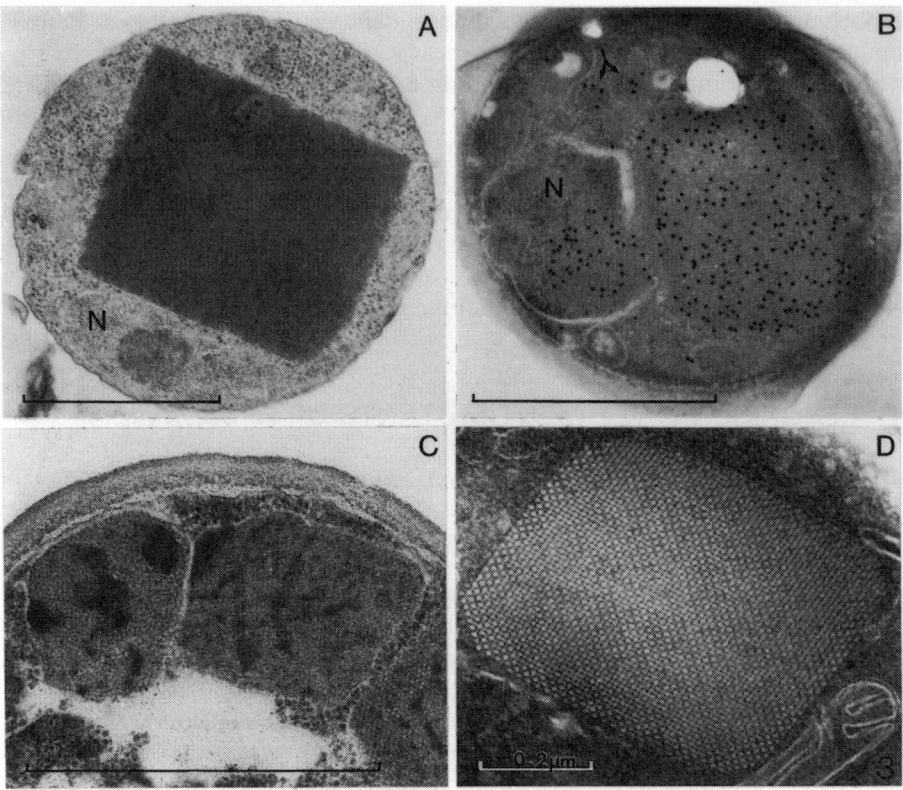

Figure 3 Fig. 3A shows the typical morphology of a *Permutant* of *H. polymorpha*, which fully lacks peroxisomal structures; when grown in chemostat cultures on glucose/methanol mixtures these cells contain large crystalloids, located in both the cytosol and the nucleus (Fig. 3A; spheroplast; glutaraldehyde/OsO$_4$)), which are composed of alcohol oxidase protein (Fig. 3D; ultrathin cryosection; uranylacetate). Pim$^-$ mutants possess small peroxisomes (Fig. 3B; arrows) which contain alcohol oxidase while bulk of the enzyme is located in the cytosolic and nucleus-bound crystalloids (protein A/gold; anti alcohol oxidase). Fig. 3C shows the peroxisomal morphology in a Pss$^-$ mutant of *H. polymorpha* which is characterized by electron dense aggregates in the crystalline matrix (spheroplast, glutaraldehyde/OsO$_4$).

pression levels of Pim$^-$- and Per$^-$-mutants are comparable. The major difference between the two classes of mutants is that in Pim$^-$-mutants, few small peroxisomes are still present. All peroxisomal proteins tested, irrespective of their inducible or constitutive nature, were located both in these organelles and in the cytosol (Fig. 3B), indicating a defect in a general major import pathway.

(c) Pss$^-$ phenotype. These mutants show normal peroxisome proliferation upon induction of cells on methanol but are characterized by abnormalities in organellar shape and matrix substructure due to improper assembly of part of

the matrix protein (Fig. 3C). Since all enzymes, involved in methanol metabolism are present and active at their correct location, we speculate that dysfunctioning of the organelles is due to a defect in peroxisome transport properties (e.g. dissipation of the proton motive force across the peroxisomal membrane [18]).

(d) Conditional mutants. i) temperature-sensitive (Ts) mutants. One group of Ts mutants show the Per⁻ phenotype (and fully lack peroxisomes) at the restrictive temperature (43°C) but have wild type properties (and contain intact peroxisomes) under permissive growth conditions (<37°C). Surprisingly, at intermediate temperatures (between 37–43°C) steady state conditions in chemostat culture have been obtained in which the cells display the Pim⁻ phenotype. ii) pH-mutants. Based on the screening procedure developed for the isolation of vacuolar pH-mutants of *S. cerevisiae* [37], different pH-mutants of *H. polymorpha* were isolated. Mutants were obtained which were unable to grow on methanol at pH 7.0 but show the normal wild type phenotype at pH 5.0. This pH-effect is specific for the utilization of methanol; growth on other substrates which require the activity of peroxisomal enzymes is not affected. In these mutants the peroxisomal proliferation and protein composition is unaltered; in cells grown at either pH the different key enzymes of methanol metabolism are peroxisome-bound. However, at pH 7.0 improperly assembled matrix protein (e.g. alcohol oxidase) was observed, whereas the enzyme was normally oligomerized in cells from cultures grown at pH 5.0. Therefore, these mutants appeared to be impaired in protein assembly rather than protein import.

4.2 Genetic analysis

As stated above, all *per* mutant phenotypes are due to monogenetic defects; this implies that the absence (or incorrect synthesis) of a single gene product may cause the absence of a complete organelle (the peroxisome). This unique property renders these mutants very attractive for a molecular analysis of peroxisome biogenesis.

All *PER* genes have been subjected to a detailed genetic analysis. Complementation and linkage analysis showed that the total *per* mutant population represented mutations in 14 genes. Figure 4 depicts the linkage relationships and map positions of the corresponding genes as deduced from random spore analysis. Twelve genes involved in peroxisome biogenesis were defined and designated *PER1* through *PER12;* ten of these genes were mapped in three separate linkage fragments. Two genes *PER8* and *PER12*, were unlinked to each other, to any of the *PER* genes indicated above (Fig. 4) or to *LEU1* and *ADE11* loci. In addition to defining the number and relative positions of *PER* genes, the combined complementation and genetic linkage analysis also made clear that Pim⁻, Per⁻ and Pss⁻ phenotypes were allele specific and not gene specific since, in at least two instances, different alleles of one gene displayed different peroxisome-defective phenotypes [38], indicating that the protein

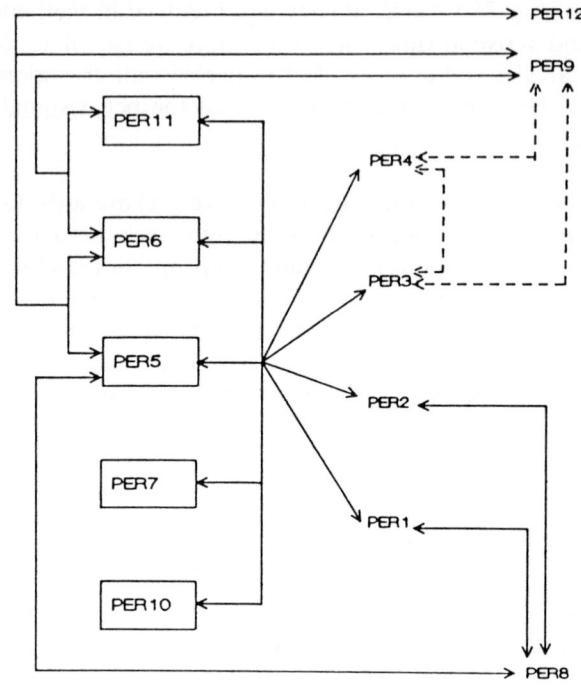

Figure 4 Interactions between products of *PER1* to *PER12* genes deduced from unlinked non-complementation data. The dotted lines indicate 'weak' interactions ('leaky' growth of the corresponding double heterozygous hybrids at restrictive temperatures). From [40].

products encoded by these genes may be part of multifunctional complexes, which participate in different aspects of peroxisome assembly/proliferation. A comparable model of functional multiplicity has been proposed to explain defects in both Golgi to vacuole sorting and vacuole segregation caused by *vac1* mutations [39].

These studies also revealed several cases of an unusual complementation behavior, namely that of unlinked noncomplementation (defined as the failure of recessive mutations in different genes to complement) between different *per* alleles, especially at low temperature, thus suggesting functional and physical links between their gene products [40].

5 FUNCTIONAL ASPECTS OF PEROXISOME METABOLISM

Physiological studies indicated that not all functions which in wild type cells are mediated by intact peroxisomes are impaired in Per$^-$ mutants of *H. polymorpha*. For instance, all mutants grew well on glucose in the presence of different organic nitrogen sources which require the activity of peroxisomal enzymes (e.g. (m)ethylamine and D-amino acids) [41]. Therefore, different peroxisomal enzymes may also effectively function in the cytosol. However, growth on other substrates is partly (ethanol) or fully impaired (methanol), despite the fact that activities of all enzymes involved in the metabolism of these compounds are present in the cytosol. Therefore, these mutants are suitable model systems to study functional aspects of peroxisomes, particularly with respect to the fundamental question of a possible advantage for the organism to compart-talize certain metabolic pathways. The results obtained so far indicate that these advantages (at least in part) may be dependent on the growth conditions. In the case of ethanol metabolism the main advantage of intact microbodies is that it enables the cell to adjust the levels of different intermediates required to generate aspartate (the final product of microbody C_2-metabolism) from isocitrate and, in particular, to prevent a drain of oxaloacetate into other metabolic pathways [42]. Methanol, however, can not be used as a carbon source by *per* mutants but may serve as an additional energy source in glucose-limited chemostat cultures [43]. These studies showed that intact, functional peroxisomes are indispensible to support growth of *H. polymorpha* on methanol (Fig. 5) and suggested that the essential functions of peroxisomes in methanol metabolism are twofold namely to facilitate i) proper partitioning of formaldehyde, generated from methanol, over the dissimilatory

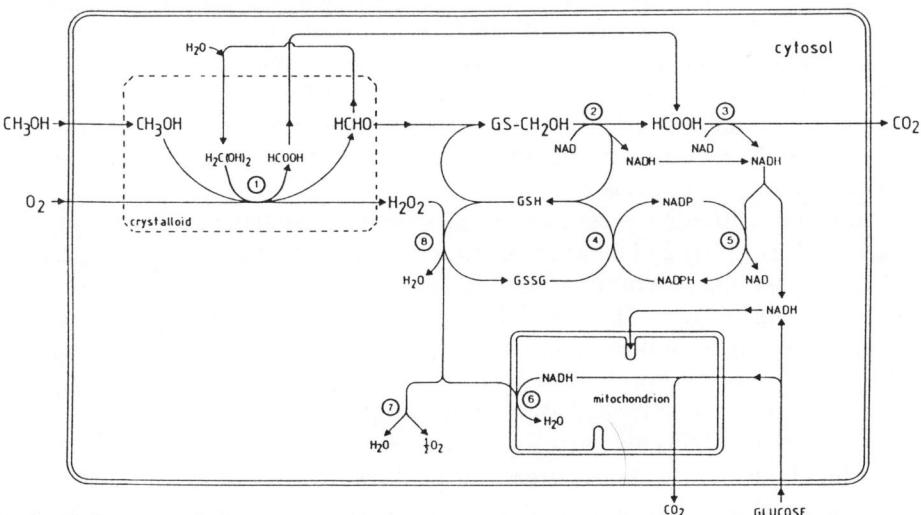

Figure 5 Hypothetical scheme of methanol metabolism in a peroxisome-deficient mutant of *H. polymorpha*. 1. alcohol oxidase (catalyzes both the oxidation of methanol and methylene glycol, the hydrated form of formaldehyde), 2. formaldehyde dehydrogenase, 3. formate dehydrogenase, 4. glutathione reductase, 5. NADH/NADPH transhydrogenase, 6. cytochrome C peroxidase, 7. catalase, 8. chemical oxidation of glutathione by H_2O_2 (reproduced from [43]).

and assimilatory pathways and ii) decomposition (by catalase) of H_2O_2 at the site where it is produced, thus preventing decomposition of hydrogen peroxide by energy consuming processes.

6 CONCLUSIONS

In the past few years much progress has been made towards the elucidation of various aspects of peroxisome biogenesis and function. Yeast have developed into very useful model organisms for these studies because they combine a number of distinct advantages: ease of growth, accessibility to both classical and molecular genetics and the possibility to manipulate peroxisome proliferation and their protein composition. Now that numerous peroxisome-deficient mutants (*per* and *pas* mutants) of different yeasts are available, a molecular analysis of gene products involved in peroxisome biogenesis and function has come within reach. Such studies are currently under way in several laboratories. Moreover, these mutants may become important as a tool for cloning *PER/ PAS* homologues from other sources (including man) by functional complementation of the corresponding yeast mutants with a heterologous genomic bank. Undoubtedly, future studies will not only continue to analyze the functional role of the various available *PER/PAS* genes in peroxisome proliferation/function but also focus on the mechanisms controlling protein import and assembly (role of cytosolic and peroxisomal chaperon(in)s, receptor/ translocation pathways). A major obstacle for a rapid expansion of this field has been the absence of reliable *in vitro* assays. Despite recent progress, the efficiencies of these systems are very low, especially for yeast. Also, establishing unambiguous import criteria akin to those developed for mitochondria (e.g. protein modification or processing) and criteria for the intactness of peroxisomes are essential in this respect.

Until now, isolated yeast peroxisomes or peroxisomal membranes are invariably leaky [20] and all attempts to restore the original *in vivo* energization of the peroxisomal membrane have so far failed, because of the presence of a pore-forming protein. The possibility to regulate this porin [23,24] has now opened the way for future studies on the possible energy dependency of various transport processes (protein import as well as transport of substrates and metabolic products/intermediates) across the peroxisomal membrane.

The purpose of this overview was to demonstrate that *Hansenula polymorpha* is a highly attractive model organism for these studies. Not only has it the advantage that it harbors various crucial peroxisomal functions which can readily be manipulated, also the availability of various mutants affected in peroxisome assembly and function [16,34,40], the identification of physical interactions between different *PER* gene products [40] in conjunction with recent major advances in the molecular genetics [44] and the extensive knowledge on the physiology/ biochemistry of the organism [7,10] renders *H. polymorpha* the organism of choice for such studies.

REFERENCES

[1] Cavalier-Smith, T. (1987) Eukaryote cell evolution. In: XIVth Int. Bot. Congress, Berlin, W. Germany (Greuter, W., Zimmer, B. and Behnke, H.D., Eds).
[2] Lazarow, P.B. and Fuyiki, Y. (1984) Biogenesis of peroxisomes. Ann. Rev. Cell Biol. 1, 489–530.
[3] Fahimi, H.D. and Sies, H. (1987). Peroxisomes in Biology and Medicine. Springer-Verlag KG, Berlin.
[4] Borst, P. (1989). Peroxisome biogenesis revisited. Biochim. Biophys. Acta **1008**, 1–13.
[5] Wanders, R.J.A., Heymans, H.S.A., Schutgens, R.B.H., Barth, P., van den Bosch, H. and Tager, J.M. (1988). Peroxisomal disorders in neurology. J. Neurol. Sci **88**, 1–39.
[6] Avers. C.J. and Federman, M. (1968) The occurrence in yeast of cytoplasmic granules which resemble microbodies. J. Cell Biol **37**, 555–559.
[7] Veenhuis, M. and Harder, W. (1989) Occurrence, proliferation and metabolic function of yeast microbodies. Yeast **5**, 517–524.
[8] Höhfeld, J., Mertens, D., Wiebel, F.F. and Kunau, W.H. (1992) Defining components required for peroxisome assembly in Saccharomyces cerevisiae. In: New Comprehensive Biochemistry, Bd.: Membrane biogenesis and Protein Targeting (Neupert, W. and Lill, R., Eds) Elsevier, New York, in press.
[9] Kunau, W.-H. and Hartig, A. (1992) Peroxisome biogenesis in Saccharomyces cerevisiae. Antonie van Leeuwenhoek, in press.
[10] Veenhuis, M. and Harder, W. (1991). Microbodies. In: The Yeasts (A.H. Rose and J.S. Harrison, Eds), Vol. 4, 2nd ed., pp. 601–653, Academic Press, London, New York.
[11] Veenhuis, M., Kram, A.M., Kunau, W.H. and Harder, W. (1990) Excessive membrane development following exposure of the methylotrophic yeast Hansenula polymorpha to oleic acid-containing media. Yeast **6**, 511–519.
[12] Veenhuis, M., van der Klei, I.J., Sulter, G. and Harder, W. (1989). Evidence for functional heterogeneity among microbodies in yeasts. Arch. Microbiol. **151**, 105–110.
[13] Waterham, H., Keizer-Gunnink, I., Goodman, J.M., Harder, W., and Veenhuis, M. (1992) Development of multi-purpose peroxisomes in the yeast Candida boidinii grown in oleic acid plus methanol limited chemostat cultures. J. Bact. **174**(12), 4057–4063.
[14] Gödecke, A., Veenhuis, M., Roggenkamp, R., Janowicz, Z.A. and Hollenberg, C.P. (1989) Biosynthesis of the peroxisomal dihydroxyacetone synthase from Hansenula polymorpha in Saccharomyces cerevisiae induces growth but not proliferation of peroxisomes. Curr. Gen. **16**, 13–20.
[15] de Hoop, M.J., Cregg, J., Keizer-Gunnik, I., Sjollema, K., Veenhuis, M. and Ab, G. (1991) Overexpression of alcohol oxidase in Pichia pastoris. FEBS Letters **291**, 299–302.
[16] Waterham, H.W., Titorenko, V., van der Klei, I.J., Harder, W. and Veenhuis, M. (1992) Isolation and characterization of peroxisomal import (Pim$^-$) mutants of Hansenula polymorpha. Yeast **8**, 961–972.
[17] Sulter, G.J., Looyenga, L., Veenhuis, M. and Harder, W. (1990). Occurrence of peroxisomal membrane proteins in methylotrophic yeasts grown under different conditions. Yeast **6**(1), 35–44.
[18] Nicolay, K., Veenhuis, M., Douma, A.C. and Harder, W. (1987) A ^{31}P NMR study of the internal pH of yeast peroxisomes. Arch. Microbiol. **147**, 37–41.
[19] Waterham, H.R., Keizer-Gunnink, I., Goodman, J.M. and Veenhuis, M. (1990) Immunocytochemical evidence for the acidic nature of peroxisomes in methylotrophic yeasts. FEBS Lett. **262**, 17–19.
[20] Douma, A.C., Veenhuis, M., Sulter, G.J., Waterham, H.R., Verheyden, K.,

Mannaerts, G.P. and Harder, W. (1990) Permeability properties of peroxisomal membranes from yeasts *Arch. Microbiol.* **153**, 490–495.

[21] van Veldhoven, P. Just, W.W. and Mannaerts, G.P. (1987) Permeability of the peroxisomal membrane to cofactors of β-oxidation. *J. Biol. Chem.* **262**, 4310–4318.

[22] Sulter, G.J., Verheyen, K., van der Veen, I.J., Mannaerts, G.P., Harder, W. and Veenhuis, M. (1993) The permeability of yeast microbody membranes is caused by an 31 kD integral membrane protein. *Yeast*, in press.

[23] Lemmens, M., Sulter, G.J., Vereecke, J., Mannearts, G.P., Veenhuis, M. and Carmeliet, E. (1993) Large conductance channels in peroxisomes and mitochondria from the yeast *Hansenula polymorpha. Biochem. Biophys. Acta*, submitted.

[24] Gould, S.J., Keller, G.A., Schneider, M., Howell, S.H., Garrard, L.J., Goodman, J.M., Distel, B., Tabak, H.F. and Subramani, S. (1990) Peroxisomal protein import is conserved between yeast, plants, insects and mammals. *EMBO J.* **9**, 85–90.

[25] Bruinenberg, P. (1988) PhD Thesis, Groningen University, The Netherlands.

[26] Hansen, H., Didion, T., Thiemann, A., Veenhuis, M. and Roggenkamp, R. (1992) Targeting sequences of the two major peroxisomal proteins in the methylotrophic yeast *Hansenula polymorpha. J. Gen. Microbiol.*, **235**, 269–278.

[27] Swinkels, B.W., Gould, S.J., Bodnar, A.G., Rachubinski, R.A. and Subramani, S. (1991) *EMBO J.* **10**, 3255–3262.

[28] Faber, K.N., Haima, P., de Hoop, M.J., Harder, W., Veenhuis, M. and Ab, G. (1993) Amine oxidase of *Hansenula polymorpha* does not require its SRL-containing C-terminal sequence for peroxisomal targeting. *Yeast*, submitted.

[29] Gödecke, A., Veenhuis, M., Roggenkamp, R., Janowicz, Z.A. and Hollenberg, C.P. (1989) Biosynthesis of the peroxisomal dihydroxy acetone synthase from *Hansenula polymorpha* in *Saccharomyces cerevisiae* induces growth but not proliferation of peroxisomes. *Curr. Gen.* **16**, 13–20.

[30] Distel, B., van der Ley, I., Veenhuis, M. and Tabak, H. (1988) Alcohol oxidase expressed under non-methylotrophic conditions is imported, assembled and enzymatically active in peroxisomes of *Hansenula polymorpha. J. Cell Biol.* **107**, 1669–1675.

[31] Wienhues, U. and Neupert, W. (1992) Protein translocation across mitochondrial membranes. *Bioassays* **14**(1), 17–23.

[32] Veenhuis, M., Douma, A., Harder, W. and Osumi, M. (1983) Degradation and turnover of peroxisomes in the yeast *Hansenula polymorpha* induced by selective inactivation of peroxisomal enzymes. *Arch. Microbiol.* **129**, 35–41.

[33] van der Klei, I.J., Harder, W. and Veenhuis, M. (1991) Selective inactivation of alcohol oxidase in two peroxisome-deficient mutants of the yeast *Hansenula polymorpha. Yeast* **7**, 813–821.

[34] Cregg, J.M., van der Klei, I.J., Sulter, G.J., Veenhuis, M. and Harder, W. (1989) Peroxisome deficient mutants of *Hansenula polymorpha. Yeast* **6**, 87–97.

[35] Sulter, G.J., Vrieling, E.G., Harder, W. and Veenhuis, M. (1993) Expression and subcellular location of peroxisomal membrane proteins in a peroxisome-deficient mutant of the yeast *Hansenula polymorpha. EMBO J.*, submitted.

[36] van der Klei, I.J., Sulter, G.J., Harder, W. and Veenhuis, M. (1991) Expression, assembly and crystallization of alcohol oxidase in a peroxisome-deficient mutant of *Hansenula polymorpha;* properties of the protein and architecture of the crystals. *Yeast* **7**, 15–24.

[37] Rothman, J.H., Howard, I. and Stevens, T.H. (1989) Characterization of genes required for protein sorting and vacuolar function in the yeast *Saccharomyces cerevisiae EMBO J.* **8**, 2057–2065.

[38] Titorenko, V., Waterham, H.R., Haima, P., Harder, W. and Veenhuis, M. (1992) Peroxisome biogenesis in *Hansenula polymorpha:* different mutations in genes, essential for peroxisome biogenesis, cause different peroxisomal mutant phenotypes. *FEMS Microbial Lett.* **95**, 143–148.

[39] Weisman, L.S. and Wickner, W. (1992) Molecular cloning of VAC1, a gene required for vacuole inheritance and vacuole protein sorting. *J. Biol. Chem.* **267**, 618–623.

[40] Titorenko, V., Haima, P., Waterham, H. R., Harder, W. and Veenhuis, M. (1993) A complex set of interacting genes controls peroxisome biogenesis in *Hansenula polymorpha*. *Proc. Nat. Acad. Sci.*, in press.

[41] Sulter, G.J., van der Klei, I.J., Harder, W. and Veenhuis, M. (1990) Expression and assembly of amine oxidase and D-amino acid oxidase in the cytoplasm of peroxisome-deficient mutants of the yeast *Hansenula polymorpha* during growth on primary amines or D-alanine as the sole nitrogen source. *Yeast* **6**, 501–507.

[42] Sulter, G.J., Schanstra, J., Harder, W. and Veenhuis, M. (1991) Ethanol metabolism in a peroxisome-deficient mutant of *Hansenula polymorpha*. *FEMS Microbial. Lett.*, **82**, 297–302.

[43] van der Klei, I.J., Harder, W. and Veenhuis, M. (1991) Methanol metabolism in a peroxisome-deficient mutant of *Hansenula polymorpha:* a physiological study. *Arch. Microbiol.* **156**, 15–23.

[44] Faber, K.N., Swaving, G.J., Faber, F., AB, G., Harder, W., Veenhuis, M. and Haima, P. (1992) Chromosomal targeting of replicating plasmids in the yeast *Hansenula polymorpha*. *J. Gen. Microbiol.* **138**, 2405–2416.

16

Structural Aspects of Methanol Oxidation in Gram-negative Bacteria

J. Frank[1], M. Janvier[2], I. Heiber-Langer[3], J.A. Duine[1], F. Gasser[2] and C. Balny[3]

[1]*Department of Microbiology and Enzymology, Delft University of Technology, Julianalaan 67, NL-2628 BC Delft, The Netherlands;* [2]*Unité de Régulation de l'Expression Génétique, Département de Biochimie et Génétique Moléculaire, Institut Pasteur, 28 rue du Dr. Roux, F-75724 Paris Cedex 15, France;* [3]*Institut National de la Santé et de la Récherche Médicale, Unité 128, BP 5051, F-34033 Montpellier Cedex, France*

1 SUMMARY

Gram-negative bacteria oxidize methanol with a periplasmic methanol dehydrogenase (MDH), a quinoprotein with pyrrolo-quinoline quinone (PQQ) as a cofactor. Methanol is oxidized by covalent catalysis, i.e. following addition to the protein-bound PQQ. It is not known yet how electrons are abstracted from methanol but they are transferred at a high rate to cytochrome c_L, the natural electron acceptor. MDH contains two types of subunits, arranged in a tetrameric $\alpha_2\beta_2$ structure.

Similar properties have been revealed for the MDH from *Methylophaga marina*, a restricted facultative methylotrophic organism, assimilating formaldehyde by the ribulose monophosphate (RuMP) pathway. In addition to methanol and methylamine, *M. marina* is also able to grow on fructose. Several mutants impaired in methanol oxidation, but still producing MDH related protein, have been isolated and it was shown that the primary defect is located either in the biosynthesis of the β-subunit and cytochrome c_L or that of PQQ.

The mutant MDH proteins are still able to bind PQQ, but this does not lead to restoration of enzymatic activity. Further characterization of these proteins suggests that PQQ binds to the α-subunit and that besides the

β-subunit, PQQ and Ca^{2+} other factors must be involved in reconstitution of an active holoenzyme.

It has been shown that electron transfer between cytochrome c_L and MDH involves electrostatic interactions, judged from the strong effect of the ionic strength on the reaction rate. High pressure experiments have shown that this process is accompanied by a relatively large volume change that is specific for the protein-protein interaction.

2 INTRODUCTION

The microbial oxidation of methanol clearly illustrates that the evolution of the capability to grow on a given substrate does not necessarily lead to a single enzyme system, the most efficient one. Indeed, both pyridine nucleotide and flavin cofactors have been adopted to perform this task and, in addition, a less universal cofactor has been adopted as well. Methylotrophic yeasts grow on methanol by virtue of an octameric flavoprotein methanol oxidase with a high molecular mass (660 kDa), tightly assembled with a catalase in peroxisomes [1]. The cofactor of this enzyme can be reversibly removed and consists of equal amounts of FAD and a FAD derivative modified in the ribityl side chain [2]. In contrast, bacteria exclusively use dehydrogenases to oxidize methanol, such as the quinoprotein methanol dehydrogenase (MDH). This enzyme is located in the periplasm of Gram-negative bacteria and it contains the unusual cofactor PQQ (2,7,9-tricarboxy-1H-pyrrolo(2,3f) quinoline-4,5-dione). Gram-positive bacteria have a different structural organisation and possibly related to this, other enzymes have emerged in these organisms; so far, a NAD-dependent MDH [3] and tetrazolium dye- and N,N'-dimethyl-4-nitrosoaniline dependent enzymes ([4], P.W. van Ophem, personal communication) have been described. Nothing is known about the nature of the cofactor of the latter two enzymes, although a possible involvement of PQQ for the latter has been suggested [5]. Apparently, these systems have evolved to a sufficiently high level of sophistication to assure the survival of the methylotrophic organisms elaborating them, but it remains worthwhile to compare the performance of the different types of enzymes by looking at their specificity for methanol, expressed as k_{cat}/K_m. This ratio has the dimension of a second order (diffusion) rate constant and therefore when k_{cat}/K_m is equal to the diffusion rate constant of the substrate, the catalytic power of the enzyme cannot be further improved. Diffusion rates of substrates in water are typically about $10^8 M^{-1} s^{-1}$, in the more viscous cytoplasm one or two orders of magnitude lower [6]. Hence, by that criterion, the quinoprotein MDH's with k_{cat}/K_m ranging from 0.6 to $3 \times 10^8 M^{-1} s^{-1}$ [7] clearly have evolved to the stage of 'perfect' enzymes [6]. A substantially lower specificity, $k_{cat}/K_m = 2\text{--}5 \times 10^5 M^{-1} s^{-1}$ is measured for methanol oxidase, but as we do not know the actual diffusion rate constant of methanol in microbodies, it might well be that this figure represents nevertheless an optimum. Dramatically lower values are reported for the methanol oxidizing enzymes in Gram-positive bacteria, ranging from 18 to 2600 $M^{-1} s^{-1}$ ([4,9], P.W. van Ophem, personal communication). Similar values are obtained for the oxidation of ethanol and this should be compared with horse liver alcohol dehydrogenase, one of the few class I ADH's capable of attacking methanol, having a $k_{cat}/K_m = 1.6 M^{-1} s^{-1}$, ten thousand times less than that for ethanol [10]. Hence, although it is hard to believe that these low k_{cat}/K_m values represent the actual catalytic power displayed *in vivo*, they nevertheless suggest that the enzymes are indeed designed for the oxidation of methanol.

Perhaps because of toxicity of the product of methanol oxidation, formaldehyde, or its relatively high redox potential, the different systems all have more or less unusual properties. For example an activator like ammonia or an amine is required for activity of quinoprotein MDH *in vitro* and it has been shown that this activator is involved in the methanol oxidation step [11]. Furthermore, a protein has been found that prevents the non-physiological, oxidation of formaldehyde by quinoprotein MDH [12]. Another example of a protein-protein interaction with MDH is the need for an activator protein in a Gram-positive thermotolerant *Bacillus* sp. to obtain full NAD-dependent MDH activity [9].

It is clear from the foregoing that a more pertinent explanation of the peculiar properties of the methanol oxidation systems will be possible only when more structural and mechanistic information is obtained, such as the three-dimensional structure that is expected to be available soon [13,14]. This paper focuses on additional structural and dynamic aspects of protein-protein and protein-cofactor interactions in a quinoprotein MDH from the Gram-negative bacterium *Methylophaga marina*.

3 STRUCTURAL ASPECTS OF METHANOL DEHYDROGENASE

3.1 *MDH from* Methylophaga marina

Methylophaga marina is a Gram-negative methylotroph able to use the C_1 compounds methanol and methylamine as a sole source of carbon and energy while the formaldehyde produced from these substrates is assimilated through the ribulose monophosphate (RuMP) pathway [15].

Organisms containing the RuMP pathway generally are obligate methylotrophs, growing only on methanol and sometimes methylamine, and are grouped in two genera designated as *Methylobacillus* [16] and *Methylophilus* [17]. *Methylophaga* clearly belongs to a different taxonomic group in view of its different GC content and the absence of DNA hybridization with the members of the other two genera [15]. Furthermore, the organism is distinct in being able to grow on the multicarbon compound fructose and this property has been exploited to isolate mutants affected in C_1 metabolism [18] to explore the factors involved in methanol oxidation in RuMP methylotrophs. Nothing is known yet about the pathway of fructose catabolism.

Methylophaga marina constitutively synthesizes a quinoprotein MDH and two cytochromes c, cytochrome c_L and cytochrome c_H [18], involved in electron transfer from MDH to cytochrome c oxidase. Like MDH from *Methylobacterium extorquens* AM1 and other methylotrophs [19] the enzyme contains two large α-subunits (65 kDa) and two rather small β-subunits (6.5 kDa), clustered in an $\alpha_2\beta_2$-tetramer with two molecules of PQQ.

3.2 Mutants of *Methylophaga marina*

Upon treatment of wild type *Methylophaga marina* with methane sulfonic acid ethyl ester or N-methyl-N'-nitro-N-nitrosoguanidine and screening for mutants impaired in methanol metabolism, three classes of mutants can be distinguished [18].

Class I mutants are impeded in the synthesis of MDH, cytochrome c_L and PQQ. Reversion to the wild type phenotype is possible and occurs in a single step, suggesting that they are affected in a single regulatory gene governing a common control of PQQ production and of the biosynthesis of MDH and cytochrome c_L.

Mutants of class II do not produce active MDH, but SDS-PAGE of a cell free extract shows a normal amount of protein at the position of the α-subunit, which is, like MDH, located in the periplasmic space. Purification and further analysis of this protein reveal that it does not contain PQQ and elutes with the same retention time from a gel filtration column as wild type MDH. Gel filtration under denaturing conditions, however, clearly shows that the small β-subunit (representing only 9% of the total molecular mass) is not present. In addition, none of these mutants produce cytochrome c_L and some of them are also impaired in PQQ biosynthesis. Spontaneous revertants of the latter category of class II mutants (grouped in class III) recover in a single step the β-subunit as well as cytochrome c_L. However, PQQ biosynthesis is not restored in this process, indicating that the PQQ deficiency of the mutants is caused by an independent mutation. Apparently, the biosynthesis of the β-subunit and cytochrome c_L are closely related, a fact that can be explained if the structural genes for MDH in *M. marina* are arranged in a similar manner as in other methylotrophs [20,21], i.e. a single operon containing *mox* F (α-subunit), *mox* J (unknown), *mox* G (cytochrome c_L) and *mox* I (β-subunit). It has been shown by van Spanning *et al.* that a single mutation in *mox* J or G in *Paracoccus denitrificans* affects both the biosynthesis of cytochrome c_L and that of the β-subunit, while normal amounts of α-subunit are produced [21], thus resulting in the phenotype typical for class II mutants. Therefore it is likely that the mutant protein found in class II mutants is in fact the α-subunit of MDH.

3.3 Properties of the mutant system

3.3.1 *Protein from class II mutants* The β-subunit of *M. marina* MDH has a strongly basic character (unpublished result), consistent with the large number of lysines found in the deduced amino acid sequences of the β-subunit in *M.*

extorquens AM1 and *P. denitrificans* [19,22]. Therefore it can be anticipated that the α-subunit isolated from class II mutants should have a more acid character than MDH. Indeed, a stronger retention in anion exchange chromatography and a different mobility in PAGE gels at pH 4.5 is observed. Most curiously, this α-subunit does not cross-react with anti-MDH serum, but, inversely, an antiserum raised against the purified α-subunit recognizes wild type MDH. Apparently, the β-subunit induces significant conformational changes in the α-subunit, leading to different epitopes.

Using the antiserum raised against the α-subunit it appears that wild type *M. marina* grown on fructose also produces small amounts of cross-reacting protein. This probably results from either a slight overproduction of α-subunits, or an inefficient assembly of α- and β-subunits [18].

Gel filtration experiments show that upon storage the α-subunits, which are originally present as dimers, spontaneously dissociate, indicating that the β-subunit plays a role in maintaining the quaternary structure. Interestingly, the monomers are still able to bind PQQ and during that process a concomitant dimerisation of the monomers takes place. A similar phenomenon is observed when α-subunits, prepared by denaturation of MDH with 6 M guanidine, are refolded in the presence of PQQ. PQQ is only found in the dimer and the resulting absorbance spectrum reflects that of PQQ but with the absorbance maxima red-shifted to 365 and 520 nm, hence different from all the known spectra of MDH. The reconstituted protein is devoid of enzymatic activity with methanol, but it still has redox properties (discussed in section 4). Thus far, recombination of the α-subunits with β-subunits extracted from wild type MDH has not been successful (unpublished results).

Cofactor induced association of subunits is a rare phenomenon. For instance, the monomer-dimer equilibrium of the $β_2$ component of tryptophan synthetase shifts to the dimer upon addition of pyridoxal phosphate [22] and Schiff-base formation occurs subsequent to dimer formation [23], suggesting a mutual interaction between cofactor and protein. Pyruvate decarboxylase from *S. cerevisiae* is another example. At alkaline pH this enzyme dissociates into monomers and thiamine pyrophosphate is an obligatory and specific cofactor for reconstitution to the oligomer [24]. This requirement has been interpreted as a conformational change induced by the cofactor, promoting subunit association [25]. Finally, *in vitro* reconstitution of NADP-nitrate reductase activity from *Neurospora crassa* is reported to involve the assembly of two subunits (covalently?) linked together by the molybdenum cofactor [26]. Since MDH also occurs in a semiquinone form [11], it seems unlikely that PQQ acts in a similar way, i.e. close to the interface between the two subunits. Instead, induction of a conformational change is a more attractive hypothesis, from which a relatively strong interaction can be inferred between cofactor and protein, perhaps explaining why PQQ is so tightly bound in MDH.

3.3.2 *ApoMDH* Many quinoprotein aldose- and alcohol dehydrogenases are synthesized as apoenzymes or can be readily converted to the apoform [27], but, unfortunately, this is not the case for the MDHs. Attempts to isolate an intact and reconstitutable apoMDH from PQQ^- mutants of *M. organophilum* have failed, probably due to an inherent instability of the apoMDH in the absence of PQQ [28]. Similarly, reconstitution of apoMDHs prepared from the MDHs of *Methylomonas methanica* and Bacterium W3A1 did not result in catalytically active holoenzymes [29,30] although in the latter case binding of the resolved cofactor by the apoenzyme could be demonstrated. Remarkably the purified W3A1 apoenzyme partly dissociates as evidenced by gel filtration chromatography, but reassociates upon addition of PQQ. Unfortunately, it is not known whether an apoenzyme prepared in that way still has its β-subunits or only consist of α-subunits, in which case the phenomenon is analogous to that mentioned in the foregoing section. Similarly, it is not clear whether the failure to observe enzymatic activity is due to the omission of Ca^{2+} in the reconstitution assay, since the presence of this cation in the catalytic site has now been established [31].

Class III PQQ^- mutants of *M. marina* can grow in methanol medium supplemented with PQQ. When grown on fructose medium in the absence of PQQ they constitute a potential source of apoMDH, especially because their cell free extracts contain a protein that cross-reacts with anti-MDH serum. Indeed, a protein containing both α- and β-subunits could be isolated. It behaves as a stable $\alpha_2\beta_2$-structure in a gel filtration column, indicating that the β-subunit is capable, like PQQ, of keeping the α-subunits together.

Methylophaga marina apoMDH stoichiometrically binds PQQ and, most unexpectedly, the enzymatically inactive holoenzyme has an absorption spectrum very similar to that of oxidized MDH prepared in the presence of cyanide, or irreversibly inhibited by cyclopropanol [18]. The explanation may be that whereas the holoenzyme is still able to bind substrate (which is ever present in the form of endogenous substrate), as a C(5)-carbonyl adduct of PQQ, it is unable to oxidize it. Ca^{2+} in the reconstitution medium has no effect, in contrast to the recovery of activity and normal absorbance spectrum that is observed when an inactive purple form of MDH isolated from fructose grown wild type cells of *M. marina*, is incubated with Ca^{2+} ions (unpublished results). This indicates that besides Ca^{2+} and PQQ additional factors are required for enzymatic activity.

4   ELECTRON TRANSFER BETWEEN MDH AND CYTOCHROME c_L

4.1   *Cytochrome c_L and the redox cycle of MDH*

Although much, indirect, evidence accumulated early, indicating that cytochrome c_L is the natural electron acceptor for MDH [7], direct proof was obtained when it was shown that methanol dependent reduction of horse heart cytochrome c only takes place in the presence of cytochrome c_L [31a], further substantiated by the finding that MDH from *Hyphomicrobium* X is rapidly oxidized by cytochrome c_L, but not by other cytochromes c, such as cytochrome c_H [32]. Similarily, electron transfer has been reported between components from other methylotrophic bacteria [33].

The redox cycle of MDH has been reviewed previously [34]. Basically, after a 2-electron reduction of MDH by substrate, the enzyme is oxidized in two 1-electron steps by either cytochrome c_L or an artificial electron acceptor. Addition of substrate to the C(5) carbonyl of PQQ (covalent catalysis) in the oxidized form of MDH then occurs, followed by electron transfer [32]. By analogy with *Hyphomicrobium* MDH, the enzyme from *M. marina* is specifically oxidized by cytochrome c_L [18].

4.2 Electron transfer

The reaction of MDH with cytochrome c_L is an electron transfer reaction where the first steps follow the minimum reaction scheme:

$$\text{MDH}_{sem} + \text{Cyt } c_L^{ox} \underset{k_{-1}}{\overset{k_1}{\rightleftarrows}} [\text{MDH}_{sem} - \text{Cyt } c_L^{ox}] \underset{k_{-2}}{\overset{k_2}{\rightleftarrows}} [\text{MDH}_{ox} - \text{Cyt } c_L^{red}]$$

with Cyt c_L^{ox}, Cyt c_L^{red}, MDH_{ox} and MDH_{sem} being the oxidized and reduced forms respectively. This mechanism has also been proposed for other electron transfer reactions [32,35–37].

To measure the rate of electron transfer between MDH and cytochrome c_L, autoreduction of the cytochrome [33] should be completely eliminated. Therefore careful purification is mandatory, since even apparently pure preparations may contain MDH as can be detected using an antiMDH serum or a sensitive test for PQQ (unpublished result).

Interestingly, the electron transfer reaction between cytochrome c_L and MDH from *Methylophaga marina* shows saturation behavior at pH 7. Assuming that the first step of the reaction is fast compared to the second ($k_1 \gg k_2$) and the backward reaction (k_{-2}) for the reduction step can be neglected, the observed rate constant can be described by:

$$k_{obs} = k_2 \cdot K_1 \cdot [\text{MDH}_{sem}]/K_1 \cdot [\text{MDH}_{sem}] + 1) \tag{1}$$

leading to $K_1 = 20 \pm 3 \cdot 10^4 \text{ M}^{-1}$ and $k_2 = 0.9 \pm 0.09 \text{ s}^{-1}$. The PQQ-α-subunit complex can also be slowly reduced by methanol at pH 9.0, but only when ammonia is present. However, a negligible rate is observed when this reduced complex is oxidized with cytochrome c_L (unpublished results).

The rate of electron transfer between cytochrome c_L and MDH decreases at high ionic strength, a behavior typical for an electrostatic type of interaction between the two proteins as shown for other systems [36,38,39]. However, at up to 20 mM NaCl the reaction rate increases with increasing ionic strength, which might be explained by assuming that extensive protein conformation and/or hydration changes occur during the binding at low salt concentration. A similar explanation has been proposed for enzyme-substrate binding in the carbamylation reaction of butyrylcholinesterase [40].

The interaction between the two proteins has also been studied as a function of temperature. The dependence of the observed reaction rate on the temperature is expressed by the Arrhenius equation, $(\delta \ln k_{obs}/\delta\ 1/T)_P = -\Delta H_{obs}‡/R$, where $\Delta H_{obs}‡$ is the activation enthalpy. The equation predicts a linear relationship between the inverse of the absolute temperature and the logarithm of the reaction rate constant, which is generally observed in practice. Most interestingly, for the reaction between cytochrome c_L and MDH a break is observed in the Arrhenius plot at 14°C and also for the reaction between cytochrome c_L and ascorbic acid, whereas a linear Arrhenius plot is obtained for the reaction between MDH and the artificial electron acceptor Wurster's blue. As discussed by Biosca et al. [41] a break in the Arrhenius plot can be due to (i) a temperature induced conformational change occurring at the protein level, (ii) a change in the rate-limiting step at a critical temperature and (iii) a modification of the solvent structure at a critical temperature. As the reaction was studied in water and over a limited temperature range, modification of the solvent structure seems unlikely, leaving the former two possibilities. In view of the narrow range of temperatures explored and the similar sharp temperature breaks in the Arrhenius plots observed for both the reduction of cytochrome c_L by MDH and by ascorbic acid, it can be postulated that a modification of the conformation of cytochrome c_L is a plausible interpretation of the non-linear Arrhenius plots observed here. Similar interpretations have been reported for a certain number of soluble enzymes [40,42]

Further support for a conformational change of cytochrome c_L during its reaction with MDH can be derived from changes in the activation volume ($\Delta V‡$) as revealed by a study of the kinetics as a function of pressure since $(\partial \ln k_{obs}/\partial P)_T = -\Delta V‡/RT$ where P is the hydrostatic pressure, T the absolute temperature and R the gas constant (82 ml atm K^{-1}, with 1 atm = 101.3 kPa). Electrophoretic analysis of MDH up to 2000 bar shows that the enzyme remains catalytically active and also that no dissociation of the subunits occurs. Large volume changes are observed for the reaction, ranging from 122 ml mol^{-1} at 4°C to 54 ml mol^{-1} at 25°C, the largest effect of the temperature on $\Delta V‡$ is seen at the higher temperature with a break point around 15°C. This supports the hypothesis that two conformations of cytochrome c_L exist.

Remarkably, $\Delta V‡$ for the reaction cytochrome c_L/MDH at 25°C is much larger than would be expected from the summation of $\Delta V‡$ values for the reactions MDH/Wurster's blue, Wurster's blue/ascorbate and ascorbate/cytochrome c_L. In contrast, $\Delta V‡$ for the corresponding reaction between cytochrome c_{HH} is in fair agreement with the calculated value, indicating that cytochrome c_L has a more specific interaction with MDH [43].

5 CONCLUSIONS

Quinoprotein methanol dehydrogenase has a complex structure, comprising two types of protein subunits, arranged in a $\alpha_2\beta_2$-structure together with Ca^{2+} and PQQ as cofactors. Hence, a multitude of interactions can be expected. Association of the α-subunits is triggered by the cofactor PQQ presumably by inducing a conformational change. This implies that a relatively strong interaction should exist between PQQ and the protein. The β-subunit has a similar interaction with the α-subunit, leading even to a significant change in the immunological response of the α-subunit. The role of Ca^{2+} is less clear, apparently it is not necessary for the binding of PQQ to α- or αβ-subunits, but it certainly plays a role in the catalytic activity of MDH.

Evidence has now been obtained that *in vitro* reconstitution of an apoMDH with PQQ and Ca^{2+} results in a blocked enzyme, while *in vivo* full activity can be restored simply by adding PQQ to the culture medium. Clearly other factors are involved, which is not surprising in view of the large number of genes implied in methanol oxidation [44].

α-Subunits recombined with PQQ exhibit weak redox properties with methanol and an artificial electron acceptor, but not with cytochrome c_L, indicating that the β-subunit is involved in electron transfer. However, cross-linking experiments have shown that cytochrome c_L interacts with the α-subunit rather than with the β-subunit [45], apparently, the conformational change imposed by the β-subunit favors also electron transfer. The results of the thermodynamic study strongly support the hypothesis of conformational changes occurring on the cytochrome c_L level and a contribution to the overall volume change due to specific interactions between the α-subunits of MDH and cytochrome c_L.

ACKNOWLEDGEMENTS

Part of this work was supported by grants from La Fondation pour la Recherche Médicale Française (C.B.), the Direction des Recherches, Etudes et Techniques (grants 89/037 and 91/1044J) (C.B.), CNRS UA 1129 (FG), the PROCOPE program (I.H.-L. and C.B.), the French Ministrère des Affaires Etrangères (J.F.), the MRT AGROBIO-ALIMENT 2002 (C.B.). I.H.-L. was supported by an ECC grant (SCI 0327C JR).

REFERENCES

[1] Harder, W. (1990) Structure/function relationships in methylotrophic yeasts. *FEMS Microbiol. Rev.* **87**, 191–200.
[2] Sherry, B. and Abeles, R.H. (1985) Mechanism of action of methanol oxidase. Reconstitution of methanol oxidase with 5-deazaflavin, and inactivation of methanol oxidase by cyclopropanol. *Biochemistry* **24**, 2594–2605.

[3] Arfman, N., Watling, E.M., Clement, W., Oosterwijk, R.J. van, Vries, G.E. de, Harder, W., Attwood, M.M. and Dijkhuizen, L. (1989) Methanol metabolism in thermotolerant *Bacillus* strains involving a novel catabolic NAD-dependent methanol dehydrogenase as a key enzyme. *Arch. Microbiol.* **152**, 280–288.

[4] Ophem, P.W. van, Euverink, G.J., Dijkhuizen, L. and Duine, J.A. (1991) A novel dye-linked alcohol dehydrogenase activity present in some Gram-positive bacteria. *FEMS Microbiol. Lett.* **80**, 57–64.

[5] Duine, J.A., Frank, J. and Berkhout, M.P.J. (1984) NAD-dependent, PQQ-containing methanol dehydrogenase: a bacterial dehydrogenase in a multienzyme complex. *FEBS Lett.* **168**, 217–221.

[6] Benner, S.A. (1989) Enzyme kinetics and molecular evolution. *Chem. Rev.* **89**, 789–806.

[7] Anthony, C. (1982) The Biochemistry of the Methylotrophs, Academic Press, New York.

[8] Hopkins, T.R. and Müller, F. (1987) Biochemistry of alcohol oxidase. In: Microbial Growth on C_1 Compounds (van Verseveld, H.W. and Duine, J.A., Eds), pp 150-157, Kluwer, Dordrecht, Netherlands.

[9] Arfman, N., Beeumen, J. van, Vries, G.E. de, Harder, W. and Dijkhuizen, L. (1991) Purification and characterization of an activator protein for methanol dehydrogenase from thermotolerant *Bacillus* spp. *J. Biol. Chem.* **266**, 3955–3960.

[10] Sheehan, M.C., Bailey, C.J., Dowds, B.C.A. and McConnell, D.J. (1988) A new alcohol dehydrogenase, reactive towards methanol, from *Bacillus stearothermophilus*. *Biochem J.* **252**, 661–666.

[11] Frank, J., Dijkstra, M., Duine, J.A and Balny, C. (1988) Kinetic and spectral studies on the redox forms of methanol dehydrogenase from *Hyphomicrobium* X. *Eur. J. Biochem.* **174**, 331–338.

[12] Long, A.R. and Anthony, C. (1991) The periplasmic modifier protein for methanol dehydrogenase in the methylotrophs *Methylophilus methylotrophus* and *Paracoccus denitrificans*. *J. Gen. Microbiol.* **137**, 2353–2360.

[13] Parker, M.W., Cornish, A., Gossain, V. and Best, D.J. (1987) Purification, crystallisation and preliminary X-ray diffraction characterisation of methanol dehydrogenase from *Methylosinus trichosporium* OB3b. *Eur. J. Biochem.* **164**, 223–227.

[14] Xia, Z.X., Hao, Z.P, Scott Matthews, F. and Davidson, V.L. (1989) Crystallization and preliminary X-ray crystallographic study of the quinoprotein methanol dehydrogenase from bacterium W3Al. *FEBS Lett.* **258**, 175–176.

[15] Janvier, M., Frehel, C., Grimont, F. and Gasser, F. (1985) *Methylophaga marina* gen. nov, sp. nov. and *Methylophaga thalassica* sp. nov., marine methylotrophs. *Int. J. Syst. Bact.* **35**, 131-139.

[16] Urakami, T. and Komogata, K. (1986) Emendation of *Methylobacillus* Yordi and Weaver 1977, a genus for methanol-utilizing bacteria. *Int. J. Syst. Bact.* **36**, 502–511.

[17] Jenkins, O., Byrom, D. and Jones, D. (1987) *Methylophilus*: a new genus of methanol-utilizing bacteria. *Int. J. Syst. Bact.* **35**, 445–448.

[18] Janvier, M., Frank, J., Luttik, M. and Gasser, F. (1992) Isolation and phenotypic characterization of methanol oxidation mutants of the restricted facultative methylotroph *Methylophaga marina*. *J. Gen. Microbiol.*, **138**, 2113–2123.

[19] Nunn, D.N., Day, D. and Anthony, C. (1989) The second subunit of methanol dehydrogenase of *Methylobacterium extorquens* AM1. *Biochem. J.* **260**, 857–862.

[20] Anderson, D.J. and Lidstrom, M.E. (1988) The *mox*FG region encodes four polypeptides in the methanol-oxidizing bacterium *Methylobacterium* sp strain AM1. *J. Bact.* **179**, 2254–2262.

[21] van Spanning R.J.M., Wansell, C.W., de Boer, T., Hazelaar, M.J., Anazawa, H., Harms, N., Oltmans, L.F. and Stouthamer, A.H. (1991) Isolation and characteriza-

tion of the *mox*J, *mox*G, *mox*I and *mox*R genes of *Paracoccus denitrificans*. Inactivation of *mox*J, *mox*G and *mox*R and the resultant effect on methylotrophic growth. *J. Bacteriol.* **173**, 6948–6961.
[22] Hathaway, G.M. (1972) Interaction of a self-associating protein with its cofactor. *J. Biol. Chem.* **247**, 1440–1444.
[23] Hathaway, G.M. and Crawford, I.P. (1970) Studies on the association of b-chain monomers of *E. coli* tryptophan synthetase. *Biochemistry* **9**, 1801–1808.
[24] Gounaris, A.D., Turkenkopf, I., Buckwald, S. and Young, A. (1971) Pyruvate decarboxylase. I. Protein dissociation into subunits under conditions in which thiamine pyrophosphate is released. *J. Biol. Chem.* **246**, 1392–1399.
[25] Gounaris, A.D., Turkenkopf, I., Civerchia, L.L. and Greenlie, J. (1975) Pyruvate decarboxylase. III. Specificity restrictions for thiamine pyrophosphate in the protein association step, sub-unit structure. *Biochim. Biophys. Acta* **405**, 492–499.
[26] Pan, S.S. and Nason, A. (1978) Purification and characterization of homogeneous assimilatory reduced nicotinamide adenine dinucleotide phosphate-nitrate reductase from *Neurospora crassa*. *Biochim, Biophys. Acta* **523**, 297–313.
[27] Duine, J.A., Frank, J. and Jongejan, J.A. (1986) PQQ and quinoproteins in microbial oxidations. *FEMS Microbiol. Lett.* **37**, 165–178.
[28] Biville, F., Mazodier, P., Gasser, F., van Kleef, M.A.G. and Duine, J.A. (1988) Physiological properties of a pyrroloquinoline quinone mutant of *Methylobacterium organophilum*. *FEMS Microbiol. Lett.* **52**, 53–58.
[29] Patel, R.N., Hou, C.T. and Felix, A. (1978) Microbial oxidation of methane and methanol: crystallization of methanol dehydrogenase and properties of holo- and apomethanol dehydrogenase from *Methylomonas methanica*. *J. Bacteriol.* **133**, 641–649.
[30] Davidson, V.L., Neher, J.W. and Cecchini, G. (1985) The biosynthesis and assembly of methanol dehydrogenase in Bacterium W3A1. *J. Biol. Chem.* **260**, 9642–9647.
[31] Adachi, O., Matsushita, K., Shinagawa, E. and Ameyama, M. (1990) Calcium in quinoprotein methanol dehydrogenase can be replaced by strontium. *Agric. Biol. Chem.* **54**, 2833–2837.
[31a] Beardmore-Gray, M., O'Keefe, D.T. and Anthony, C. (1983) The methanol-cytochrome-*c* oxidoreductase activity of methylotrophs. *J. Gen. Microbiol.* **129**, 923–933.
[32] Frank, J., Dijkstra, M., Balny, C., Verwiel, P.E.J. and Duine, J.A. (1989) Methanol dehydrogenase: Mechanism of action. *Antonie van Leeuwenhoek* **56**, 25–34.
[33] Dijkstra, M., Frank, J., Duine, J. A. (1989) Studies on electron transfer from methanol dehydrogenase to cytochrome c_L, both purified from *Hyphomicrobium* X. *Biochem. J.* **257**, 87–94.
[34] Anthony C. (1992) The c-type cytochromes in methylotrophic bacteria. *Biochim. Biophys. Acta* **1099**, 1–15.
[35] Duine, J.A. and Frank, J. (1990) The role of PQQ and quinoproteins in methylotrophic bacteria. *FEMS Microbiol. Rev.* **87**, 221–226.
[36] van Eldik, R. (1991) Pressure dependence of inorganic electron-transfer reactions in solution. *High Press. Res.* **6**, 251–259.
[37] Heiber-Langer, I, Hooper, A. B. and Balny, C. (1992) Pressure modulation of cytochrome-to-cytochrome electron-tranfer. Models and enzyme reactions. *Biophys. Chem.* **43**, 265–277.
[38] Heremans, K. (1982) High-pressure effects on proteins and other biomolecules. *Ann. Rev. Biophys. Bioeng.* **11**, 1–21.
[39] Cusanovich, M. A., Meyer, T. E. and Tollin, G. (1988) c-Type cytochromes: oxidation-reduction properties. In: Heme Proteins 7 (Eichhorn, G.L. and Marzilli, L.G., Eds) pp. 37–91, Elsevier, New York.
[40] Masson, P. and Balny, C. (1986) Thermodynamic arguments for temperature-

induced cryptic conformational change of human plasma cholinesterase. *Biochim. Biophys. Acta.* **874**, 90–98.
[41] Biosca, J.A., Travers, F., Hillaire, D., Barmann, T. E. (1984) Cryoenzymic studies on myosin subfragment 1: perturbation of an enzyme reaction by temperature and solvent. *Biochemistry* **23**, 1947–1955.
[42] Massey, V., Curti, B. and Ganther H. (1966) A temperature-dependent conformational change in D-amino acid oxidase and its effects on catalysis. *J. Biol. Chem.* **24**, 2347–2357.
[43] Heiber-Langer, Cléry, C., Frank, J., Masson, P. and Balny, C. (1992) Interaction of cytochrome c_L with methanol dehydrogenase from *Methylophaga marina* 42: Thermodynamic arguments for conformational change. *Eur. Biophys. J.* **21**, 241–250.
[44] Lidstrom, M.E. (1990) Genetics of carbon metabolism in methylotrophic bacteria. *FEMS Microbiol. Rev.* **87**, 431–436.
[45] Cox, J.M., Day, D.J. and Anthony, C. (1992) The interaction of methanol dehydrogenase and its electron acceptor, cytochrome c_L, in methylotrophic bacteria. *Biochim. Biophys. Acta* **1119**, 97–106.

17
Methanol Dehydrogenase and Cytochrome Interactions

C. Anthony, H.T.C. Chan, J.M. Cox and I.W. Richardson

SERC Centre for Molecular Recognition, Biochemistry Department, University of Southampton, Southampton SO9 3TU, UK

1 SUMMARY

This review covers recent work on the quinoprotein methanol dehydrogenase (MDH) from Gram-negative bacteria and on its interaction with cytochrome c_L, its specific electron acceptor. The reaction depends upon electrostatic interactions between lysyl residues on the α subunit of MDH and carboxylate groups on cytochrome c_L. The same domain on cytochrome c_L that 'docks' with MDH is also involved in docking with cytochrome c_H. The usual structure of MDHs is a an $\alpha_2\beta_2$ tetramer containing 2 molecules of PQQ plus a single atom of calcium whose insertion requires the products of the *Mox A, K* and *L* genes. The calcium is directly or indirectly involved in binding PQQ at the active site. It is not the site of action of inhibitory chelating agents such as EGTA; these act by binding to MDH so as to prevent effective docking with cytochrome c_L.

2 INTRODUCTION

This review covers aspects of our recent work, published since the *1989 C_1-Symposium* [1], on the interaction of the quinoprotein methanol dehydrogenase (MDH) and its specific electron acceptor cytochrome c_L. MDH is responsible for oxidation of methanol in all Gram-negative bacteria growing on methane or methanol; it is part of a periplasmic electron transport chain in

Microbial Growth on C_1 Compounds
© Intercept Ltd, PO Box 716, Andover, Hampshire SP10 1YG, UK

which the cytochrome c_L reacts with a typical Class I c-type cytochrome (cytochrome c_H) which is the substrate for the membrane oxidase (either cytochrome aa_3 or cytochrome co):

MeOH → MDH → Cytochrome c_L → Cytochrome c_H → Oxidase

More complete reviews have been published recently, of the dehydrogenase [2,3], of quinoprotein structure [4], of electron transport from quinoproteins [5,6] and of the c-type cytochromes of methylotrophs [7].

When the small β subunit of MDH was first described [8], the most remarkable feature of its sequence was the predicted secondary structure for the C-terminal half of the protein; this suggested a well-defined amphipathic helix with a pattern of positively charged lysyl residues facing away from a hydrophobic side of the helix. It was suggested that these are probably involved in the 'docking' of MDH with carboxylates on the specific electron acceptor, cytochrome c_L [8].

This review considers the following questions relating to this process: Is the interaction between MDH and cytochrome c_L primarily between oppositely charged domains on the two proteins? Are the lysyl residues on MDH and carboxylates on cytochrome c_L involved? Is the β subunit specifically involved in this interaction? Does the domain on cytochrome c_L that interacts with MDH also interact with lysyl residues on cytochrome c_H? Does the calcium ion present in MDH have any role in these interactions?

3 MDH 'DOCKS' WITH CYTOCHROME c_L BY ELECTROSTATIC INTERACTIONS

When protein-protein interactions are electrostatic in nature they are affected by the ionic strength of the surrounding medium. This has been clearly demonstrated with MDH and cytochrome c_L from *Methylobacterium extorquens* [9], *Methylophilus methylotrophus* [9] and *Acetobacter methanolicus* [10]; 50% inhibition was caused by ionic strengths between 0.01 and 0.04 for a wide range of salts. Figure 1 demonstrates that the effect of high ionic strength is to decrease the affinity of the two proteins for each other.

4 INTERACTION IS BETWEEN LYSINES ON MDH AND CARBOXYLATES ON CYTOCHROME c_L

This was demonstrated by chemically modifying lysyl or carboxylate residues on MDH and cytochrome c_L [9,10]. The modified MDH retained activity in the dye-linked assay system showing that the active site for reaction with substrate had not been altered. It was shown that reagents that changed the charge on lysines led to inactive MDH whereas those that modified MDH with retention of charge had relatively little affect (Table 1). The inhibition by

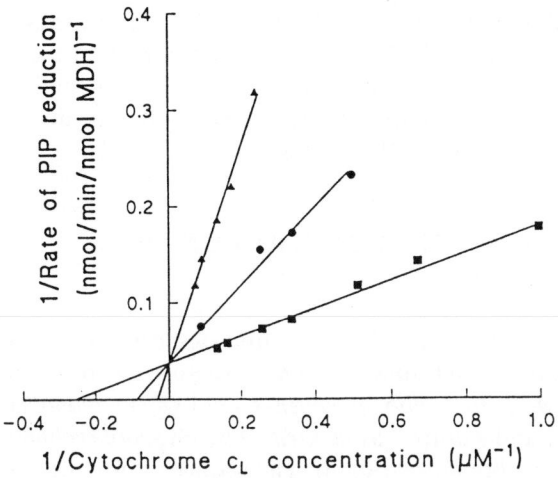

Figure 1 The effect of NaCl on the affinity of MDH for cytochrome c_L. MDH of *Mb. extorquens* (2 µM) was assayed with 2,6-dichlorophenolindophenol (PIP) as terminal electron acceptor as described in [9]. ■, no NaCl, K_m, 3 µM; ●, 25 Mm-NaCl, K_m, 8.6 µM; ▲, 50 mM-NaCl, K_m, 30 µM. Taken with permission from [9].

reagents specific for arginine residues suggests that these may also be involved. When cytochrome c_L was modified with lysine-modifying reagents, its activity was retained, but those reagents that modified carboxyl groups led to greatly diminished activity. The fact that modification of fewer than 7% of the lysines on MDH led to complete loss of activity demonstrates that those lysines that are most amenable to modification are also those that are involved in the interaction with cytochrome c_L. After modification of the MDH from *Mp. methylotrophus* with trinitrobenzene sulphonate (TNBS) (Table 1), it was se-

Table 1 Modification of lysine residues on MDH

Reagent	Organism	No. of residues changed	Change	Activity (%) with cyt c_L	Activity (%) with dye
SPDP	*Mb. extorquens*	7	+ → 0	0	77
SPDP	*Mp. methylotrophus*	7	+ → 0	9	123
TNBS	*Mb. extorquens*	4	+ → 0	0	82
TNBS	*Mp. methylotrophus*	9	+ → 0	9	93
TNBS	*A. methanolicus*	9	+ → 0	0	100
IT+DTDP	*Mp. methylotrophus*	9	+ → +	91	99
IT+DTDP	*A. methanolicus*	5	+ → +	125	111

MDH was modified and activity measured with cytochrome c_L or dye (PES/PIP)
SPDP, N-Succinimydyl-(pyridyldithio)-propionate
TNBS, Trinitrobenzenesulphonate; IT+DTDP, Iminothiolane+Dithiodipyridine

parated into its subunits by SDS-gel filtration; this demonstrated that 94% of the TNBS was associated with the α subunit, suggesting that the lysines whose modification leads to loss of activity are on this subunit and not on the β subunit [9,10], thus suggesting that this small subunit does not have a predominant role in docking with cytochrome c_L.

5 THE STRUCTURE AND FUNCTION OF THE β SUBUNIT OF MDH

When the small β subunit of MDH was first described it was shown to be present as part of an $\alpha_2\beta_2$ tetramer [8]. This structure has now been confirmed for MDHs from *Methylobacillus glycogenes* [11], *A. methanolicus* [10], *Mp. methylotrophus* [9] and *Paracoccus denitrificans*, *Hyphomicrobium* and *Methylophaga marina* (our unpublished data). The sequences for the β subunits of 4 MDHs from very different types of methylotroph are shown in Figure 2. It can be seen that the sequence in the N-terminal region of the β subunit is more conserved than the C terminus; of particular note are the conserved cysteines (involved in a cystine bridge). The protein sequence predicted from the gene sequence from *P. denitrificans* [12] showed that 4 out of 5 of the lysines we had predicted to be important in reaction with cytochrome c_L were conserved, thus apparently strengthening our argument. However, the sequences of the β subunits from *A. methanolicus* [10] and *Mp. methylotrophus* (our unpublished data) show that none of these lysines is completely conserved (Figure 2), indicating that they cannot have a special role in docking with the cytochrome c_L. This is consistent with the observations described in the previous section, that the α subunit is more likely to be involved in this process.

6 THE α SUBUNIT OF MDH DOCKS WITH CYTOCHROME c_L AS SHOWN BY CHEMICAL CROSS-LINKING

An additional approach to determine which subunit of MDH interacts with cytochrome c_L is to cross-link the two proteins chemically during interaction and then to determine which subunit is cross-linked to the cytochrome (determined by haem-staining and sizing of cross-linked peptides by SDS-PAGE). When conventional cross-linking reagents were used, having relatively long 'arms', different results were obtained using proteins from *Mb. extorquens* and *Mp. methylotrophus* [9]. A better approach is to use the so-called zero-length, two-stage, cross linking reagents [13]. In this system the carboxyl group on one protein is modified and this modified protein is attacked by an unmodified lysine on the second (unmodified) protein forming a cross-linker arm consisting of a peptide bond between the original carboxyl and lysyl residues. Only groups that come very close together are able to react in this system. The results of such experiments were unambiguous; the subunit that is involved in docking

Methanol dehydrogenase and cytochrome interactions 225

```
      !!!       !!    !  !!!        !       !  !!!!  !! !!  !   !                  !!
a]  YDGTKCKAAGNCWEPKPGFPEKIAGSKYDPKHDPKELNKQADSIKQMEERNKKRVENFKKTGKFEYDVAKISAN
            +                        +           +      ++      +  +          +
b]  YDGQTCKEAGNCWEAKPGYPEKIAGSKYDPKHDPVELNKQEQAIKAMDERNAARIANAKSSGNFVFDVKK
c]  YDGANCKAPGTCWEPKPDYPAKVEGSKYDPQHDPAELSKQGESLAVMDARYEWRVWNMKKTGKFEYDVKKIDGYGDETKAPPAD
d]  AYDGTHCKKPGVCWEPQPGYPEQLVGVKYDPHFDPAELGVQxESxxxMNARNEARTAYFILTGTWEEDVNQIPK
```

Figure 2 Amino acid sequences of β subunits of MDH. (a) *Mb. extorquens*; (b) *Mp. methylotrophus*; (c) *P. denitrificans*; (d) *A. methanolicus*. + indicates lysyl residues originally proposed to be involved in interaction with cytochrome c_L. ! indicates conserved residues. References to the literature sources are provided in the text.

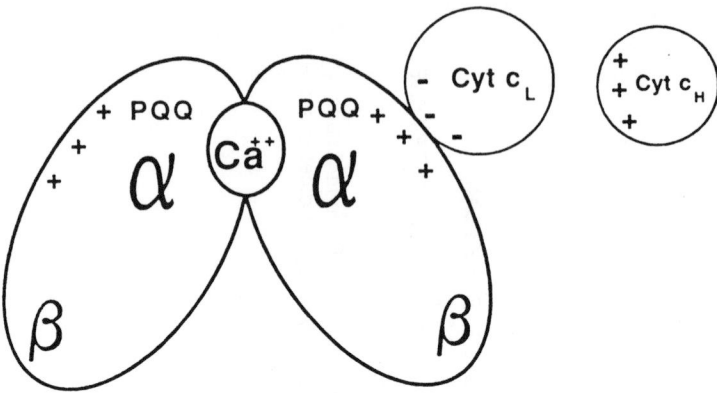

Figure 3 The interaction of MDH, cytochrome c_L and cytochrome c_H. The three proteins interact by reversible electrostatic interactions by way of carboxylates on cytochrome c_L and lysyl residues on the large subunit of MDH and on cytochrome c_H. The same domain on cytochrome c_L is involved in docking with both MDH and cytochrome c_H.

with cytochrome c_L is the α subunit. This was demonstrated using proteins from four different methylotrophs: *A. methanolicus* [10], *Mb. extorquens* [9], *Mp. methylotrophus* [9] and *Methylophaga marina* (our unpublished data). Using a similar technique it was also shown that the same site on cytochrome c_L that is involved in docking with MDH is also involved in the next reaction in the electron transport chain – with the class I cytochrome c (cytochrome

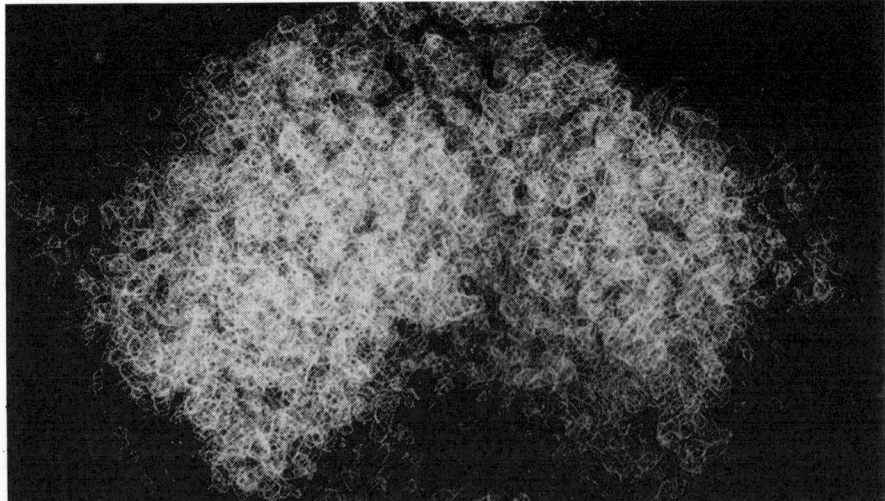

Figure 4 A 3Å electron density map of the $\alpha_2\beta_2$ tetramer of MDH of *Mb. extorquens* (M. Ghosh, K. Harlos, C.C.F. Blake, I.W. Richardson, and C. Anthony, unpublished data). Information on the crystals and methods used is provided in [37].

c_H) [9]. By sequencing of the cross-linked peptides produced by hydrolysis of the cross-linked proteins, it was shown that the same lysyl residues of cytochrome c_H are involved in reaction with both the cytochrome c_L and with the terminal oxidase. These experiments also confirmed that the three proteins do not at any stage form a single multicomponent electron transport complex. Figure 3 provides an illustration of the interaction of the 3 proteins; the general shape of the MDH is based on the preliminary X-ray structure from *Mb. extorquens* (Figure 4).

The demonstration that the β subunit is not involved in docking with cytochrome c_L raises the question of its function. No sequences having similarity to this subunit are found in sequence databases, and there are no regions of similarity to the β subunit of MDH in other quinoproteins. This would appear to rule out a role in binding the PQQ prosthetic group, although it is worth noting that the PQQ in MDH is more tightly bound than in other quinoproteins and this may be correlated with the presence of the β subunit.

7 THE ROLE OF CALCIUM IN THE STRUCTURE AND FUNCTION OF MDH

It has been known for some time that calcium or magnesium ions are necessary for reconstitution of activity using apoenzymes of glucose dehydrogenase or alcohol dehydrogenase plus PQQ [14,15]. More recently, the soluble alcohol dehydrogenase from *Pseudomonas aeruginosa* and the soluble glucose dehydrogenase from *Acinetobacter calcoaceticus* have been shown to contain calcium ions [16,17], and calcium was recently reported in the MDH of the newly described methylotroph *Methylobacillus glycogenes* [11]. We have now demonstrated that there is a single Ca^{2+} per $\alpha_2\beta_2$ tetramer of the MDHs from *Mb. extorquens* [18], *Mp. methylotrophus* [18], *P. denitrificans* [18], *Hyphomicrobium* X [18] and *Methylophaga marina* [36].

We have recently gained some understanding of the role of this calcium, and of the process of MDH biosynthesis, from an investigation of the MoxA, K and L mutants of *Mb. extorquens* [18]. These mutants map separately from the operon that codes for the α and β subunits of MDH and cytochrome c_L [19,20]. MDH isolated from the methanol oxidation mutants Mox A, K and L was identical to the enzyme isolated from wild-type bacteria with respect to molecular size, subunit configuration, isoelectric point, N-terminal amino acid sequence and stability in denaturing conditions. The most obvious difference in these mutant MDHs was in their absorption spectra, indicating that the prosthetic group was modified PQQ, or PQQ in a different redox state, or bound differently in the enzyme. It was subsequently shown that the PQQ was normal PQQ but was present exclusively in the oxidized state (2 moles PQQ/$\alpha_2\beta_2$ tetramer); and this was consistent with the absence of a semiquinone signal in the epr spectra. Although in the oxidized form, the spectra of these enzymes were not the same as the spectrum of the oxidized wild type enzyme,

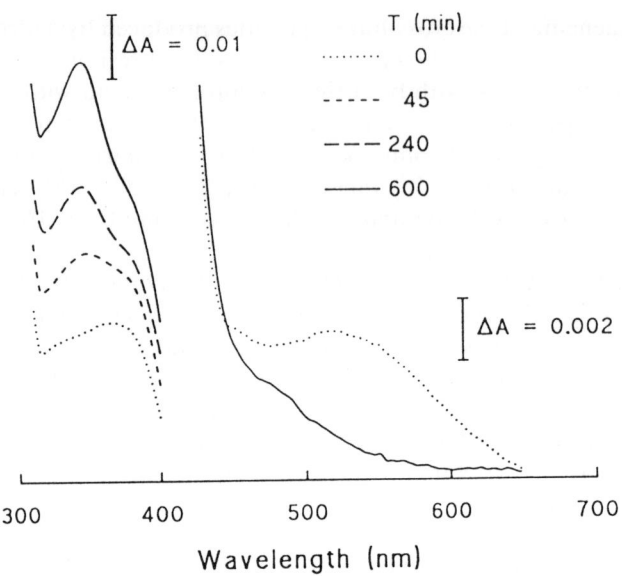

Figure 5 Absorption spectra of MDH from the MoxA⁻ mutant strain during reconstitution with calcium ions. MDH (0.7 μM) was incubated with 500 μM calcium chloride in 100 mM-tris (pH 9) at 20°C [18].

and the mutant enzymes were unable to react with cyclopropanol. This suggested that the PQQ must be bound differently in the active site. We eventually demonstrated that these MDHs contain no calcium. Remarkably, incubation of MDH from the mutants in calcium salts led to irreversible time-dependent reconstitution of full activity concomitant with restoration of a spectrum corresponding to that of fully reduced normal MDH (Figure 5). We concluded that the reduced enzyme was produced, instead of the usual wild-type semiquinone form, because after formation of the active enzyme by calcium addition, the oxidized PQQ became reduced by endogenous substrate on the enzyme.

We have recently obtained similar results with MDH from a wild-type organism, the marine methylotroph *Methylophaga marina* [36]. A violet-red form of MDH was occasionally produced which was completely inactive and lacked Ca^{2+}. Although not identical, its spectrum was similar to that of the inactive MDH from MoxA, K and L mutants of *Methylobacterium extorquens*. The inactive form of the *M. marina* enzyme was restored to full activity by incubation with Ca^{2+}. These results, obtained with an enzyme from a wild-type organism confirmed that Ca^{2+} is not absolutely required for binding PQQ to MDH but that it *is* required (directly or indirectly) for maintaining the correct configuration of PQQ in the active enzyme. If a single Ca^{2+} is bound per $\alpha_2\beta_2$ tetramer it is difficult to see how the Ca^{2+} can be involved directly in a reaction mechanism in which the Ca^{2+} is bound directly to PQQ. The presence of a

```
MDH   M.ext   477- GGTMATAGDL VFYGTLDGY L-KARDSDTG DL-LWKFKIP SGAIGYPMT YTHKGTQYVA IYYGVGG
MDH   M.org   477- GGTLATAGDL VFYGTLDGY L-KARDSDTG DL-LWKFKIP SGAIGYPMT YTHKGTQYVA IYYGVGG
MDH   P.den   476- GGTMATAGGL TFYVTLDGF I-KARDSDTG EL-LWKFKLP SGVIGHPMT YKHDGRQYVA IMYGVGG

ADH   A.ace   486- GGILATGGDL LFQGLANGE F-HAYDATNG SD-LYKFDAQ SGIIAPPMT YSVNGKQYVA VEVGWGG

GDHA  E.col   713- GGPISTAGNV LFIAATADN YLRAYNMSNG EK-LWQGRLP AGGQATPMT YEVNGKQYVV ISAGGHG
GDHA  A.cal   719- GGSISTAGNV MFVGATQDN YLRAFNVTNG KK-LWEARLP AGGQATPMT YEINGKQYVV IMAGGHG
GDHB  A.cal   137- GLPSSKDHQS GRLVIGPDQ KIYYTIGDQG RNQLAYLFLP NQAQHTPTQ QELNGKDYHT YMGKVLR
```

Figure 6 The putative PQQ-binding region on quinoproteins. The methanol dehydrogenases (α subunits) are from *Mb. extorquens* [4,21], *Mb. organophilum* [22] and *P. denitrificans* [23]. The glucose dehydrogenases (GDHA and GDHB) are from *Acinetobacter calcoaceticus* [26,27] and *E. coli* [28], and the alcohol dehydrogenase is from *Acetobacter aceti* [24,25]. These sequences are discussed extensively in [4].

single atom of calcium per $\alpha_2\beta_2$ tetramer suggests that the calcium might be located at the interface between the two $\alpha\beta$ dimers.

Analysis of the protein sequences for a number of quinoprotein MDHs [4,21–23], alcohol dehydrogenase [24,25] and glucose dehydrogenases [26–28] has identified a region that might be involved as a PQQ-binding region [4] as indicated in Figure 6. Although there is no close similarity to any known calcium binding sequences, it remains possible that this region might be involved in binding both PQQ and calcium.

8 THE ROLE OF THE *MOX A, K* AND *L* GENES IN THE BIOSYNTHESIS OF MDH

In the MoxA, K and L mutants the structural genes *moxF* and *moxI* code for the normal α and β subunits, which are present in the normal $\alpha_2\beta_2$ tetrameric configuration.

The *moxA, K* and *L* genes are clearly implicated in the insertion of calcium into MDH. Figure 7 summarizes 3 possible roles for the proteins coded by these genes. A key feature of the 3 models is that they are based on the observation that PQQ is tightly bound to the $\alpha_2\beta_2$ tetramer before insertion of Ca^{2+}; and that the conformation in the absence of calcium is sufficiently similar to that of the holoenzyme for this to be formed when the concentration of calcium in the medium is artificially high. The concentration of free calcium in the periplasm where the MDH is assembled is presumably the same as that in the surrounding growth medium. The concentration in normal growth media is at least 20 μM. Incubation of pure MDH from the MoxA mutant in this concentration of calcium led to reconstitution of active MDH but only 25% of maximum activity was achieved and the rate was extremely slow (10% reconstitution after 2 days). The rate with higher concentrations was very much greater and full activity could be achieved (100% activity in 1 h in 10 mM calcium). The simplest description of the role of the MoxA, K and L proteins might therefore be in providing a high local concentration of calcium in the

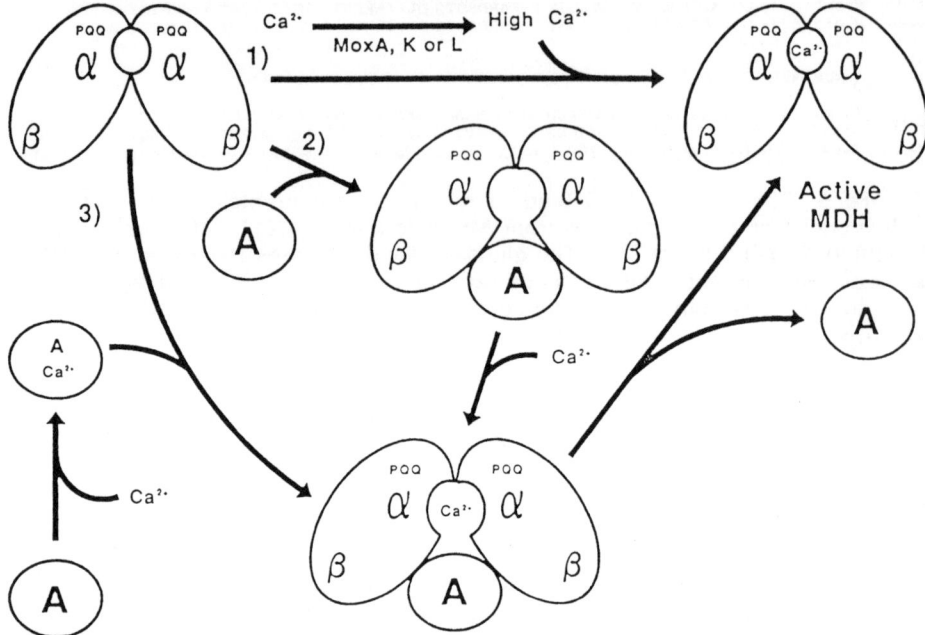

Figure 7 The role of the *moxA*, *K* and *L* gene products in MDH assembly. The three possible roles summarized here are for the MoxA protein; an identical role could be described for MoxK and/or MoxL proteins. 1. the protein is involved merely in maintaining a high concentration of calcium in the periplasm; 2. the MoxA protein stabilises the $\alpha_2\beta_2$ tetramer together with 2 molecules of PQQ in a conformation able to bind low concentrations of calcium, after which the MoxA protein dissociates; 3. this proposal is the same as the second, except that the MoxA protein carries the calcium ion to its location. Taken with permission from [18].

periplasm. However, as calcium flux in and out of the periplasm is likely to be rapid (by way of porins in the outer membrane), it is difficult to conceive of such a mechanism. Alternative functions of the proteins are suggested in Figure 7. These include a calcium-binding role in which the protein carries calcium to the active site and is then released. Alternatively, the MoxA protein (for example) might be involved in stabilizing a conformation of MDH that is then able to bind calcium, after which the MoxA protein is released and does not form part of the final structure. In this respect these proteins would be fulfilling a molecular chaperone function (see [29]).

9 THE INHIBITION OF MDH BY EDTA AND RELATED CHELATING AGENTS

In the first description of methanol oxidation by whole cells of *Mb. extorquens* [30] it was shown that 0.1 mM-EDTA was a specific inhibitor. Subsequently four mechanisms of action have been suggested: binding to MDH [30], binding to cytochrome c_L [31], removal of ions required for binding electron transport proteins together [32,33], inhibition of electron transfer [9]. The demonstration that MDH requires a Ca^{2+} ion for activity clearly suggests a fifth possibility – that EDTA and related inhibitors act by removal of this essential ion from the enzyme, and we have now investigated this [34]. Using proteins from *Mb. extorquens* it was shown that EGTA is a potent competitive inhibitor of electron transfer between MDH and cytochrome c_L ($I_{50} = 2.5\ \mu M$). After treatment with a high concentration of EGTA, the inhibition was completely reversed by passage through a gel filtration column in Ca^{2+}-free buffer. However, this treatment did not remove the tightly bound Ca^{2+} present in the MDH. It might thus be assumed that the EGTA acts by chelation of calcium in the MDH, causing inhibition, but not removing the calcium. This was shown not to be the case by using indo-1, a fluorescent analogue of EGTA, which changes its fluorescence properties on chelation of Ca^{2+} [35]. We found that indo-1 bound tightly to MDH in a 1:1 ratio but it was not able to bind to cytochrome c_L. This reaction of MDH with indo-1 was prevented by EGTA, showing, as expected, that indo-1 and EGTA compete for the same site. However the binding of indo-1 was identical (1 mole/mole MDH) in the MDH from the calcium-free MDH purified from the MoxA mutant. It was thus concluded that EGTA inhibits methanol oxidation by binding to lysyl or arginyl residues on MDH thus preventing docking with cytochrome c_L. It is pleasing to note that this is the mechanism (binding to MDH) first proposed some years ago in our first paper on methanol oxidation by *Mb. extorquens* [30].

ACKNOWLEDGEMENT

We thank the SERC for financial support.

REFERENCES

[1] Anthony, C. (1990) The oxidation of methanol in Gram-negative bacteria. *FEMS. Microbiol. Rev.* **87**, 209–214.
[2] Anthony, C. (1986) The bacterial oxidation of methane and methanol. *Adv. Microbial. Physiol.* **27**, 113–210.
[3] Anthony, C. (1993) Methanol dehydrogenase in Gram-negative bacteria. In: Principles and Applications of Quinoproteins (Davidson, V.L., Ed.) pp. 17–45, Marcel Dekker, New York.
[4] Anthony, C. (1992) The structure of bacterial quinoprotein dehydrogenases. *Int. J. Biochem.* **24**, 29–39.
[5] Anthony, C. (1988) Quinoproteins and energy transduction. In: Bacterial Energy Transduction (Anthony, C., Ed.), pp. 293–316, Academic Press, London.
[6] Anthony, C. (1993) The role of quinoproteins in bacterial energy transduction. In: Principles and Applications of Quinoproteins (Davidson, V.L., Ed.), pp. 223–244, Marcel Dekker, New York.
[7] Anthony, C. (1992) The c-type cytochromes of methylotrophic bacteria. *Biochim. Biophys. Acta* **1099**, 1–15.

[8] Nunn, D.N., Day, D.J. and Anthony, C. (1989) The second subunit of methanol dehydrogenase of *Methylobacterium extorquens* AM1. *Biochem. J.* **260**, 857–862.
[9] Cox, J.M., Day, D.J. and Anthony, C. (1992) The interaction of methanol dehydrogenase and its electron acceptor, cytochrome c_L, in the facultative methylotroph *Methylobacterium extorquens* AM1 and in the obligate methylotroph *Methylophilus methylotrophus*. *Biochim. Biophys. Acta* **1119**, 97–106.
[10] Chan, H.T.C. and Anthony, C. (1991) The interaction of methanol dehydrogenase and cytochrome c_L in an acidophilic methylotroph, *Acetobacter methanolicus*. *Biochem. J.* **280**, 139–146.
[11] Adachi, O., Matsushita, K., Shinagawa, E. and Ameyama, M. (1990) Purification and properties of methanol dehydrogenase and aldehyde dehydrogenase from *Methylobacillus glycogenes*. *Agric. Biol. Chem.* **54**, 3123–3129.
[12] van Spanning, R.J.M., Wansell, C.W., de Boer, T., Hazelaar, M.J., Hideharu, A., Harms, N., Oltmann, L.F. and Stouthamer, A.H. (1991) Isolation and characterisation of the *moxJ*, *moxG*, *moxI* and *moxR* genes of *Paracoccus denitrificans*: Inactivation of *moxJ*, *moxG* and *moxR* and the resultant effect on methylotrophic growth. *J. Bacteriol.* **173**, 6948–6961.
[13] Grabarek, Z. and Gergely, J. (1990) Zero-length crosslinking procedure with the use of active esters. *Anal. Biochem.* **185**, 131–135.
[14] Shinagawa, E., Matsushita, K., Adachi, O. and Ameyama, M. (1989) Formation of the apo-form of quinoprotein alcohol dehydrogenase from *Gluconobacter suboxydans*. *Agric. Biol. Chem.* **53**, 1823–1828.
[15] Groen, B.W., van Kleef, M.A.G. and Duine, J.A. (1986) Quinohaemoprotein alcohol dehydrogenase apoenzyme from *Pseudomonas testosteroni*. *Biochem. J.* **234**, 611–615.
[16] Mutzel, A. and Gorisch, H. (1991) Quinoprotein ethanol dehydrogenase: preparation of the apo-form and reconstitution with pyrroloquinoline quinone and Ca^{2+} or Sr^{2+} ions. *Agric. Biol. Chem.* **55**, 1721–1726.
[17] Geiger, O. and Gorisch, H. (1989) Reversible thermal inactivation of the quinoprotein glucose dehydrogenase from *Acinetobacter calcoaceticus*. *Biochem. J.* **261**, 415–421.
[18] Richardson, I.W. and Anthony, C. (1992) Characterization of mutant forms of the quinoprotein methanol dehydrogenase lacking an essential calcium ion. *Biochem. J.* **287**, 709–715.
[19] Nunn, D.N. and Lidstrom, M.E. (1986) Isolation and complementation analysis of 10 methanol oxidation mutant classes and identification of the methanol dehydrogenase structural gene of *Methylobacterium* sp. strain AM1. *J. Bacteriol.* **166**, 581-590.
[20] Nunn, D.N. and Lidstrom, M.E. (1986) Phenotypic characterisation of 10 methanol oxidation mutant classes of *Methylobacterium* sp. strain AM1. *J. Bacteriol.* **166**, 591–597.
[21] Anderson, D.J., Morris, C.J., Nunn, D.N., Anthony, C. and Lidstrom, M.E. (1990) Nucleotide sequence of the *Methylobacterium extorquens* AM1 *moxF* and *moxJ* genes involved in methanol oxidation. *Gene.* **90**, 173–176.
[22] Machlin, S.M. and Hanson, R.S. (1988) Nucleotide sequence and transcriptional start site of the *Methylobacterium organophilum* XX methanol dehydrogenase structural gene. *J. Bacteriol.* **170**, 4739–4747.
[23] Harms, N., de Vries, G.E., Maurer, K., Hoogendijk, J. and Stouthamer, A.H. (1987) Isolation and nucleotide sequence of the methanol dehydrogenase structural gene from *Paracoccus denitrificans*. *J. Bacteriol.* **169**, 3969–3975.
[24] Inoue, T., Sunagawa, M., Mori, A., Imai, C., Fukuda, M., Takagi, M. and Yano, K. (1990) Possible functional domains in a quinoprotein alcohol dehydrogenase from *Acetobacter aceti*. *J. Ferment. Bioeng.* **70**, 58–60.

[25] Inoue, T., Sunagawa, M., Mori, A., Imai, C., Fukuda, M., Takagi, M. and Yano, K. (1989) Cloning and sequencing of the gene encoding the 72-Kilodalton dehydrogenase subunit of alcohol dehydrogenase from *Acetobacter aceti*. *J. Bacteriol.* **171**, 3115–3122.

[26] Cleton-Jansen, A.-M., Goosen, N., Odle, G. and van de Putte, P. (1988) Nucleotide sequence of the gene coding for quinoprotein glucose dehydrogenase from *Acinetobacter calcoaceticus*. *Nucleic Acids Res.* **16**, 6228.

[27] Cleton-Jansen, A.-M., Goosen, N., Vink, K. and van de Putte, P. (1989) Cloning, characterization and DNA sequencing of the gene encoding the Mr 50,000 quinoprotein glucose dehydrogenase from *Acinetobacter calcoaceticus*. *Mol. Gen. Genet.* **217**, 430–436.

[28] Cleton-Jansen, A.-M., Goosen, N., Fayet, O. and van de Putte, P. (1990) Cloning, mapping, and sequencing of the gene encoding *Escherichia coli* quinoprotein glucose dehydrogenase. *J. Bacteriol.* **172**, 6308–6315.

[29] Ellis, R.J. (1990) Molecular chaperones—the plant connection. *Science* **250**, 954–959.

[30] Anthony, C. and Zatman, L.J. (1964) The methanol-oxidizing enzyme of *Pseudomonas* sp. M27. *Biochem. J.* **92**, 614–621.

[31] Frank, J., Dijkstra, M., Balny, C., Verwiel, P.E.J. and Duine, J.A. (1989) Methanol dehydrogenase: mechanism of action. In: PQQ and Quinoproteins, (Jongejan, J.A. and Duine, J.A., Eds), pp. 13–22, Kluwer Academic Publishers, Dordrecht.

[32] Carver, M.A., Humphrey, K.M., Patchett, R.A. and Jones, C.W. (1984) The effect of EDTA and related chelating agents on the oxidation of methanol by the methylotrophic bacterium *Methylophilus methylotrophus*. *Eur. J. Biochem.* **138**, 611–615.

[33] Carver, M.A. and Jones, C.W. (1984) The role of c-type cytochromes in the terminal respiratory chain of the methylotrophic bacterium *Methylophilus methylotrophus*. *Arch. Microbiol.* **139**, 76–82.

[34] Chan, H.T.C. and Anthony, C. (1992) The mechanism of inhibition by EDTA and EGTA of methanol oxidation by methylotrophic bacteria. *FEMS Microbiol. Lett.* **96**, 231–234.

[35] Grynkiewicz, G., Poenie, M. and Tsien, R.Y. (1985) A new generation of Ca^{2+} indicators with greatly improved fluorescence properties. *J. Biol. Chem.* **260**, 3440–3450.

[36] Chan, H.T.C. and Anthony, C. (1992) Characterisation of a red form of methanol dehydrogenase from the marine methylotroph *Methylophaga marina*. *FEMS Microbiol. Lett.* **97**, 293–298.

[37] Ghosh, M., Harlos, K., Blake, C.C.F., Richardson, I.W. and Anthony, C. (1992) Crystallization and preliminary crystallographic investigation of methanol dehydrogenase from *Methylobacterium extorquens* AM1. *J. Mol. Biol.* **228**, 302–305.

18
Genetics of Methanol Oxidation in *Paracoccus denitrificans*

Nellie Harms

Department of Microbial Physiology, Biological Laboratory, Vrije Universiteit, De Boelelaan 1087, 1081 HV Amsterdam, The Netherlands

1 SUMMARY

A genomic DNA fragment of 10.7 kb containing 8 *mox* genes and the 5' region of a 9th gene has been isolated from *Paracoccus denitrificans*. *moxF*, *moxI*, and *moxG*, are the structural genes and encode the large and small subunits of methanol dehydrogenase (MDH) and its electron acceptor, cytochrome c_{551i}, respectively. In the same transcription direction, three additional genes *moxJ*, *moxR*, and *moxS*, have been found that encode proteins of unknown function. Upstream of *moxF*, three regulatory genes, *moxZ*, *moxY*, and *moxX*, which are transcribed in a direction opposite to *moxF*, have been found. Computer alignment analysis revealed that the gene products of *moxY* and *moxX* have homology with sensor and regulator proteins of the two-component regulatory systems. The gene product of *moxZ* showed no homology with any protein sequenced thus far. Mutants with marked and unmarked mutations in the *moxX*, *Y* and *Z* regulatory genes have been isolated and characterized. Mutants with a mutation in *moxX* did not grow on methanol, did not synthesize MDH and lacked MDH activity. Expression of a *moxF-lacZ* transcriptional fusion gene was only observed in strains with a functional *moxX*. This indicates that the *moxX* gene product is a positively acting regulator which is required for the expression of the structural *mox* genes. A mutant with an unmarked deletion in *moxY* showed a phenotype comparable to the wild type. This indicates that although *moxY* is part of the two-component regulatory system that is involved in activation of the structural *mox* genes it is not indispensable for methanol oxidation. It seems that, at least in *moxY* mutants, another sensor is able to activate *moxX*. A mutant with an unmarked deletion in *moxZ* showed reduced growth on methanol as compared with the wild type. MDH activities and

---

Microbial Growth on C_1 Compounds
© Intercept Ltd, PO Box 716, Andover, Hampshire SP10 1YG, UK

expression of *moxF-lacZ* transcriptional fusion genes were also lower than in the wild type. This indicates that in addition to the two-component regulatory system, a third protein, *moxZ*, is involved in expression of the structural *mox* genes.

2 INTRODUCTION

Paracoccus denitrificans is a versatile gram-negative bacterium that is able to grow under a large variety of growth conditions. The organism can grow heterotrophically both under aerobic and anaerobic conditions with either nitrate, nitrite, nitric oxide or nitrous oxide as a terminal electron acceptor. In addition, it can grow autotrophically with either hydrogen, thiosulfate or reduced C_1-compounds such as methanol, methylamine or formate as an energy source. Methanol and methylamine are oxidized *via* formaldehyde and formate to carbon dioxide, which is subsequently fixed by ribulose bisphosphate carboxylase, one of the key enzymes of the ribulose bisphosphate pathway.

The oxidation of methanol in *P. denitrificans* is catalyzed by methanol dehydrogenase (MDH). This enzyme is located in the periplasm [1] and contains the cofactor pyrrolo-quinoline-quinone (PQQ). The enzyme has a molecular mass of about 150 kDa [2] and is comprised of two large and two small subunits with molecular masses of 66.6 kDa and 9.4 kDa, respectively. Electrons originating from the oxidation of methanol are transferred to the electron transport chain at the level of cytochrome *c*. The primary electron acceptor for MDH is the inducible soluble cytochrome c_{551i} with a molecular mass of 17.1 kDa [3, 4].

The synthesis of MDH in *P. denitrificans* is strictly regulated [5]. A catabolite repression mechanism represses synthesis in cells grown either on a heterotrophic substrate or on a mixture of a heterotrophic substrate and methanol. Derepression, resulting in low amounts of MDH, has been found in cells grown under carbon limitation in chemostat cultures. Activation of MDH synthesis has been found in cells grown on methanol, methylamine or choline, resulting in amounts of MDH up to 15% of total cell protein. These substrates have a common feature in that formaldehyde is formed during their metabolism. It has been suggested therefore that a product induction mechanism, with formaldehyde as an effector, is involved in activation of MDH synthesis [5]. In addition, a post-translational control mechanism has been found that regulates the amount of MDH synthesized and its activity.

In *P. denitrificans* a large variety of *c*-type cytochromes have been found [6, 7]. Two of these, cytochrome c_{551i} and cytochrome c_{553i}, are only induced in cells grown on methanol, methylamine and choline and have therefore a similar regulation pattern as MDH. Cytochrome c_{551i} is the electron acceptor for MDH. The function of cytochrome c_{553i} is still unknown.

This chapter will summarize what is known about the genes involved in methanol oxidation and the function of their products in *P. denitrificans*.

3 ORGANIZATION OF mox GENES

A DNA fragment of 10.7 kb has been isolated from *P. denitrificans*. The nucleotide sequence of this DNA fragment has been determined and revealed 8 *mox* genes and the 5' region of a 9th one (see Figure 1) ([3,8]; Harms, unpublished results). *MoxFJGIR* and *S* are transcribed in the same direction. *MoxFJGI* are tightly linked and appear to form an operon [3,6]. *MoxZY* and

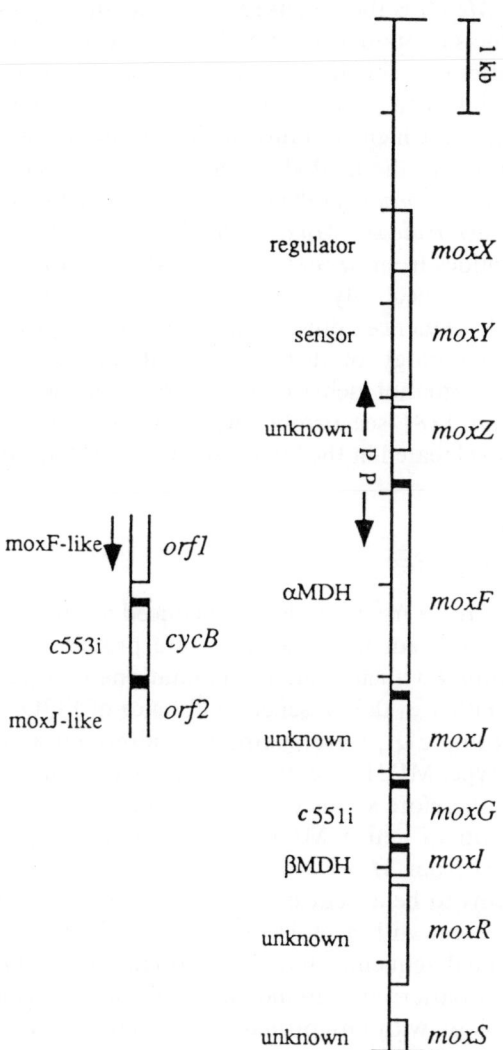

Figure 1 Organization of the *mox* gene cluster of *Paracoccus denitrificans*. Open boxes indicate open reading frames. The black boxes indicate signal sequences. The direction of the arrows indicate the direction of transcription.

X have a transcription direction opposite to that of *moxF* and also appear to form an operon (Harms, unpublished results). Mutants with either an insertion of a kanamycin resistance marker or an unmarked deletion in 7 of these genes have been isolated and characterized.

3.1 *Structural* mox *genes*

MoxF and *moxI* are the structural genes encoding the large and small subunit of MDH, respectively. *MoxG* is the structural gene encoding cytochrome c_{551i}. These three gene products contain at their N-terminus a stretch of amino acids that shows characteristics of a signal sequence, which confirms the periplasmic location of these proteins. The primary structure of the cytochrome shows a large hydrophobic domain, which contains the heme-binding site [3]. *MoxFG* and *I* have been identified in the methylotrophs *Methylobacterium extorquens* AM1 and *Methylobacterium organophilum* XX [9–15]. Alignment of the amino acid sequences of *P. denitrificans* MoxF, MoxI and MoxG [3, 8] and the corresponding polypeptides from *M. extorquens* AM1 [15, 16, 17] revealed 76, 60 and 50% homology, respectively. Comparison of the deduced amino acid sequence of *moxF* with sequences present in international protein and DNA data banks revealed homology of this protein with quinoproteins such as glucose dehydrogenase, ethanol dehydrogenase, but also with an unknown protein in *P. denitrificans* [18] (see also below). It has been suggested therefore that the PQQ cofactor is located in the large subunit of MDH. For an overview see Anthony [7].

3.2 *Accessory* mox *genes*

MoxJ is located downstream from *moxF*. The deduced amino acid sequence of this gene predicts a signal sequence and a mature protein with a molecular mass of 28.1 kDa. A mutant with an unmarked mutation in this gene was unable to grow on methanol, either in the presence or absence of PQQ. Both subunits of MDH and cytochrome c_{551i} were synthesized although at reduced levels compared to the wild type. MDH activity was not detected in this mutant. The gene product of *moxJ* therefore seems to be a periplasmic located protein that is involved in formation of active MDH [3]. Alignment of *MoxJ* from *P. denitrificans* with *MoxJ* from *M. extorquens* AM1 [15] revealed a homology of 41%. This gene seems to be absent in *M. organophilum* XX [10].

MoxR encodes a protein with a molecular mass of 36.9 kDa. This protein does not contain a signal sequence and is hydrophilic in nature. Within the amino acid sequence, a pattern was found that has characteristics of an ATP binding site. No homology with any protein sequenced thus far was observed. However, in the sequence downstream from *moxI* of *M. extorquens* AM1, the start of a gene was found that encodes a protein which showed 50% homology with *MoxR* of *P. denitrificans* [3]. A mutant with a mutation in *moxR* was unable to grow on methanol. MDH and cytochrome c_{551i} were synthesized as in the wild type, but MDH activity was not detected. It has been suggested that *moxR* has a function in regulation of formation of active MDH [3].

Downstream from *moxR* the 5' region of *moxS* has been identified [3]. A mutation in this gene resulted in the inability of *P. denitrificans* to grow on methanol, and MDH activity was absent (H. Yang, unpublished results). This indicates that this gene is involved in methanol oxidation. Whether *moxS* is also present in other methylotrophs is not known.

Upstream of *moxF*, three regulatory genes, *moxZ*, *Y* and *X*, have been identified. Characteristics of these genes and their function in the regulation of MDH synthesis will be described in section 4.

3.3 cycB *and accessory genes*

On a separate chromosomal fragment a gene, *cycB*, encoding cytochrome c_{553i} has been identified [18]. Upstream of this gene, part of an open reading frame is present, the deduced amino acid sequence of which shows homology with *moxF* and other quinoproteins. Downstream of *cycB* the 5' region of another open reading frame has been identified that shows homology with *moxJ*. A mutant with a mutation in *cycB* was able to grow on methanol, although with a reduced maximum specific growth rate compared to the wild type [18]. This result indicates that this gene cluster is somehow involved in methanol oxidation. However, the function of the gene cluster is still unclear. It has been suggested that the gene upstream of *cycB* encodes a PQQ dependent dehydrogenase with unknown substrate specificity. Studies are in progress to isolate the complete *cycB* gene cluster and to characterize this DNA fragment in more detail.

4 REGULATION OF *mox* GENE EXPRESSION

The regulatory genes, *moxZ*, *Y* and *X*, encode proteins of 16.4 kDa, 48.2 kDa and 24.5 kDa, respectively. The gene product of *moxZ* showed no significant homology with any known protein sequenced thus far. However, *MoxY* and *MoxX* showed homology with the protein histidine kinases and the response regulators forming the two-component regulatory systems. The three regulatory *mox* genes were analyzed by characterization of mutants, complementation analysis, and promoter studies with *moxF-lacZ* transcriptional gene fusions (see Table 1). This analysis has led to the model as presented in Figure 2.

The levels of β-galactosidase encoded by the *moxF-lacZ* transcriptional fusions generally correlated with MDH activity [19]. β-Galactosidase activity was found in cells grown on methylamine, but was absent in cells grown on glucose. Mutants with a mutation in *moxX* did not grow on methanol, did not synthesize MDH and consequently lacked MDH activity. Similar characteristics were also found for mutants that contain an insertion of a kanamycin resistance gene in either *moxZ* or *moxY*. Addition of *moxZ* or *moxZY*, *in trans*,

Table 1 Characteristics of regulatory mox mutants

Strains	Relevant genotype[a]	Growth on methanol	MDH[b] activity	MDH/cyt c_{551i} synthesis	Promoter activity[c]	
					Methylamine	Glucose
PD1222	wt	+++	1810	+	2250	30
PD0721	Z::Km	−	2	−	89	15
PD0821	Y::Km	−	2	−	nt	nt
PD0721/pMA11[d]	Z::Km/Z	−	2	nt	nt	nt
PD0721/pNH30[d]	Z::Km/Z	nt	nt	nt	87	10
PD0921	X::Km	−	2	−	nt	nt
PD 0721/pNH56	Z::Km/ZY	−	nt	nt	nt	nt
PD0721/pWR124[e]	Z::Km/ZY	nt	nt	nt	136	<1
PD0741	ΔZ	+	440	+	750	20
PD0841	ΔY	+++	1870	+	nt	nt

wt = wild type; nt = not tested; cyt c_{551i} = cytochrome c_{551i}
[a] Z = moxZ; Y = moxY; X = moxX::Km = insertion of kanamycin resistance marker
[b] in nmol DCPIP per min per mg protein. Cultures were grown on methylamine.
[c] β-galactosidase activities are expressed in arbitrary (Miller) units [25]. Promoter activities were measured in cells harboring plasmid pCP112 (= pMP190 [26] derivative carrying the region between moxF and moxZ) or another plasmid as indicated. Cultures were grown either on methylamine or glucose as indicated.
[d] pMA11 = pEG400 [27] derivative containing the 5' region of moxY, moxZ, both promoters and the 5' region of moxF; pNH30 = pMP190 containing the same insert as pMA11.
[e] pNH56 = pEG400, containing moxY, moxZ, both promoters and the 5' region of moxF; pWR124 = pMP190 containing the same insert as pNH56.

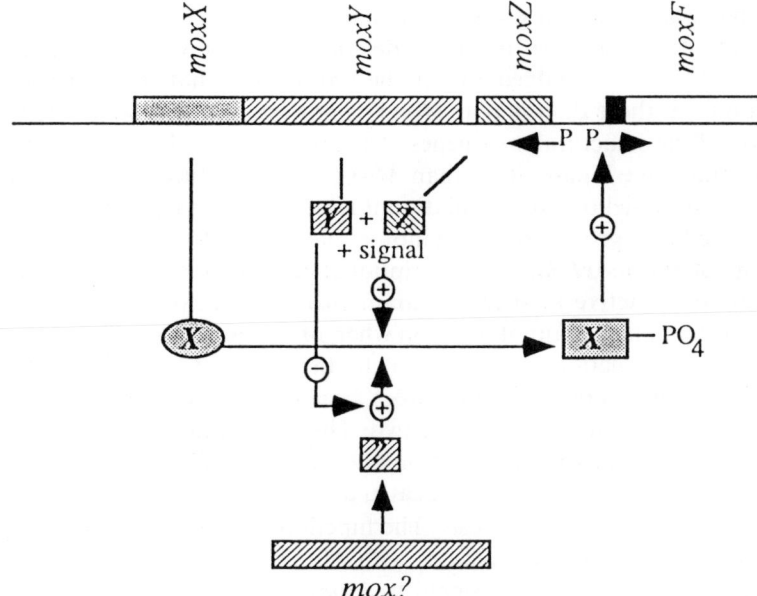

Figure 2 Model for the regulation of the structural *mox* genes in *Paracoccus denitrificans*. The *moxX* gene product code for a putative transcriptional activator. The *moxY* and *moxZ* and the hypothetical *mox?* gene product may be involved in modulating MoxX activity. +, positive regulation. Details are given in the text.

to the *moxZ* mutant PD0721 also resulted in the same phenotype. These latter strains showed little or no expression of *moxF-lacZ* transcriptional fusion genes. Addition of *moxX*, in trans, to one of the mutants restored growth, MDH was synthesized and β-galactosidase activities were observed (data not shown). These data indicate that the *moxX* gene product is a positively acting regulator which is required for activation of the *moxF* promoter. Mutants with an unmarked deletion in *moxY* showed a phenotype comparable to the wild-type. Mutants with an unmarked deletion in *moxZ*, showed reduced growth on methanol, reduced MDH activity and reduced expression of the transcriptional fusion. Addition of *moxZ*, in trans, did not result in changes of this phenotype. These data indicate that *moxY* and *moxZ* are not indispensable for methanol oxidation.

Two-component regulatory systems consist of at least two proteins, a histidine protein kinase and a response regulator. The histidine protein kinase detects an environmental signal and transduces this signal *via* phosphorylation reactions to the response regulator. A phosphorylated regulator protein is able to bind DNA after which transcription is activated [20, 21]. This mechanism for two-component regulatory systems, applied to *MoxY* and *MoxX*, predicts that loss of *MoxY* and *MoxX* function should virtually eliminate C_1-dependent regulation of *moxF* gene expression. Indeed a mutation in *moxX* resulted in a mutant that was unable to grow on methanol and to synthesize MDH. However, a mutant with a large 'in frame' deletion in *moxY* was indistinguishable from the wild type as judged by all the parameters that were measured. A conclusion from these data is that *MoxY* is not essential for *MoxX*-dependent expression of the structural *mox* genes. An explanation is that at least in *moxY* mutants, a histidine kinase other than *MoxY* can transduce a signal to *MoxX*. Following this model of cross regulation it is surprising that a deletion in *moxZ* resulted in reduced growth on methanol, reduced MDH activities and reduced expression of the *moxF-lacZ* transcriptional fusion. It seems that the cross regulation is not active in such a mutant probably because the hypothetical histidine protein kinase cannot function when *MoxY* is present. The characteristics of the *moxZ* deletion mutant might be the result of absence of *MoxZ* or of reduced expression of *moxY* and *moxX* because of downstream effects of the mutation or of a combination of the two. The observation that this mutation could not be complemented by addition of *moxZ in trans* strongly suggests that the reduced expression of the genes located downstream plays an important role in the reduction of MDH synthesis. The function of *MoxZ* remains unclear. It is conceivable that *MoxZ* cooperates with *MoxY* in sensing an environmental signal and transduction of this signal to *MoxX*.

The signal that is sensed by *MoxY* is at present unknown. Computer analysis of *MoxY* predicts that the protein contains two membrane spanning regions, bordering a more hydrophilic domain that is probably located at the periplasmic site of the inner membrane. This may indicate that the stimulus which is detected by MoxY is present in the periplasm. In earlier studies it has been shown that MDH synthesis in *P. denitrificans* is not induced by its substrate, methanol [5]. It has been postulated that a product induction mechanism, with formaldehyde as an effector, is involved in regulation of MDH synthesis. It is likely, therefore, that formaldehyde, which is formed in the periplasm, is the signal that is detected by *MoxY*.

So far the *moxF* promoter is the only promoter that was found to be activated by *MoxX*. The regulation of expression of the *cycB* gene cluster is not controlled by the *MoxY-MoxX* pair, since in all *moxZY* and *X* mutants, grown on methylamine, cytochrome c_{553i} was found (H. Yang and N. Harms, unpublished).

5 CONCLUSIONS

The studies summarized here indicate that at least 9 genes are involved in oxidation of methanol. Correct expression of MDH in *P. denitrificans* involves several processes. At the transcriptional level, a repression and an activation mechanism has been discovered. So far no genes involved in the repression mechanism have been identified. Activation of the structural *mox* genes was found to be regulated by a two-component regulatory system together with a third regulatory protein. Once the *mox* genes are transcribed and translated, the two subunits have to be transported to the periplasm and assembled with the cofactor. So far two genes, *moxJ* and *moxR*, have been identified that are involved in these post-translational processes. Certainly more *P. denitrificans mox* genes have to be identified and characterized. In *M. extorquens* AM1 and *M. organophilum* XX, for instance, up to 17 *mox* genes have been described, 4 of which are involved in PQQ biosynthesis [10, 22].

ACKNOWLEDGEMENTS

I thank my coworkers Dr Rob van Spanning for the many fruitful discussions and helpful suggestions and Dr Hui Yang for providing results prior to publication.

REFERENCES

[1] Alefounder, P.R. and Ferguson, S.J. (1981) A periplasmic location for methanol dehydrogenase from *Paracoccus denitrificans*: implications for proton pumping by cytochrome aa_3. Biochem. Biophys. Res. Comm. **948**, 778–784.
[2] Bamforth, C.W. and Quayle, J.R. (1978) Aerobic and anaerobic growth of *Paracoccus denitrificans* on methanol. Arch. Microbiol. **119**, 91–97.
[3] Van Spanning, R.J.M., Wansell, C.W., De Boer, T., Hazelaar, M.J., Anazawa, H., Harms, N., Oltmann, L.F. and Stouthamer, A.H. (1991) Isolation and characterization of the *moxJ, moxG, moxI and moxR* genes of *Paracoccus denitrificans*. Inactivation of the *moxJ, moxG, and moxR* genes and the resultant effect on methylotrophic growth. J. Bacteriol. **173**, 6948–6961.
[4] Long, A.R. and Anthony, C. (1991) Characterization of the periplasmic cytochromes c of *Paracoccus denitrificans*: identification of the electron acceptor for methanol dehydrogenase, and description of a novel cytochrome c heterodimer. J. Gen. Microbiol. **137**, 415–425.
[5] De Vries, G.E., Harms, N., Maurer, K., Papendrecht, A. and Stouthamer, A.H. (1988) Physiological regulation of *Paracoccus denitrificans* methanol dehydrogenase synthesis and activity. J. Bacteriol. **170**, 3731–3737.
[6] Harms, N. and Van Spanning, R.J.M. (1991) C_1 metabolism in *Paracoccus denitrificans*: Genetics of *Paracoccus denitrificans*. J. Bioenerg. Biomembr. **23**, 187–210.
[7] Anthony, C. (1992) The c-type cytochromes of methylotrophic bacteria. Biochim. Biophys. Acta **1099**, 1–15.
[8] Harms, N., De Vries, G.E., Maurer, K., Hoogendijk, J. and Stouthamer, A.H. (1987) Isolation and nucleotide sequence of the methanol dehydrogenase structural gene from *Paracoccus denitrificans*. J. Bacteriol. **169**, 3969–3975.
[9] Anderson, D.J. and Lidstrom, M.E. (1988) The *moxFG* region encodes four polypeptides in the methanol-oxidizing bacterium *Methylobacterium* sp. strain AM1. J. Bacteriol. **170**, 2254–2262.
[10] Bastien, C., Machlin, S., Zhang, Y., Donaldson, K. and Hanson, R.S. (1989) Organization of genes required for the oxidation of methanol to formaldehyde in three type II methylotrophs. Appl. Environ. Microbiol. **55**, 3124–3130.
[11] Machlin, S.M. and Hanson, R.S. (1988) Nucleotide sequence and transcriptional start site of the *Methylobacterium organophilum* XX methanol dehydrogenase structural gene. J. Bacteriol. **170**, 4739–4747.

[12] Machlin, S.M., Tam, P.E., Bastien, C.A. and Hanson, R.S. (1988) Genetic and physical analyses of *Methylobacterium organophilum* XX genes encoding methanol oxidation. *J. Bacteriol.* **170**, 141–148.
[13] Nunn, D.N. and Lidstrom, M.E. (1986) Phenotypic characterization of 10 methanol oxidation mutant classes in *Methylobacterium* sp. strain AM1. *J. Bacteriol.* **166**, 591–597.
[14] Nunn, D.N. and Lidstrom, M.E. (1986) Isolation and complementation analysis of 10 methanol oxidation mutant classes and identification of the methanol dehydrogenase structural gene of *Methylobacterium* sp. strain AM1. *J. Bacteriol.* **166**, 581–590.
[15] Anderson, D.J., Morris, C.J., Nunn, D.N., Anthony, C. and Lidstrom, M.E. (1990) Nucleotide sequence of the *Methylobacterium extorquens* AM1 *moxF* and *moxJ* genes involved in methanol oxidation. *Gene* **90**, 173–176.
[16] Nunn, D.N. and Anthony, C. (1988) The nucleotide sequence and deduced amino acid sequence of the cytochrome c_L gene of *Methylobacterium extorquens* AM1, a novel class of *c*-type cytochrome. *Biochem. J.* **256**, 673–676.
[17] Nunn, D.N., Day, D. and Anthony, C. (1989) The second subunit of methanol dehydrogenase of *Methylobacterium extorquens* AM1. *Biochem. J.* **60**, 857–862.
[18] Ras, J., Reijnders, W.N.M., Van Spanning, R.J.M., Harms, N., Oltmann, L.F. and Stouthamer, A.H. (1991) Isolation, sequencing and mutagenesis of the gene encoding cytochrome c_{553i} of *Paracoccus denitrificans* and characterization of the mutant strain. *J. Bacteriol.* **173**, 6971–6979.
[19] De Vries, G.E., Harms, N., Hoogendijk, J. and Stouthamer, A.H. (1989) Isolation and characterization of *Paracoccus denitrificans* mutants with increased conjugation frequencies and pleiotropic loss of a n(GATC)n DNA-modifying property. *Arch. Microbiol.* **152**, 52–57.
[20] Stock, J.B., Ninfa, A.J. and Stock, A.M. (1989) Protein phosphorylation and regulation of adaptive responses in bacteria. *Microbiol. Rev.* **53**, 450–490.
[21] Ninfa, A.J. (1991) Protein phosphorylation and the regulation of cellular processes by the homologous two-component regulatory systems of bacteria. In: Genetic Engineering. (Setlo, J.K., Ed.), pp. 39–72. Plenum Press, New York.
[22] Lidstrom, M.E. (1990) Genetics of carbon metabolism in methylotrophic bacteria. *FEMS Microbiol. Rev.* **87**, 431–436.
[23] Miller, J.H. (1972) Experiments in Molecular Genetics, pp.352–355, Cold Spring Harbor, N.Y., Cold Spring Harbor Laboratory.
[24] Gerhus, E., Steinrücke, P. and Ludwig, B. (1990) *Paracoccus denitrificans* cytochrome c_1 gene replacement mutants. *J. Bacteriol.* **172**, 2392–2400.
[25] Spaink, H.P., Okker, R.J.H., Wijffelman, C.A., Pees, E. and Lugtenberg, B.J.J. (1987) Promoters in the nodulation region of the *Rhizobium leguminosarum* Sym plasmid pRL1J1. *Plant Mol. Biol.* **9**, 27–39.

19
The Methanol-oxidizing Enzyme Systems in Gram-positive Methylotrophic Bacteria

Leonid V. Bystrykh, Nico Arfman and Lubbert Dijkhuizen

Department of Microbiology, University of Groningen, Kerklaan 30, 9751 NN Haren, The Netherlands

1 SUMMARY

In recent years it has become evident that Gram-positive methylotrophic bacteria employ a novel type of alcohol dehydrogenase for methanol oxidation. All thermotolerant, methanol-utilizing strains of *Bacillus methanolicus* investigated were found to possess an NAD-dependent methanol dehydrogenase (MDH), which is strongly stimulated by a specific (activator) protein. No NAD-dependent MDH activity could be detected in the methanol-utilizing bacteria *Amycolatopsis methanolica* and *Mycobacterium gastri*. Instead, methanol oxidation in these organisms resulted in concomitant reduction of N,N'-dimethyl-4-nitrosoaniline (NDMA). The corresponding enzymes are designated as methanol:NDMA oxidoreductases (MNO).

Analysis of the quaternary protein stuctures of the purified *B. methanolicus* MDH (subunit M_r 43 kDa) and the *A. methanolica* and *M. gastri* MNO enzymes (subunit M_r 49–50 kDa) by electron microscopy and image processing revealed similar decameric structures with five-fold symmetry. The three proteins are also similar with respect to their metal composition (Zn^{2+}- and Mg^{2+}-ions) and the presence of bound pyridine nucleotide cofactors. The primary amino acid sequences of these three enzymes share a high degree of identity and show that they belong to the Type III alcohol dehydrogenases (ADH).

In addition to the methanol:NDMA oxidoreductase activity of MNO, also dye (dichlorophenol indophenol, DCPIP and 3-[4,5-dimethylthiazol-2-yl]-2,5-diphenyl-tetrazolium bromide, MTT)-linked methanol dehydrogenase activities can be detected reproducibly in crude extracts of *A. methanolica*. These dye-linked methanol dehydrogenases appear to represent the overall activities of multienzyme systems. The biochemistry of methanol oxidation in Gram-positive bacteria is complex and both MDH of *B. methanolicus* and MNO of *A. methanolica* and *M. gastri in vivo* require additional proteins, most likely participating in the transfer of reducing equivalents from NAD(P)H cofactors to NAD-coenzyme and/or to the electron transport chain.

2 INTRODUCTION

The initial oxidation of methanol to formaldehyde in methanol-utilizing microorganisms can be catalyzed by at least three different types of enzymes. Methylotrophic yeasts employ alcohol oxidase (EC 1.1.3.13), a FAD-containing peroxisomal enzyme, which catalyzes the oxidation of methanol into formaldehyde and hydrogen peroxide, using oxygen as electron acceptor [1,2]. Methanol oxidation in Gram-negative bacteria involves a periplasmic quinoprotein methanol dehydrogenase (EC 1.1.99.8), containing pyrroloquinoline quinone (PQQ) as prosthetic group, and special *c*-type cytochromes for transfer of reducing equivalents towards the electron transport chain [3–5]. Methanol dehydrogenase activity *in vivo* is most likely influenced by additional factors, e.g. a low molecular weight factor X from *Hyphomicrobium* X [6], and a so-called modifier protein, detected in several Gram-negative methylotrophs [7]. In contrast to the well-known enzymology and genetics of methanol oxidation in methylotrophic yeasts and Gram-negative methylotrophic bacteria, little is known about the enzymes of primary methanol oxidation in Gram-positive methylotrophs. Studies with thermotolerant *B. methanolicus* strains [8,9], *A. methanolica* [10] and *M. gastri* [11] have shown that these organisms, which lack a periplasmic space, do not possess a PQQ-containing methanol dehydrogenase. Evidence is emerging now that these organisms employ a novel cytoplasmic methanol dehydrogenase containing bound NAD(P)H.

3 METHANOL DEHYDROGENASE FROM *Bacillus methanolicus* C1

NAD-dependent MDH constitutes up to 40% of total soluble protein in cells of *B. methanolicus* strain C1 grown under methanol-limiting conditions in continuous cultures at low dilution rates [12,13]. NAD-dependent ADH enzymes (EC 1.1.1.1) display little, if any, activity with methanol. MDH from *B. methanolicus* oxidizes C_1-C_4 primary alcohols as well as 1,3-propanediol and further biochemical and electron microscopic studies revealed that its structural, kinetic, and mechanistic properties are clearly different from general NAD-dependent ADH enzymes. In contrast to the common dimeric and tetrameric ADH enzymes, MDH consists of ten identical subunits of M_r 43 kDa, arranged in a 'sandwich' of two pentagonal rings [14]. Each subunit contains 1 zinc and 1–2 magnesium ions. Zinc is commonly found in the active site of ADH enzymes but the presence of magnesium has not been reported before. In addition, each MDH subunit contains a tightly (but non-covalently) bound NAD(H) molecule. UV-spectrophotometry indicates that this NAD(H) molecule is redox active, participating in the alcohol oxidation reaction [15]. Activity of MDH strictly requires exogenous NAD (coenzyme) in addition to bound NAD(H) (cofactor). No incorporation of label in MDH protein was observed in experiments with radioactively labeled coenzyme NAD, indicating that these two NAD(H) species are not exchanged during catalysis. NAD thus appears to play a dual role in the MDH reaction, with cofactor NAD acting as primary electron acceptor and coenzyme NAD being responsible for reoxidation of the NADH cofactor in a transhydrogenase type of reaction. MDH displays a ping-pong type reaction mechanism, which implies that the alcohol substrate and coenzyme NAD bind sequentially to the enzyme. This is consistent with a mechanism involving a temporary deposit of reducing equivalents at the MDH-bound cofactor. In contrast, NAD-ADH enzymes lacking a cofactor obey a reaction mechanism which involves simultaneous binding of alcohol and NAD substrates (ternary substrate complex) [16].

The affinities of MDH for alcohol substrates and exogenous NAD, as well as the V_{max} values, are strongly increased by a soluble M_r 50 kDa activator protein (consisting of two M_r 27 kDa subunits). The activation process strictly requires the presence of exogenous NAD and Mg^{2+} ions. Activation may result in a 40-fold overall increase in the methanol turnover rate of MDH at physiological methanol concentrations [17]. The purified activator protein does not contain a cofactor, but is able to bind 1 molecule of NAD(H) per subunit. The activator protein can form a (loose) protein complex with MDH, but the interaction only occurs in the presence of exogenous NAD plus magnesium ions, indicating that these low molecular weight components are part of the MDH-activator protein complex. The activator protein changes the steady state kinetics of the MDH reaction from ping-pong to Theorell-Chance type. These observations, in combination with the structural and kinetic properties of MDH, have led to the postulation that the activator protein facilitates the reoxidation of the bound NADH cofactor [15]. This model implies that the NAD-activator protein complex is a more favorable electron acceptor than free NAD and that the transhydrogenase reaction, resulting in reoxidation of the NADH cofactor, is the rate-limiting step in alcohol oxidation.

The MDH structural gene of *B. methanolicus* strain C1 has been cloned and sequenced [18]. The deduced MDH amino acid sequence was found to share significant identity with *Zymomonas mobilis* ADH2, *Saccharomyces cerevisiae* ADH4, *Escherichia coli* ADHE and 1,2-propanediol oxidoreductase and *Clostridium acetobutylicum* ADH1. The latter enzymes do not oxidize methanol and contain zinc or iron. Since the MDH related enzymes clearly differ from the horse liver ADH-type and the *Drosophila* ADH-type [19], we have adopted the designation Type III for this group of enzymes [18].

The N-terminal amino acid sequence of the activator protein revealed no significant similarity with any published protein sequences. Recently we have succeeded in cloning the *B. methanolicus* gene encoding the activator protein (H. Kloosterman, J.W. Vrijbloed, L. Dijkhuizen, unpublished data). The precise identity of the activator protein however remains to be determined.

4 METHANOL DEHYDROGENASES FROM *Amycolatopsis methanolica* AND *Mycobacterium gastri*

No NAD- or PQQ-dependent MDH activity could be detected in the methanol-utilizing bacteria *Amycolatopsis methanolica* and *Mycobacterium gastri*. Instead, methanol oxidation in these organisms resulted in concomitant reduction of p-nitroso-N,N'-dimethylaniline (NDMA) as an artificial electron acceptor [20]. Duine et al. [21] previously reported that *A. methanolica* possesses low activity levels of a dye (DCPIP)-linked methanol dehydrogenase which is strictly dependent on the presence of exogenous NAD. However, accumulation of free NADH could not be detected with this system, which suggested that NADH is not in the usual free form, but remains enzyme associated. NDMA is known to reoxidize pyridine nucleotides which are tightly bound to the active centers of dehydrogenases [22,23]. The methanol dehydrogenases present in *A. methanolica* and *M. gastri* subsequently were purified to homogeneity and characterized using NDMA as the coupling dye [20]. Both enzymes use methanol as well as formaldehyde as substrates. As the *in vivo* electron acceptor of these enzymes remains unknown, they are referred to as methanol:NDMA oxidoreductase (MNO) rather than methanol dehydrogenase. MNO failed to oxidize methanol in the presence of the artificial electron acceptors DCPIP or MTT. No evidence was obtained for the presence of an activator-like protein in these organisms and addition of *B. methanolicus* activator protein to crude extracts did not result in appearance of NAD-dependent MDH activity. The enzymatic properties of the MNO enzymes differ from *B. methanolicus* MDH in various other aspects as well [20]. The MNO enzymes not only catalyze the NDMA-linked oxidation of methanol and formaldehyde but also NADH-dependent formaldehyde reductase and formaldehyde dismutase [24] reactions (yielding formaldehyde and formate from methanol). On the other hand, *B. methanolicus* MDH displayed NAD-dependent MDH and NADH-dependent formaldehyde reductase activities but was inactive with NDMA; it also failed to catalyze the formaldehyde dismutase reaction.

Analysis of the quaternary protein stuctures of the purified *A. methanolica* and *M. gastri* MNO enzymes (subunit M_r 49–50 kDa) by electron microscopy and image processing [25] revealed that these proteins are strikingly similar to *B. methanolicus* MDH (i.e. decameric structures displaying five-fold symmetry). The three proteins are also similar with respect to their metal composition (Zn^{2+}- and Mg^{2+}-ions) and the presence of a bound pyridine nucleotide cofactor (NADPH in case of both MNO enzymes). Finally, analysis of the

N-terminal and internal peptide amino acid sequences of MNO, and comparison with *B. methanolicus* MDH, showed that the three enzymes also share a high degree of identity at the primary amino acid sequence level and all three clearly belong to the Type III ADH enzymes.

5 DYE-LINKED METHANOL DEHYDROGENASE ACTIVITIES IN *Amycolatopsis methanolica*

Crude extracts of *A. methanolica* were reported to possess low dye-linked (PMS/DCPIP) methanol dehydrogenase activity, which was measured at pH 7.0 in the absence of ammonium salts [26]. Duine *et al.* [21] reported the presence of DCPIP-linked (without PMS) methanol dehydrogenase activity which was measured at pH 9.0. This system strictly required the presence of ammonium salts and NAD for activity but accumulation of free NADH could not be detected. A tentative characterization revealed that this MDH forms part of a rather loose multienzyme complex, together with an NAD-dependent formaldehyde dehydrogenase and an NADH dehydrogenase [21]. A similar NAD-dependent methanol dehydrogenase activity was demonstrated in crude extracts of methanol-grown cells of the actinomycete strain 381 [27]. Only recently, we have been able to develop reproducible assay conditions for the DCPIP-dependent methanol dehydrogenase activity in *A. methanolica*.

Meanwhile, van Ophem *et al.* [28] reported the presence of a tetrazolium dye (MTT)-linked methanol dehydrogenase activity in extracts of methanol-grown cells of *A. methanolica* which is stably maintained and can be assayed reproducibly. The activity of this MTT-dependent activity strongly increased with increasing ionic strength of the buffer solution as well as with increasing amounts of extract. These observations are taken to indicate that activity depends on association of some essential components. MTT-dependent activity was completely lost following chromatographic fractionation of crude extracts. Activity could be restored however by combining separate fractions of the eluate, indicating that this MTT-linked methanol dehydrogenase represents the overall activity of a multienzyme system.

6 CONCLUSIONS

The identification and characterization of structurally similar, NAD(P)H-containing methanol dehydrogenases from *B. methanolicus*, *A. methanolica* and *M. gastri* shows that a new and uniform type of methanol-oxidizing enzyme is employed by these Gram-positive methylotrophs. All enzymes involved in methanol oxidation investigated to date possess a tightly bound cofactor (FAD in yeast alcohol oxidase; PQQ in quinoprotein MDH; NAD in *Bacillus* MDH; NADP in the MNO enzymes), suggesting that a temporary deposit for reduction equivalents is a prerequisite for methanol-converting alcohol dehydrogenases.

Under *in vivo* conditions both MDH of *B. methanolicus* and MNO of *A. methanolica* and *M. gastri* appear to be associated with additional proteins. Our current knowledge indicates that these proteins participate in the reoxidation of the NAD(P)H cofactors, most likely resulting in transfer of reducing equivalents to NAD coenzyme and/or to the electron transport chain.

Further biochemical investigations are certainly required in order to obtain a full understanding of the nature of the methanol oxidation pathways in Gram-positive, methanol-utilizing bacteria.

REFERENCES

[1] Woodward, J.R. (1990) Biochemistry and applications of alcohol oxidase from methylotrophic yeasts. In: Advances in Autotrophic Microbiology and One-Carbon Metabolism, Vol. I (Codd, G.A., Dijkhuizen, L. and Tabita, F.R., Eds) pp. 193–225, Kluwer Academic Publishers, Dordrecht.
[2] Müller, F., Hopkins, T.R., Lee, J. and Bastiaens, P.I.H. (1992) Methanol oxidase. In: Chemistry and Biochemistry of Flavoenzymes. Vol. III (Müller, F., Ed.) pp. 95-119, CRC Press, Boca Raton.
[3] Anthony, C. (1982) The Biochemistry of Methylotrophs, Academic Press, London.
[4] Anthony, C. (1986) Bacterial oxidation of methane and methanol. *Adv. Microb. Physiol.* **27**, 113–210.
[5] de Vries, G.E., Kües, U. and Stahl, U. (1990) Physiology and genetics of methylotrophic bacteria. *FEMS Microbiol. Rev.* **75**, 57–102.
[6] Dijkstra, M., Frank, J. and Duine, J.A. (1988) Methanol oxidation under physiological conditions using methanol dehydrogenase and a factor isolated from *Hyphomicrobium* X. *FEBS Lett.* **227**, 198–202.
[7] Long, A.R. and Anthony, C. (1991) The periplasmic modifier protein for methanol dehydrogenase in the methylotrophs *Methylophilus methylotrophus* and *Paracoccus denitrificans. J. Gen. Microbiol.* **137**, 2353–2360.
[8] Dijkhuizen, L., Arfman, N., Attwood, M.M., Brooke, A.G., Harder, W. and Watling, E.M. (1988) Isolation and initial characterization of thermotolerant methylotrophic *Bacillus* strains. *FEMS Microbiol. Lett.* **52**, 209–214.
[9] Arfman, N., Dijkhuizen, L., Kirchhof, G., Ludwig, W., Schleifer, K.H., Bulygina, E.S., Chumakov, K.M., Govorukhina, N.I., Trotsenko, Y.A., White, D. and Sharp, R.J. (1992) *Bacillus methanolicus* sp. nov., a new species of thermotolerant, methanol-utilizing, endospore-forming bacteria. *Int. J. Syst. Bacteriol.* **42**, 439–445.
[10] de Boer, L., Dijkhuizen, L., Grobben, G., Goodfellow, M., Stackebrandt, E., Parlett, J.H., Whitehead, D. and Witt, D. (1990) *Amycolatopsis methanolica* sp. nov., a facultatively methylotrophic actinomycete. *Int. J. Syst. Bacteriol.* **40**, 194–204.
[11] Kato, N., Miyamoto, N., Shimao, M. and Sakazawa, C. (1988) 3-Hexulose phosphate synthase from a new facultative methylotroph, *Mycobacterium gastri* MB19. *Agric. Biol. Chem.* **52**, 2659–2661.
[12] Arfman, N., Watling, E.M., Clement, W., van Oosterwijk, R.J., de Vries, G.E., Harder, W., Attwood, M.M. and Dijkhuizen, L. (1989) Methanol metabolism in thermotolerant methylotrophic *Bacillus* strains involving a novel catabolic NAD-dependent methanol dehydrogenase as a key enzyme. *Arch. Microbiol.* **152**, 280–288.
[13] Arfman, N., de Vries, K.J., Moezelaar, H.R., Attwood, M.M., Robinson, G.K., van Geel, M. and Dijkhuizen, L. (1992) Environmental regulation of alcohol metabolism in thermotolerant methylotrophic *Bacillus* strains. *Arch. Microbiol.* **157**, 272–278.

[14] Vonck, J., Arfman, N., de Vries, G.E., Van Beeumen, J., Van Bruggen, E.F.J. and Dijkhuizen, L. (1991) Electron microscopic analysis and biochemical characterization of a novel methanol dehydrogenase from the thermotolerant *Bacillus* sp. C1. *J. Biol. Chem.* **266**, 3949–3954.

[15] Arfman, N. (1991) Methanol metabolism in thermotolerant Bacilli. Ph.D. thesis, University of Groningen, The Netherlands.

[16] Cook, P.F. and Bertagnolli, B.L. (1987) Kinetics of pyridine nucleotide-utilizing enzymes. In: Pyridine Nucleotide Coenzymes, Part A, (Dolphin, D., Avramović, O. and Poulson, R., Eds) pp.405–448, Wiley, New York.

[17] Arfman, N., Van Beeumen, J., de Vries, G.E., Harder, W. and Dijkhuizen, L. (1991) Purification and characterization of an activator protein for methanol dehydrogenase from thermotolerant *Bacillus* spp. *J. Biol. Chem.* **266**, 3955–3960.

[18] de Vries, G.E., Arfman, N., Terpstra, P. and Dijkhuizen, L. (1992) Cloning, expression and sequence analysis of the *Bacillus* strain C1 NAD-dependent methanol dehydrogenase gene. *J. Bacteriol.* **174**, 5346–5353.

[19] Jörnvall, H., Persson, M. and Jeffery, J. (1981) Alcohol and polyol dehydrogenases are both divided into two protein types, and structural properties cross-relate the different enzyme activities within each type. *Proc. Natl. Acad. Sci. USA* **78**, 4226–4230.

[20] Bystrykh, L.V., Govorukhina, N.I., van Ophem, P.W., Hektor, H., Dijkhuizen, L. and Duine, J.A. (1993) Methanol:NDMA oxidoreductase and formaldehyde dismutase activities in Gram-positive bacteria oxidizing methanol. *J. Gen. Microbiol.*, in press.

[21] Duine, J.A., Frank, J. and Berkhout, M.P.J. (1984) NAD-dependent, PQQ-containing methanol dehydrogenase: a bacterial dehydrogenase in a multienzyme complex. *FEBS Lett.* **168**, 217–221.

[22] Dunn, M.F. and Bernhard, S.A. (1971) Rapid kinetic evidence for adduct formation between the substrate analog p-nitroso-N,N'-dimethylaniline and reduced nicotinamide-adenine dinucleotide during enzymic reduction. *Biochemistry* **10**, 4569–4575.

[23] Kovář, J., Šimek, K., Kučera, I. and Matyska, L. (1984) Steady-state kinetics of horse-liver alcohol dehydrogenase with a covalently bound coenzyme analogue. *Eur. J. Biochem.* **139**, 585–591.

[24] Kato, N., Yamagami, T., Shimao, M. and Sakazawa, C. (1986) Formaldehyde dismutase, a novel NAD-binding oxidoreductase from *Pseudomonas putida* F61. *Eur. J. Biochem.* **156**, 59–64.

[25] Bystrykh, L.V., Vonck, J., van Bruggen, E.F.J., Van Beeumen, J., Samyn, B., Govorukhina, N.I., Arfman, N., Duine, J.A. and Dijkhuizen, L. (1993) Electron microscopic analysis and structural characterization of novel NADP(H)-containing methanol:NDMA oxidoreductases from the Gram-positive methylotrophic bacteria *Amycolatopsis methanolica* and *Mycobacterium gastri* MB19. *J. Bacteriol.*, in press.

[26] Kato, N., Tsuji, K., Tani, Y. and Ogata, K. (1975) Utilization of methanol by an actinomycete. In: Microbial Growth on C_1 Compounds (The Organizing Committee, Eds), pp. 91–98. The Society of Fermentation Technology, Tokyo.

[27] Eshraghi, S., Hancock, I.C. and Williams, E. (1990) A methylotrophic thermotolerant actinomycete with a NAD-dependent methanol dehydrogenase. In: Abstracts of the 6th Int. Symp. Microbial Growth on C_1 Compounds, Göttingen, Abstract p.105.

[28] Ophem, P.W. van, Euverink, G.J., Dijkhuizen, L. and Duine, J.A. (1991) A novel dye-linked alcohol dehydrogenase activity present in some Gram-positive bacteria. *FEMS Microbiol. Lett.* **80**, 57–64.

20
Overview of the Current State of Methylotroph Taxonomy

Peter N. Green

NCIMB Ltd, 23 St Machar Drive, Aberdeen AB2 1RY, Scotland, UK

1 SUMMARY

The contribution made by a polyphasic approach to the taxonomy of methylotrophic bacteria is examined. The present taxonomic status of the obligate methylotrophs, based largely upon phenotypic and chemotaxonomic data, is discussed in some detail. The impact of some new genotypic data on our perception of the taxonomic structure within this group of organisms is also examined.

A selected number of more recently described facultative methylotrophs are reviewed as is the phylogeny of some of these organisms.

2 INTRODUCTION

Many present day taxonomists favour a polyphasic approach to the study of taxonomy and systematics. This requires a dual approach in which there is a balance between detailed phenotypic and chemotaxonomic information on the one hand, and genotypic features on the other. Ideally, this genotypic information should complement and supplement the other data and contribute to the overall phylogeny of the organisms examined.

In the case of methylotrophic bacteria, and the obligate methylotrophs in particular, the restricted nutritional abilities of most of these organisms, have resulted in less data of the former type being available on which a sound taxonomy can be constructed. In addition, genotypic data, especially those involving 5S and 16S ribosomal RNA, have only recently begun to show relationships within and between the various groups of bacteria which share the common feature of C_1 metabolism. In many cases this information has confirmed existing taxonomic structures, whereas in others it has shown relationships where none was thought to exist, or conversely that phenotypically similar strains were more distantly related than previously envisaged.

Thus almost inevitably, with a group of organisms such as the methylotrophs, their taxonomy is in a state of semi-permanent flux where one piece of information may advance the cause, while another may complicate or confuse our understanding of, or ability to circumscribe, a particular group of organisms.

In this chapter, an attempt is made to describe briefly the current taxonomic status of certain groups of methylotrophic bacteria and to examine trends or similarities which exist within this complex array of microbes.

2.1 *Groups of convenience*

Figure 1 illustrates the major groupings that most workers use to subdivide methylotrophic bacteria. Basically these organisms are split into two broad sections: those which utilize only C_1 compounds (obligate methylotrophs) and those which can utilize, in addition to C_1, compounds, multicarbon substrates (facultative methylotrophs). An intermediate group which can grow on a restricted range of multi-carbon compounds is also recognized by some workers. The obligate methylotrophs are further sub-divided into those which can utilize methanol and methylated amines (but not methane), and those which can utilize methane (the methanotrophs). The facultatively methylotrophic bacteria encompass a wide variety of organisms some of which bear little or no phylogenetic relationship to one another.

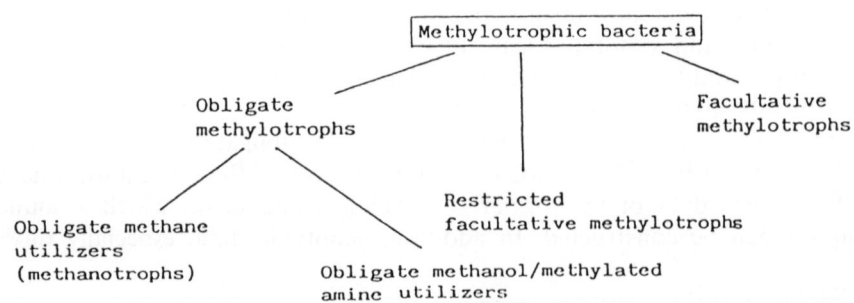

Figure 1 Schematic diagram of methylotrophic bacteria.

3 OBLIGATE METHYLOTROPHS

3.1 *Methanotrophs*

Published descriptions of obligate methane utilizing bacteria were relatively rare prior to 1970, when Whittenbury et al. [1] isolated and described a large number of strains. The tentative classification scheme which they devised for these organisms still exists as the cornerstone on which present day taxonomy is based. Initially two groups were proposed. One contained 'type I' organisms which were rod shaped, were shown to contain bundles of vesicular discs when viewed in thin section, used the ribulose monophosphate (RUMP) pathway for C_1 assimilation, had an incomplete tricarboxylic acid (TCA) cycle and had predominant fatty acids with a 16 carbon chain length. 'Type II' organisms were often morphologically different from their type I counterparts, contained paired peripheral membranes in thin section, used the serine pathway for C_1 assimilation, had a complete TCA cycle and had predominant fatty acids with an 18 carbon chain length. A third group of methanotrophs ('type X') was subsequently proposed to support strains of *Methylococcus capsulatus*: the only obligate methanotroph of that time, shown to be capable of autotrophic CO_2 fixation (see Table 1).

Type I bacteria contained organisms tentatively named as *Methylomonas* and *Methylobacter* by Whittenbury, while type II bacteria were assigned to either *Methylosinus* or *Methylocystis*. Although not all of these have been formally proposed as validated genera [1,2], they often tend to be viewed as such in the literature. Useful taxonomic features which may help to differentiate between the various methane utilizing bacteria are discussed in detail elsewhere [1,2].

One such feature in particular: phospholipid composition [3], is useful in distinguishing between some of the 'genera' described above. For example, type II organisms *Methylosinus* and *Methylocystis* contain methylated derivatives of phosphatidylethanolamine in addition to phosphatidylcholine. These phospholipids are absent from the pool of the type I *Methylomonas* and *Methylobacter* strains. Also bacteria grouped by Whittenbury and his colleagues as *Methylomonas* differ from those grouped as *Methylobacter* by the lack of cardiolipin which is present in the latter in substantial amounts. Type II methanotrophs (*Methylocystis* and *Methylosinus*) can be differentiated by a 2-D chromatogram of their phospholipids. Strains belonging to *Methylosinus* contain only trace amounts of phosphatidylcholine, one of the major components of *Methylocystis*. Other workers [2–7] have shown that phospholipid fatty acid composition among methanotrophs is another useful distinguishing tool, possibly even at the sub-'generic' level.

While further studies, such as PAGE of whole cell proteins, broadly supported Whittenbury's groupings, it was almost inevitable that as more chemotaxonomic and in particular, genotypic data became available, these groupings would be challenged as a clearer picture of their heterogeneity emerged. The

Table 1 Primary groupings of methanotrophic bacteria[a]

Characteristic	Type I	Type X	Type II
Morphology	Straight rod or coccobacillus	Coccus	Straight, curved or pear-shaped rod
Membrane arrangement			
Bundles or vesicular disks	+	+	−
Paired peripheral membranes	−	−	+
Motility	V	−	V
Resting stage	Azotobacter-type cyst	Azotobacter-type cyst	Lipid cyst or terminal exposure
Rosette formation	−	−	+ (most strains)
Major carbon assimilation pathway	Rump	Rump	Serine
Autotrophic CO_2 fixation	−	+	−
Complete TCA cycle	−	−	+
Nitrogenase	−	+	+
Isocitrate dehydrogenase[a]			
NAD^+ and $NAD(P)^+$ specific	+	−	−
NAD^+ specific	−	+	−
$NAD(P)^+$ specific	−	−	+
Glucose-6-dehydrogenase	+($NADP^+$-specific)	+($NADP^+$-specific)	−[c]
6-Phosphogluconate dehydrogenase	+($NADP^+$-specific)	+($NADP^+$-specific)	−
Predominant fatty acid carbon-chain length	16	16	18
Growth at 45°C	V[b]	+	−
Mol% G+C of DNA	50–54	62.5	61.7–63.1

[a] Not all strains classifiable into groups I and II have been shown to possess all the biochemical characteristics outlined in this scheme.
[b] V = variable.
[c] During growth on methane.

taxonomic logic in the naming of some newly isolated methanotrophs and the re-naming of existing strains (mainly as *Methylococcus* spp.) did little to help the non-taxonomists understanding of these organisms. Such was the confusion created by these proposals, that Romanovskaya *et al.* [8] proposed that some of the '*Methylococcus*' strains, along with some *Methylobacter* strains should be transferred to a new genus; *Methylovarius*. This genus was not however adequately circumscribed and was never validated. More recently, Bowman *et al.* [9] have isolated two new orange-pigmented species of *Methylomonas* (*M. fodinarum* and *M. aurantiaca*) which have mol % G+C values in the mid to high 50s and fix carbon via the RUMP cycle. During their studies, they also examined a number of new isolates alongside a range of existing type cultures. Specifically, they examined the DNA base compositions, genome molecular weights and nucleotide distributions among these strains [10]. While they found that the type II strains formed separate 'generic' clusters according to genome molecular weights, the same strains formed two overlapping clusters when nucleotide distributions and DNA base compositions were examined. Type I methanotrophs, on the other hand, were shown to form three separate clusters based on both nucleotide distribution and genome size analysis. The first group contained the new orange pigmented isolates and the pink *Methylomonas* strains *M. methanica* and *M. rubra*. A second cluster contained low G+C strains cited as *Methylococcus whittenburyi*, *Methylococcus bovis*, *Methylococcus vinelandii*, *Methylococcus luteus*, *Methylococcus ucrainicus*, *Methylomonas albus* and *Methylomonas pelagica*. The third and final cluster contained the high G+C strains *Methylomonas gracilis*, along with *Methylococcus capsulatus*, *Methylococcus thermophilus* and other moderately thermophilic strains examined. However, it is equally possible to postulate, from this and other data, that at least five separate genera may exist within the type I methanotrophs (see Fig. 2); namely: (1) an emended genus *Methylomonas* containing the two pink pigmented species (*M. methanica* and *M. rubra*); although even this may have to be reviewed later as some recent 16S cataloguing data suggest there may not be a close phylogenetic link between *M. rubra* and *M. methanica* (R.S. Hanson, pers. comm.); (2) the two new orange-pigmented species (*M. fodinarum* and *M. aurantiaca*) could prove to be sufficiently different from the pink isolates to merit generic status on their own; (3) an emended genus *Methylobacter* containing, at present, all the other low (48–56 mol %) G+C strains listed above; although the spread of G+C values (8%) does not indicate a particularly homogeneous group. (The members of this group would not be too dissimilar from those proposed for *Methylovarius*; although there is disparity among the mol % G+C values of the organisms involved between the data of Bowman *et al.* and that of Romanovskaya and co-workers.) ; (4) *Methylococcus*, containing only the type X *M. capsulatus* strains and related isolates of Bowman *et al.* [10]; and (5) the two brown pigmented thermophilic or thermotolerant species (*M. themophilus* and *M. gracilis*) which, as with the first two groups, may be sufficiently unrelated to the *M. capsulatus* cluster to merit separate genus status (see Fig. 2).

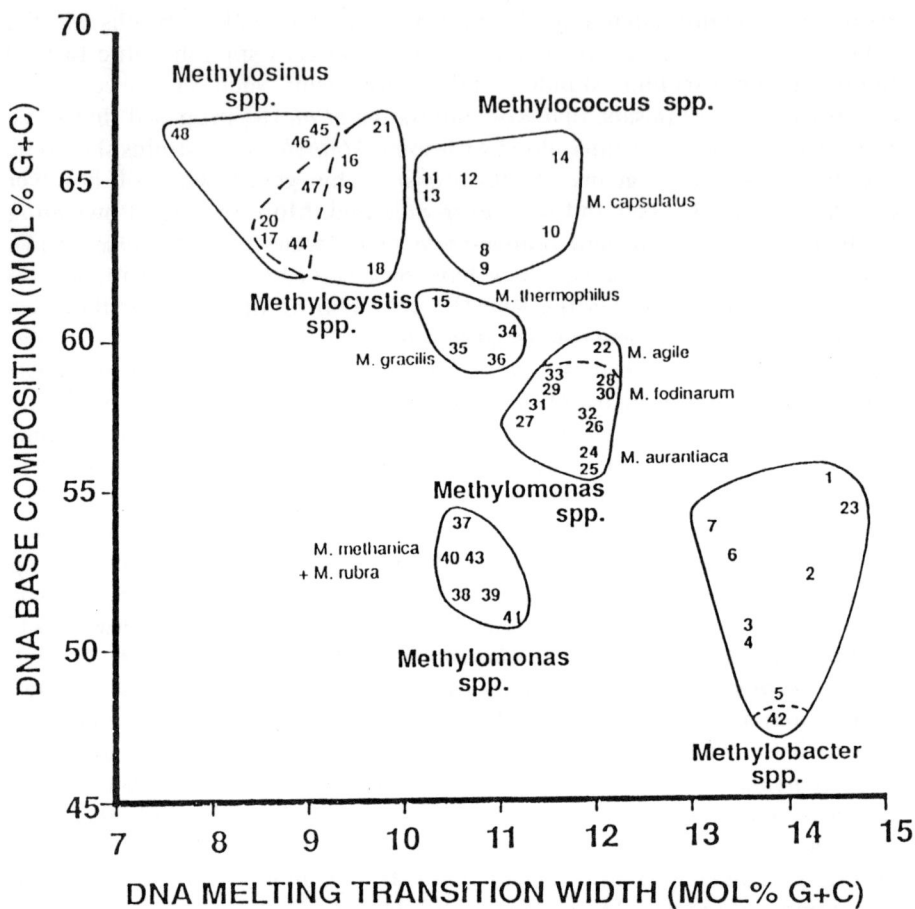

Figure 2 Similarity map of obligate methanotrophic bacteria based on DNA base compositions and nucleotide distributions. Modified from [10].

While I think it would be premature to make any such proposals formally, the taxonomy of the obligate methanotrophs is without doubt at an exciting new crossroads.

3.2 *Obligate methylotrophs which do not utilize methane*

Before moving on to the obligate methanol and methylated amine utilizing bacteria, we should first consider the restricted facultative methylotrophs. These were a group of organisms studied by Colby and Zatman [11] who coined the names 'type L or type M restricted facultative' for organisms which had a restricted nutritional capability. The type L or 'less restricted' organisms are viewed by some workers as facultative methylotrophic heterotrophs with a restricted capability. The type M organisms, in contrast, have a much more limited nutritional spectrum; many of them being able to utilize only glucose and/or fructose in addition to C_1 compounds. It will be interesting to see, if as one would expect, these isolates are much more closely related phylogenetically to what are considered to be the obligate methanol/methylated amine utilizers.

If, therefore, for the sake of convenience, we consider Colby and Zatman's type M restricted facultative strains to be part of the obligate methanol utilizing group of organisms, we find that there currently exist four validly published genera, viz: *Methylophilus* [12], *Methylobacillus* [13], *Methylovorus* [14] and *Methylophaga* [15]; where previously all these organisms were accommodated in a single genus: *Methylobacillus*. I think it is widely accepted that this group of phenotypically similar organisms is quite heterogeneous, but so far they have not been studied in sufficient detail to be able both to distinguish and identify all the taxa present.

Table 2 shows some of the features presently available which may help to differentiate between the four genera presently described, and at the same time, illustrates the lack of characteristics which the bench bacteriologist can easily use to differentiate between these organisms; some groups of which can only be identified on the basis of DNA:DNA homologies [16] or the electrophoretic mobilities of specific enzymes [17]. All these organisms are Gram-negative motile or non-motile rods which use the RUMP pathway as their major route for C_1 assimilation; have straight chain $C_{16:0}$ and monounsaturated $C_{16:1}$ as their major fatty acids and have ubiquinone with 8 isoprene units as their major isoprenoid quinone [18]. Of the four currently recognized genera, one (*Methylophaga*) contains organisms of marine origin, which have a low (43–49) mol % G+C value, an obligatory requirement for Na^+ and Mg^{2+} ions and are auxotrophic for vitamin B_{12}. Of the other three genera, strains belonging to *Methylophilus* utilize glucose and have phospholipids which contain PG and PE (see Table 2) along with two unidentified glycolipids; while strains belonging to *Methylobacillus*, do not utilize glucose and have DPG and cardiolipin as well as PG and PE and one unidentified glycolipid as their phospholipid components. (See Green [1] for more detailed information.)

4 FACULTATIVE METHYLOTROPHS

Facultative methylotrophs are an extremely varied and cosmopolitan group of organisms. Unlike the obligate strains discussed, they do not form an identifiable taxonomic entity. Indeed in many cases the ability of these bacteria to utilize C_1 compounds has been an overweighted taxonomic feature. Nevertheless, there are some examples where specific groups of organisms can be distinguished as discreet taxa where all the members have methylotrophy in common (e.g. *Methylobacterium*). Some of these taxa are completely unrelated to one another while others share common physiological, biochemical or morphological properties; and perhaps a close phylogenetic relationship.

Table 2 Obligate methanol/methylated amine utilizers: some differentiating features

Organism	Phospholipids				(mol %)	Glucose utilization	Fructose utilization	Requirement for Na$^+$ and Mg^{++}	Auxotrophic for B$_{12}$	Branched C$_{17}$ fatty acids	6 PGA dehydrogenase (NADP)	Ammonia assimilation
	DPG	PG	PE	GL								
METHYLOPHILUS					50–53							
M. methylotrophus	−	+	+	A + B		+	V	−	−	+	+	Glutamate cycle
M. glucoseoxidans						+	−	−	−			
METHYLOBACILLUS					50–56							
M. glycogenes	+	+	+	A		−	V	−	−	−	+	Glutamate dehydrogenase
M. fructoseoxidans						−	+	−	−			
METHYLOVORUS					56–57							
M. glucosostrophus	+	+	+	C		+	−	−	−	−	−	Glutamate cycle
METHYLOPHAGA					43–49							
M. marina							+	++	+			
M. thalassica							+	++	+			

Organism feature	P. denitrificans	P. alcaliphilus	P. aminovorans	P. kocurii	P. aminophilus
G + C content (mol %)	65–68	64–66	67–68	71	63
Urease	+	–	–	+	–
Growth factors	–	biotin	thiamine	thiamine	thiamine
growth at pH6	+	–	+	–	+
growth at pH9	–	+	W	–	W
Utilization of:					
D-xylose	–	+	–	–	+
D-mannose	+	+	+	NT	–
D-fructose	+	+	+	–	–
Trehalose	+	–	–	–	–
D-sorbitol	+	+	+	–	–
D-Mannitol	+	+	+	–	–
Inositol	+/–	+/–	–	–	–
Methanol	+/–	+/–	–	–	–
Methylamine	–	–	+	+	+
Dimethylamine	–	–	+	+	+
Trimethylamine	–	–	+	+	+
Trimethylamine –N–oxide	–	–	+	+	+
Formamide	–	–	W	NT	W
Methylformamide	–	–	+	NT	+
dimethylformamide	–	–	+	NT	+
TMAH[a]	NT	NT	NT	+	NT
hydrogen	+	–	–	NT	–
Nitrogen sources:					
KNO₃	+	+	–	–	W
(NH₄)₂SO₄	+	+	+	+/–	+
Glutamate	+	NT	NT	–	NT

[a] TMAH = tetramethylammonium hydroxide.
NT = not tested; W = weak.

Before looking at some of these relationships we should briefly examine a few of the more recently described facultative methylotrophs. (Readers should consult Green [1] for a more comprehensive review.) Increasingly, the organisms of interest being isolated are 'extremophile' bacteria which can survive and replicate in environments of extreme pH or temperature. *Acidomonas methanolica*, as the name implies contains acidophilic organisms which can utilize methanol and grow at pH 2.2–5.5. These organisms, thought to be related to *Acetobacter* are discussed in more detail elsewhere [19] as indeed are another group of extremophiles, the thermophilic *Bacillus* strains isolated by Dijkhuizen and his colleagues [20] and also reviewed by Arfman in this book. *Paracoccus*, a rapidly expanding genus containing facultative methylotrophs (Table 3), also contains one extremophile, *P. alcaliphilus* [21] which, unlike other members of the genus, can grow at pH 9.5. This organism grows on methanol but not on any of the other C_1 substrates examined. Other species of this genus, which are described by Urakami *et al.* [22] are shown in Table 3. Some of these organisms utilize N-methylformamide (MF) and N-N-dimethyl-formamide (DMF).

Organisms recently classified as *Pseudomonas aminovorans* (which includes strains *Ps.* MS and *Ps.* MA, both of which grow on methylated amines) have recently be reclassified in a new genus *Aminobacter* [23]. Two new species have been proposed, *A. aganoensis* and *A. niigataensis*, of which the latter can also utilize MF and DMF as sole carbon source, while *A. aganoensis* can utilize tetra- methylammonium hydroxide.

Aminobacter, in common with several other genera of facultative methylotrophs, share a number of features. One such feature is budding or asymmetric growth. Indeed this particular morphological feature extends to the following genera, all of which contain methylotrophic strains: *Methylobacterium*, *Blastobacter*, *Ancylobacter*, *Xanthobacter*, *Rhizobium*, *Agrobacterium*, *Methylomicrobium* [24], *Hyphomicrobium*, *Nitrobacter* and *Rhodopseudomonas*. A close phylogenetic relationship is thus likely to exist between at least a number of these organisms. This relationship is further strengthened, in some instances, by other mutual characteristics. In particular, members of several of the above genera contain large amounts of straight-chain monounsaturated $C_{18:1}$ as their principal cellular fatty acid; and have a ubiquinone with 10 isoprene units as their major isoprenoid component [25]. Additionally, smaller groups of strains which may or may not be closely related share other features such as nitrogen fixation (*Xanthobacter*, *Azospirillum* and some members of the *Rhizobiaceae*). Strains of *Xanthobacter*, *Blastobacter*, *Ancylobacter*, *Paracoccus*, *Nitrobacter*, *Aminobacter*, *Acidomonas*, *Rhodopseudomonas* and *Thiobacillus* have also been shown to assimilate C_1 compounds autotrophically, while *Methylobacterium*, *Hyphomicrobium*, *Methylomicrobium* and *Aminobacter* strains use the serine pathway.

Another noteworthy and possibly ancestral link concerns that of bacteriochlorophyll (Bchl) a, which is present in members of the genus *Methylobacterium* [24], suggestive of a common root with some of the photosynthetic organisms.

Interestingly, many of the facultative methylotrophs discussed in this section can be recovered in rRNA superfamily IV (*sensu* De Ley) which is equivalent to Woese's α sub-class within the Proteobacteria. Focusing in on the α-2 sub-division of this group of organisms, it is likely that most, if not all members, are budding organisms (L.B. Perry, pers. com., [26]) which contain 27-hydroxyoctacosanoic acid as a group specific fatty acid [27] chemical marker.

5 CONCLUSIONS

While there is clearly still much work to be done before we have a complete picture of the phylogenetic relationships of all the methylotrophic organisms which have been described, it is clear that methylotroph taxonomy is now entering an exciting new era. Much of this is due to several recent studies involving ribosomal RNA sequencing [28–30], which have allowed us an insight into the phylogeny of these organisms (see the chapters by Hanson and Bulygina in this book). Of particular satisfaction is that, in many cases, these genotypic characterizations have broadly agreed with taxonomic structures achieved by more 'traditional' methods, thus vindicating the value and reliability of polyphasic taxonomy.

More rapid methodologies in the future will surely lead us in new directions, such as more refined chemotaxonomic analysis, and the development of oligonucleotide probes with their many different applications; and to discovery of new strains, confirmation of existing taxa and the quantification of members (both culturable and non-culturable) in microbial consortia.

ACKNOWLEDGEMENTS

I wish to pay due credit to the work of L.B. Perry, whose meticulous attention to morphological detail over many years resulted in the phylogenetic link between polar growth (budding) and many of the members of Woese's α-class Proteobacteria which are discussed in this chapter.

REFERENCES

[1] Green, P.N. (1992) Taxonomy of methylotrophic bacteria. In: Methane and Methanol Utilizers. Biotechnology Handbooks 5 (Murrell, J.C. and Dalton, H., Eds), pp. 23–77, Plenum Press, New York.
[2] Hanson, R.S., Netrusov, A.I. and Tsuji, K. (1992) The obligate methanotrophic bacteria *Methylococcus*, *Methylomonas* and *Methylosinus*. In: The Prokaryotes, 2nd edn. (Balows, A., Truper, H.G., Dworkin, M., Harder, W. and Schleifer, K-H., Eds), pp. 2350–2364, Springer-Verlag, New York.
[3] Galchenko, V.F. and Andreev, L.V. (1984) Taxonomy of obligate methanotrophs. In: Microbial Growth on C_1-Compounds (Crawford, R.L. and R.S., Eds), pp. 269–275, American Society for Microbiology, Washington, DC.

[4] Bowman J.P., Skerratt, J.H., Nichols, P.D. and Sly, L.I. (1991) Phospholipid fatty acid and lipopolysaccharide fatty acid signature lipids in methane-utilizing bacteria. *FEMS Microbiol. Ecol.* **85**, 15–22.
[5] Guckert, J.B., Ringelberg, D.B., White, D.C., Hanson, R.S. and Bratina, B.J. (1991) Membrane fatty acids as phenotypic markers in the polyphasic taxonomy of methylotrophs within the Proteobacteria. *J. Gen. Microbiol.* **137**, 2631–2641.
[6] Urakami, T. and Komagata, K. (1987) Cellular fatty acid composition with special reference to the existence of hydroxy fatty acids in Gram-negative methanol-methane- and methylamine-utilizing bacteria. *J. Gen. Appl. Microbiol.* **33**, 135–165.
[7] Romanovskaya, V.A., Malashenko, Y.R. and Grischenko, N.I. (1980) Diagnosis of methane-oxidizing bacteria by numerical methods based on cell fatty acid composition. *Mikrobiologiya* **49**, 969–975.
[8] Romanovskaya, V.A. (1984) *Methylovarius* gen. nov., a new genus. *Mikrobiologiya* **53**, 777–784.
[9] Bowman, J.P., Sly, L.I., Cox, J.M. and Hayward, A.C. (1990) *Methylomonas fodinarum* sp. nov. and *Methylomonas aurantiaca* sp. nov.: two closely related type 1 obligate methanotrophs. *Syst. Appl. Microbiol.* **13**, 279–287.
[10] Bowman, J.P., Sly, L.I. and Hayward, A.C. (1991) Contribution of genome characteristics to the assessment of taxonomy of obligate methanotrophs. *Int. J. Syst. Bacteriol.* **41**, 301–305.
[11] Colby, J. and Zatman, L.J. (1975) Tricarboxylic acid cycle and related enzymes in restricted facultative methylotrophs. *Biochem. J.* **148**, 505–511.
[12] Jenkins, O., Byrom, D. and Jones, D. (1987) *Methylophilus*: a new genus of methanol utilizing bacteria. *Int. J. Syst. Bacteriol.* **37**, 446–448.
[13] Yordy, J.R. and Weaver, T.L. (1977) *Methylobacillus*: a new genus of obligately methylotrophic bacteria. *Int. J. Syst. Bacteriol.* **27**, 247–255.
[14] Govorikhina, N.I. and Trotsenko, Y.A. (1991) *Methylovorus*—a new genus of restricted facultatively methylotrophic bacteria. *Int. J. Syst. Bacteriol.* **41**, 158–162.
[15] Janvier, M., Frehel, G., Grimont, F. and Gasser, F. (1985) *Methylophaga marina* gen. nov. sp. nov. and *Methylophaga thalassica* sp. nov., marine methylotrophs. *Int. J. Syst. Bacteriol.* **35**, 131–139.
[16] Urakami, T., Tamaoka, J. and Komagata, K. (1985) DNA base composition and DNA-DNA homologies of methanol utilizing bacteria. *J. Gen. Appl. Microbiol.* **27**, 243–253.
[17] Urakami, T. and Komagata, K. (1981) Electrophoretic comparison of enzymes in the Gram negative methanol utilizing bacteria. *J. Gen. Appl. Microbiol.* **27**, 381–403.
[18] Urakami, T. and Komagata, K. (1979) Cellular fatty acid composition and coenzyme Q system in Gram negative methanol utilizing bacteria. *J. Gen. Appl. Microbiol.* **25**, 343–360.
[19] Urakami. T., Tamaoka, J., Suzuki, K-I, and Komagata, K. (1989) *Acidomonas* gen. nov. incorporating *Acetobacter methanolicus* as *Acidomonas methanolica* comb. nov. *Int. J. Syst. Bacteriol.* **39**, 50–55.
[20] Dijkhuizen, L., Arfman, N., Attwood, M.M., Brook, A.G., Harder, W. and Watling, E.M. (1988) Isolation and initial characterization of thermotolerant methylotrophic *Bacillus* strains. *FEMS Microbiol. Lett.* **52**, 209–214.
[21] Urakami, T., Tamaoka, J., Suzuki, K-I, and Komagata, K. (1990). *Paracoccus alcaliphilus* sp. nov., an alkaliphilic and facultatively methylotrophic bacterium. *Int. J. Syst. Bacteriol.* **39**, 116–121.
[22] Urakami, T., Araki, H., Oyanagi, H., Suzuki, K-I. and Komagata, K. (1990) *Paracoccus aminophilus* sp. nov. and *Paracoccus aminovorans* sp. nov., which utilize N,N-dimethylformamide. *Int. J. Syst. Bacteriol.* **40**, 287–291.

[23] Urakami, T., Araki, H., Oyanagi, H., Suzuki, K-I, and Komagata, K. (1992) Transfer of *Pseudomonas aminovorans* (den Dooren de Jong 1926) to *Aminobacter* gen. nov. as *Aminobacter aminovorans* comb. nov. and description of *Aminobacter aganoenis* sp. nov. and *Aminobacter niigataensis* sp. nov. Int. J. Syst. Bacteriol. 42, 84–92.

[24] Govorukhina, N.I., Doronina, N.V., Andreev, L.V. and Trotsenko, Y.A. (1989) *Methylomicrobium: a new genus of facultative methylotrophic bacteria.* Microbiology 58, 260–266.

[25] Komagata, K. (1989) Taxonomy of facultative methylotrophs. In: Aerobic Photosynthetic Bacteria (Harashima, K., Shiba, T. and Murata, N., Eds), pp. 27–38, Springer-Verlag, Berlin.

[26] Woese, C.R., Stackebrandt, E., Weisburg, W.E., Paster, B.J., Madigan, M.T., Fowler, V.J., Hahn, C.M., Blanz, P., Gupta, R., Nealson, K.H. and Fox, G.E. (1984) The phylogeny of purple bacteria: The alpha subdivision. Syst. Appl. Microbiol. 5, 315–326.

[27] Ramadas Bhat, U., Carlson, R.W., Busch, M. and Mayer, H. (1991) Distribution and phylogenetic significance of 27-hydroxy-octacosanoic acid in lipopolysaccharides from bacteria belonging to the alpha-2 subgroup of *Proteobacteria*. Int. J. Syst. Bacteriol. 41, 213–217.

[28] Ando, S., Kato, S-I., and Komagata, K. (1989) Phylogenetic diversity of methanol-utilizing bacteria deduced from their 5S ribosomal RNA sequences. J. Gen. Appl. Microbiol. 35, 351–361.

[29] Bulygina, E.S., Galchenko, V.F., Govorakhina, N.I., Netrusov, A.I., Nikitin, D.I., Trotsenko, Y.A. and Chumakov, K.M. (1990) Taxonomic studies on methylotrophic bacteria by 5S ribosomal RNA sequencing. J. Gen. Microbiol. 136, 441–446.

[30] Tsuji, K., Tsien, H.C., Hanson, R.S., Depalma, S.R., Scholtz, R. and Laroche, S. (1990) 16S ribosomal RNA sequence analysis for phylogenetic relationship among methylotrophs. J. Gen. Microbiol. 36, 1–10.

21
Taxonomy of Thermotolerant Methylotrophic Bacilli

N. Arfman and L. Dijkhuizen

Department of Microbiology, University of Groningen, Kerklaan 30, 9751 NN Haren, The Netherlands

1 SUMMARY

The generic position of fourteen strains of Gram-positive bacteria able to use methanol as growth substrate has been determined. All are obligately aerobic, thermotolerant organisms, able to grow at temperatures of 35–60°C. Nine of the strains produce oval spores at a subterminal to central position in slightly swollen rod-shaped cells. DNA-DNA hybridization studies, 5S rRNA sequence analysis and physiological characteristics revealed that all 14 strains cluster as a well-defined group and form a distinct new genospecies. Analysis of the 16S rRNA and 5S rRNA sequences indicated that this new species is distinct from *Bacillus brevis* but closely related to *B. firmus* and *B. azotoformans*. The name proposed for this new species is *Bacillus methanolicus*. The type strain, strain PB1, has been deposited as NCIMB 13113.

2 INTRODUCTION

Bacteria able to grow on methanol at elevated temperatures are of technological interest for single cell protein production, solvent degradation in aerobic thermophilic biotreatment processes, and fermentative production of amino acids [1–3]. Initial attempts to isolate such methanol-utilizing strains were often fraught with difficulties, particularly when using conventional isolation procedures involving batch cultivation and methanol-agar plating techniques [4]. In recent years, however, several groups have successfully employed continuous

---

Microbial Growth on C_1 Compounds
© Intercept Ltd, PO Box 716, Andover, Hampshire SP10 1YG, UK

Table 1 General physiological properties of thermotolerant methanol-utilizing *Bacillus* strains

Growth temperature range	35–60°C
Optimal growth temperature (T_{opt})	±55°C
t_d on methanol at 50–55°C (at T_{opt})	40–80 min
pH range for growth	6.5–9
pH optimum for growth	7.0–7.5
Molar growth yield	16–18 gram dry weight/mol methanol
Assimilation pathway	ribulose monophosphate pathway
Methanol oxidation	NAD-dependent methanol dehydrogenase
Methanol tolerance	1.5 M
Growth substrates	methanol, glucose, mannitol
Substrates not used for growth	succinate, cellobiose
Acid production from	glucose, mannitol
No acid production from	glycerol, glycogen, inulin, lactose, salicin, starch, sucrose, i-erythritol, inositol, xylose, galactose, adonitol
Substrates not hydrolyzed	casein, hippurate
Nitrogen sources	ammonia, glutamine

All of the methanol-utilizing *Bacillus* strains display catalase and oxidase activity. None of the strains is capable of reducing nitrate or nitrite. None of the strains is able to grow in the presence of 5% NaCl.

culture techniques for the isolation of pure cultures of *Bacillus* strains growing rapidly on methanol at 55°C [2–6]. These organisms appear to be widespread and ubiquitous in nature, with isolates originating from different soil samples, aerobic (thermophilic) waste water treatment systems, and volcanic hot springs. The *Bacillus* strains are strongly resistance to high methanol concentrations and the molar growth yield on methanol at the optimum growth temperatures in methanol-limited chemostats is among the highest reported (Table 1) [4].

Enzyme analysis revealed that all isolates employ a novel NAD-dependent methanol dehydrogenase (MDH) for methanol oxidation and the ribulose monophosphate (RuMP) pathway for formaldehyde assimilation [7,8]. Synthesis of methanol dehydrogenase and hexulose-6-phosphate synthase, the key enzyme of the RuMP pathway, are controlled differently in *Bacillus* strain C1 [8]. Relatively high MDH levels and low HPS levels occur in cells grown at low dilution rates in methanol-limited continuous cultures, in cells grown on glucose in batch or continuous cultures, or in cells which have reached the end of the exponential growth phase. Transfer of such cells into media containing relatively high methanol concentrations immediately results in accumulation of toxic formaldehyde [8]. With hindsight, it is clear now that this phenomenon has seriously hampered batch and methanol-agar plating attempts to isolate pure cultures [4].

On the basis of a number of phenotypical tests Al-Awadhi *et al.* [3] concluded that 7 of their isolates were *Bacillus brevis* strains. A further isolate was a sheathed filamentous Gram-positive endospore-forming obligately aerobic bacterium that could not be allocated to any presently described genus. In a recent investigation, 14 organisms including most of Al-Awadhi's isolates were further characterized [9]. The results of this study are summarized and discussed in the next sections.

3 MOLECULAR SYSTEMATIC AND CHEMOTAXONOMIC ANALYSIS

3.1 16S rRNA sequence analysis

The 16S rRNA primary structure of strain C1 (NCIMB 13114) was analyzed (deposited under the EMBL Data Library accession number X64465) and compared with published sequences of bacilli [10–13]. Based on a matrix of phylogenetic distances, a phylogenetic tree was constructed (Figure 1). The combined 16S rRNA data show that *Bacillus* strain C1 is more closely related to *B. azotoformans* and *B. firmus* than to *B. brevis*, as originally thought [3].

3.2 DNA-DNA hybridization studies and G+C content

The mol % G+C content of the DNAs of strains C1, PB1, AR2, TS1, TS4, S2, 4(55), WM5.1 and KA was determined as 48–50%. DNA-DNA hybridization studies showed that the methanol-utilizing *Bacillus* strains are closely related. The DNA similarity values obtained at optimal hybridization conditions were 60% and higher.

3.3 5S rRNA sequence analysis

5S rRNA sequences of the methylotrophic bacteria are shown in Figure 2. These sequences (deposited under the EMBL Data Library accession numbers Z11816 to Z11827), compared with published data on *Bacillus* species [14], confirm that the methanol-utilizing *Bacillus* strains cluster as a well-defined group, separate from the mesophilic trimethylamine-utilizing *Bacillus* strain S2A1 and non-methylotrophic organisms such as *B. stearothermophilus*. The 5S rRNA data also provide further evidence that the methylotrophic bacilli are closely related to *B. firmus* and *B. megaterium*, but not to *B. brevis*.

3.4 Cell wall and lipid analysis

All methylotrophic *Bacillus* strains tested contain *meso*-diaminopimelic acid as cell wall diamino acid. This murein type is also present in cell walls of *B. subtilis*, *B. pumilus*, *B. megaterium*, *B. cereus*, *B. firmus*, *B. fastidiosus*, as well as *B. brevis* [15]. Lipid analysis indicated the presence of squalene and phosphatidylethanolamine in strain C1. Hopanoids [16] could not be detected.

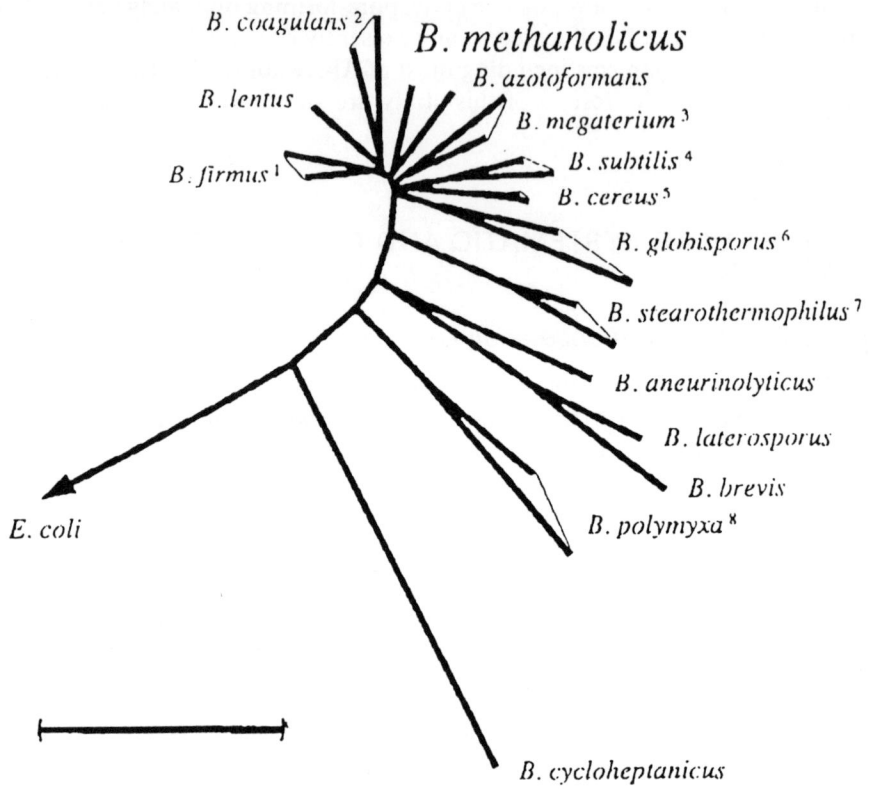

Figure 1 Distance matrix tree showing the relationships between the methanol-utilizing *Bacillus methanolicus* strain C_1 and other bacilli. The phylogenetic distances were calculated including those alignment positions which are invariant in at least 40% of the entire set of 16S rRNA sequences. Groups of more closely related bacilli whose relationships are supported by distance as well as parsimony analyses are indicated by triangles. These groups are: 1: *Bacillus firmus, B. benzoevorans, B. circulans*; 2: *B. coagulans, B. acidoterrestris, B. badius, B. smithii*; 3: *B. megaterium, B. fastidiosus, B. maroccanus, B. psychosaccharolyticus, B. simplex*; 4: *B. subtilis, B. amyloliquefaciens, B. atrophaeus, B. lautus, B. lentimorbus. B. licheniformis. B. popillae, B. pumilus*; 5: *B. cereus, B. anthracis, B. medusa, B. mycoides, B. thuringiensis*; 6: *B. globisporus, B. fusiformis, B. insolitus, B. pasteurii, B. psychrophilus, B. sphaericus*; 7: *B. stearothermophilus, B. kaustophilus, B. thermoglucosidasius*; 8: *B. polymyxa, B. amylolyticus, B. alvei, B. azotofixans, B. gordonae, B. larvae, B. macquariensis, B. macerans, B. pabuli, B. pulvifaciens*. The bar represents 0.05 K_{nuc}.

The fatty acid profile of strain C1 consists mainly of 13-methyltetradecanoic acid (*iso*-C_{15}; 27% of total fatty acid composition), 12-methyltetradecanoic acid (*anteiso*-C_{15}; 16%), 14-methylpentadecanoic acid (*iso*-C_{16}; 12%), 13-methylpentadecanoic acid (*anteiso*-C_{16}; 13%), 15-methylhexadecanoic acid (*iso*-C_{17}; 4%), 14-methylhexadecanoic acid (*anteiso*-C_{17}; 14%).

```
PB1   NCIMB 13113              :   CCUGGUGGCG*AUGGCGAGAAGGUCACACCCGUUCCCAUACCGAACACGGAAGUUAAGCU
C1    NCIMB 13114              :   ----------*-------------------------------------------------
AR2                            :   ----------*-------------------------------------------------
TS1                            :   YY--------*----------------------------------------G--------
TS4                            :   YY--------*----------------------------------------G--------
S1    NCIMB 12524              :   ----------*-------------------------------------------------
S2    NCIMB 12528              :   YY--------*----------------------------G-----------G--------
WM5.1 NCIMB 12527              :   YY--------*----------------------------R--------------------
4(55) NCIMB 12523              :   ----------*------------A------------------------------------
KA    NCIMB 12529              :   YY--------*U---------------------------G--------------------
40M                            :   YY--------*----------------------------G--------------------
S2A1                           :   YYY-----AUU*A-----AG----U--------------C--------A-CC-AG--A--
B. firmus ATCC 14575           :   U-Y-------*--A----------------------------------------------
B. megaterium KM               :   U--R------*--A----AG----------------------------------------
B. subtilis 168                :   U------AU-*-------AG-------------------G--------------------
B. stearothermophilus 799      :   ---A----U-*--A---GAGG--AA--------------C-------------------C
B. brevis ATCC 8185            :   U------AU-*------GAGG--A--------------------------CC-------C


PB1                            :   UCUCAGCGCCGAUGGUAGUUGGGGCAUU*GCCCCUGCGAGAGUAGGACGCUGCCAGGC
C1                             :   ----------------------------*--------------------U--------
AR2                            :   ----------------------------*-----------------------------
TS1                            :   ----------------------------*-------Y---------------------
TS4                            :   ----------------------------*--------------------U--------
S1                             :   ----------------------------*-----------------------------
S2                             :   ----------------------------*-----------------------------
WM5.1                          :   ----------------------------*-----------------------------
4(55)                          :   ----------------------------*-----------------------------
KA                             :   ----------------------------*-----------------------------
40M                            :   ----------------------------*--------------------U--------
S2A1                           :   CU----------------------GUC-*C------U-----C-------C--UC--G
B. firmus                      :   --------------------R---GC-G*U------U-----------------R---
B. megaterium                  :   CU-U------R------------A-U--*-U-----K------------U---Y----
B. subtilis                    :   CU-----------------C----GU--*C------U-----------AUC-------
B. stearothermophilus          :   CUC----------------------CAGC--------A--------U------U----
B. brevis                      :   CUC-------A------C---CUC-GCAG-GAG-C-G------------UC-------
```

Figure 2 Alignment of 5S rRNA sequences of methylotrophic bacteria (determined in this study) and *B. firmus* ATCC 14575, *B. megaterium* KM, *B. subtilis* 168, *B. stearothermophilus* 799, *B. brevis* ATCC 8185 [14]. Only the nucleotides that differ from the *Bacillus* strain PB1 sequence are shown. Alignment gaps are indicated as *; Y = C or U; R = A or G; K = G or U.

4 MORPHOLOGY AND PHYSIOLOGICAL CHARACTERISTICS

4.1 Colony morphology

All strains formed circular colonies on tryptone soya broth agar (TSBA) after two days of incubation, in most cases with rough surfaces and crenated, undulating edges. Colonies of strains S2 and KA were markedly different, with smooth surfaces and convex edges [3].

4.2 Cell and spore morphology

Colonies contained non-motile, rod-shaped Gram-positive cells with dimensions of 0.5–2.0 μm in diameter and 1–6 μm in length (in case of exponentially growing cells in methanol batch culture). Strain KA formed filamentous cells during all growth stages; some other isolates only formed filamentous cells towards the end of growth, e.g. in the centers of TSBA colonies. In methanol-limited continuous cultures of strains C1 and AR2, the majority of cells were present as short chains of rod-shaped cells. Reduced growth rates caused the formation of strongly helical filaments in both these cultures. Similar cellular structures could be detected in colonies of all the isolates grown on mannitol mineral agar for 2 days. Under these conditions, however, they constituted only a minority of the population. With the exception of strains C1, TS1, TS2, WM5.2 and TFB, cells of all strains sporulated on TSBA and/or on mannitol mineral agar. These cells were swollen and possessed oval spores at a subterminal to central position.

4.3 Physiological characteristics

All of the methanol-utilizing isolates studied were obligately aerobic and able to grow in various media at temperatures of 35–60°C. These endospore-forming isolates may therefore be assigned to the genus *Bacillus* [17]. The physiological characteristics shared by all methanol-utilizing *Bacillus* strains are summarized in Table 1. In addition, the entire group of methylotrophic *Bacillus* strains appears to be able to synthesize poly-β-hydroxybutyrate (PHB) as a storage material.

Determination of the average linkage (S_{SM}/UPGMA) of the methanol utilizing *Bacillus* strains, based on 68 morphological and physiological characters, revealed that all isolates are linked at >80% similarity, which indicates that the strains are closely related.

5 CONCLUSIONS

The molecular systematic, chemotaxonomic and phenotypic data confirm that all 14 strains examined are closely related and belong to the genus *Bacillus*.

Until the internal heterogeneity of this group of organisms has been studied in more detail, all 14 isolates will be considered as belonging to a single new species, *Bacillus methanolicus* (me.tha'noli.cus. M. L. n. *methanolicum*, methanol; M. L. masc. adj. *methanolicus*, relating to methanol). The type strain is *Bacillus methanolicus* strain PB1 NCIMB 13113.

The thermotolerant methylotrophic test strains can readily be separated from the mesophilic *B. firmus* on the basis of growth temperature; growth at pH 6.0; methanol utilization; acid production from glycerol, inulin, mannitol, and sucrose; and hydrolysis of casein. *B. brevis* differs with respect to methanol utilization; acid production from inulin, salicin and sucrose; and hydrolysis of casein. *B. azotoformans* can be distinguished by its ability to grow anaerobically and to produce N_2 from nitrate or nitrite, whereas it is unable to grow at temperatures above 50°C or to utilize glucose [18].

ACKNOWLEDGEMENT

These investigations were financially supported by the Biotechnology Action Programme of the Commission of European Communities Contract BAP-0267-NL and the Programme Committee for Industrial Biotechnology (The Netherlands).

REFERENCES

[1] Dijkhuizen, L., Hansen, T.A. and Harder, W. (1985) Methanol, a potential feedstock for biotechnological processes. *Trends Biotechnol.* 3, 262–267.
[2] Schendel, F.J., Bremmon, C.E., Flickinger, M.C., Guettler, M. and Hanson, R.S. (1990) L-Lysine production at 50°C by mutants of a newly isolated and characterized methylotrophic *Bacillus* sp. *Appl. Environ. Microbiol.* 56, 963–970.
[3] Al-Awadhi, N., Egli, T., Hamer, G. and Wehrli, E. (1989) Thermotolerant and thermophilic solvent-utilizing methylotrophic, aerobic bacteria. *Syst. Appl. Microbiol.* 11, 207–216.
[4] Dijkhuizen, L., Arfman, N., Attwood, M.M., Brooke, A.G., Harder, W. and Watling, E.M. (1988) Isolation and initial characterization of thermotolerant methylotrophic *Bacillus* strains. *FEMS Microbiol. Lett.* 52, 209–214.
[5] Brooke, A.G., Watling, E.M., Attwood, M.M. and Tempest, D.W. (1989) Environmental control of metabolic fluxes in thermotolerant methylotrophic *Bacillus* strains. *Arch. Microbiol.* 151, 268–273.
[6] Govorukhina, N.I. and Trotsenko, Y.A. (1989) Isolation and characterization of a thermotolerant methanol-utilizing *Bacillus* strain. Abstr. 6th Int. Symp. Microbial Growth on C_1 Compounds, Abstract p.108.
[7] Arfman, N., Watling, E.M., Clement, C., van Oosterwijk, R.J., de Vries, G.E., Harder, W., Attwood, M.M. and Dijkhuizen, L. (1989) Methanol metabolism in thermotolerant methylotrophic *Bacillus* strains involving a novel catabolic NAD-dependent methanol dehydrogenase as a key enzyme. *Arch. Microbiol.* 152, 280–288.
[8] Arfman, N., de Vries, K.J., Moezelaar, H.R., Attwood, M.M., Robinson, G.K., van Geel, M. and Dijkhuizen, L. (1992) Environmental regulation of alcohol

metabolism in thermotolerant methylotrophic *Bacillus* strains. *Arch. Microbiol.* **157**, 272–278.

[9] Arfman, N., Dijkhuizen, L., Kirchhof, G., Ludwig, W., Schleifer, K.-H., Bulygina, E.S., Chumakov, K.M., Govorukhina, N.I., Trotsenko, Y.A., White, D. and Sharp, R.J. (1992) *Bacillus methanolicus* sp. nov., a new species of thermotolerant, methanol-utilizing, endospore-forming bacteria. *Int. J. Syst. Bacteriol.* **42**, 439–445.

[10] Ash, C., Farrow, J.A.E., Dorsch, M., Stackebrandt, E. and Collins, M.D. (1991) Comparative analysis of *Bacillus anthracis*, *Bacillus cereus*, and related species on the basis of reverse transcriptase sequencing of 16S rRNA. *Int. J. Syst. Bacteriol.* **41**, 343–346.

[11] Ash, C., Farrow, J.A.E., Wallbanks, S. and Collins, M.D. (1991) Phylogenetic heterogeneity of the genus *Bacillus* revealed by comparative analysis of small-subunit-ribosomal RNA sequences. *Lett. Appl. Microbiol.* **13**, 202–206.

[12] Rössler, D., Ludwig, W., Schleifer, K.H., Lin, C., McGill, T.J., Wisotzkey, J.D., Jurtshuk, Jr., P. and Fox, G.E. (1991). Phylogenetic diversity in the genus *Bacillus* as seen by 16S rRNA sequencing studies. *Syst. Appl. Microbiol.* **14**, 266–269.

[13] Wisotzkey, J.D., Jurtshuk, Jr., P. and Fox, G.E. (1990) PCR amplification of 16S rDNA from lyophilized cell culture facilitates studies in molecular systematics. *Curr. Microbiol.* **21**, 325–327.

[14] Wolters, J. and Erdmann, V.A. (1988) Compilation of 5S rRNA and 5S rRNA gene sequences. *Nucl. Acids Res.* **16**, r1–r70.

[15] Schleifer, K.-H. and Kandler, O. (1972) Peptidoglycan types of bacterial cell walls and their taxonomic implications. *Bacteriol. Rev.* **36**, 407–477.

[16] Ourisson, G., Rohmer, M. and Poralla, K. (1987) Prokaryotic hopanoids and other polyterpenoid sterol surrogates. *Ann. Rev. Microbiol.* **41**, 301–333.

[17] Priest, F.G., Goodfellow, M. and Todd, C. (1988) A numerical classification of the genus *Bacillus*. *J. Gen. Microbiol.* **134**, 1847–1882.

[18] Pichinoty, F., de Barjac, H., Mandel, M. and Asselineau, J. (1983) Description of *Bacillus azotoformans* sp. nov. *Int. J. Syst. Bacteriol.* **33**, 660–662.

22

Systematics of Gram-negative Methylotrophic Bacteria Based on 5S rRNA Sequences

E.S. Boulygina[1*], K.M. Chumakov[1**] and A.I. Netrusov[2]

[1]*Institute of Microbiology of the Russian Academy of Sciences, Prosp. 60-let Oktyabrya 7, Moscow 117811, Russia* and [2]*Microbiology Department, Moscow State University, Moscow 119899, Russia*

1 SUMMARY

Nucleotide sequences of 5S ribosomal RNA (rRNA) isolated from 80 strains of Gram-negative methylotrophic bacteria have been determined. Comparative analysis of methanotrophs, obligate and facultative methylotrophs showed that these bacteria constitute separate branches within the *Proteobacteria* and are genetically distinct from each other. The phylogenetic relationships within each group of methylotrophs are discussed as well as the origin of methylotrophy.

2 INTRODUCTION

Up to now, the taxonomy of methylotrophic bacteria has been based on different pheno- and chemo-taxonomic analysis. These two approaches have resolved the taxonomic status of some groups of methylotrophs [1,2,3,4] but for many of these bacteria their status remains unclear.

---

*Present address: Gray Freshwater Biological Institute, University of Minnesota, P.O. Box 100, Navarre, MN 55392, USA.
**Present address: FDA Center for Biologics Evaluation and Research, 8800 Pockville Pike, Bldg. 29A, Bethesda, MD 20892, USA.

To solve this problem, we tried to use one of the simplest and the quickest methods of gene systematics, the comparative analysis of 5S rRNA sequences. More than one hundred strains and species of different groups of Gram-negative bacteria have been analyzed using the 'maximum topological similarity' method of phylogenetic tree reconstruction [5,6].

3 SYSTEMATICS OF OBLIGATE METHANOTROPHS

More than one hundred strains belonging to five main 'genera' of methanotrophs have been described, but just a few representatives of these groups are included in the Approved List of Bacterial Names [7,8]. We analyzed 28 strains of methanotrophs [9] representatives of which are represented in Figure 1. To avoid the confusion over nomenclature we used the same bacterial names that have been suggested by Whittenbury and colleagues [10]. The present tree shows that the phylogenetic relations within this group confirm the Whittenbury scheme as well as the taxonomic studies of Galchenko [11]. All of these strains are divided into three major groups according to the classification mentioned above. Subgroups of these, corresponding to the five described 'genera' of methanotrophs, are also clearly differentiated. It should be noted that sequence comparisons revealed heterogeneity within the genus *Methylomonas*. The nucleotide sequences of *M. methanica* and *M. rubra* differ significantly from *M. agile* and *M. albus*. Therefore, the taxonomy of *Methylomonas* requires further study.

As a whole, the obligate methane-oxidizing bacteria form two distinct groups. Comparison with other published data [12] shows that type I methanotrophs

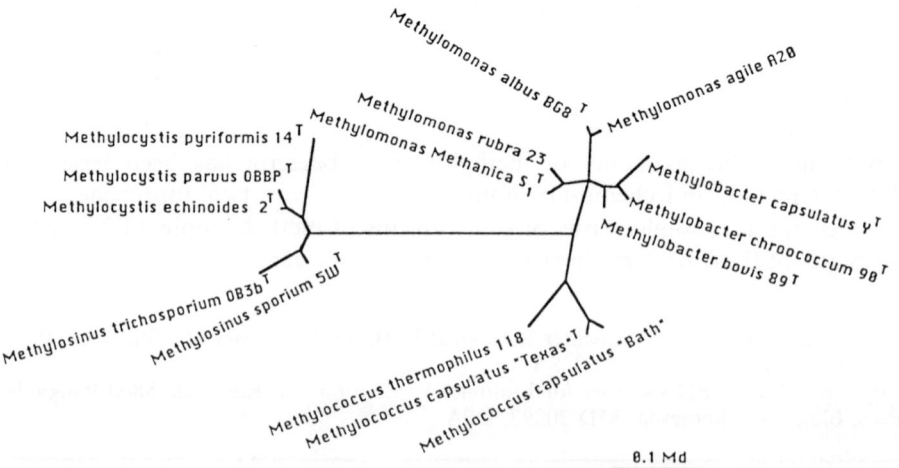

Figure 1 Tentative phylogenetic tree of representative strains of methanotropns based on 5S rRNA sequence analysis.

belong to the gamma-subgroup of the *Proteobacteria* [13,14] and type II methanotrophs belong to the alpha-subgroup (Figure 4). Thus, these two groups of methanotrophic bacteria should be placed into separate families. The differences in their phenotypic features support this suggestion.

4 TAXONOMY OF OBLIGATE METHYLOTROPHS

All terrestrial strains of obligate and restricted facultative methylotrophic bacteria were divided into three genera: *Methylobacillus* [4, 15], *Methylophilus* [3] and *Methylovorus* [16]. In spite of the diversity of the described strains of these bacteria only single species are validly published in each genus.

Eighteen strains of obligate methylotrophs were analyzed in our current study, some of which were previously assigned to genera such as *Methanomonas, Methanolomonas, Methylomonas, Pseudomonas, Protaminobacter*. Our results showed that all of these strains were organized into three tight clusters corresponding to the genera *Methylobacillus, Methylophilus* and *Methylovorus* (Figure 2). Therefore, we propose that all of these microorganisms should be accommodated in the three latter genera. DNA-DNA hybridization studies [17] support this conclusion. Moreover, both our data and other phenotypic characteristics indicate that some of these strains merit separate species status. Although our data do not clearly distinguish between strains of *Methylovorus*

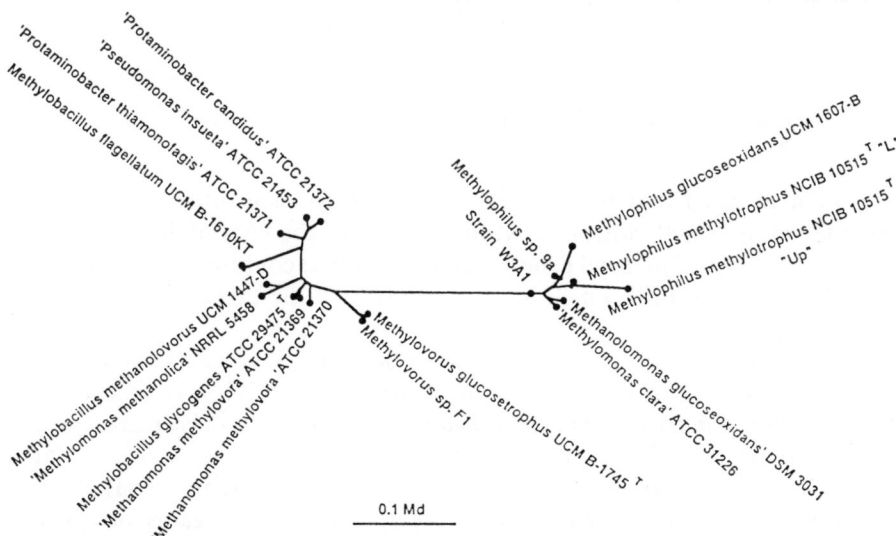

Figure 2 Tentative phylogenetic tree of obligate and restricted facultative methylotrophs according to 5S rRNA sequencing. Nucleotide sequences of *'Methanolomonas glucoseoxidans'* DSM 3031, *'Methylomonas clara'* ATCC 31226 and *'Methanomonas methylovora'* ATCC 21369 were taken from published data base [12].

and *Methylobacillus*; other chemotaxonomic data suggests that they are indeed different genera [16].

The comparison of nucleotide sequences of obligate methylotrophs with the published data bank [12] revealed an evolutionary relationship between these bacteria and representatives of the beta-subgroup of *Proteobacteria* (Figure 4). Nevertheless, a high level of genetic divergence indicates the early separation of this group of microorganisms from a common ancestor. At the same time these analyses show that obligate methylotrophs and methanotrophs differ as significantly from each other as from facultative methylotrophs.

5 SYSTEMATICS OF FACULTATIVE METHYLOTROPHIC BACTERIA

Facultative Gram-negative methylotrophs are represented by a wide spectrum of bacteria belonging to such genera as *Acetobacter*, *Acidomonas*, *Blastobacter*, *Methylobacterium*, *Pseudomonas* and some others. It is easier to characterize these microorganisms than the obligate methylotrophs, although some of these genera contain both methylotrophic and non-methylotrophic species. Moreover, the taxonomic position of many such bacteria still remains obscure.

The phylogenetic analysis that we performed showed that different representatives of facultative methylotrophs are divided into distinct branches within the alpha-subgroup of *Proteobacteria* (Figure 3).

5.1 *Taxonomic analysis of genus* Methylobacterium

Comparison of 5S rRNA sequences of representatives of the genus *Methylobacterium* revealed that they are closely related to each other but not to the other facultative methylotrophs. At the same time, the level of divergence of 5S rRNA is evidence of internal genetic heterogeneity. It confirmed the DNA-DNA hybridization data [18] which have demonstrated the existence of more than 8 species of this genus.

The 5S rRNA analysis also showed a high level of genetic relationship between strain *Pseudomonas* PP, unidentified strain KI-4, and species of the genus *Methylobacterium*. Strain *Pseudomonas* PP is closely related to *M. extorquens* on the basis of its phenotypic features, but our study, as far as DNA-DNA homology analysis (unpublished data by Govorukhina), revealed that this strain is quite distinct from previously studied strains of *Methylobacterium*. Unfortunately, not all the type strains and species isolated have been included in our work. However, we compared the strains that were most closely related on the basis of their phenotypic features. Therefore, we believe that our data is quite reliable.

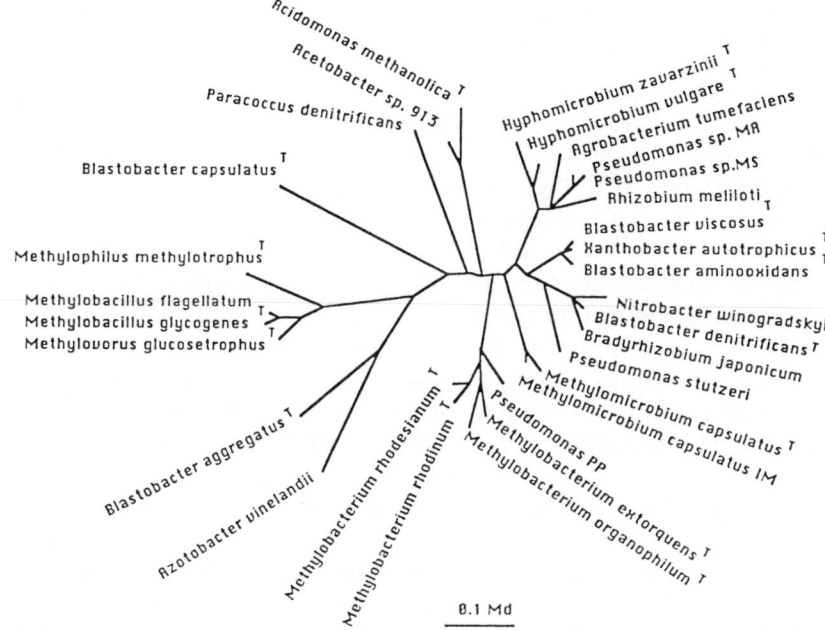

Figure 3 Phylogenetic tree reflecting evolutionary relationship between the major genera of facultative methlyotrophs. Some non-methlyotrophic bacteria are shown for comparative purposes. Nucleotide sequences of *Agrobacterium tumefaciens*, *Azotobacter vinelandii*, *Nitrobacter winogradskyi* and *Paraococcus denitrificans* were taken from published data base [12].

5.2 Phylogenetic analysis of acidophilic methylotrophs

Recently a new genus, *Acidomonas*, has been suggested for methylotrophic strains of the genus *Acetobacter* [19]. 5S rRNA sequence analysis have been done for *Acidomonas methanolicus*, some other methylotrophic strains of *Acetobacter* not included in the study mentioned above, and nonmethylotrophic strains of *Acetobacter* and *Gluconobacter* [20]. These data showed that non-methylotrophic strains of *Acetobacter* proved to be phylogenetically more closely related to *Gluconobacter* strains than to methylotrophic strains of the genera *Acetobacter* and *Acidomonas*. This finding, as well as the range of phylogenetic diversity within this group, substantiate the earlier proposal to place methylotrophic strains of *Acetobacter* into the new genus *Acidomonas* and shows, that in this case the feature of methylotrophy has significant taxonomic weight.

5.3 Determination of the taxonomic status of some budding methylotrophic bacteria

The taxonomy of *Pseudomonas*-like strains, that use the serine pathway of C_1-assimilation and are capable of budding cell division, such as *Pseudomonas aminovorans* MA and MS is unclear. It was shown earlier on the basis of DNA: rRNA hybridization that *Pseudomonas aminovorans* is closely related to the *Rhizobium-Agrobacterium* complex [21].

The results of our study substantiated the phylogenetic relationship of *Pseudomonas aminovorans* MA and MS with this complex. The value for mutation distances were approximately the same between these strains and the genera *Rhizobium* and *Agrobacterium*. Other representatives of the family *Rhizobiaceae* (*Bradyrhizobium japonicum*) are phylogenetically remote from all of these genera. Thus, our data confirmed the close relationship of budding *Pseudomonas*-like methylotrophic strains to the *Rhizobium-Agrobacterium* complex.

The name *Hyphomicrobium* was attributed to the group of budding and appendaged bacteria in Bergey's Manual of Systematic Bacteriology [22], but the taxonomic position of this genus is still unclear. Our analysis revealed that strains of *Hyphomicrobium* were organized into a tight cluster with *Pseudomonas aminovorans* MA and MS and the *Rhizobium-Agrobacterium* complex.

The genus *Methylomicrobium* contains methylotrophic bacteria that utilize C_1-compounds via the serine pathway, have a budding-like type of cell division and possess an unusual cellular fatty acid: cis-methyloctadecene acids (cMe-18:1w7) [23]. However, the taxonomic position of this genus has been unclear. 5S rRNA sequence analysis indicates a remote phylogenetic relationship between *Methylomicrobium* and such genera as *Hyphomicrobium*, *Rhizobium*, *Nitrobacter* and some species of *Blastobacter*.

5.4 Taxonomy of the genera Blastobacter and Xanthobacter

Methylotrophic species are also represented in the genera *Xanthobacter* and *Blastobacter*. The former genus is included in Section 4 of Bergey's Manual of Systematic Bacteriology [22] with 'Gram-negative aerobic rods and cocci'. The latter is included with 'budding and/or appendaged bacteria'. However, neither of these genera is assigned to any of the described families in Bergey's Manual of Systematic Bacteriology [22].

In our study we tried to detect systematic positions of these genera. A tentative phylogenetic tree (Figure 3) shows that the species of the genus *Blastobacter* examined were split into different clusters. *Bl. aminooxidans* and *Bl. viscosus* proved to be very closely related to *Xanthobacter autotrophicus*. Comparison of the main morphological and physiological features indicates that the basic difference between them is the mode of cell division and the

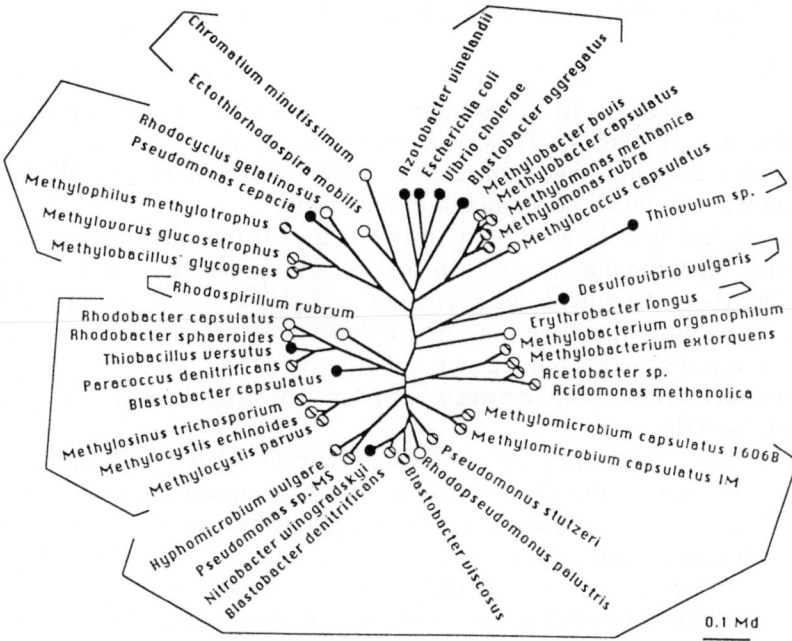

Figure 4 Tentative phylogenetic tree, based on the 5S rRNA sequences comparison of representatives of phototrophs (filled circles), methylotrophs (striped circles) and heterotrophs (open circles). 5S rRNA nucleotide sequences of phototrophs and the most of heterotrophs were taken from published data base [12].

ability to fix atmospheric nitrogen [24–26]. However, all of these strains can grow autotrophically in an atmosphere of H_2, CO_2 and O_2; utilize C_1-compounds *via* ribulosebisphosphate pathway and have very similar mol% G+C values.

Comparison of 5S rRNA sequences of three other species of the *Blastobacter* with the published data bank [12] revealed that *Bl. denitrificans* is phylogenetically similar to *Nitrobacter winogradskyi*. At the same time *Bl. capsulatus* is remotely related to representatives of such genera as *Rhodobacter* and *Caulobacter*, which belong to the alpha-3 subgroup of the *Proteobacteria*. *Bl. aggregatus* proved to be more closely related to photobacteria, which are microorganisms belonging to the gamma-subgroup of the *Proteobacteria*. Therefore our analysis revealed a substantial phylogenetic divergence within *Blastobacter* that should lead to the reclassification of species of this genus into four different groups. It still remains unclear which of these species should be retained as members of the genus *Blastobacter*.

6 EVOLUTIONARY ASPECTS OF THE METHYLOTROPHY

As a whole, our analysis showed that the group of methylotrophic bacteria is quite heterogeneous, with representatives in most of the main phylogenetic lines of eubacteria (Figure 4). Thus, it seems that the ability to use one-carbon compounds does not have sufficient taxonomic weight to classify all methylotrophic bacteria into a single taxon of high rank.

The phylogenetic heterogeneity of the methylotrophs revealed by our data combined with the differences in their one-carbon utilization pathways seems to tell us about the polyphyletic origin of methylotrophy, i.e. its appearance in evolution at more than one time. The independent loss of this feature at later time in various lines might occur and this is suggested by the very close evolutionary relationship between some methylotrophic and non-methylotrophic bacteria.

The fact that methylotrophs are phylogenetically related to bacteria in three main subgroups of the *Proteobacteria*, which comprise most of eubacterial physiological types, probably puts the methylotrophy at the same level with autotrophy and heterotrophy; perhaps like a transition type of metabolism.

It is therefore evident that 5S rRNA analysis alone is not enough, and that the task of forming a phylogenetic structure for these organisms requires using phenotypic as well as genotypic methods. Thus, our results should be considered as preliminary. Nevertheless, the lack of significant differences in the results of both 5S and 16S rRNA sequence analysis [27] as well as the various chemotaxonomic studies makes our conclusions quite useful for rapid phylogenetic screening.

7 CONCLUSION

5S rRNA sequence analysis revealed phylogenetic relationships between main groups of methylotrophic bacteria and allowed us to determine the taxonomic position of some of these bacteria. Comparisons of our data with the other pheno- and chemotaxonomic studies makes this analysis quite useful for systematic purposes.

ACKNOWLEDEGMENTS

We gratefully thank N.A. Govorukhina, Yu.A. Trotsenko, V.F. Galchenko, D.I.Nikitin, O.M. Goulikova for supplying the strains used in this study; R.S. Hanson, P.N. Green and T. Barta for helpful discussion and comments on the manuscript.

REFERENCES

[1] Patt, T.E., Cole, G.C. and Hanson, R.S. (1976). *Methylobacterium*, a new genus of facultatively methylotrophic bacteria. *Int. J. Syst. Bacteriol.* **26,** 226–229.
[2] Green, P.N. and Bousfield, I.J. (1983) Emendation of *Methylobacterium* Patt et al., 1976; *Methylobacterium rhodos* (Heumann, 1962); *Methylobacterium radiora* (Ito and Iizuka, 1971); *Methylobacterium mesophilica* (Austin and Goodfellow, 1979). *Int. J. Syst. Bacteriol.* **33,** 875–877.
[3] Jenkins, O., Byrom, D. and Jones, D. (1987) *Methylophilus:* a new genus of methanol-utilizing bacteria. *Int. J. Syst. Bacteriol.* **37,** 446–448.
[4] Urakami, T. and Komagata, K. (1986) Emendation of *Methylobacillus* Yordy and Weaver 1977, a genus for methanol-utilizing bacteria. *Int. J. Syst. Bacteriol.* **36,** 502–511.
[5] Chumakov, K.M. and Yushmanov, S.V. (1988) Maximum topological similarity priciple in molecular systematics. *Mol. Gen. Microbiol. Virol.* **3,** 3–9 (in Russian).
[6] Yushmanov, S.V. and Chumakov, K.M. (1988) Algorithms for building maximum topological similarity phylogenetic trees. *Mol. Gen. Microbiol. Virol.* **3,** 9–15 (in Russian).
[7] Approved Lists of Bacterial Names (1989) (Skerman, V.B.D., McGowan, V. and Sneath, P.H.A., Eds), American Society for Microbiology, Washington, DC.
[8] Index of the Bacterial and Yeast Nomenclatural Changes (1989) (Moore, W.E. and Moore, L.V.H., Eds), American Society for Microbiology, Washington, DC.
[9] Boulygina, E.S. (1991) Systematic of Gram-negative methylotrophic bacteria based on 5S rRNA sequence analysis. Ph.D. thesis, Inst. of Microbiol. Russ. Acad. Sci., Moscow, Russia.
[10] Whittenbury, R., Phillips, K.S. and Wilkinson, J.F. (1970) Enrichment, isolation and some properties of methane-utilizing bacteria. *J. Gen. Microbiol.* **61,** 205–218.
[11] Galchenko, V.F., Andreev, L.V. and Trotsenko, Yu.A. (1986) Taxonomy and Identification of Obligate Methanotrophic Bacteria. (Kalakoutskii, L.V., Ed.), pp. 64, 88, Pushchino, Acad. Press (in Russian).
[12] Wolters, J. and Erdman, V.A. (1988) Compilation of 5S rRNA and 5S rRNA gene sequences. *Nucl. Acids Res.* **16,** suppl., r1-r70.
[13] Woese C.R. (1987) Bacterial evolution. *Microbiol. Rev.* **51,** 221–271.
[14] Woese C.R., Stackebrandt, E., Macke, T.J. and Fox, G. E. (1985) A phylogenetic definition of the major eubacterial taxa. *Syst. Appl. Microbiol.* **6,** 143–151.
[15] Yordy, J.R. and Weaver, T.Y. (1977) *Methylobacillus*, a new genus of obligate methylotrophic bacteria. *Int. J. Syst. Bacteriol.* **27,** 247–255.
[16] Govorukhina, N.I. and Trotsenko, Yu.A. (1991) *Methylovorus*, a new genus of restricted facultatively methylotrophic bacteria. *Int. J. Syst. Bacteriol.* **41,** 158–162.
[17] Boulygina, E.S., Govorukhina, N.I., Netrusov, A.I., Trotsenko, Yu.A. and Chumakov, K.M. (1992) Comparative studies on 5S rRNA sequences and DNA-DNA hybridization of obligately and restricted facultatively methylotrophic bacteria. *Syst. Appl. Microbiol.*, in press.
[18] Hood, D.W., Dow, C.S. and Green, P.N. (1987) DNA:DNA hybridization studies on the pink-pigmented facultative methylotrophs. *J. Gen. Microbiol.* **133,** 709–720.
[19] Urakami, T., Tamaoka, J., Suzuki, K. and Komagata, K. (1989) *Acidomonas* gen. nov., incorporating *Acetobacter methanolicus* as *Acidomonas methanolica* comb. nov. *Int. J. Syst. Bacteriol.* **39,** 50–55.
[20] Bulygina, E.S., Gulikova, O.M., Dikanskaya, E.M., Netrusov, A.I., Turova, T.P. and Chumakov, K.M. (1992) Taxonomic studies of the genera *Acidomonas, Acetobacter and Gluconobacter* by 5S ribosomal RNA sequencing. *J. Gen. Microbiol.*, in press.

[21] De Smedt, J. and De Ley, J. (1977) Intra- and intergeneric similarities of *Agrobacterium* ribosomal ribonucleic acid cistrons. *Int. J. Syst. Bacteriol.* **27**, 222–240.
[22] Bergey's Manual of Systematic Bacteriology (1984) (Krieg, N.R. and J.G. Holt, Eds). The Williams & Wilkins Co., Baltimore.
[23] Doronina, N.V. and Govorukhina, N.I. (1987). Current state of taxonomy of methylotrophic bacteria. In: Biochemestry and Physiology of Methylotrophs (Trotsenko, Yu.A., Ed.), pp. 85–94. Pushchino, Acad. Press (in Russian).
[24] Loginova, N.V. and Trotsenko, Yu.A. (1979) *Blastobacter viscosus*, a new species of methanol-utilizing facultative autotrophic bacteria. *Microbiology* **48**, 644–651 (Engl. Transl. Microbiologiya).
[25] Doronina, N.V., Govorukhina, N.I. and Trotsenko, Yu.A. (1983) *Blastobacter aminooxidans*–a new species of bacteria growing autotrophically on methylated amines. *Mikrobiologiya* **52**, 709–715 (in Russian).
[26] Wiegel, J., Wilke, D., Baumgarten, J. and Schlegel, G. (1978) Transfer of the nitrogen-fixing hydrogen bacterium *Corynebacterium autotrophicum* Baumgarten et al. to *Xantobacter* gen. nov. *Int. J. Syst. Bacteriol.* **28**, 573–581.
[27] Tsuji, K., Tsien, H.C., Hanson, R.S., DePalma, S.R., Scholtz, R. and Laroche, S. (1990) 16S ribosomal RNA sequence analysis for determination of phylogenetic relationship among methylotrophs. *J. Gen. Microbiol.* **136**, 1–10.

23
Phylogeny and Ecology of Methylotrophic Bacteria

R.S. Hanson, B.J. Bratina and G.A. Brusseau

Department of Microbiology and Gray Freshwater Biological Institute, University of Minnesota, P.O. Box 100, Navarre, MN 55392, USA

1   SUMMARY

Methane-oxidizing bacteria (methanotrophs), a subset of bacteria that utilize single-carbon compounds (methylotrophs), have attracted considerable attention because of their abilities to degrade low molecular weight halogenated hydrocarbons and their participation in global carbon cycles. The small number of phenotypic characteristics of obligate methylotrophs has limited taxonomic studies. Sequencing of ribosomal RNAs has been employed to define more clearly the evolutionary relationships between methylotrophs. This analysis reveals that methylotrophs utilizing the serine pathway for formaldehyde assimilation were found in two clusters within the α-subgroup of the Proteobacteria. The serine pathway methanotrophs form one cluster, while the pink-pigmented facultative methylotrophs that did not utilize methane form another phylogenetically coherent group. The methylotrophs that employ the ribulose monophosphate pathway for formaldehyde assimilation are a more diverse group found within the β- and γ-subgroups of Proteobacteria.

Conventional microbiological methods for enumeration and identification of methanotrophs in environmental samples and bioreactors have been inadequate. Phylogenetic probes useful for characterizing populations of methylotrophs in mixed culture were synthesized after analyzing sequence data. These probes, as well as gene probes, were used to study the occurrence and relative abundance of methanotrophs in bioreactors and freshwater lakes.

Microbial Growth on C_1 Compounds
© Intercept Ltd, PO Box 716, Andover, Hampshire SP10 1YG, UK

2 INTRODUCTION

Methylotrophic bacteria are a diverse group of microbes that utilize one-carbon compounds more reduced than carbon dioxide [1]. In addition, these bacteria assimilate carbon at the oxidation level of formaldehyde [1,2]. Methanotrophs are a subset of methylotrophs that are able to use methane as a carbon and energy source.

In 1970, Whittenbury et al. [3] described the properties of more than 100 new isolates of methanotrophic bacteria. These isolates were classified into five groups on the basis of morphology, fine structure of intracytoplasmic membranes and types of resting stages formed. The proposed groups were named *Methylosinus*, *Methylocystis*, *Methylomonas*, *Methylobacter* and *Methylococcus*. Subsequent studies led to the separation of methanotrophs into three groups [4,5]. Type I included those bacteria that utilized the ribulose-monophosphate pathway (RuMP) for formaldehyde assimilation while bacteria included in Type II are those bacteria that employed the serine pathway for formaldehyde assimilation. Another group, Type X, was added to accommodate *Methylococcus capsulatus*, a methanotrophic bacterium that contained enzymes of the RuMP pathway, to a lesser extent enzymes of the serine pathway and enzymes of the Calvin Cycle for CO_2 fixation.

The separation of methanotrophic bacteria into these three groups was further justified because of differences in growth temperatures, phospholipid fatty acid composition, DNA base composition, and other biochemical characteristics [5–9]. Subsequent studies including comparisons of 5S rRNA [10] and 16S rRNA sequences [11] have also supported the separation of methanotrophic bacteria into these three major groups and 5 or more subgroups at the generic level [10–12].

Methanotrophic bacteria are known to play an important role in global carbon cycles. Methane is an abundant gas in the atmosphere, its concentration has been increasing at a rate of 1–3% per year [13,14] and has reached a concentration of 1.7 parts per million in the atmosphere. Because methane absorbs infrared irradiation more efficiently than carbon dioxide, it may contribute significantly to global warming [14]. Whalen and Reeburgh [16,17], and Harriss [15] have shown that tundra and wetland environments, known to produce methane when flooded, are sinks for atmospheric methane during dry periods. This phenomenon may result in a significant reduction in atmospheric methane concentrations if global warming does occur. Deserts are also net sinks for methane [18]. However, nothing is known of the types and numbers of methanotrophs in these terrestrial environments, although they are believed to consume over 10^{13} g of methane per year [18].

Methanotrophs capable of expressing soluble methane monooxygenase (sMMO) oxidize some halogenated low molecular weight hydrocarbons at rates significantly higher than other aerobic bacteria which contain different non-specific oxygenases [19–23].

Because of the impact of methanotrophs in biotechnology [20–24] and the chemistry of the earth's atmosphere, studies of their ecology are important. For example, an ideal bioremediation system for destruction of trichloroethylene (TCE), vinyl chloride (VC), chloroform and related toxic or carcinogenic compounds found in groundwaters may employ *in situ* bioremediation using indigenous methanotrophic bacteria. It will be necessary to stimulate growth of those species of methanotrophs capable of degrading target chemicals. In order to model these processes, it will be necessary to follow the activity, biomass and other characteristics of the microbes grown in aquifers. It is difficult to determine the numbers of specific methanotrophic bacteria in an ecosystem using conventional microbiological methods because of poor colony forming efficiencies of cells and the absence of selective media or easily identifiable characteristics that permit the enumeration of separate genera and species of methanotrophs [25,26]. Therefore, we have developed oligonucleotide signature probes that will detect specific groups of methanotrophs in environmental samples and gene probes that can be used to distinguish between different species and isolates.

3 PHYLOGENETIC RELATIONSHIPS AMONG METHYLOTROPHIC BACTERIA BASED ON 16S rRNA SEQUENCE COMPARISONS

The sequences of 16S rRNAs extracted from 25 Gram-negative methylotrophic bacteria have been determined. Over 90% of all nucleotides in 16S rRNAs from each bacterium were unambiguously determined. A previously published analysis of the sequences of 16S rRNAs from twelve Gram-negative methylotrophic bacteria indicated that Type I methylotrophs can be classified in the β- and γ-subdivision of the purple bacteria (class Proteobacteria) [11]. A comparison of the sequences of 16S rRNAs from eleven pure cultures of RuMP pathway methylotrophs indicated that those bacteria capable of utilizing methane formed 2 or 3 loosely related clusters (Fig. 1b). One cluster included *Methylomonas rubra*, *Methylomonas luteus*, *Methylomonas* A4 and *Methylomonas albus* BG8. *Methylomonas methanica* and *Methylococcus capsulatus* were distantly related to each other and to the other RuMP pathway methantrophs named above. We have not yet completed the sequences of 16S rRNA from a sufficient number of RuMP pathway methanotrophs presently listed under the genus *Methylobacter* to draw conclusions about their relationships to each other or to other methylotrophs.

Those bacteria that utilize the RuMP pathway for formaldehyde fixation but do not utilize methane as a source of carbon and energy are not closely related to Type I methanotrophs (Fig. 1b). Those bacteria in this group from which 16S rRNAs have been isolated and sequenced, with the exception of *Methylophilus methylotrophus*, are closely related to each other (Fig. 1b). *M. methylotrophus* is more closely related to the RuMP pathway methylotrophs that do not utilize methane than to Type I methanotrophs.

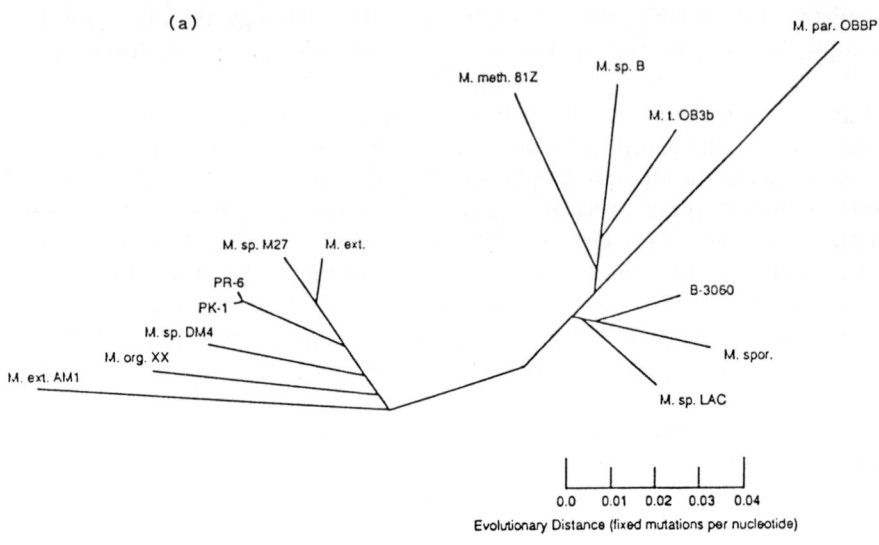

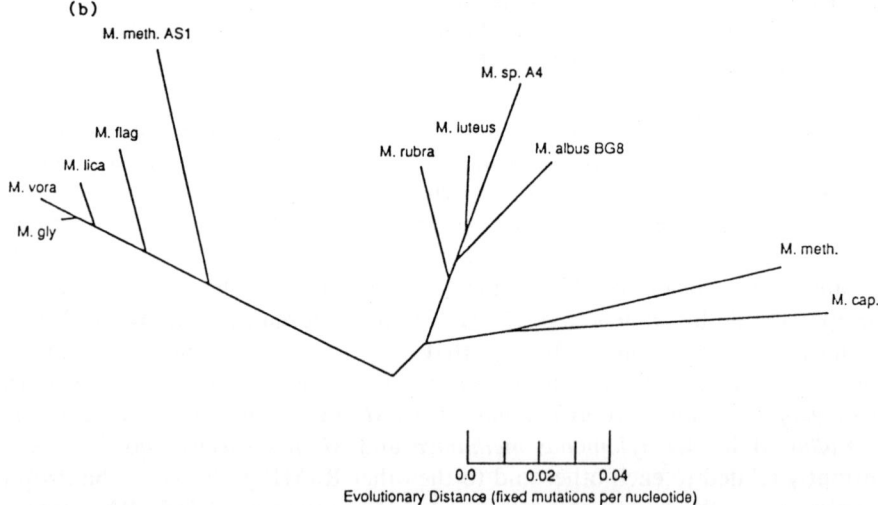

Figure 1 Unrooted phylogenetic trees showing the relationships among methylotrophic bacteria. (a) The serine pathway methylotrophs. (b) The RuMP pathway methylotrophs. The abbreviations used are defined in Table 1.

The pink pigmented facultative methylotrophs that employ the serine pathway for formaldehyde fixation are tightly clustered within the α-subgroup of the Proteobacteria (Fig. 1a).

The serine methanotrophs form a phylogenetically coherent assemblage within the α-2-subgroup of the Proteobacteria. They are, as a group, more closely related to each other than to any serine pathway methylotrophs that do not utilize methane (Fig. 1a).

4 IDENTIFICATION OF TARGET SEQUENCES FOR DESIGN OF HYBRIDIZATION PROBES

Specific hybridization probes that are complementary to target (signature) sequences of phylogenetically related bacteria, individual species and subspecies of microorganisms have been fabricated [27–32]. The target sequences are unique (insofar as it is possible to determine uniqueness using a sampling of 16s rRNA sequences) to the species that signature probes are designed to detect. The probes have been end-labelled with ^{32}P-PO_4 or fluorescent dyes. Radioactively labelled probes have been employed to detect rRNAs of target species in RNA extracted from naturally occurring communities of microorganisms and bound to nitrocellulose or nylon membranes [27–32]. The probes specifically hybridize to the rRNAs of the target species if the hybridization and washing conditions are sufficiently stringent [29,31]. Signature probes, labelled with fluorescent dyes, have been employed as hybridization probes to specifically detect cells of target microbes fixed to microscope slides [27–32]. Cells of several bacteria and some fungi that were treated with buffered paraformaldehyde and dehydrated by alcohol treatment were permeable to fluorescently labelled oligonucleotide signature probes. The probes hybridized with intracellular rRNA molecules. Positive hybridization reactions were detected by epifluorescence microscopy.

A target sequence that was uniquely present in serine pathway methylotrophs and another that was uniquely present in RuMP pathway methylotrophs have been identified (Figs 2 and 3).

Probe 9α that is complementary to the target sequence of serine pathway methylotrophs hybridized to membrane bound RNA extracted from all serine pathway methylotrophs (Table 1 and Fig. 2) [27]. Probe 9α did not hybridize to RNAs extracted from RuMP methylotrophs or to RNAs extracted from a limited number of other Eubacteria [27].

Similarly, probe 10γ that is complementary to the target sequence of RuMP pathway methylotrophs hybridized only to membrane bound RNA extracted from RuMP pathway methylotrophs [27].

Probes 9α and 10γ labeled with fluorescent dyes were similarly specific when used as a probe to detect cells fixed to microscope slides [27].

We have completed several more sequences since our previous study [11] and the data (Figs 2 and 3) confirm that the target sequence for Probe 9α is present in all serine pathway methylotrophs. Further examination has revealed potential target sequences that can be used to design probes to distinguish between serine pathway methanotrophs and the serine pathway methylotrophs that do not utilize methane. We have not yet rigorously tested the specificity of potential signature probes that may distinguish between methanotrophs that utilize methane and those that do not. It is possible that more than one probe may be necessary to specifically detect RuMP pathway methanotrophs and methylotrophs because of their phylogenetic diversity. Probe 10γ did not hybridize to membrane bound RNA of all species of RuMP pathway methylotrophs we have tested. For reasons that are not clear, this probe did not hybridize to RNA

A. SERINE PATHWAY METHANOTROPHS

Target sequence-9-alpha	GGUUCGG	AAUAACUCAGG		
Methylosinus trich. OB3b	UCGGUUCGG	AAUAACUCAGG	GAAA	*
Methylosinus sp B	UCGGUUCGG	AAUAACUCAGG	GAAA	*
M. methanica 81Z	UCGGUUCGG	AAUAACUCAGG	GAAA	*
M. sporium	UCGGUUCGG	AAUAACUCAGG	GAAA	*
Methylosinus LAC	UUGGUUCGG	AAUAACUCAGG	GAAA	*
Methylocystis parv/s	UCGGUUCGG	AAUAACUCAGG	GAAA	*
Strain B-3060	UCGGUUCGG	AAUAACUCAGG	GAAA	ND

B. SERINE PATHWAY METHYLOTROPHS THAT DO NOT USE METHANE

Methylobacterium org XX	UCGGUUCGG	AAUAACUCAGG	GAAA	*
M exroquens	UCGGUUCGG	AAUAACUCAGG	GAAA	*
M. extorquens AM1	UCGGUUCGG	AAUAACUCAGG	GAAA	*
Methylobacterium DM4	UCGGUUCGG	AAUAACUCAGG	GAAA	*
Methylobacterium M27	UCGGUUCGG	AAUAACUCAGG	GAAA	*
Methylobacterium PK-1	UCGGUUCGG	AAUAACUCAGG	GAAA	*
Methylobacterium PR-6	UCGGUUCGG	AAUAACUCAGG	GAAA	*

C. RUMP PATHWAY METHYLOTROPHS

M. capsulatus Bath	UUCUGGGGG	AAUAACUCGGG	GAAA
M. flagellatum	UAAUGGGGG	- AUAACUAGUC	GAAA
M. methylophilus AS1	UCGUGGGGG	A- CAACUAGUC	GAAA
M. methanolica	UAAUGGGGG	A- NAACUAGUC	GAAA
M. methylovora	UAAUGGGGG	AN- AACUAGUC	GAAA
M. glycogenes	UAAUGGGGG	AW-AACUAGUC	GAAA
M. rubra	UAGUGGGGG	AU- AACGUGGG	GAAA
Methylomonas A4	UAGUGGGGG	ACGAACUUGGG	GAAA
M. albus BG8	UAGUGGGGG	NC - AACGUGGG	GAAA
M. methanica	UGGUGGGGG	AU- AACUNGGG	GAAA
M. luteus	UAGUGGGGG	AC- AACUUGGG	GAAA
Methylobacter vinel.	UAGUGGGgg	u C- AACUUGGG	GAAA
Methylobacter bovis	UAGUGGGgg	a C- AACUUGGG	GAAA
Lake Mendota clone -1	UAGUGGGGG	AU- AACCCGGG	GAAA
Lake Mendota clone-2	AAGUGGGGG	AU- AACCCGGG	GAAA

Figure 2 Target sequence for serine pathway methylotrophs. Signature probes are complementary to target sequences. The target sequence 9α is at positions 136–154 of the *E. coli* 16S rRNA sequence. The sequences followed by (*) are present in the RNAs of the organisms indicated and hybridize with the signature probe 9α.

extracted from *M. methanica* in several attempts, although the sequence data indicate that it should (Fig. 3).

5 COLORIMETRIC ASSAY FOR METHANE MONOOXYGENASE

We have developed a colorimetric assay for the rapid detection of sMMO in whole cells or extracts of methanotrophic bacteria [19]. The assay detected the conversion of naphthalene to 1- or 2-naphthol by sMMO. The products of the reaction were reacted with tetrazoitized *o*-dianisidine to produce violet-colored derivatives. Only those methanotrophs that utilized the serine pathway for formaldehyde assimilation that were grown under copper-limited conditions synthesized sufficient sMMO to produce detectable metabolites under the assay conditions employed [19]. The assay has proven to be a convenient and rapid means of determining the presence of sMMO in cell suspensions and colonies

A. NON METHYLOTROPHIC BACTERIA

E. coli (190-224)	AGACCA	AAGAGGGGGACCUU	CGGGCCUCCUGCCA
A. tumefaciens	UACGCC	CCUACGGGGAAAG	AUUUAUCGGGGUA

B. RuMP PATHWAY METHYLOTROPHS

Target seq-10 gamma	Gg A	AAGCGGGGGAuCUU	CGG	
M. meth AS1	GGGGGA	AAGCGGGGGAUCUU	CGGACCA- ACGUU	*
M. flagellatum	GGGGGA	AAGCGGGGGAUCUU	CGGACCU- ACGUU	*
M. methanica	GGUGGA	AAGCGGGGGAUCUU	CGGACCUCGCGCA	*
M. luteus	GGGGCA	AAGCGGGGGACCUU	CGGGCCUCGCGCU	*
M rubra	CGGGUA	AAGCUGGGGACCUU	CGGGCCUGGCGCU	*
M. capsulatus	GGAGGA	AAGCGGGGGANCUU	CGGACCUCGCGCA	*
Methylobact. vinel.	GGGGgA	AAGCGGGGGANCUU	CGGGUUUGCGCUA	*
Methylobact. bovis	GGGGNA	AAGCGGGGGANCUU	CGGGCCUGCGCUA	*
Lake Mendota clone-1	GGAGGA	AAGCGGGGGAUCUU	CGGACCUCGCGCU	*
Lake Mendota clone-2	GGAGGA	AAGCGGGGGAUCUU	CGGACCUCGCGUU	*

C. SERINE PATHWAY METHANOTROPHS

M. trich OB3b	CGCc uU	AA-- GGGGGAAAG--	AUUUAUUCGCGAAA
Methylosinus LAC	CGCCUU	UA-- GGGGGAAAG--	AUUUAUUCGCGAAA
Methylosinus sp B	CGC--- U	UC-- GGGGGAAAG--	AUUUAUUCGCGAAA
M. methan 81Z	CGCCUA	Uc-- GGGGGAAAG--	AUUUAUUCGCGAAA
M. sporium	CGC--- U	Uc -- GGGGNAAAG--	AUUUAUUCGCGAAA
M. parvus OBBP	CGCCUA	UU-- GGGGGAAAG--	AUUUAUUCGCGAAA
Strain B-3060	CGUGCG	AG--AGCAGAAAG--	AUUUAUNCGCGAAA

D. SERINE PATHWAY METHYLOTROPHS THAT DO NOT UTILIZE METHANE

M. organoph XX	CGCCCA	CAAAGGGGAAAGA	UU---UAUCGCG- AAA
M. extorquens	CGC--CU	UU--U- GGGGAAAGG	Uu---UACYGCGGAAG
M. extorquens AM1	CGC--CU	UU--U- GGGGAAAGG	UU---UACUGCG- AAG
M. sp M 27	CGCCCU	UU--U- GGGGAAAGG	UU---UACUGCG- AAG
M. sp DM4	CGCNCU	UU--U- GGGGAAAGG	UU---UAUCGCG- AAG
Isolate PK1	CGCCCU	UU--U- GGGGAAAGG	UU---UACUGCG- AAG
Isolate PR6	CGCCCU	UU--U- GGGGAAAGG	UU---UACUGCG- AAG

Figure 3 Target sequences for RuMP pathway methylotrophs and serine pathway methanotrophs. The target sequence 10γ is at positions 191-211 of the E. coli 16S rRNA sequence. The sequences with * indicate that membrane bound RNA from these organisms hybridized with the signature probe 10γ [12,27].

on agar media. The assay should be useful for the detection of mutants on agar plates that are unable to produce sMMO and mutants in which the regulation of sMMO synthesis is altered.

6 GENE PROBES USEFUL FOR THE DETECTION OF METHYLOTROPHIC BACTERIA

The genes encoding methanol dehydrogenase (MDH) are highly conserved in Gram-negative methylotrophic bacteria [12,25]. A cloned 2.5 kb gene fragment containing the open reading frame encoding the MDH large subunit and approximately 500 bp of an upstream sequence hybridized to DNA purified from all of 18 Gram-negative methylotrophs examined [12,25].

Table 1 Methylotrophs used in this study

Organism	Abbreviations used in Figures 1a and 1b
A. Methanotrophs That Utilize the RuMP Pathway for Formaldehyde Assimilation (γ-Subclass of Proteobacteria).	
Methylococcus capsulatus Bath	*M. cap.*
Methylomonas sp A4	*M. sp.* A4
Methylomonas albus BG8	*M. albus* BG8
Methylomonas luteus	*M. luteus*
Methylomonas methanica	*M. meth.*
Methylomonas rubra	*M. rubra*
B. Methylotrophs That Utilize the RuMP Pathway for Formaldehyde Assimilation But Do Not Utilize Methane (β-Subclass of Proteobacteria).	
Methylobacillus flagellatum	*M. flag*
Methylobacillus glycogenes	*M. gly*
Methylomonas methanolica	*M. lica*
Methylomonas methylovora	*M. vora*
Methylophilus methylotrophus AS1	*M. meth.* AS1
C. Methanotrophs That Utilize The Serine Pathway for Formaldehyde Assimilation (α-Subclass of The Proteobacteria).	
Methylosinus trichosporium OB3b	*M.t.* OB3b
Methylosinus sp. B	*M. sp.* B
Methylosinus sp. LAC	*M. sp.* LAC
Methylosinus methanica 81Z	*M. meth.* 81Z
Methylosinus sporium	*M. spor.*
Methylocystis parvus OBBP	B-3060
D. Facultative Methylotrophs That Utilize The Serine Pathway For Formaldehyde Assimilation But Do Not Utilize Methane (α-Subclass of the Proteobacteria).	
Methylobacterium organophilum XX	*M. org.* XX
Methylobacterium extorquens AM1	*M. ext.* AM1
Methylobacterium extorquens	*M. ext*
Methylobacterium sp. DM4	*M. sp.* DM4
Methylobacterium sp. M27	*M. sp.* M27
Methylobacterium sp. PK-1	PK-1
Methylobacterium sp. PR-6	PR-6

Some properties of methylotrophs not described in previous publications [3,6–11]. Isolate B-3060 was obtained from V.A. Romanovskaya, Institute for Microbiology and Virology, Kiev, Russia. *Methylomonas* sp. A4 is a marine methanotroph obtained from M. Lidstrom, California Institute of Technology, Pasadena, CA. *Methylosinus* sp. LAC was isolated from a bioreactor and was described in the publication by L. Alvarez-Cohen, et al. [40]. *Methylobacterium* strains PK-1 and PR-6 were isolated from bean plants by S. Hirano, University of Wisconsin, Madison, WI. *Methylobacterium* sp. DM4, utilizes dichloromethane as well as methanol and methylamine. *M. organophilum* strain XX was grown on methane when initially isolated [42]. The strain used in this study does not grow on methane.

A cloned 2.2 kb DNA fragment from *Methylosinus trichosporium* OB3b that encoded the B-component of sMMO (*mmoB*), the γ subunits of the sMMO (*mmoZ*), orfY and a portion of the component C (*mmoC*) gene [33–36] has been cloned [36]. This cloned DNA fragment hybridized to DNA purified from *M. trichosporium* OB3b, *Methylosinus methanica* 81Z, *Methylosinus* sp. B, *Methylosinus sporium* and *Methylosinus* sp. LAC. All of these bacteria are closely related serine pathway methanotrophs that rapidly degrade TCE (Fig. 2) [37]. The cloned DNA fragment did not hybridize to DNA from methylotrophs that employed the RuMP pathway for formaldehyde assimilation and did not rapidly degrade TCE [36].

Highly conserved gene probes like the cloned 2.5 kb MDH gene are useful for detecting methylotrophs in environmental samples. When DNA from a sample is digested with an infrequently cutting restriction endonuclease and the restriction fragments are separated by pulsed field gel electrophoresis, transferred by Southern blotting to membranes and hybridized with a conserved gene probe, useful information can be obtained. Because different bacteria produce restriction fragments differing in size, the lengths of the hybridizing fragments can be compared to hybridizing fragments from known cultures. This information together with other data can be used to determine if bioreactors inoculated with a specific culture maintain that strain over long periods of time. The data can also be used to determine persistence and distribution of strains in an environment.

The number of restriction fragments that hybridize to a conserved gene probe can also be used to define the complexity of populations of Gram-negative methylotrophs in a sample. If one assumes (or determines by experimentation) that infrequently cutting restriction enzymes do not recognize restriction sites within the target genes, a single fragment will be produced from DNA of each bacterium in the population. The number of fragments produced will provide an indication of the number of different methylotrophs present in sufficient numbers to permit their detection. The strength of the hybridization signals for each fragment will reflect the relative populations of different methylotrophs.

The 2.2 kb sMMO gene probe described above is more specific because it selectively detects those methanotrophic bacteria with sufficiently conserved sMMO genes. The production of sMMO is apparently restricted to a limited number of methanotrophs [36,37].

7 STUDIES OF A CONSORTIUM OF BACTERIA CAPABLE OF RAPIDLY DEGRADING TRICHLOROETHYLENE AND CHLOROFORM

Alvarez-Cohen and McCarty [38–40] described a mixed culture of bacteria enriched with methane and oxygen from aquifer material from Moffett Field Naval Air Station, Mountain View, California. This mixed culture was grown in a bioreactor with methane as a carbon and energy source, and it rapidly oxidized TCE and chloroform. Although grown with the addition of unsterile media and air, the culture has been stable for several years. We wished to characterize the dominant methanotroph(s) in the mixed culture and to determine if the methanotroph produced sMMO under the conditions used to maintain the culture.

The most abundant bacterial cells in the mixed culture were crescent-shaped bacteria that bound the fluorescently labeled signature probe 9α when fixed to microscope slides. The dominant 5S rRNA was sequenced and was shown to be 93.5% homologous to the sequence of 5S rRNA from *Methylosinus trichosporium* OB3b. The sequence of the 16S rRNA of this bacterium was found to be 92 and 94% homologous to the 16S rRNA sequences of *M. trichosporium* OB3b and *M. sporium* respectively (Fig. 1a). A pure culture of a methane-utilizing bacterium was isolated from the bioreactor. The cells were indistinguishable in morphology from the most abundant cells in the mixed culture. The pure culture and cells in the bioreactor produced sMMO as evidenced by their ability to oxidize naphthalene to naphthol. Proteins in extracts of bioreactor cells and the pure culture were separated by polyacrylamide electrophoresis and were transferred to nitrocellulose membranes by Western blotting. Both extracts contained proteins indistinguishable in molecular weights that crossreacted on Western blots with antibodies prepared against sMMO components of *M. trichosporium* OB3b. The pure culture also oxidized TCE when grown under copper-limiting conditions [40].

The *Ase* 1 DNA fragments from the bioreactor pure and mixed cultures, separated in agarose gels by pulsed field electrophoresis and hybridized to the 2.2 kb sMMO gene probe described above, were identical in size [40]. The size of the *Ase* 1 DNA fragments from the bioreactor pure and mixed cultures differed in size from *Ase* 1 restriction fragments that hybridized to the sMMO gene probe produced by digestion of DNAs extracted from our collection of pure cultures.

We concluded that a serine pathway methylotroph phylogenetically related to *M. trichosporium* and *M. sporium* produced sMMO and was responsible for the oxidation of TCE and chloroform in the bioreactor [40].

8 ATTEMPTS TO CHARACTERIZE METHANOTROPHIC BACTERIA IN TWO FRESHWATER LAKES

Biomass was collected from approximately 1.5–2.5 l of water from different depths in Lake Minnetonka, Minnetonka, MN, on 10th September 1990. Total RNA was purified from each sample and bound to nylon membranes as previously described [27]. The membrane-bound RNA was hybridized with ^{32}P-labeled signature probes specific for eubacteria, RuMP methylotrophs (probe 10γ) and serine pathway methylotrophs (probe 9α). After hybridization with each probe the portions of the blots containing each RNA sample were cut from the blot and the radioactivity of the bound, radiolabeled probe was

Table 2 Distribution of eubacteria and methanotrophic bacteria in Lake Minnetonka, Minnesota

Depth from surface (m)	Dissolved oxygen concentration (ppm)	Temperature (°C)	cpm ^{32}P-Probe bound per μg filterbound RNA	
			Eubacterial probe	Probe 10γ
1	8.2	23.0	4.1×10^3	ND
5	7.8	23.5	8.6×10^3	11
6	8.0	23.5	1.1×10^4	ND
7.25	2.0–3.0	22.0	1.7×10^4	150
7.5	0.5–1.0	21.5	2.0×10^4	175
8.0	0.3	20.5	*	*

ND Below limits of detection by the methods employed.
*Not tested.

determined. The results (Table 2) show that the eubacterial probe hybridized to filter-bound RNA from all depths sampled in the epiliminion (0–7 meters) and the metalimnion (7–8 meters). Probe 10γ hybridized only to RNA extracted from metalimnion samples (Table 2). Insufficient RNA was available from hypolimnion samples to accurately estimate hybridization to either probe. Probe 9α did not hybridize to any of the RNAs extracted from water column samples.

Biomass (approximately 5g wet weight) was collected from 100–200 l of Lake Minnetonka using a tangential flow filtration device [12]. Extracts were prepared from biomass samples and were used to assay for key enzymes of the RuMP and serine pathways for formaldehyde fixation. Extracts from Lake Minnetonka metalimnion samples contained hexulose-6-phosphate synthase (0.2–0.6 μmol hexulose-6-phosphate formed min^{-1} (mg protein^{-1}) but did not contain detectable amounts of hydroxypyruvate reductase [12].

These data indicate that Type I methanotrophs are the dominant methanotrophs within the metalimnion of Lake Minnetonka. Methane oxidation has been shown to occur most rapidly within the metalimnion of eutrophied freshwater lakes during summer stratification [25,26,40–44]. The specific activities of the eubacterial and 10γ oligonucleotide probes used for the hybridization reactions were approximately equal. The data in Table 2 indicate that the population of all eubacteria is approximately 2 orders of magnitude greater than that of the Type I methanotrophic bacteria within the metalimnion of Lake Minnetonka.

RNA prepared from biomass samples from the metalimnion of Lake Mendota, Madison, WI also hybridized strongly with probe 10γ and weakly with probe 9α [12]. DNA isolated from metalimnion samples was amplified by polymerase chain reaction using 16S rRNA universal primers [12]. The amplified DNA was cloned into *E. coli* and sequenced. The 16S rDNA sequence homologous to

16S rRNA was determined [12]. The 16S rRNA contained a target sequence identical to other RuMP pathway methylotrophs (Fig. 3).

9 METHYLOTROPHS ASSOCIATED WITH PLANTS

We examined the ability of some free-floating plants to oxidize methane and TCE. Samples were obtained from wetlands and ponds in the University of Minnesota Landscape Arboretum and near Lake Minnetonka. Plant material samples (1 g wet weight) were placed into 15 ml serum bottles with 5 ml of 0.01 M potassium phosphate buffer pH 7.2. The atmosphere contained 200 μl of ^{14}C-methane (7.3×10^6 dpm). Heat-killed controls were employed to determine background levels of $^{14}CO_2$. Triplicate samples of each plant species were tested. After 24 hours the amount of methane converted to carbon dioxide was determined as previously described [41]. Of the plants tested. *Lemna minor* (little duckweed) samples oxidized methane at higher rates than other plant species (Table 3). *Lemna minor* is a small (approximately 1 cm in diameter) floating plant that forms a confluent layer of growth on the surface of wetland ponds during the summer months. These wetlands produce methane in sediments which diffuses to the pond's surface.

The ability of the duckweed plants and associated bacteria to oxidize TCE was examined. Plant material (1 g) and 1 ml of biomass, removed by collecting the sediment after washing 180 g of plants by vigorous shaking in 180 ml of water, filtration to remove plants, centrifugation at 10,000 **g** for 10 min and resuspension in 18 ml of water, were tested for TCE oxidation. The plants or biomass washed from plants were added to a 10 ml glass serum bottles capped with teflon lined rubber septa. The duckweed plants oxidized 0.40 nmol TCE per 24 h per gram tissue. Biomass washed from plants also oxidized TCE (0.3 nmol TCE per 24 h per gram). Membrane-bound RNA extracted from the plants hybridized to probes 9α and 10γ as well as to a eubacterial probe.

Pink pigmented facultative methylotroph bacteria (PPFMs) are abundant on the surfaces of plant leaves [46–48]. The 16S rRNAs purified from two

Table 3 Oxidation of methane by aquatic plants

Plant tested	$^{14}CO_2$ Produced (μmol/24 h/g biomass)
Wofia (Mixed Watermeal)	1.2 ± 0.15
Lemna minor (Little Duckweed)	24.6 ± 11.3
Lemna trisula (Star Duckweed)	7.0 ± 0.5
Spirodela polyrhiza (Big Duckweed)	0.73 ± 0.65
Riccia	1.6 ± 1.1
Ricciocarpus	3.5 ± 4.1

strains of PPFMs isolated from bean leaves by S. Hirano (University of Wisconsin, Madison, WI) were sequenced. Sequence comparisons indicated that these bacteria (isolates PK-1 and PR-6) are closely related to other PPFMs (Fig. 1a).

10 DISCUSSION

The 16S rRNAs of several methylotrophic bacteria have been sequenced. Our sequence data base has been limited by our ability to obtain well characterized pure cultures of methylotrophs, particularly methane utilizing bacteria that could be grown in the laboratory. The ribosomal RNAs of some bacteria have been difficult to sequence because the primers used for sequencing with reverse transcriptase may not be suitable and perhaps because of secondary structures and methylated bases within some rRNA molecules that cause premature termination of the chain elongation reaction.

The sequence information available indicates that the serine pathway methylotrophs that do not use methane and methanotrophs form two phylogenetically coherent assemblages within the α-2 subdivision of the *Proteobacteria*. The data do not permit a decision concerning the separation of the serine pathway methanotrophs into two genera. We are sequencing 16S rRNAs from more cultures that are presently classified as *Methylocystis* species.

It is reasonably clear that the pink pigmented facultative methylotrophs (PPFMs) are closely related to each other. The PPFMs are ubiquitous and have been isolated as airborne microbes, from soil, water and sediment samples, from leaves and nodules of plants, rice grains and hospital environments [7,25,26]. Several PPFMs isolated from different sources are closely related to each other (Fig. 1a). The 16S rRNAs of other non-pigmented serine pathway methylotrophs that do not use methane remain to be sequenced.

Because the serine pathway methanotrophs and methylotrophs that do not utilize methane form two distinct phylogenetic assemblages, it should be possible to develop signature probes that will distinguish between these groups of bacteria in natural samples.

The methylotrophs that employ the RuMP pathway for formaldehyde assimilation are less closely related to each other than are the methylotrophs that employ the serine pathway for formaldehyde assimilation. The RuMP methanotrophs, *M. capsulatus* and *M. methanica*, are not closely related to other Type I methanotrophs. The RuMP pathway methylotrophs that do not utilize methane form a separate cluster within the β-subdivision of the *Proteobacteria*. It should be possible when other 16S rRNA sequences from RuMP pathway methylotrophs are available to develop signature probes for the detection of several different groups of RuMP methylotrophs. We are sequencing 16S rRNAs of bacteria presently classified as members of the genus *Methylobacter*.

Although the number of sequences of 16S rRNAs available is not yet sufficient for the selection of the best signature probes for studies of the distribution of these methylotrophic bacteria in natural samples, the potential of this approach has been demonstrated [27–32]. The phylogenetic diversity of the methylotrophs is both an advantage and a disadvantage. The diversity should permit the selection of target sequences unique to each phylogenetic assemblage of methylotrophs. On the other hand, the number of sequences required to develop the necessary probes is relatively large and some of these bacteria are difficult to grow and are not easily obtained in pure cultures.

The signature probes we have employed have proven useful for detection of methanotrophs in bioreactor samples and samples taken from freshwater lakes. The signature probes, gene probes and other techniques have been employed to demonstrate that a methanotroph related to *Methylosinus* species has been maintained over long periods of time in a bioreactor employed for studies of the biodegradation of TCE and chloroform at Stanford University [39]. This methanotroph synthesized soluble MMO under the conditions used to operate the bioreactor. Signature probes and enzymatic assays have been employed in studies of the distribution of methanotrophs in freshwater lakes [12]. In highly eutrophied lakes like Lake Minnetonka, Minnesota and Lake Mendota, Madison, Wisconsin, methane oxidation is confined to the metalimnion during summer stratification [25,41–45]. Probe 10γ hybridized to filter bound RNA from metalimnion samples but not samples from the epilimnion of Lake Minnetonka [12]. The presence of RuMP pathway methanotrophs in the metalimnion was further suggested by the presence of hexulose phosphate synthase but not hydroxypyruvate reductase in extracts of biomass concentrated from water collected from within the metalimnion of Lake Minnetonka.

We have employed gene probes, antibodies against soluble MMO and a sensitive colorimetric assay for sMMO to detect bacteria capable of synthesizing this enzyme. There is a correlation between the ability of cells to synthesize sMMO and their ability to rapidly degrade TCE and other low molecular weight halogenated hydrocarbons [19,37]. Studies by DiSpirito et al. [49], Henry and Grbic-Galic [50] and Little et al. [51] have provided good evidence that cells containing only the particulate MMO also oxidize TCE but at rates two to three orders of magnitude lower than methanotrophs capable of synthesizing sMMO that are grown under conditions which permit its synthesis [19,20,22]. Because a limited number of bacteria have been examined for their ability to produce sMMO, it is not wise to conclude that the synthesis of this enzyme is restricted to Type II methanotrophs of the genus *Methylosinus* and the Type X methanotroph *Methylococcus capsulatus*.

In cases where 16S rRNA or 16S rDNA were amplified from nucleic acids extracted from environmental samples, previously uncultured organisms were shown to be present [52]. It is probable that the methanotrophs in nature may prove to be different from those in culture collections [12].

ACKNOWLEDGEMENTS

The research reported in this manuscript was supported by a grant to R.S. Hanson from The U.S. Department of Energy (DE-FGO2-88ER13862), a grant from U.S. Department of Energy through Oak Ridge National Laboratories and The University of Tennessee (UTN ORA 3546.04/SUB DOE) and a grant from the National Science Foundation (BSR 8903833). We are grateful to L. Alvarez Cohen and P.C. McCarty for permitting us to include data from a collaborative project on methanotrophs involved in TCE and chloroform biodegradation.

REFERENCES

[1] Anthony C. (1982) The Biochemistry of Methylotrophs, Academic Press, New York.
[2] Zatman, L. (1981) A search for patterns in methylotrophic pathways. In: Microbial Growth on C1 Compounds (Dalton, H., Ed.), pp. 42–54, Heyden, London.
[3] Whittenbury, R., Phillips, K.C. and Wilkinson, J.F. (1970) Enrichment, isolation and some properties of methane utilizing bacteria. J. Gen. Microbiol. **61**, 205–218.
[4] Whittenbury, R. and Dalton, H. (1981) The methylotrophic bacteria. In: The Procaryotes (Starr, M.P., Stolp, H., Trüper, H.G., Balows, A. and Schlegel, H.G., Eds), pp. 894–902, Springer-Verlag, Berlin.
[5] Whittenbury, R. and Krieg, N.R. (1984) Methylococcaceae fam. nov. In: Bergey's Manual of Determinative Bacteriology, Vol. 1, pp. 256–262. Williams & Wilkins, Baltimore.
[6] Galchenko, V.F. and Andreev, L.V. (1984) Taxonomy of obligate methanotrophs. In: Microbial Growth on C1 Compounds (Crawford, R.L. and Hanson, R.S., Eds), pp. 269–275. American Society of Microbiology, Washington, DC.
[7] Green, P.N. (1992) Taxonomy of methylotrophic bacteria. In: Methane and Methanol Utilizers (Murrell, J.C. and Dalton, H., Eds), pp. 23–84, Plenum Press, New York.
[8] Hanson, R.S., Netrusov, A.I. and Tsuji, K. (1991) The obligate methanotrophic bacteria *Methylococcus*, *Methylosinus*, and *Methylomonas*. In: The Procaryotes, Vol. II (Balows, A., Truper, H.G., Dworkin, M., Harder, W. and Schliefer K., Eds), pp. 2350–2364, Springer-Verlag, New York.
[9] Guckert, J.B., Ringelberg, D.B., White, D.C., Hanson, R.S. and Bratina, B.J. (1991) Membrane fatty acids as phenotypic markers in the polyphasic taxonomy of methylotroph within the proteobacteria. J. Gen. Microbiol. **137**, 2631–2641.
[10] Boulygina, E.S., Galchenko, V.F., Govorukhina, N.I., Netrusov, A.I., Nikitin, D.I., Trotsenko, Y.A. and Chumakov, K.M. (1990) Taxonomic studies on methylotrophic bacteria by 5S ribosomal RNA sequencing. J Gen Microbiol. **136**, 441–446.
[11] Tsuji, K., Tsien, H.C., Hanson, R.S., DePalma, S.R., Scholtz, R. and LaRoche, S. (1990) 16S ribosomal RNA sequence analysis for determination of phylogenetic relationshops among methylotrophs. J. Gen. Microbiol. **136**, 1–10.
[12] Bratina, B.J. (1992) PhD Thesis, University of Minnesota.
[13] Crutzen, P.J. and Graedel, T.E. (1986) The role of atmospheric chemistry in environment development interactions. In: Sustainable Development of the Biosphere (Clark, W.C. and Munn R.E., Eds), Cambridge University Press, Cambridge, MA.

[14] Graedel, T.E. and Crutzen, P.J. (1989) The changing atmosphere. Sci. Am. 261, 136–143.
[15] Harriss, R.C., Gorham, E., Sebacher, D.I., Bartlett, K.B. and Flebbe, P.A. (1985) Methane flux from northern peatlands. Nature, Lond. 315, 652–654.
[16] Whalen, S.C. and Reeburgh, W.S. (1988) A methane flux time series for tundra environments. Global Biogeochem. Cycles 2, 399–409.
[17] Whalen, S.C. and Reeburgh, W.S. (1990) Consumption of atmospheric methane by tundra soils. Nature, Lond. 346, 160–162.
[18] Streigel, R.G., McCannaughey, T.A., Thorstenson, D.C., Weeks, E.P. and Woodward, J.C. (1992) Consumption of atmospheric methane by desert soils. Nature, Lond. 357, 145–147.
[19] Brusseau, G., Tsien, H.C., Hanson, R.S. and Wackett, L.P. (1990) Optimization of trichloroethylene oxidation by methanotrophs and the use of a colorimetric assay to detect soluble methane monooxygenase activity. Biodegradation 1, 19–29.
[20] Oldenhuis, R., Vink, R.L.J.M., Janssen, D.B. and Witholt, B. (1989) Degradation of chlorinated aliphatic hydrocarbons by Methylosinus trichosporium OB3b expressing soluble methane monooxygenase. Appl. Environ. Microbiol. 55, 2819–2862.
[21] Oldenhuis, R., Oedzes, J.Y., Waarde, J.J.v.d. and Janssen, D.B. (1991) Kinetics of chlorinated hydrocarbon degradation by Methylosinus trichosporium OB3b and toxicity of trichloroethylene. Appl. Environ. Microbiol. 57, 7–14.
[22] Tsien, H.C., Brusseau, G.A., Hanson, R.S. and Wackett, L.P. (1989) Biodegradation of trichloroethylene by Methylosinus trichosporium OB3b. Appl. Environ. Microbiol. 55, 3155–3161.
[23] Wackett, L.P., Brusseau, G.A., Householder, S.R. and Hanson, R.S (1989) Survey of microbial oxygenase: trichloroethylene degradation by propane-oxidizing bacteria. Appl. Environ. Microbiol. 55, 2960–2964.
[24] Large, P. and Bamforth, C.W. (1988) Methylotrophy and Biotechnology. Copublished by Longman and John Wiley, New York.
[25] Hanson, R.S. (1980) Ecology and diversity of methylotrophic organisms. In: Advances in Applied Microbiology, Vol. 26 (Perlman, D., Ed.), pp. 3–39, Academic Press, New York.
[26] Hanson, R.S. and Wattenburg. E. (1991) Ecology of methylotrophic bacteria. In: Biology of Methylotrophs (Goldberg, I. and Rokem, J.S., Eds), pp. 325–348, Butterworth Publishers.
[27] Tsien, H.C., Bratina, B.J., Tsuji, K. and Hanson, R.S. (1990) Use of oligonucleotide signature probes for identification of physiological groups of methylotrophic bacteria. Appl. Environ. Microbiol. 56, 2858–2865.
[28] Amann, R.I., Stromley, J., Devereaux, R., Key, R. and Stahl, D. (1992) Molecular and microscopic identification of sulfate-reducing bacteria in multispecies biofilms. Appl. Environ. Microbiol. 58, 614–623.
[29] Giovannoni, S.J., DeLong, E.F., Olsen, G.J. and Pace, N.R. (1988) Phylogenetic group-specific oligodeoxynucleotide probes for identification of single microbial cells. J. Bacteriol. 170, 720–726.
[30] DeLong, E.F., Wickham, G.S. and Pace, N.R. (1989) Phylogenetic stains: ribosomal RNA-based probes for the identification of single cells. Science 243, 1360–1363.
[31] Amann, R.I., Krumholz, L. and Stahl, D.A. (1990) Fluorescent oligonucleotide probing of whole cells for determinative, phylogenetic, and environmental studies in microbiology. J. Bacteriol. 172, 762–770.
[32] Stahl, D.A., Flesher, B.A., Mansfield, H.R. and Montgomery, L. (1988) Use of phylogenetically based hybridization probes for studies of ruminal microbial ecology. Appl. Environ. Microbiol. 54, 1079–1084.

[33] Tsuji, K., Tsien, H.C., Bratina, B., Brusseau, G., Bastein, C., Zhang, Y., Donaldson, K., Machlin, S. and Hanson, R.S. (1990) Genetic and biochemical studies of methylotrophic bacteria. In: Gas, Oil and Coal Biotechnology (Akin, I.C. and Smith, J., Eds), pp. 445–462, Inst. of Gas Technol., Chicago, Illinois.

[34] Cardy, D.L.N., Laidler, V., Salmond, G.P.C. and Murrell, J.C. (1991) The methane monooxygenase gene cluster of *Methylosinus trichosporium*: cloning and sequencing of the mmoC gene. *Arch. Microbiol.* **156**, 477–483.

[35] Cardy, D.L.N., Laidler, V., Salmond, G.P.C. and Murrell, J.C. (1991) Molecular analysis of the methane monooxygenase gene cluster of *Methylosinus trichosporium* OB3b. *Molecular Microbiol.* **5**, 335–342.

[36] Murrell, J.C. (1992) The genetics and molecular biology of obligate methane-oxidizing bacteria. In: Methane and Methanol Utilizers (Murrell, J.C. and Dalton, H., Eds), pp. 115–148, Plenum Press, New York.

[37] Tsien, H.C. and Hanson, R.S. (1992) Soluble methane monooxygenase component B gene probe for identification of methanotrophs that rapidly degrade trichloroethylene. *Appl. Environ. Microbiol.* **58**, 953–960.

[38] Alvarez-Cohen, L. and McCarty, P.L. (1991) Effects of toxicity, aeration and reductant supply on trichloroethylene transformation by a mixed methanotrophic culture. *Appl. Environ. Microbiol.* **57**, 228–235.

[39] Alvarez-Cohen, L. and McCarty, P.L. (1991) Product toxicity and cometabolic competitive inhibition modeling of chloroform and trichloroethylene transformation by methanotrophic resting cells. *Appl. Environ. Microbiol.* **57**, 1031–1037.

[40] Alvarez-Cohen, L., McCarty, P.L., Boulygina, E., Hanson, R.S., Brusseau, G.A. and Tsien, H.C. (1992) Characterization of a methane-utilizing bacterium from a bacterial consortium that rapidly degrades trichloroethylene and chloroform. *Appl. Environ. Microbiol.* **58**, 1886–1893.

[41] Harrits, S. and Hanson, R.S. (1980) Stratification of aerobic methane-oxidizing organisms in Lake Mendota, Madison, Wisconsin. *Limnol. Oceanogr.* **5**, 412–421.

[42] Patt, T.E., Cole, G.C., Bland, J. and Hanson, R.S. (1974) Isolation of bacteria that grow on methane and organic compounds as sole sources of carbon and energy. *J. Bacteriol.* **120**, 955–964.

[43] Rudd, J.W.M., Furutani, A., Flett, R.J. and Hamilton, R.D. (1976) Factors controlling methane oxidation in shield lakes: The role of nitrogen fixation and oxygen concentration. *Limnol. Oceanogr.* **21**, 357–364.

[44] Rudd, J.W.M. and Hamilton, R.D. (1978) Methane cycling in a eutrophic shield lake and its effects on whole lake metabolism. *Limnol. Oceanogr.* **23**, 337–355.

[45] Rudd, J.W.M., Hamilton, R.D. and Campbell, N.E.R. (1974) Measurement of microbial oxidation of methane in lake water. *Limnol. Oceanogr.* **19**, 519–524.

[46] Corpe, W.A. and Basile, D.V. (1982) Methanol-utilizing bacteria associated with green plants. *Devel. Industrial Microbiol.* **23**, 483–493.

[47] Corpe, W.A. and Rheem, S. (1989) Ecology of methylotrophic bacteria on living leaf surfaces. *FEMS Microb. Ecol.* **62**, 143–250.

[48] Hirano, S.S. and Upper, C.D. (1991) Bacterial community dynamics. In: Microbial Ecology of Leaves (Andrews, J.H. and Hirano, S.S., Eds), pp. 271–294, Springer-Verlag, New York.

[49] DiSpirito, A.A., Gulledge, J., Shiemke, A.K., Murrell, J.C., Lidstrom, M.E. and Krema, C.L. (1992) Trichloroethylene oxidation by membrane-associated methane monooxygenase in Type I, Type II and Type X methanotrophs. *Biodegradation* **21**, 151–164.

[50] Henry, S.M. and Grbic-Galic, D. (1990) Effect of mineral media on trichloroethylene oxidation by aquifer methanotrophs. *Microb. Ecol.* **20**, 151–169.

[51] Little, C.D., Polumbo, A.V., Herbes, S.E., Lidstrom, M.E., Tyndall, R.L. and P.J. Gilmer (1988). Trichloroethylene oxidation by a methane-utilizing bacterium. *Appl. Environ. Microbiol.* **54**, 951–956.

[52] Ward, D.M., Weller, R. and Bateson, M.M. (1990) 16S rRNA sequences reveal numerous uncultured microorganisms in a natural community. *Nature, Lond.* **345**, 63–65.

24
Ecophysiological Characteristics of Obligate Methanotrophic Bacteria and Methane Oxidation *in situ*

Gary M. King

Darling Marine Center, University of Maine, Walpole, ME 04573, USA

1 SUMMARY

Most of the obligate methane-oxidizing bacteria (MOB) described to date are neutrophilic mesophiles that grow optimally in dilute media. Kinetic analyses generally indicate that bacterial methane uptake occurs by transport systems with a $K_m > 1 \mu M$. These and other properties of MOB are inconsistent with characteristics of methane oxidation *in situ*. The inconsistencies indicate a need for greater attention to the ecophysiological characteristics of isolates and the design of enrichment and isolation schemes which emphasize ecologically relevant parameters (e.g., low temperature, limited and diverse substrate availability, low water potential).

2 INTRODUCTION

Aerobic bacterial methane oxidation is an ancient process, perhaps not long removed from the advent of oxygenic photosynthesis. Consequently, sufficient time has elapsed for the obligate methane-oxidizing or methanotrophic bacteria (MOB) to adapt to the numerous environments containing both oxygen and methane. Although the conditions encountered by the earliest MOB are unknown, the diversity of habitats suitable for MOB has undoubtedly increased during the evolution of the biosphere, and MOB now occur in a wide variety of habitats. A review of the recent literature suggests that most of the known MOB are mesophilic, neutrophilic, obligate aerobes that grow optimally at relatively high water potentials [1].

Microbial Growth on C_1 Compounds
© Intercept Ltd, PO Box 716, Andover, Hampshire SP10 1YG, UK

Is the apparently limited diversity of MOB a product of biochemical constraints on the pathways of methane oxidation and assimilation? Is MOB diversity the product of ecological constraints associated with microenvironments where methane and oxygen coexist? Alternatively, is MOB diversity as currently known constrained by the sampling and enrichment schemes in use? Though the former questions are certainly more challenging, they are beyond the scope of this essay, and perhaps not easily addressed in any case. However, it is possible to address limitations imposed by the nature of the observations made to date. In particular, sufficient data are available to facilitate a comparison of the physiological characteristics of MOB and the characteristics of methane-oxidation as it occurs *in situ*. Such a comparison provides a means for determining whether existing cultures possess traits consistent with those that might be expected on the basis of field data. For example, if existing cultures are unable to consume methane at unamended atmospheric concentrations, one might conclude that this process is facilitated by as yet unknown MOB.

This brief essay considers field and laboratory observations in an attempt to compare the known properties of obligate methanotrophs and the 'ecophysiological' characteristics of methane oxidation. Most, though not all, methanotrophs have been isolated following procedures adapted from Whittenbury et al. [1–3], whose methods have formed the basis for the numerous enrichments that have provided the material for research on the microbiology of methane oxidation. Though even the earliest reports of MOB enrichment and isolation indicate a sensitivity to ecological problems, the characterization of MOB has often (and justifiably) emphasized taxonomic criteria rather than criteria relevant for understanding ecological behavior or function. This emphasis on taxonomy, and observations which suggest that most bacteria *in situ* remain unisolated, naturally provokes questions about the extent to which the MOB in culture represent the active methane consumers in various aquatic and soil ecosystems.

3 ATMOSPHERIC METHANE CONSUMPTION BY SOILS AND MOB

Atmospheric methane consumption provides a suitable subject for comparing the physiological diversity of MOB with characteristics of the process *in situ*. Atmospheric methane uptake by soils contributes significantly to the global methane budget; it occurs almost ubiquitously from the arctic and sub-arctic tundra to tropical savannahs and even arid deserts [4–8]. MOB have been isolated from at least some of these habitats, especially in temperate regions [e.g., 2]. Several of the isolates belong to well-characterized genera and species, including *Methylomonas, Methylosinus* and *Methylococcus*. However, it is unclear if these isolates can account for the oxidation observed in soils. For example, no known isolates appear to consume atmospheric methane. We have observed atmospheric methane consumption by a variety of strains, among

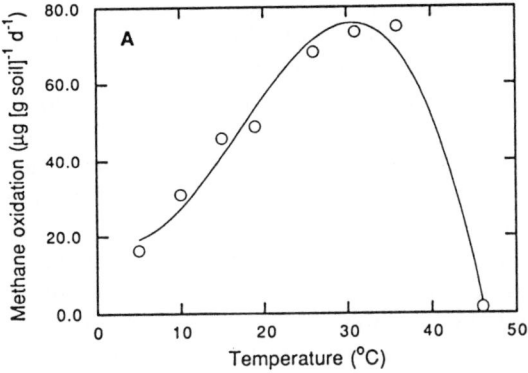

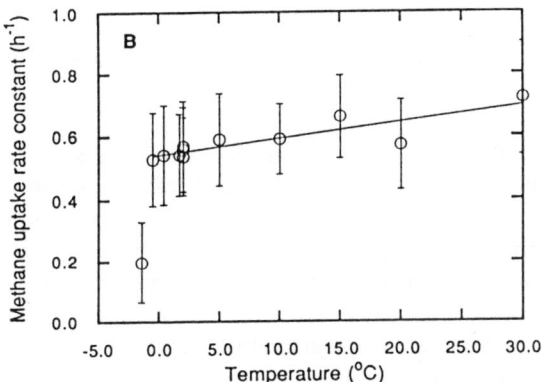

Figure 3 A. Response of methane oxidation in landfill soils at initial concentrations of about 10,000 ppm to changes in incubation temperature; redrawn from Whalen et al. [17]. B. Response of methane oxidation in intact soil cores from a temperate forest with initial methane concentrations of 1.7 ppm to changes in incubation temperature; data are mean uptake rate constants from quintuplicate determinations and vertical bars are ±1 standard deviation. From King and Adamsen [16].

responses of soil and cultures. Though present, the mesophiles isolated from soil may be inactive *in situ* at low temperature and low methane concentrations. The presence of two types of methanotrophs as defined by temperature responses is analogous to suggestions by Bender and Conrad [10], who defined populations on the basis of kinetic data. Psychrotrophic MOB with high affinity uptake systems could account for atmospheric methane consumption, while mesophilic bacteria with low affinity uptake systems could account for activity at high methane concentrations and temperatures >20–30°C. The latter types of MOB might dominate systems such as landfills [17], while the former dominate soils exposed to low methane concentrations.

Regardless of the type(s) of MOB responsible for atmospheric methane consumption, the low Q_{10} values reported to date [6, 15] suggest only a modest direct sensitivity of soil methane oxidation to anticipated mean annual or maximum seasonal temperature changes associated with global climate change. However, indirect effects of climate change, especially as related to soil moisture regimes, could have a more dramatic impact. For instance, drier conditions may convert some wetlands from net methane emitters to methane consumers (see Harriss et al. [18] for an example of changes due to decreases in water tables); drier soils may enhance soil-atmosphere gas exchange and thus increase soil methane oxidation. Assuming that soil water regimes and land use practices are no less optimal than at present, soil methane oxidation should increase slightly in importance as a loss term in the atmospheric methane budget due to increased temperature alone; more significant increases might be expected as a consequence of changes in precipitation patterns.

5 WATER POTENTIAL AND METHANE OXIDATION IN CULTURES AND SOILS

The effects of soil water on methane oxidation are expressed through two basic parameters, the soil void volume (gas space) and soil water potential. The former obviously determines in part gas exchange between the soil and the atmosphere. Increasing soil water content decreases the void volume with a consequent decrease in gas exchange rates. Several studies indicate that water contents of 10–20% (g water [g dry weight soil]$^{-1}$) are optimal for oxidation, with decreasing rates at higher or lower values (Fig. 4) [16, 17]. While some consistency exists among the published data, water contents from diverse soils are not easily compared without additional information, e.g. bulk density.

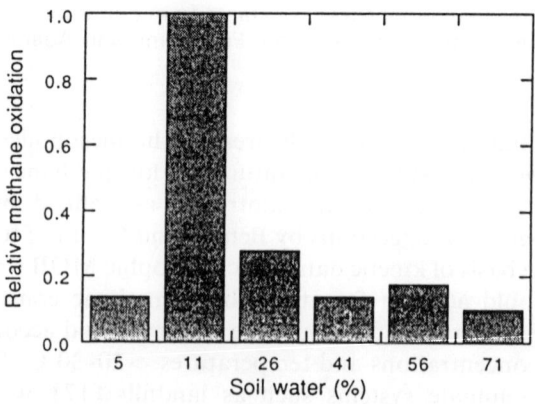

Figure 4 Relative response of methane oxidation in landfill soils to changes in soil water contents. Redrawn from Whalen et al. [17].

The second parameter, soil water potential, is perhaps as important as the soil void volume. Soil water potentials are the sum of gravitational, osmotic, pressure and matric potentials (see Griffin [19] for a pertinent review). Of these, the matric potential often dominates the total water potential, at least with respect to conditions experienced by the soil microflora. Matric potentials, sometimes referred to as a matric suction, arise from the interaction of water at the surface of soil particles. Matric potentials are lower in clays than in sands at the same volumetric or gravimetric water content. Matric potentials also show a pronounced hysteresis, with potentials varying as a function of the direction of water content change (wetting or drying). Matric potentials can play a critical role in soil metabolism, since the potential of cytoplasmic water in microbial cells must equilibrate with the potential of the cell's local environment.

Although the relationship between microbial activity and water potential is generally well-described (see reviews by Griffin [19] and Brown [20]), relatively little is known about the specific responses of MOB to water stress. Preliminary studies in our laboratory indicate that neither *M. rubra* nor *Ms. trichosporium* OB3b tolerate particularly low water potentials. These organisms do not grow or oxidize methane at total water potentials less than about -2 MPa in media containing NaCl. The minimum potentials for growth are somewhat higher in media containing polyethylene glycol 200 (PEG) or sucrose (King, unpub. observations). This is notable because neither PEG nor sucrose are transportable solutes. As a result, they provide models for predicting responses to matric potential, since adjustments by the cells must be based on the transport of salts in the medium. This is analogous to the response of cells in soils where adjustment of cellular water potential is also dependent on solute transport for turgor pressure regulation.

Soil matric potentials can fall to values less than the minima tolerated by *M. rubra* and *Ms. trichosporium* OB3b. Gravimetric water contents of 10–20%, at which active methane consumption has been reported [16–17], are approximately equivalent to water potentials of about -10^{-4} to -10^{-2} MPa in sands but -1 to -10 MPa in clays. The relatively high water potentials in sands could facilitate the recently reported atmospheric methane consumption in desert soils, even at low water contents [8]. The modest water potentials in sandy systems might even select for MOB with physiological characteristics similar to those of many extant cultures, but significantly different from the MOB active in clays or loams.

The MOB active in soils with a high clay content may possess a degree of xerotolerance atypical of most known cultures. Several MOB [e.g., 13, 14] tolerate water potentials comparable to that of seawater (about 0.54 M NaCl, equivalent to a potential of -2.5 MPa), but many tolerate only more modest potentials. Somewhat surprisingly, the marine isolates characterized thus far do not tolerate high ionic strengths (low water potentials) and, with the exception of requiring NaCl, their upper limits for salt tolerance are similar to those of isolates from non-saline systems [13, 14]. Two exceptional reports describe tolerance for 5–7% NaCl for salt lake isolates [21, 22].

Although these limits fall within the range of soil water potentials at which methane oxidation occurs, there are several reasons that the observed growth minima overestimate the potential for MOB to tolerate soil water stress. First, the potentials are derived from media containing transportable salts. As noted earlier, cells in media with non-transportable solutes appear less tolerant of water stress. In some respects, this is a curious observation, since sodium ions disrupt protein tertiary structure and inhibit metabolic activity in many organisms when present at seawater concentrations [23, 24]. Perhaps the energetics of ion accumulation in dilute media are a greater limitation than changes in protein conformation. It is also notable that the addition of glycine betaine, a compatible solute, does not significantly improve the response of *M. rubra* or *Ms. trichosporium* to NaCl (King, unpub. observations); this result contrasts with the response of many other eubacteria [25–27].

Second, it appears that salt tolerance has always been assayed in the presence of high methane concentrations. In such circumstances, cellular energy sources probably do not limit solute transport and thus turgor regulation. At methane concentrations more typical of soils, substrate availability may substantially limit turgor regulation. The response of *Ms. trichosporium* OB3b to variations in methane at modest salt concentrations illustrates the point. Similar oxidation rates occur at elevated methane partial pressures (e.g. 1000 ppm) and NaCl concentrations of 0.1 and 120 mM; however, at 10 ppm methane, rates decrease by about 70% for the higher salt level (King, unpub. observations). Sub-optimal values for other growth parameters, e.g., temperature and pH, may also limit the ability of MOB to respond to water stress. In addition, water stress, low pH, temperature and substrate concentrations, all of which are common features of soil environments, very likely act synergistically to decrease the V_{max} per cell for methane uptake. As indicated earlier, low V_{max} values per cell, in conjunction with typical K_ms ($>1\mu M$), probably preclude a significant role for currently known MOB in atmospheric methane oxidation [9].

6 SIGNIFICANCE OF NON-METHANE SUBSTRATES AND FACULTATIVE METHANOTROPHS

While typical *in situ* regimes of temperature, pH, water potential and methane probably constrain the activities of MOB substantially, the availability of substrates other than methane might have a positive effect. For example, simultaneous uptake of methane and even small amounts of methanol, methylated amines or more complex substrates (organic acids, amino acids, sugars, etc.) could provide sufficient energy or cell carbon to satisfy maintenance requirements and ameliorate stresses resulting from sub-optimal conditions. Unfortunately, very little is known about the uptake of non-methane substrates by MOB under *in situ* regimes. It is certainly clear that the 'classical' MOB do not grow on substrates other than methane, methanol or formaldehyde [1]. However, this does not preclude a role for other substrates in growth or catabolism, particularly when methane is present at nanomolar to micromolar concentrations. Stimulation of growth by glucose, malate, acetate and yeast extract in several MOB [2, 12] indicates that some form of co-metabolism or multi-substrate utilization may have important ecological implications. The role of non-methane substrates in soils as well as other systems deserves particular attention since methoxylated sugars (e.g. pectins) in plant tissues could provide a relatively abundant source of methanol.

Even if soil MOB utilize only methane, sufficient evidence exists to propose an important role for non-methane substrates. Several different microorganisms, including 3 genera of yeasts, *Methylobacterium organophilum*, and a *Mycobacterium* spp. oxidize methane and catabolize a wide diversity of additional substrates [28–31]. These facultative methylotrophs could co-metabolize methane while other substrates provided their primary energetic and carbon demands. Unfortunately, the capacity of these organisms for atmospheric methane oxidation is unknown. Likewise, their responses to key environmental variables have not been adequately explored. Current uncertainties about atmospheric methane oxidation by MOB and the postulated role of soils in the global methane budget justify greater attention to a group of organisms which have largely been ignored. It is especially notable that the various species of *Methylobacterium* grow at temperatures < 10°C, resist dessication and utilize trace concentrations of organics [31]. The behavior of these organisms as well as MOB under *in situ* conditions needs particular attention.

7 CONCLUSIONS

Much about the ecophysiology of methanotrophs and the process of methane oxidation *in situ* remains unknown. Inconsistencies between properties of cultures and *in situ* activities suggest that as yet unisolated, perhaps novel methanotrophs or facultative methanotrophs dominate atmospheric methane consumption by soils. Novel bacteia and possibly fungi may dominate methane oxidation in other systems as well, including marine and freshwater water columns, the rhizospheres of aquatic plants and animal-bacterial symbioses (e.g. in mytilids). Enrichment, isolation and characterization of such organisms should emphasize ecologically relevant regimes of pH, temperature, water potential and substrate availability. Interactions among key parameters and the consequences of low methane concentrations merit particular attention. Since the total global rate of methane oxidation very likely exceeds the net annual global flux of methane to the atmosphere [32], the importance of analyzing the behavior of methane-utilizing bacteria under *in situ* conditions cannot be over-emphasized. At a minimum, results of these studies will aid in predicting variations in the dynamics of methane as a consequence of global climate change.

ACKNOWLEDGEMENTS

The author acknowledges support from NSF BSR-9107315 and NASA NAGW-1428. Helpful discussion and input from P. Roslev, A. Adamsen and Drs S. Schnell, N. Iversen and R.S. Hanson are appreciated.

REFERENCES

[1] Hanson, R.S., Netrusov, A.I. and Tsuji, K. (1990) The obligate methanotrophic bacteria *Methylococcus*, *Methylomonas*, and *Methylosinus*. In: The Prokaryotes, 2nd ed. (Balows, A., Trüper, H.G., Dworkin, M., Harder, W. and Schleifer, K.H., Eds), pp. 2350–2364, Springer-Verlag, New York.
[2] Whittenbury, R., Phillips, K.C. and Wilkinson, J.F. (1970) Enrichment, isolation and some properties of methane-utilizing bacteria. *J. Gen. Microbiol.* **61**, 205–218.
[3] Whittenbury, R. and Dalton, H. (1981) The methylotrophic bacteria. In: The Pokaryotes (Starr, M.P., Stolp, H., Trüper, H.G., Balows, A. and Schlegel, H.G. Eds), pp. 894–902, Springer-Verlag, New York.
[4] Keller, M., Mitre, M.E. and Stallard, R.F. (1990) Consumption of atmospheric methane in tropical soils of Central Panama. *Glob. Biogeochem. Cyc.* **4**, 21–27.
[5] Whalen, S.C., and Reeburgh, W.S. (1990) Consumption of atmospheric methane to sub-ambient concentrations by tundra soils. *Nature, Lond.* **346**, 160–162.
[6] Crill, P.M. (1991) Seasonal patterns of methane uptake and carbon dioxide release by a temperate woodland soil. *Glob. Biogeochem. Cyc.* **5**, 319–334.
[7] Mosier, A., Schimel, D., Valentine, D., Bronson, K. and Parton, W. (1991) Methane and nitrous oxide fluxes in native, fertilized and cultivated grasslands. *Nature, Lond.* **350**, 330–332.
[8] Striegl, R.G., McConnaughey, T.A., Thorstenson, D.C., Weeks, E.P. and Woodward, J.C. (1992) Consumption of atmospheric methane by desert soils. *Nature, Lond.* **357**, 142–145.
[9] Conrad, R. (1984) Capacity of aerobic microorganisms to utilize and grow on atmospheric trace gases (H_2, CO, and CH_4). In: Current Perspectives in Microbial Ecology (Klug, M.J. and Reddy, C.R., Eds), pp. 461–467. American Society for Microbiology Publications, Washington, DC.
[10] Bender, M. and Conrad, R. (1993) Kinetics of methane oxidation in oxic soils. *Chemosphere*, in press.
[11] Hutton, W.E. and Zobell, C.E. (1949) The occurrence and characteristics of methane-oxidizing bacteria in marine sediments. *J. Bacteriol.* **58**, 463–473.
[12] Hazeu, W. (1975) Some cultural and physiological aspects of methane-utilizing bacteria. *Antonie van Leeuwenhoek J. Serol.* **41**, 121–131.
[13] Sieburth, J. McN., Johnson, P.W., Eberhardt, M.A., Sieracki, M.E., Lidstrom, M. and Laux, D. (1987) The first methane-oxidizing bacterium from the upper mixing layer of the deep ocean, *Methylomonas pelagica* sp. nov. *Curr. Microbiol.* **14**, 285–293.
[14] Bowman, J.P., Sly, L.I. and Cox, J.M. (1990) *Methylomonas fodinarum* sp. nov. and *Methylomonas aurantiaca* sp. nov.: two closely related type I obligate methanotrophs. *Sys. Appl. Microbiol.* **13**, 279–289.
[15] King, G.M. and Adamsen, A.P.S. (1992) Effects of temperature on methane oxidation in a forest soil and pure cultures of the methanotroph, *Methylomonas rubra*. *Appl. Environ. Microbiol*, in press.
[16] Adamsen, A.P.S. and King, G.M. (1993) Methane consumption in temperate and sub-arctic forest soils: rates, vertical zonation, and responses to water and nitrogen. *Appl. Environ. Microbiol.*, in press.

[17] Whalen, S.C., Reeburgh, W.S. and Sandbeck, K.A. (1990) Rapid methane oxidation in a landfill cover soil. *Appl. Environ. Microbiol.* **56**, 3405–3411.
[18] Harriss, R.C., Sebacher, D.I. and Day, F.P., Jr. (1982). Methane flux in the Great Dismal Swamp. *Nature, Lond.* **297**, 673–674.
[19] Griffin, D.M. (1981) Water and microbial stress. *Adv. Microb. Ecol.* **5**, 91–136.
[20] Brown, A.D. (1990) Microbial Water Stress Physiology: Principles and Applications, 313 pp. John Wiley, Chichester.
[21] Heyer, J., Malaschenko, Y., Berger, U. and Budkova, E. (1984) Verbreitung methanotropher Bakterien. *Z. für All. Mikrobiol.* **24**, 724–744.
[22] Malaschenko, Y.R. (1976) Isolation and characterization of new species (thermophilic and thermotolerant ones) of methane utilizers. In: Microbial Production and Utilization of Gases (Schlegel, H.G. Gottschalk, G., and Pfennig, N., Eds) pp. 293–300. Goltze KG, Göttingen.
[23] Yancey, P.H., Clark, M.E., Hand, S.C., Bowlus, R.D. and Somero, G.N. (1982) Living with water stress: evolution of osmolyte systems. *Science* **217**, 1214–1222.
[24] Somero, G.N. (1986) Protons, osmolytes, and fitness of internal milieu for protein function. *Am. J. Physiol.* **251**, R197–R213.
[25] Galinski, E.A. and Trüper, H.G. (1982) Betaine, a compatible solute in the extremely halophilic phototrophic bacterium *Ectothiorhodospira halochloris*. *FEMS Microbiol. Lett.* **13**, 357–360.
[26] Le Rudulier, D. and Bouillard, L. (1983) Glycine betaine, an osmotic effector in *Klebsiella pneumoniae* and other members of the *Enterobacteriaceae*. *Appl. Environ. Microbiol.* **46**, 152–159.
[27] Landfald, B. and Strøm, A.R. (1986) Choline-glycine betaine pathway confers a high level of osmotic tolerance in *Escherichia coli*. *J. Bacteriol.* **165**, 849–855.
[28] Patt, T.E., Cole, G.C. and Hanson, R.S. (1974) Isolation of bacteria that grow on methane and organic compounds as sole sources of carbon and energy. *J. Bacteriol.* **120**, 955–964.
[29] Wolf, H.J. and Hanson, R.S. (1979) Isolation and characterization of methane-utilizing yeasts. *J. Gen. Microbiol.* **114**, 187–194.
[30] Reed, W.M. and Dugan, P.R. (1987) Isolation and characterization of the facultative methylotroph *Mycobacterium* ID-Y. *J. Gen. Microbiol.* **133**, 1389–1395.
[31] Green, P.N. (1990). The genus Methylobacterium. In: The Prokaryotes (Balows, A., Trüper, H.G., Dworkin, M., Harder, W. and Schleifer, K.H., Eds), 2nd ed. pp. 2342–2349. Springer-Verlag, New York.
[32] King, G.M. (1993) Ecological aspects of methane oxidation, a key determinant of global methane dynamics. *Adv. Microb. Ecol.*, in press.

25
Methanotroph–Invertebrate Symbioses in the Marine Environment: Ultrastructural, Biochemical and Molecular Studies

Colleen M. Cavanaugh

Department of Organismic and Evolutionary Biology, Harvard University, 16 Divinity Avenue, Cambridge, MA 02138, USA

1 SUMMARY

Symbiotic associations involving methanotrophs have been suggested to occur for a number of marine invertebrates. Here I review the evidence indicating that species from at least two phyla (Mollusca and Pogonophora) harbor intracellular symbionts capable of using reduced C_1 compounds such as methane as their main carbon and energy source.

2 INTRODUCTION

The discovery of thriving communities of invertebrates at deep-sea hydrothermal vents gave us our first insights into a new set of ecosystems based on chemosynthesis. A major contributor to the base of the food-chain was sulfur-oxidizing chemoautotrophic bacteria found within the invertebrates (see reviews [1,2]). Symbioses between chemoautotrophs and invertebrates are now well described, occuring in species from five different phyla and from habitats ranging from coastal reducing sediments to deep-sea vents. It appears that these symbioses provide the hosts a nutritional source and the symbionts simultaneous access to substrates from both reducing and oxidizing environments. The benefits afforded the symbiotic partners suggest that similar symbioses could have evolved between eukaryotic hosts and other types of bacteria which are both dependent on substrates at oxic-anoxic interfaces and capable of synthesizing C_3 compounds from C_1 compounds.

Indeed, we and other researchers have recently presented data suggesting that methane-utilizing bacteria (methanotrophs) exist in symbiotic association with certain deep-sea invertebrates [3–6]. Methanotrophs are a subset of methylotrophs uniquely capable of deriving both their carbon and energy from methane (see review [7]). In these associations, the bacterial endosymbionts are hypothesized to provide the animal host with an internal source of nutrition via production of organic compounds from methane. As in chemoautotrophic symbioses, these symbionts afford the host the opportunity to utilize carbon, nitrogen, and energy sources otherwise unavailable to metazoans.

Here I review the data supporting the hypothesis that 'methanotrophic symbioses' exist. Although the putative bacterial symbionts have not yet been grown in pure culture, their nature is being elucidated by studies on symbiont and host ultrastructure, physiology, biochemistry, and stable isotope signatures in the intact symbioses. Cellular and molecular biological approaches, which have been successfully used to characterize other 'unculturable' symbionts and organelles, are currently being applied in my laboratory to study the taxonomic and evolutionary relationships of the putative symbionts to free-living bacteria. These studies, coupled with microscopical observations, have led to the intriguing possibility that some of the animals harbor two different types of symbionts. Guided by the diverse metabolic potential of known free-living methanotrophs, we are gaining an understanding of these symbioses in terms of metabolic and genetic capabilities of the symbionts as well as partner interactions and the cycling of energy and nutrients at oxic/anoxic interfaces of marine ecosystems.

3 ANIMALS AND HABITATS

Symbiotic associations with methanotrophs have been described for three species of invertebrates: two new species (as yet unnamed) of deep-sea mytilid mussels (Family Mytilidae) collected in the Gulf of Mexico and the pogonophoran tubeworm *Siboglinum poseidoni* from the central Skagerrak (species, habitats, and references listed in Table 1). The two mytilids, referred to here as FL mytilid and LA mytilid, are related to the hydrothermal vent mussel, *Bathymodiolus thermophilus* (R.D. Turner, pers. comm.). *S. poseidoni* is characterized along with the other species in the phylum Pogonophora by its complete lack of mouth and gut. As in chemoautotrophic symbioses, the bacterial symbionts are found in the bivalve gills and in the pogonophoran trophosome, an organ which extends through the trunk of the animal.

All of these animals occur in reducing sediments where methane and oxygen (from ambient deep-sea water) co-occur (Table 1). The Gulf of Mexico mytilids

Table 1 Marine invertebrate–methanotroph symbioses. Species list and collection site data including sediment porewater concentrations of potential symbiont energy sources (methane, sulfide, ammonia)

Phylum	Mollusca		Pogonophora
Species	FL Mytilid[a]	LA Mytilid[a]	*Siboglinum poseidoni*
Collection site	Florida Escarpment	Louisiana slope	Central Skagerrak
Depth (m)	3266	400–900	300
Temp (°C)	4.4	7	5.5
CH_4[b]	≤10 mM	0.04–66 μM	10 to 100 μM
H_2S	>3 mM	≤3 mM	u[c]
NH_3	3–4 mM	≤1.6 mM	nd
References	[5,9,14,20]	[4,10–12]	[6,13]

nd = not determined; u = undetectable.
[a] The FL mytilid and LA mytilid are new species (as yet unnamed) of mussels (Family Mytilidae). The names 'FL mytilid' and 'LA mytilid', as they are referred to in this paper, correspond to the temporary names 'Seep mytilid Va' (the Florida Escarpment mussel) and 'Seep mytilid Ia' (the common Bush Hill/Louisiana Slope mussel) designated pending formal description (R.G. Gustafson, R.A. Lutz, and R.D. Turner, pers. comm).
[b] Porewater methane concentrations are likely low estimates due to degassing prior to analysis.
[c] In Skaggerak sediments, free sulfide was not detected in top 20 cm. Bound sulfide may be present but was not determined [6].

are found in habitats which are referred to as 'cold seeps'. The FL mytilid is associated with hypersaline seeps at the base of the Florida Escarpment [5,8]. Up to 10 mM methane, which is presumably produced and transported within the Florida platform, is detectable in these sediment porewaters [9]. The LA mytilid lives in areas of hydrocarbon seeps and gas hydrates on the Louisiana slope in the Gulf of Mexico [4]. Mytilid density is significantly correlated with water methane levels [10,11] and they are observed living in sediments from which gases are bubbling [12]. *S. poseidoni* is found on the Danish slope of the central Skagerrak [6], the posterior portion of their tube buried in sediment and the anterior portion lying on the surface. While porewater methane concentrations appear low in these sediments, possibly due to degassing of sediment cores, 'shadow zones' are found in this area of the Skagerrak indicative of deep sediment gas reserves [6].

Finding methanotrophic symbionts in these animals was surprising since all other pogonophorans and vent mytilids investigated to date contain chemoautotrophic symbionts within their tissues [2]. Indeed, at both of the Gulf of Mexico sites sulfide concentrations are also elevated (Table 1), and symbioses with sulfur-oxidizing chemoautotrophs have been described from each of the sites [12–14]. Ammonia, another potential chemoautotroph energy source, also occurs in high concentrations in the Gulf of Mexico sediments (Table 1). Efforts to determine the nature of the symbionts observed in the tissues of these animals has thus focused on both autotrophic and methanotrophic metabolism.

4 METHANOTROPH–INVERTEBRATE SYMBIOSES: SUPPORTING EVIDENCE

4.1 Symbiont ultrastructure

The intracellular symbionts may be observed in the mytilid gills (Fig. 1) and in the pogonophoran trophosome tissue (Fig. 2) by transmission electron microscopy (TEM). These symbionts, which appear to be Gram-negative bacteria, contain distinctive stacked intracytoplasmic membrane arrays. Such membranes are seen in several types of bacteria including Type I or Type X methanotrophs [15]. These symbionts are coccoid- to oval-shaped and range in diameter from 0.75 to 2.0 μm (Table 2). The gills of the FL mytilids also contain smaller (~ 0.4 μm diameter) coccoid- to rod-shaped Gram-negative symbionts which lack internal membranes (Fig. 1A,B). The ultrastructural arrangement of the symbionts within the gills and trophosome is similar to that observed for chemoautotrophic bivalve and pogonophoran symbioses. All of the symbionts are intracellular, contained in vacuoles surrounded by a peribacterial membrane thought to be derived from the host cell (Figs 1, 2). Gill bacteriocytes are interspersed with symbiont-free intercalary cells. Structures resembling lysosomal residual (myelin-like) bodies are observed in the basal region of the mytilid bacteriocytes (Fig. 1A,C), suggesting that digestion of symbionts occurs in this region.

Attempts to culture methanotrophs from the FL mytilid have been unsuccessful (C.M. Cavanaugh and M.E. Lidstrom, unpubl. observations). Bacteria containing membrane stacks like those of the *S. poseidoni* symbionts have been enriched in mineral media with methane from both the pogonophoran tissue and from Skagerrak sediment [6]. However, pure cultures of these bacteria were not obtained, and it is not yet known whether these bacteria are the symbionts.

4.2 Enzyme activities

The occurrence of key enzymes from methanotrophic and autotrophic pathways have served as diagnostic indicators of the metabolic capabilities of the symbionts observed in the tissues of these animals. Initial tests for methane monooxygenase (MMO) which catalyzes the oxidation of methane to methanol, were negative for frozen tissues of the FL mytilid [5]. This result is not conclusive, however, because MMO activity is quite labile in free-living methanotrophs [16]. However, methane utilization has been demonstrated in fresh tissues of all three species (see section 4.3).

Other key enzyme activities considered diagnostic of both dissimilatory and assimilatory methylotrophic pathways are detected in the symbiont-containing tissues of these animals (Table 2). Methanol dehydrogenase (MeDH), an enzyme considered diagnostic of methylotrophic bacteria [7], was detected in extracts of symbiont-containing tissues for all three species. Estimating that bacterial protein accounts for about 20% of the total, the activities are comparable to those of free living methylotrophs (60 to 600 nmol min^{-1} (mg protein)$^{-1}$) [17]. Hexulose phosphate synthase (HPS), the key enzyme of the ribulose monophosphate carbon assimilation pathway of Type I and Type X strains, was detected at high levels in the FL mytilid and *S. poseidoni*. Enzymes of the serine cycle carbon assimilation pathway of Type II strains, hydroxypyruvate reductase and serine glyoxylate aminotransferase were either not detectable or present at variable concentrations, but were never found together in the FL mytilid [5].

The key enzyme of the Calvin-Benson cycle, ribulose-1,5-bisphosphate carboxylase (RuBPCase), was not detectable in the FL mytilid [5,14], but low activities were detected in extracts of the LA mytilid gills [18]. This enzyme is considered diagnostic of autotrophic metabolism and is also detected in Type X methanotrophs [7].

4.3 Physiology

Symbiont-containing tissues of freshly collected specimens of both mytilids and *S. poseidoni* have been shown to utilize methane by one or two different methods (Table 2). Incorporation and oxidation of ^{14}C-labelled methane has been demonstrated for all three species [6,18; C.M. Cavanaugh, unpub. data]. Rates of methane utilization in FL gill tissue homogenates averaged 13 ± 5 nmol min^{-1} mg protein^{-1} (n = 3). The label was incorporated into acid stable compounds and oxidized to CO_2 in a ratio of $\sim 1:2$ (C.M. Cavanaugh, unpub. data), similar to the ratio measured for free-living methanotrophs [17]. Methane consumption by the Gulf of Mexico mytilids has also been directly measured using gas chromatography [4,19; C.M. Cavanaugh and B. Tilbrook, unpub. data]. Methane consumption is limited to the gill tissues, is dependent on oxygen, and is inhibited by acetylene and formalin. The LA mytilid exhibited maximal consumption rates at 250 to 300 μM oxygen and 200 to 300 μM methane [19].

In an attempt to determine whether a second symbiont type occurs in the mytilid gills, thiosulfate or sulfide stimulation of CO_2 fixation was tested. This activity was not detectable in the FL mytilid (C. M. Cavanaugh, unpub. data) but could be demonstrated for LA mytilids for up to two days after capture [18]. The authors suggest that this could be due to either a second symbiont or to contaminating bacteria on the surface of freshly collected mussels, since this activity was not maintained in captivity [18].

4.4 Stable isotopes

Stable isotopes, used to infer an organism's food source, indicate that methane plays a major role in the nutrition of at least three of these species (Table 2) [12,13,20]. Stable nitrogen and sulfur isotope values of both Gulf of Mexico mytilids also differ from tubeworm-chemoautotroph symbioses collected from the same sites [12,20], suggesting that either the sources of assimilated nitrogen or the biochemical fractionation pathways differ between these organisms.

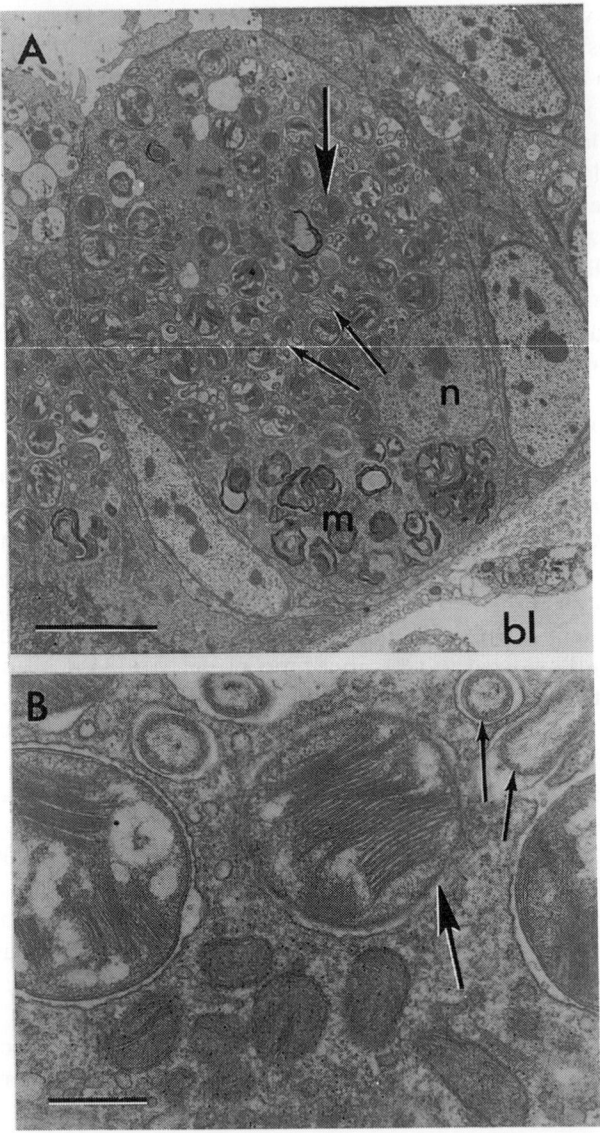

Figure 1 Transmission electron micrographs of FL mytilid (A,B) and LA mytilid (C,D) gills showing intracellular Gram-negative bacterial symbionts. Symbionts of both mytilids contain stacked intracytoplasmic membranes (large arrows); the FL mytilid contains a second smaller symbiont which lacks internal membranes (small arrows). A,C. Low magnification of transverse section of gill filaments showing

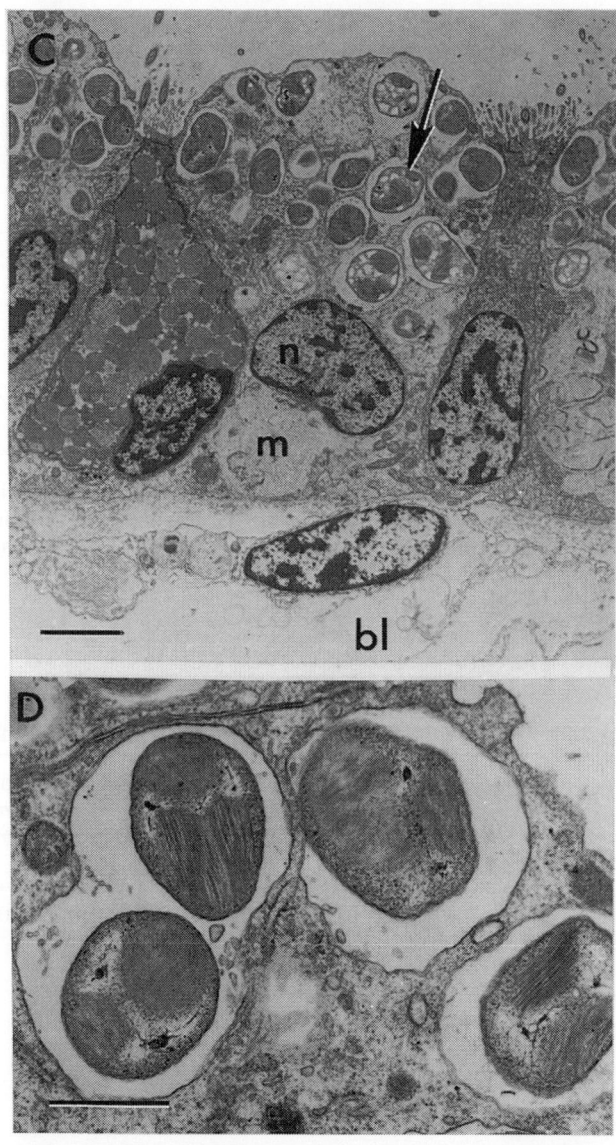

symbionts contained within membrane-bound vacuoles in epithelial cells (bacteriocytes) interspersed by symbiont-free intercalary cells. Scale bar = 5 μm. m, myelin-like residual bodies; n, bacteriocyte nucleus; bl, blood space. B,D. Higher magnification showing detail of symbionts. Scale bar = 0.5 μm. Figs C and D courtesy of C.R. Fisher.

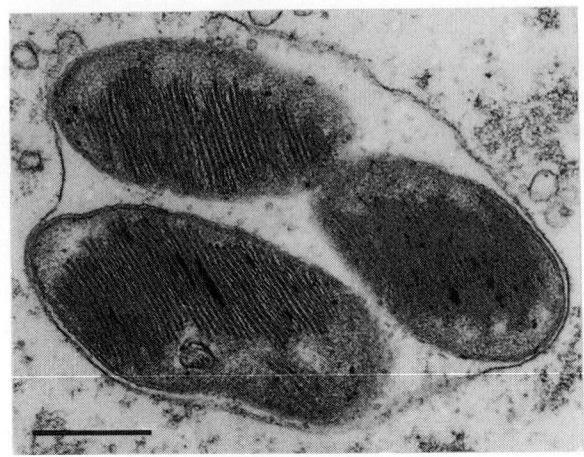

Figure 2 Transmission electron micrograph of intracellular bacterial symbionts within the trophosome of *Siboglinum poseidoni*. Gram-negative symbionts containing intracytoplasmic membranes are found within membrane bound vacuoles. Scale bar = 0.5 µm. Courtesy of H.J. Flügel.

Table 2 Marine invertebrate–methanotroph symbioses: supporting evidence

| Phylum | Mollusca | | Pogonophora |
Species	FL Mytilid	LA Mytilid	*Siboglinum poseidoni*
Bacterial symbionts			
Morphology	coccus rod	coccus	coccus/oval
Diameter (µm)	1.6 0.4	0.75–2.0	0.75–1.0
Intracytoplasmic membranes	+ −	+	+
Enzyme activities[a]			
MeDH	4–163	2.7–4.4	4–46
HPS	132–314	nd	36–287
RuBPcase	u	0.12	nd
Methane utilization[b]	+	+	+
Stable isotope[c], $\delta^{13}C$ (‰)			
Tissues	−74.3 ± 2.0	−40.1 to −57.6	−73.6, −78.3
CH$_4$	−61.1 to −93.8	−41.2	−72.9, −80.3
References	[5,9,14,20]	[4,12,18,19]	[3,6,13]

nd = not determined; u = undetectable.

[a] Activities expressed as nmol min^{-1} (mg protein)$^{-1}$. Protein estimated as 15% of wet weight. MeDH, methanol dehydrogenase; HPS, hexulose phosphate synthase; RuBPcase, ribulose-1,5-bisphosphate carboxylase oxygenase.

[b] Methane utilization determined using gas chromatography and/or ^{14}C-labeled methane.

[c] Isotope values are reported relative to the standard Pee Dee Belemite (PDB) using the standard delta notation where $\delta^{13}C = [(^{13}C:^{12}C \text{ Sample}/^{13}C:^{12}C \text{ Standard})-1] \times 10^3$.

The stable carbon isotope ratios detected in tissues of all three species are closely correlated with the stable isotopic composition of the methane source where the samples were collected (Table 2). The carbon isotope composition of the FL mytilid and *S. poseidoni* is extremely depleted ($\partial^{13}C = -73$ to -78%) [6,9,20] reflecting biogenic methane sources, while the LA mytilid tissue values ($\partial^{13}C = -40.1$ to -57.6%) reflect thermogenic methane sources [12]. These results imply that methane is a major source of carbon for these animals, presumably mediated via symbiont utilization.

4.5 Phylogeny

We have analyzed the phylogenetic relationship of putative methanotrophic bacteria observed in the tissues of the FL and LA mytilids by comparison of 16S ribosomal RNA sequences (D.L. Distel and C.M. Cavanaugh, unpub. data). Unlike other chemoautotrophic symbioses examined to date [21,22; D. L. Distel, unpub. data] the FL mytilid showed strong evidence of the presence of more than one abundant 16S rRNA gene. By cloning and sequencing, two distinct sequences were recovered from the specimens, while a single sequence was detected in the LA mytilid. These results are in agreement with ultrastructural data showing two symbiont morphotypes in the FL mussel and only one symbiont morphotype in LA specimens.

Preliminary phylogenetic analysis places all of the mytilid symbionts into two distinct clusters within the gamma subdivision of the Proteobacteria. One of the 16S rRNA genes found in the FL mytilid is closely related to the chemoautotrophic symbiont of the hydrothermal vent mussel, *Bathymodiolus thermophilus*. The other FL mytilid 16S rRNA sequence clusters in a very distinct group which also contains the single LA mytilid 16S rRNA sequence. The latter cluster appears to be affiliated with known free-living Type I methanotrophs, including *Methylomonas rubra* (see R. S. Hanson, this volume).

Preliminary results demonstrate that we can successfully perform *in situ* hybridizations allowing us to identify, with single-bacterial-cell resolution, the location of the cells which contain the identified 16S rRNA sequences. Thus we can determine if these sequences truly reflect multiple symbiont species within a single host tissue.

5 DISCUSSION

The data presented here support the hypothesis that methanotrophic bacteria exist in symbiosis with certain invertebrates collected from a variety of marine environments. To date the major evidence supporting this hypothesis is the co-occurrence in specific tissues of (a) subcellular inclusions resembling Gram-negative bacteria with Type I intracytoplasmic membranes and (b) enzymatic or physiological activities typically associated with methanotrophic metabolism. The stable isotope data further indicate that methane is a major nutritional carbon source for the cold seep animals with assimilation presumed to be mediated by their intracellular symbionts. Our preliminary phylogenetic analysis of the mytilid symbionts further supports this hypothesis by indicating that each of the mytilids contains at least one symbiont which clusters with free-living Type I methanotrophs (D.L. Distel and C.M. Cavanaugh, unpub. data).

The possibility that two different symbionts exist in the FL mytilid is suggested based on two lines of evidence: TEM observations of two different symbiont 'morphotypes' within single host cells, and phylogenetic analysis indicating a second dominant 16S rRNA gene sequence in DNA extracted from the symbiont-containing gills. The occurrence of different types of symbionts in a single host cell is common in protists, but to our knowledge has not been observed in metazoans [23,24]. This is presumably due to the many adaptations of both host and symbiont necessary for successful uptake and maintenance of a prokaryote within a eukaryotic cell [25,26]. When two or more intracellular symbionts are known to occur in metazoans, they are typically restricted to different tissues or cell types, e.g. primary and secondary symbionts found in aphid mycetocytes and sheath cells respectively [27]. A single bacterial species may also display multiple morphologies within a single eukaryotic cell, e.g. bacteria of the genus *Chlamydia* [28] and the rickettsia *Coxiella burnetti* [29].

It is not clear yet whether the two morphotypes observed in the mytilid gills are (1) two different species or (2) represent developmental stages of a single species. The occurrence of membranes in free-living methanotrophs is variable depending on culture conditions [16,30]; therefore the latter explanation could hold. However, our 16S rRNA sequence analysis provides evidence that two different bacterial species occur in the FL mytilid gills (D.L. Distel and C.M. Cavanaugh, unpub. data). Physiological and enzymatic data support the hypothesis that at least some of the symbionts found in these mytilids and pogonophora are capable of utilizing methane. If two symbionts do in fact occur within the same tissues, the metabolism of the second type remains obscure. If phylogeny is any indication of metabolism, it would appear that these animals harbor chemoautotrophs similar to the sulfur-oxidizers, as found in the vent mytilid *B. thermophilus* [2,21]. Other possibilities (not specifically indicated by phylogenetic data) are nitrifiers, or methanol-utilizing autotrophs which are commonly found in co-culture with methanotrophs [31]. RuBPCase activity is extremely variable in *B. thermophilus* gills, ranging from undetectable to activities an order of magnitude lower than those measured in other vent invertebrate-chemoautotroph symbioses due to abundant protease activity [18; D. Nelson, pers. comm.] and thus may be present but not detectable in these mytilids. Alternatively a second species may be present but not metabolically active. *In situ* localization of enzymes using antibodies and probes for genes unique to methanotrophy and to chemoautotrophy will help resolve the metabolic and genetic capabilities of the two types of symbionts observed in the FL mytilid.

The importance of the symbionts in the nutrition of the host and the degree of integration and co-evolution between host and symbionts appears variable between the different species. As in all pogonophoran symbioses, the methanotrophic symbionts are contained within a specialized tissue, the trophosome, which is primarily devoted to housing bacteria, whereas the mytilid symbionts are contained within epithelial cells interspersed among symbiont-free cells in the gills. *S. poseidoni* completely lacks mouth and gut, and is therefore presumed dependent on its symbionts as a major source of nutrition, whereas both the mytilids still retain a functioning gut. While stable carbon isotopes indicate methane is an important carbon source for all of the cold seep species, studies on the LA mytilid have shown it to be capable of filter feeding [32]. We recently observed that the gut of freshly collected FL mytilids was full of particulate material (C.M. Cavanaugh, unpub. observation). Unlike symbioses between dinoflagellates and corals or sea anemones, in which carbon compounds are translocated from symbiont to host [23], ultrastructural observations suggest that the symbionts in the mytilids are being 'farmed', i.e., they are digested in the basal region of the cell (Fig. 1A,C), with organic compounds presumably being translocated to the host via the blood space.

Although relatively few methanotrophic symbioses are known, we continue to examine new environments for additional examples. Indeed, examination of a new species of mytilid (as yet unnamed; Family Mytilidae) collected from hydrothermal vents along the Mid-Atlantic Ridge (Snakepit Site; depth ~3500 m [33]) revealed surprising results (C.M. Cavanaugh, C.O. Wirsen and H.W. Jannasch, in press). Unlike all other vent bivalves, which contain sulfur-oxidizing chemoautotrophic symbionts, the MAR mussel contains two distinct morphological types of symbionts like the FL mytilid, including one with Type I intracytoplasmic membranes. Activities of methanol dehydrogenase were detected in gill extracts, while RuBPCase was not. Stable carbon isotope values of the MAR mytilid ($\partial^{13}C = -32.5, -32.7\%$) fall within the range of values reported for Pacific vent bivalve-chemoautotroph symbioses, but do not preclude the use of vent-derived methane reported for East Pacific Rise vents to be isotopically heavy ($\partial^{13}C = -15$ to -17%) [34] relative to biogenically produced methane. Methane is present in end member fluids at the MAR Snakepit site in similar concentrations (50–100 μM) to vents along the East Pacific Rise [34,35]. While tests for methane utilization by the MAR mytilid must await fresh specimens, the co-occurrence of bacteria containing Type I membranes and MeDH activity suggest this animal also harbors methanotrophs, extending the occurrence of methanotrophic symbioses to include hydrothermal vents.

Two other new species of mytilids from the Gulf of Mexico have also been suggested to harbor methanotrophic symbionts on the basis of symbiont ultrastructure, enzyme activities, and stable isotopes (C.R. Fisher, pers. comm.). Although the extent of methanotrophic symbioses is not known, isotopically light carbon values (indicative of biogenically produced methane) have been noted for other deep-sea invertebrates [20,36]. Lower Cretaceous fossil communities, including serpulid worm tubes and bivalves, have also been described from isotopically light ($\partial^{13}C = -25$ to -50%) carbonates, suggesting that methane-based symbioses may be quite ancient [37].

Associating with a eukaryote may best be viewed as an adaptation by methanotrophs to 'bridge' oxic-anoxic interfaces, thus assuring the simultaneous supply of necessary substrates typically available only from anoxic and oxic environments. Such symbioses appear to yield benefits for both partners – an internal source of organic carbon for the host and substrate accessibility for the symbiont. Given the fairly common environmental circumstances of methane release to aerobic waters, we suggest the search for methanotroph-invertebrate associations in marine, as well as freshwater, habitats should be widened.

6 CONCLUSION

Bacteria containing complex intracytoplasmic membranes, observed in the tissues of two mytilids and one tubeworm, are concluded to be methanotrophs on the basis of biochemical and physiological data. Preliminary phylogenetic analysis, indicating the mytilid symbionts cluster with free-living methanotrophs, further supports this conclusion. We are using various molecular approaches to determine whether the two symbiont morphotypes observed in the FL mytilid are developmental stages of a single species or are two different bacteria.

ACKNOWLEDGEMENTS

I thank Daniel Distel and Philip Gschwend for helpful discussion and critical reading of the manuscript, Richard Hanson for sharing unpublished sequence data, and Hans Flügel and Chuck Fisher for providing micrographs. This work was supported by the Office of Naval Research grant N00014-91-J-1489 and National Science Foundation grant DCB 8718799.

REFERENCES

[1] Cavanaugh, C.M. (1985) Symbiosis of chemoautotrophic bacteria and marine invertebrates from hydrothermal vents and reducing sediments. In: Hydrothermal Vents of the Eastern Pacific: An Overview (M.L. Jones, Ed.), pp. 373–388, Bulletin of the Biological Society of Washington, Washington, DC.

[2] Fisher, C. (1990) Chemoautotrophic and methanotrophic symbioses in marine invertebrates. Rev. Aquat. Sci. 2, 399–436.

[3] Southward, A.J., Southward, E.C., Dando, P.R., Rau, G.H., Felbeck, H. and Flügel, H. (1981) Bacterial symbionts and low $^{13}C/^{12}C$ ratios in tissues of

Pogonophora indicate unusual nutrition and metabolism. *Nature (Lond.)* **293**, 616–620.

[4] Childress, J.J., Fisher, C.R., Brooks, J.M., Kennicutt, M.C.I., Bidigare, R. and Anderson, A.E. (1986) A methanotrophic marine molluscan (Bivalvia, Mytilidae) symbiosis: mussels fueled by gas. *Science* **233**, 1306–1308.

[5] Cavanaugh, C.M., Levering, P.R., Maki, J.S., Mitchell, R. and Lidstrom, M.E. (1987) Symbiosis of methylotrophic bacteria and deep-sea mussels. *Nature (Lond.)* **325**, 346-348.

[6] Schmaljohann, R. and Flügel, H.J. (1987) Methane-oxidizing bacteria in pogonophora. *Sarsia* **72**, 91–98.

[7] Anthony, C. (1986) Bacterial oxidation of methane and methanol. *Adv. Microb. Physiol.* **27**, 113–209.

[8] Paull, C.K., Hecker, B., Commeau, R., Freeman-Lynde, R.P., Neumann, C., Corso, W.P., Golubic, S., Hook, J.E., Sikes, E. and Curray, J. (1984) Biological communities at the Florida Escarpment resemble hydrothermal vent taxa. *Science* **226**, 965–967.

[9] Martens, C.S., Chanton, J.P. and Paull, C.K. (1991) Biogenic methane from abyssal brine seeps at the base of the Florida Escarpment. *Geology* **19**, 851–854.

[10] MacDonald, I.R., Boland, G.S., Baker, J.S., Brooks, J.M., Kennicutt, M.C.I. and Bidigare, R.R. (1989) Gulf of Mexico hydrocarbon seep communities. II. Spatial distribution of seep organisms and hydrocarbons at Bush Hill. *Mar. Biol.* **101**, 235–247.

[11] MacDonald, I.R., Callender, W.R., Burke, R.A., Jr., McDonald, S.J. and Carney, R.S. (1990) Fine-scale distribution of methanotrophic mussels at a Louisiana cold seep. *Prog. Oceanog.* **24**, 15–24.

[12] Brooks, J.M., Kennicutt II, M.C., Fisher, C.R., Macko, S.A., Cole, K., Childress, J.J., Bidigare, R.R. and Vetter, R.D. (1987) Deep-sea hydrocarbon seep communities: evidence for energy and nutritional carbon sources. *Science* **238**, 1138–1142.

[13] Schmaljohann, R., Faber, E., Whiticar, M.J. and Dando, P.R. (1990) Co-existence of methane- and sulphur-based endosymbioses between bacteria and invertebrates at a site on the Skaggerak. *Mar. Ecol. Prog. Ser.* **61**, 119–124.

[14] Cary, C., Fry, B., Felbeck, H. and Vetter, R.D. (1989) Multiple trophic resources for a chemoautotrophic community at a cold water brine seep at the base of the Florida Escarpment. *Mar. Biol.* **100**, 411–418.

[15] Hanson, R.S., Netrusov, A.I. and Tsuji, K. (1992) The obligate methanotrophic bacteria *Methylococcus*, *Methylomonas*, and *Methylosinus*. In: The Prokaryotes, 2nd ed. (A. Balows et al., Eds), pp. 2350–2364, Springer-Verlag, Berlin.

[16] Prior, S.D. and Dalton, H. (1985) The effect of copper ions on membrane content and methane monooxygenase activity in methanol-grown cells of *Methylococcus capsulatus* (Bath). *J. Gen. Microbiol.* **131**, 155–163.

[17] Anthony, C. (1982) The Biochemistry of Methylotrophs, Academic Press, London.

[18] Fisher, C.R., Childress, J.J., Oremland, R.S. and Bidigare, R.R. (1987) The importance of methane and thiosulfate in the metabolism of the bacterial symbionts of two deep-sea mussels. *Mar. Biol.* **96**, 59–71.

[19] Kochevar, R.E., Childress, J.J., Fisher, C.R. and Minnich, E. (1992) The methane mussel: roles of symbiont and host in the metabolic utilization of methane. *Mar. Biol.* **112**, 389–401.

[20] Paull, C.K., Jull, A.J.T., Toolin, L.J. and Linick, T. (1985) Stable isotope evidence for chemosynthesis in an abyssal seep community. *Nature (Lond.)* **317**, 709–711.

[21] Distel, D.L., Lane, D.J., Olsen, G.J., Giovannoni, S.J., Pace, B., Pace, N.R., Stahl, D.A. and Felbeck, H. (1988) Sulfur-oxidizing bacterial endosymbionts: analysis of phylogeny and specificity by 16S rRNA sequences. *J. Bacteriol.* **170**, 2506–2510.

[22] Eisen, J.A., Smith, S.W. and Cavanaugh, C.M. (1992) Phylogenetic relationship of chemoautotrophic bacterial symbionts of *Solemya velum* Say (Mollusca:

Bivalvia) determined by 16S rRNA gene sequence analysis. *J. Bacteriol.* **174**, 3416–3421.
[23] Smith, D.C. and Douglas, A.E. (1987) Biology of Symbiosis, Edward Arnold, London.
[24] Jeon, K.W. (1983) Intracellular Symbiosis, Academic Press, New York.
[25] Moulder, J.W. (1985) Comparative biology of intracellular parasitism. *Microbiol. Rev.* **49**, 298–337.
[26] Falkow, S. (1990) The 'Zen' of bacterial pathogenicity. *The Bacteria* **11**, 3–9.
[27] Gutnick, D.L. (1992) Prokaryotic symbionts of the aphid. In: The Prokaryotes, 2nd ed. (A. Balows *et al.*, Eds), pp. 3907–3913, Springer-Verlag, Berlin.
[28] Fields, P.I. and Barnes, R.C. (1992) The genus *Chlamydia*. In: The Prokaryotes, 2nd ed. (A. Balows *et al.*, Eds), pp. 3691–3709, Springer-Verlag, Berlin.
[29] Williams, J.C., Weiss, E. and Dasch, G.A. (1992) The genera *Coxiella*, *Wolbachia*, and *Rickettsiella*. In: The Prokaryotes, 2nd ed. (A. Balows *et al.*, Eds), pp. 2471–2484, Springer-Verlag, Berlin.
[30] Collins, M.L.P., Buchholz, L.A. and Remsen, C.C. (1991) Effect of copper on *Methylomonas albus* BG8. *Appl. Environ. Microbiol.* **57**, 1261–1264.
[31] Hanson, R.S. (1980) Ecology and diversity of methylotrophic bacteria. *Adv. Appl. Microbiol.* **26**, 3–39.
[32] Page, H.M., Fisher, C.R. and Childress, J.J. (1990) Role of filter-feeding in the nutritional biology of a deep-sea mussel with methanotrophic symbionts. *Mar. Biol.* **104**, 251–257.
[33] Thompson, G., Humphries, S.E., Schroeder, B. and Sulanowska, M. (1988) Active vents and massive sulfides at 26°N (TAG) and 23°N (Snakepit) on the Mid-Atlantic Ridge. *Can. Mineral.* **26**, 697–711.
[34] Welhan, J.A. and Craig, H. (1983) Methane, hydrogen and helium in hydrothermal fluids at 21°N on the East Pacific Rise. In: Hydrothermal Processes at Seafloor Spreading Centers (P. Rona *et al.*, Eds), pp. 391–409. Plenum Press, New York.
[35] Lilley, M.D., Baross, J.A. and Gordon, L.I. (1983) Reduced gases and bacteria in hydrothermal fluids: The Galapagos spreading center and 21°N East Pacific Rise. In: Hydrothermal Processes at Seafloor Spreading Centers (P. Rona *et al.*, Eds), pp. 411–449. Plenum Press, New York.
[36] Kulm, L.D. *et al.* (1986) Oregon subduction zone: venting, fauna, and carbonates. *Science* **231**, 561–566.
[37] Beauchamp, B., Krouse, H.R., Harrison, J.C., Nassichuk, W.W. and Eluik, L.S. (1989) Cretaceous cold-seep communities and methane-derived carbonates in the Canadian Arctic. *Science* **244**, 53–56.

26

L-Phenylalanine Synthesis by the Facultative RuMP Cycle Methylotroph *Amycolatopsis methanolica*

L. Dijkhuizen, G.J.W. Euverink, G.I. Hessels and J.W. Vrijbloed

Department of Microbiology, University of Groningen, Kerklaan 30, 9751 NN Haren, The Netherlands

1 SUMMARY

Studies of the aromatic amino acid biosynthetic pathway in various microorganisms have revealed a complex pattern of organization and regulation, involving isoenzymes, enzyme complexes, bi- or even penta-functional proteins, and feedback control of enzymes at the activity and/or synthesis level [1]. Currently we are engaged in a detailed physiological, biochemical and genetic analysis of this pathway in the actinomycete *Amycolatopsis methanolica*. This Gram-positive bacterium is a very versatile methanol-utilizing organism, employing the ribulose monophosphate (RuMP) cycle of formaldehyde fixation (fructose bisphosphate aldolase cleavage variant). The results of our studies with *A. methanolica* show that the organism is readily amenable to the extensive physiological and genetic manipulations required for the construction of amino acid overproducing strains [2].

2 INTRODUCTION

A unique feature of methylotrophic micro-organisms is their ability to synthesize a C_3-metabolite from C_1-units. Once such a compound has been produced, the synthesis of all cell constituents (polysaccharides, proteins, nucleic acids and lipids) further proceeds *via* the general pathways of intermediary metabolism, as found in other organisms. Methanol is a commodity chemical of high purity. It is an attractive feedstock for fermentation processes which aim to produce bulk chemicals. There is also a strong interest in applying the unique metabolic pathways in methylotrophs to open up new, or alternative ways for the production of fine chemicals such as amino acids [3,4].

Methanol-utilizing bacteria employing the fructose bisphosphate aldolase cleavage variant of the ribulose monophosphate (RuMP) cycle of formaldehyde assimilation are potential vehicles for fermentative overproduction of aromatic amino acids [5–7]. In these organisms the precursors for the shikimate pathway, erythrose-4-phosphate (E4P) and phosphoenolpyruvate (PEP), are intermediate and end product, respectively of the RuMP pathway (Fig. 1). Transketolase activity in wild type *Escherichia coli* is relatively low and clearly limits the supply of E4P for biosynthesis of aromatic amino acids in overproducing strains [8]. The pentose phosphate pathway enzymes (transaldolase, transketolase, pentose phosphate epimerase and pentose phosphate isomerase) also function in the RuMP cycle (rearrangement phase enzymes) and are generally present at very high activity levels in these methylotrophic bacteria. Mutant studies

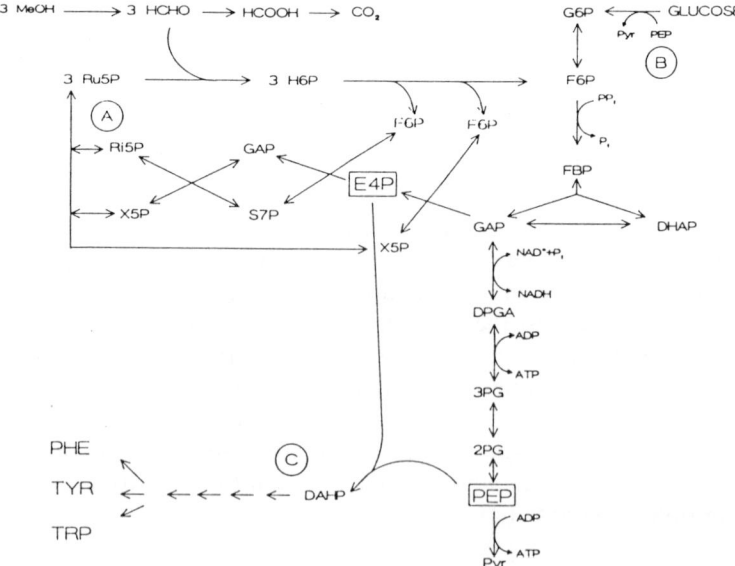

Figure 1 Schematic representation of methanol (A) and glucose (B) metabolism, and synthesis of aromatic amino acids (C) in RuMP cycle methylotrophic bacteria.

with the Gram-positive bacterium *Arthrobacter* P1 have shown that a second, C_1-inducible transaldolase isoenzyme contributes to this high activity. It appears likely that a complete set of pentose phosphate pathway isoenzymes becomes induced during growth of this organism on C_1-compounds [9]. The Gram-negative RuMP cycle bacteria investigated, however, are obligate methylotrophs and not amenable to the extensive physiological and genetic manipulations required for strain development [10]. The isolation of auxotrophic mutants of these organisms has met with considerable difficulty and their sensitivity to growth inhibition by amino acid analogs is low. Facultative methylotrophs can be found abundantly among organisms employing the Calvin cycle, the serine pathway, or the xylulose monophosphate (XuMP) cycle for the assimilation of C_1-compounds. Only in recent years, however, have we succeeded in identifying and characterizing a number of versatile RuMP cycle bacteria. These facultative RuMP cycle methylotrophs are found almost exclusively among Gram-positive bacteria. Representatives are various bacilli, coryneform bacteria, and actinomycete species [11,12]. In the following sections the possible use of one of these organisms, *Amycolatopsis methanolica*, for overproduction of aromatic amino acids (e.g. L-phenylalanine) is reviewed.

3 ORGANISM

The actinomycete *Amycolatopsis methanolica* was initially labeled *Streptomyces* sp. 239 [13], then *Nocardia* sp. 239 [14]. Recent developments in the taxonomy of actinomycetes allowed its proper identification [15]. Chemosystematic studies showed that the organism has a wall chemotype IV (meso-diaminopimelic acid, arabinose and galactose present). Unlike representatives of the genus *Nocardia*, cell walls of the organism are devoid of mycolic acids. Further chemotaxonomic and morphological data, and a comparison of 16S rRNA

Table 1 Properties of wild type *Amycolatopsis methanolica*

+	Utilization of methanol as a substrate
+	Carbon assimilation via the RuMP cycle
+	FBP aldolase cleavage variant of the RuMP cycle
+	Facultative methylotroph
+	Growth in simple mineral salts media
+	No special growth conditions required
+	Existing fermentation experience with actinomycetes
+	Sensitivity to a variety of amino acid analogs
+	Methods for the isolation of genetically stable mutants
−	No genetic system available
−	Phenylalanine/phenylpyruvate/quinate catabolism
−	Low growth rate on methanol
−	Glucose catabolite repression of methanol assimilation

sequences, identified the organism as a member of the genus *Amycolatopsis*. On the basis of a variety of biochemical and microbiological tests it was concluded that the organism forms the nucleus of a new species, *Amycolatopsis methanolica* [15]. At the outset of our studies the organism clearly possessed some undesirable properties and we were faced with a general lack of methods for its physiological and genetic manipulation (Table 1). Most of these problems have been solved meanwhile and it is now clear that, at least in principle, *A. methanolica* is a suitable strain for the development of a process for fermentative production of aromatic amino acids. The main emphasis at the moment is on a detailed characterization of the pathways and enzymes involved in phenylalanine biosynthesis (and degradation), and methanol utilization, and the development of further genetic techniques for *A. methanolica*.

4 DEGRADATIVE ENZYMES

Amycolatopsis methanolica is a very versatile bacterium, able to grow on phenylalanine, tyrosine, and quinate. The presence of these catabolic pathways may seriously interfere with attempts to overproduce aromatic amino acids *via* the shikimate pathway. The catabolism of the amino acids proceeds *via* (4-hydroxy)phenyl-pyruvate and (4-hydroxy)phenylacetate and the pathways merge at the level of homogentisate. Deamination of phenylalanine and tyrosine is catalyzed by an inducible NAD-dependent phenylalanine dehydrogenase, and tyrosine aminotransferase, respectively [16,17]. Following diepoxyoctane treatment, mutants blocked in either phenylalanine dehydrogenase or phenylpyruvate decarboxylase were isolated. These mutants, however, were still able to grow on phenylalanine, albeit at strongly reduced growth rates. Only double mutants blocked in both enzymes completely failed to catabolize phenylalanine [16]. Quinate is initially converted into dehydroquinate (DHQ) and dehydroshikimate, intermediates of the shikimate pathway, and further catabolized via protocatechuate and the β-ketoadipate pathway. Enzyme studies and mutant analysis clearly showed that only a single DHQ dehydratase, functioning both in quinate catabolism and aromatic amino acid biosynthesis, is present in *A. methanolica* [18].

5 BIOSYNTHETIC ENZYMES

Using various procedures, a set of more than 200 aromatic amino acid auxotrophic mutants has been isolated. Characterization of these mutants revealed that synthesis of L-phenylalanine and L-tyrosine proceeds *via* single pathways, involving phenylpyruvate and L-arogenate as intermediates, respectively. DHQ dehydratase mutants were also invariably blocked in DHQ synthase, suggesting common control elements or gene clustering [18]. No mutants were obtained in 3-deoxy-D-arabinoheptulosonate 7-phosphate

(DAHP) synthase, chorismate mutase and aromatic amino acid aminotransferase (AAAT), suggesting the presence of isoenzymes, as has been reported for various other organisms. Enzyme purification studies confirmed the presence of several AAAT enzymes; the most dominant of these AAAT enzymes was active with both prephenate and phenylpyruvate. Only a single DAHP synthase enzyme species could be detected in wild type *A. methanolica*. This enzyme is sensitive to cumulative feedback inhibition by all three aromatic amino acids. Partially purified enzyme showed apparent K_i values of 3, 160 and 180 μM for L-tryptophan, L-phenylalanine and L-tyrosine, respectively. The aromatic amino acids displayed competitive inhibition with respect to E4P. L-Tryptophan and E4P showed uncompetitive and competitive inhibition towards PEP, with apparent K_i values of 11 and 530 μM, respectively. Chorismate mutase functions in L-phenylalanine and L-tyrosine biosynthesis. The activity of the single chorismate mutase detectable in extracts of the wild type organism was inhibited by both L-phenylalanine and L-tyrosine (apparent K_i values of 60 and 35 μM, respectively). The activity of prephenate dehydratase, an enzyme specifically involved in L-phenylalanine biosynthesis, was inhibited by L-phenylalanine (apparent K_i value of 10 μM) and stimulated by L-tyrosine (activator constant of 10 μM). Anthranilate synthase, the first enzyme in the L-tryptophan specific branch, was strongly inhibited by L-tryptophan (apparent K_i value of 5 μM). Addition of the aromatic amino acids, either separately or in combinations, did not result in significant repression of the synthesis of these enzymes [19].

Amycolatopsis methanolica is very sensitive to inhibition of growth by various amino acid analogs. o-Fluoro- and p-fluorophenylalanine also inhibited the activities of chorismate mutase and prephenate dehydratase *in vitro* [20]. Efficient methods for the isolation of analog-resistant mutants of *A. methanolica* have been developed subsequently. Many analog-resistant mutant strains had become unable to grow on L-phenylalanine as carbon source and most likely had lost phenylalanine (analog) transport systems. Several mutants were found to possess either a chorismate mutase or a prephenate dehydratase enzyme which had become completely insensitive to L-phenylalanine (analog) inhibition. Some prephenate dehydratase mutants were still activated by tyrosine, while others had become insensitive to both phenylalanine and tyrosine.

6 METHANOL AND GLUCOSE METABOLISM

The pathways for methanol and glucose utilization in *A. methanolica*, and their regulation, have been studied in detail [21–23]. Protein purification studies revealed that glucose metabolism is regulated at the level of a PPi-dependent phosphofructokinase, phosphoglycerate mutase and pyruvate kinase. No evidence was obtained for the involvement of a PEP-dependent sugar phosphotransferase system for glucose uptake. Mixed substrate experiments in batch cultures with glucose plus methanol resulted in simultaneous utilization of these substrates. The presence of glucose repressed synthesis of the RuMP cycle enzymes hexulose phosphate synthase (HPS) and hexulose phosphate isomerase (HPI), and methanol was only utilized as an energy source. Similar results were found following addition of formaldehyde (fed-batch system) to a culture growing on glucose. The synthesis of enzymes involved in methanol dissimilation and assimilation in *A. methanolica* thus appears to be regulated differently. Methanol and/or formaldehyde induce the synthesis of these enzymes, but under carbon-excess conditions their inducing effect on HPS and HPI synthesis is overruled completely by glucose. Repression of HPS and HPI was of minor significance following addition of methanol to glucose-, acetate- and ethanol-limited chemostat cultures. The strong repressive effect of glucose on methanol metabolism makes it unlikely that the presence of methanol as an additional substrate will enhance the intracellular availability of E4P and PEP, precursors for aromatic amino acid biosynthesis. Nevertheless, utilization of methanol as an energy source may still allow a further increase in the flow of glucose-carbon towards biosynthetic rather than energy-generating processes. An alternative approach would be the isolation of mutants that have lost glucose repression and constitutively express HPS and HPI.

7 GENETICS

A. methanolica was found to possess a 13.3 kb plasmid (pMEA300) which is able to integrate into the chromosome at a specific site. A derivative plasmid, carrying the thiostrepton resistance gene (pMEA301), was used for the development of a whole cell transformation procedure [24] with the plasmid-deficient *A. methanolica* strain WV1. Further detailed analysis of pMEA300 has resulted in identification of plasmid fragments encoding integration, conjugation and replication functions, allowing construction of *E. coli–A. methanolica* shuttle vectors in further work. Cloning of amino acid biosynthetic genes is currently being attempted by complementation of the available *A. methanolica* mutants, using a gene bank of chromosomal DNA of *A. methanolica*.

8 CONCLUSIONS

The data obtained provide a firm basis for the rational construction of amino acid overproducing strains of *A. methanolica*, using methanol or mixtures of glucose and methanol (to enhance generation of the precursors E4P and PEP) as feedstocks. Conceivably, this necessitates the stepwise isolation of mutants blocked in phenylpyruvate catabolism and the tyrosine-specific biosynthetic pathway, combined with attempts to introduce phenylalanine feedback inhibition deregulated DAHP synthase, chorismate mutase and prephenate dehydratase mutant enzymes. Further metabolic reprogramming of *A. methanolica* will involve cloning and over-expression of shikimate pathway enzymes to increase the metabolic flux to the end product L-phenylalanine, combined with studies on substrate uptake and product efflux.

ACKNOWLEDGEMENT

These studies were supported by grant no. GBI81.1510 from the Netherlands Technology Foundation (STW) which is subsidized by the Netherlands Organization for the Advancement of Pure Research (NWO).

REFERENCES

[1] Bentley, R. (1990) The shikimate pathway - a metabolic pathway with many branches. *Crit. Rev. Biochem. Mol. Biol.* **25**, 307–384.
[2] de Boer, L. (1990) L-phenylalanine and methanol metabolism in the facultative methylotroph *Amycolatopsis methanolica (Nocardia* sp. 239). Ph.D. Thesis, University of Groningen.
[3] Dijkhuizen, L., Hansen, T.A. and Harder, W. (1985) Methanol, a potential feedstock for biotechnological processes. *Trends in Biotechnol.* **3**, 262–267.
[4] Leak, D.J. (1992) Biotechnological and applied aspects of methane and methanol utilizers. In: Methane and Methanol Utilizers (Murrell, J.C. and Dalton, H., Eds), pp. 245–279, Plenum Press, New York.
[5] Morinaga, Y. and Hirose Y. (1984) Production of metabolites by methylotrophs. In: Methylotrophs: Microbiology, Biochemistry, and Genetics (Hou, C.T., Ed.), pp. 107–118, CRC Press, Boca Raton, Florida.
[6] Minoda, Y. (1986) Raw materials for amino acid fermentation. In: Biotechnology of Amino Acid Production. Progr. Industrial Microbiol. (Aida, K., Chibata, I., Nakayama, K., Takinami, K. and Yamada, H., Eds), Vol. 24, pp. 51-66. Kodansha Ltd., Tokyo; Elsevier, Amsterdam.
[7] Suzuki, M., Berglund, A., Unden, Å. and Heden, C.G. (1977) Aromatic amino acids production by analogue-resistant mutants of *Methylomonas methanolophila* 6R. *J. Ferment. Technol.* **55**, 466–475.
[8] Draths, K.M., Pompliano, D.L., Conley, D.L., Frost, J.W., Berry, A., Disbrow, G.L., Staversky, R.J. and Lievense, J.C (1992) Biocatalytic synthesis of aromatics from D-glucose: The role of transketolase. *J. Am. Chem. Soc.* **114**, 3956–3962.
[9] Levering, P.R. and Dijkhuizen, L. (1986) Regulation and function of transaldolase isoenzymes involved in sugar and one-carbon metabolism in the ribulose monophosphate cycle methylotroph *Arthrobacter* P1. *Arch. Microbiol.* **144**, 116–123.
[10] Holloway, B.W. (1984) Genetic techniques for methylotrophs. In: Microbial Growth on C_1 Compounds (Crawford, R.L. and Hanson, R.S., Eds), pp. 215–220, American Society for Microbiology, Washington DC.
[11] Dijkhuizen, L. and Arfman, N. (1990) Methanol metabolism in thermotolerant methylotrophic *Bacillus* species. *FEMS Microbiol. Rev.* **87**, 215–220.
[12] Dijkhuizen, L., Levering, P.R. and de Vries, G.E. (1992) The physiology and biochemistry of aerobic methanol-utilizing Gram-negative and Gram-positive bacteria. In: Methane and Methanol Utilizers (Murrell, J.C. and Dalton, H., Eds), pp. 149–181, Plenum Press, New York,
[13] Kato, N., Tsuji, K., Tani, Y. and Ogata, K. (1974) A methanol-utilizing *Actinomycete. J. Ferment. Technol.* **52**, 917–920.
[14] Hazeu, W., de Bruyn, J.C. and van Dijken, J.P. (1983) *Nocardia* sp. 239, a facultative methanol utilizer with the ribulose monophosphate pathway of formaldehyde fixation. *Arch. Microbiol.* **135**, 205–210.
[15] de Boer, L., Dijkhuizen, L., Grobben, G., Goodfellow, M., Stackebrandt, E., Parlett, J.H., Whitehead, D. and Witt, D. (1990) *Amycolatopsis methanolica* sp. nov., a facultatively methylotrophic actinomycete. *Int. J. Syst. Bacteriol.* **40**, 194–204.

[16] de Boer, L., Harder, W. and Dijkhuizen, L. (1988) Phenylalanine and tyrosine metabolism in the facultative methylotroph *Nocardia* sp. 239. *Arch. Microbiol.* **149**, 459–465.
[17] de Boer, L., van Rijssel, M., Euverink, G.J. and Dijkhuizen, L. (1989) Purification, characterization and regulation of a monomeric L-phenylalanine dehydrogenase from the facultative methylotroph *Nocardia* sp. 239. *Arch. Microbiol.* **153**, 12–18.
[18] Euverink, G.J.W., Hessels, G.I., Vrijbloed, J.W., Coggins, J.R. and Dijkhuizen, L. (1992) Purification and characterisation of a dual function 3-dehydroquinate dehydratase from *Amycolatopsis methanolica*. *J. Gen. Microbiol.* **138**, 2449–2457.
[19] de Boer, L., Vrijbloed, J.W., Grobben, G. and Dijkhuizen, L. (1989) Regulation of aromatic amino acid biosynthesis in the ribulose monophosphate cycle methylotroph *Nocardia* sp. 239. *Arch. Microbiol.* **151**, 319–325.
[20] de Boer, L., Grobben, G., Vrijbloed, J.W. and Dijkhuizen, L. (1990) Biosynthesis of aromatic amino acids in *Nocardia* sp. 239: effects of amino acid analogues on growth and regulatory enzymes. *Appl. Microbiol. Biotechnol.* **33**, 183–189.
[21] de Boer, L., Euverink, G.J., van der Vlag, J. and Dijkhuizen, L. (1990) Regulation of methanol metabolism in the facultative methylotroph *Nocardia* sp. 239 during growth on mixed substrates in batch- and continuous cultures. *Arch. Microbiol.* **153**, 337–343.
[22] Bystrykh, L.V., Arfman, N. and Dijkhuizen, L. (1993) The methanol oxidizing enzyme systems in Gram-positive methylotrophic bacteria. This volume.
[23] Euverink, G.J.W., van der Vlag, J., Hessels, G.I., Hondmann, D., Visser, J. and Dijkhuizen, L. (1993) Enzymes of glucose and methanol metabolism in the facultative RuMP cycle methylotroph *Amycolatopsis methanolica*. *J. Gen. Microbiol.*, in press.
[24] Madon, J. and Hütter, R. (1991) Transformation system for *Amycolatopsis (Nocardia) mediterranei:* Direct transformation of mycelium with plasmid DNA. *J. Bacteriol.* **173**, 6325–6331.

27

Application of Methanotrophic Oxidations for the Bioremediation of Chlorinated Organics

Lisa Alvarez-Cohen

Department of Civil Engineering, 726 Davis Hall, University of California, Berkeley, CA 94720, USA

1 SUMMARY

The use of bioremediation for the destruction of hazardous organics is emerging as a promising and economical treatment alternative. A group of the most common hazardous organics, chlorinated solvents, can be degraded cometabolically by methane oxidizing bacteria. While the application of bioremediation for the degradation of organics which can be used by micro-organisms as a primary substrate is becoming more widespread, bioremediation based upon cometabolic transformations is still relatively undeveloped.

The application of methane oxidizers for bioremediation presents a range of challenges unique to the cometabolic process. Significant among these are the difficulties and kinetic limitations associated with the administration of methane, which is necessary to grow and maintain the cells and is only slightly soluble in water. Additionally, significant kinetic limitations are imposed by the toxicity of the cometabolic transformation products, the presence of additional substrates which compete for the active enzyme site and the generation of required reducing equivalents.

A model which incorporates mass transfer, cell growth, product toxicity, and competitive inhibition is used along with experimentally derived parameters to predict the transformation potential of methanotrophic reactor systems. Modelling results indicate that overall bioremediation efficiency can be effectively improved by maximizing the mass transfer rate of methane and oxygen to the cells thereby increasing the methanotrophic growth rate, and that product toxicity and competitive inhibition significantly effect degradation efficiency.

2 INTRODUCTION

Bioremediation has been used for the destruction of organic contaminants in both groundwater and surface water applications. Unlike the most commonly applied physical treatment processes which merely transfer contaminant from the water phase to the solid or gas phase, bioremediation has the potential for resulting in total contaminant destruction by transformation to harmless end products. To date, the most common application of bioremediation has been for removal of hydrocarbon contamination. This is logical since most hydrocarbons are easily biodegraded and can serve as a primary carbon and energy source for microbial growth. However, for compounds which cannot serve as primary substrates for micro-organisms, such as many of the halogenated aliphatics, biological remediation has not been widely applied [1]. The concept of applying methanotrophic cometabolism for bioremediation of halogenated aliphatics was proposed by Wilson and Wilson [2], after their work with soil columns suggested that an enriched methanotrophic population was most likely responsible for the observed trichloroethylene (TCE) degradation. This conclusion was later supported by studies conducted with pure methanotrophic cultures [3,4] and expanded to include degradation of a broad range of halogenated aliphatics by both mixed [5-7] and pure [8-10] methane oxidizing cultures. Additionally, studies conducted with the purified enzyme responsible for methane oxidation have conclusively shown that methane monooxygenase (MMO) is also responsible for the oxidation of a range of chlorinated aliphatics [11,12].

3 ENGINEERING CHALLENGES

Typical reactor configurations for bioremediation are designed to grow the microbial population on the contaminant of interest, accomplishing contaminant removal and microbial growth in one step. The use of an alternate substrate to grow and maintain a microbial population which will subsequently remove the contaminant of interest has not yet been commonly applied in bioremediation. Specific contaminants of interest for application of methanotrophic bioremediation include the most common halogenated groundwater contaminants: trichloroethylene (TCE), 1,1,1 trichloroethane (TCA), chloroform (CF), dichloroethylene (DCE) isomers, vinyl chloride (VC), and other halogenated aliphatics. Although the fully chlorinated aliphatics (tetrachloroethylene, carbon tetrachloride) are also common groundwater contaminants, they have not proved to be subject to methanotrophic oxidation [5,8,9,11], either due to steric interference by the chlorine atoms or due to the oxidized nature of the highly chlorinated molecules.

Questions regarding cometabolic degradation kinetics and reaction optimization techniques still exist. Since methanotrophic cell growth is tied to consumption of an alternate substrate (methane), while contaminant degradation is tied to successful enzyme production, the dual goals of optimizing contaminant removal and optimizing cell growth may not be mutually compatible. Additionally, applicability of methanotrophs for use in bioremediation processes will be limited by several important factors: product toxicity associated with halogenated organics, competitive inhibition between methane and the cometabolites as well as among mixtures of cometabolites, and low aqueous solubility of methane and oxygen. Each of these limitations will be discussed in detail below.

3.1 Product toxicity

There is a growing body of evidence that although moderate concentrations (μM) of halogenated organics do not exhibit substrate toxicity to methanotrophs, their transformation products can be extremely toxic [10,13–15]. Experiments conducted with purified soluble methane monooxygenase (sMMO) have shown enzyme inactivation resulting from TCE and DCE oxidations [11,12]. Further, whole cell studies have shown that in addition to specific sMMO inactivation, a general cellular inactivation results from TCE oxidation. Experiments using whole cells of *Methylosinus trichosporium* OB3b showed nonspecific binding of the TCE transformation products to cellular proteins [10]. Studies conducted with mixed consortia of methane oxidizers showed that diminished formate mineralization by cells which had transformed TCE or CF compared to cells which had been exposed to either compound in the absence of transformation indicated that TCE and CF themselves were not toxic, but that their transformation products caused overall cellular inactivation [14]. Additional batch studies [8,9,15], and reactor studies [16–18], have shown evidence for product toxicity associated with a range of chlorinated aliphatics. Methanotrophic cell activity was found to decrease as a linear function of TCE and CF oxidation [10,14]. Two studies which have quantified this cell inactivation have reported transformation capacities of 0.27 mg TCE/mg cells for a pure culture of *Methylosinus trichosporium* OB3b at 30°C in the presence of excess formate [10] and 0.08 mg TCE/mg cell and 0.015 mg CF/mg cell for a mixed culture at 21°C with formate added and 0.04 mg TCE/mg cell and 0.008 mg CF/mg cell without formate [14,19].

3.2 Competition for the active sites

Since the methanotrophic degradation of chlorinated organics is catalyzed by the same enzyme as the primary growth and energy reaction, it is reasonable to expect the growth and degradation reaction kinetics to be governed by competitive inhibition. That is, the molecules of methane and chlorinated organics will compete for the active site on the MMO causing the rates of both reactions to be inhibited. The simultaneous presence of multiple chlorinated organics would also be expected to cause mutual inhibition. Reports by researchers using pure cultures of *Methylosinus trichosporium* OB3b support this conclusion by showing competitive inhibition between methane and TCE [4,8,10]. Additional studies with mixed methane oxidizing cultures showed methane competitive inhibition with TCE, and 1,1,1 TCA [20–23].

Further, when more than one chlorinated organic is present, competition for the active enzyme site would be expected to result in competitive inhibition kinetics for degradation of each compound. Palumbo et al. [24] reported significantly decreased rates of TCE transformation due to competitive inhibition by 1,1 DCE, cis- and trans-1,2 DCE, and tetrachloroethylene (PCE). Competitive inhibition between TCE and 1,1,1 TCA [21], and between TCE and CF have also been reported [14].

3.3 Reducing equivalents

Methanotrophic oxidations catalyzed by MMO require NADH as a cosubstrate to supply necessary reducing equivalents [25]. Although NADH is consumed by the preliminary step in the oxidation of methane, it is subsequently regenerated as the reaction proceeds. Methanotrophs are also capable of regenerating NADH in the absence of methane from endogenous energy reserves and from the mineralization of formate [26]. Since MMO-catalyzed cometabolic oxidations consume NADH without regenerating it, reactions occurring in the absence of primary substrate may require an exogenous supply of reducing equivalents. Formate, whose mineralization is not catalyzed by MMO, can serve as a noncompetitive source of reducing equivalents [27,28]. Several researchers have shown significantly increased TCE degradation rates with formate addition for both pure and mixed methane oxidizing cultures [4,8,13–15,24]. Oldenhuis et al. [8] reported that 1 mM formate was required for complete degradation of 0.2mM TCE by *Methylosinus trichosporium* OB3b cells in the absence of methane. In reactor studies, researchers using methane-fed trickle-filters showed that replacing the methane with formate induced an increased TCE removal rate which was significant but short lived [29]. Short-term formate addition to the second stage of a two-stage recycle reactor also caused increased TCE removal [18]. Although reactor configurations have been proposed to take advantage of the formate induced accelerated bioremediation [4,8,30], it is important to recognize that formate is degraded by many heterotrophic bacteria and therefore could result in the methanotrophs being out-competed by other organisms.

3.4 Mass transfer limitations

Since both the electron donor and electron acceptor (methane and oxygen) are gaseous at ambient temperature, they must be dissolved into aqueous phase in

Table 1 Physical properties of compounds at 20°C

	Solubility in H_2O[a] (mg/l)	Henry's constant[b] (−)	Saturated gas phase concentration[c] (mg/l)
Methane	24	28	670
Oxygen	43	30	1,350
TCE	1,100	0.30	415
CF	8,000	0.11	1,027
VC	2,700	0.90	8,520
1,1 DCE	400	0.86	2,640
c1,2 DCE	800	0.12	864
1,1,1 TCA	4,400	0.54	716

[a] Adapted from [31,32,33].
[b] Computed from [34,35].
[c] Computed from vapor pressures given in [32,33].

order to become available to cells. Both methane and oxygen are only slightly soluble in water at 20°C as shown in Table 1. Additionally, the characteristically high Henry's coefficient (a dimensionless constant representing the ratio of a compound's equilibrium gas phase and liquid phase concentrations) of both compounds indicates that they preferentially partition into the gas phase. The stoichiometry of methane oxidation and cell production indicates that depending on the fraction of methane utilized for cell synthesis, each mole of methane oxidized will require 1–2 moles of oxygen. Consequently, an appropriate methanotrophic feed mixture would contain 33–50% methane and pure oxygen or 9–17% methane and air. Since solubilities are directly related to gas partial pressures, this would result in maximum aqueous methane concentrations of 7.9–12 mg/l with pure oxygen or 2.2–4.1 mg/l with air.

In order to determine the microbial growth limitations imposed by this solubility range, let us consider the steady-state operation of a suspended growth methanotrophic bioreactor where cell growth is limited by the supply of methane. The transfer rate of compounds from the headspace to the liquid phase of the reactor can be expressed by the following [36]:

$$r_{Li} = [k_L a]_i \left(\frac{S_{hi}}{H_i} - S_i \right) \qquad (1)$$

where:

r_{Li} = rate of compound i transfer from gas to liquid (mg/l/d)
$[k_L a]_i$ = gas/liquid mass transfer coefficient for compound i (d^{-1})
S_i = steady-state aqueous concentration of i (mg/l)
S_{hi} = concentration of compound i in the headspace (mg/l)
H_i = Henry's law coefficient for compound i (−)

The driving force for the mass transfer is the concentration gradient between the gas and liquid phases. This force is an inverse function of the Henry's coefficient, indicating that the mass transfer rate of methane and oxygen into solution would be much slower than that of the other cometabolic substrates listed in Table 1. The mass transfer coefficient is a function of the liquid and gas diffusion properties of the compound as well as the physical properties of the reactor and mixing system [36]. For a reactor equipped with typical commercial-size mechanical aerators, $[k_L a]$ for oxygen into water ranges from 1–20 h^{-1}, although higher transfer rates are achievable at the bench scale [37]. Since the diffusion properties of oxygen and methane are similar [34], the mass transfer coefficients for both compounds will fall within the same range. Therefore, the optimistic value of 20 h^{-1} is used for the mass transfer of methane in the following calculations.

The maximum rate of methane transfer can be computed using Eqn. 1 and the concentrations for methane/air and methane/oxygen mixtures given above. As a result of the methane limitation, the maximum rate of cell growth within a suspended reactor can be computed by multiplying the maximum methane transfer rate by the cell yield. Using a cell yield of 0.81 g/g, the average of 14 reported methanotrophic cell yields [38], and assuming a highly agitated commercial reactor ($k_L a = 20$ h^{-1}), the maximum rate of methanotrophic cell growth in a suspended growth bioreactor is estimated at 3100–4700 mg/l/d for a methane/oxygen feed mixture and 860–1600 mg/l/d for a methane/air mixture. Although these rates have been computed for a full-scale suspended growth reactor operated under the most favorable circumstances and neglecting cell decay, the resulting required detention times to build up an appropriate cell concentration for organics removal would be long for full scale application, on the order of one day or greater.

Reactor configurations which increase the cell detention time using high recycle or by providing a support matrix for biofilm accumulation could possibly be used to mitigate the slow methane mass transfer. However, for degradation of chlorinated organics, these configurations would result in the accumulation of cells inactivated by the product toxicity described above. Additionally, the lower Henry's coefficients and higher solubilities characteristic of the chlorinated aliphatics listed in Table 1 suggest that delivery of the cometabolites to the cells would not be subject to similar mass transfer limitations, and hence would cause a diminished methane uptake due to competitive inhibition. It has therefore been proposed by several researchers that the growth reaction and the degradation reaction be carried out in separate vessels, avoiding competitive inhibition and allowing each reaction to be optimized individually [8,18,21,30].

An additional option for significantly increasing the methane and oxygen uptake rates is to promote substrate uptake more directly from the gas phase. The saturated gas phase concentrations of both methane and oxygen are 25 times greater than the aqueous saturation concentrations (Table 1). Therefore, cells grown in a gas-phase biofilm reactor where substrates become available after diffusing through only a thin liquid film may be much more efficient for methanotrophic growth. Speitel and McLay [16] reported limited success using a gas-phase bioreactor for the removal of TCE and DCE from a gas stream, but were unable to sustain long-term activity presumably because of product toxicity. Strandberg et al. [39] reported success growing and maintaining a mixed consortium of methane oxidizers on solid media in a partially saturated reactor with significant removal of low concentrations of TCE and trans-1,2 DCE from a liquid stream, and Strand et al. [40] estimated that a gas-sparged methanotrophic biofilm reactor could achieve 99% reduction of μM levels of TCE and TCA with a 12 hour detention time.

4 KINETIC MODELING

In order to design reactor configurations to effectively use methanotrophic cometabolism for the degradation of organics, we must be capable of modeling the associated cometabolic reaction kinetics. The microbial degradation rate of organics can typically be modeled using the following expression of Monod kinetics [36]:

$$r_i = \frac{k_i S_i X}{K_{si} + S_i} \quad (2)$$

where:

r_i = degradation rate of compound i (mg/l)
S_i = concentration of compound i (mg/l)
k_i = maximum degradation rate of compound i (mg i/mg cells-d)
K_{si} = half velocity constant for i (mg/l)
X = active microbial concentration (mg/l).

However, the characteristics of methanotrophic oxidations outlined above require a modified kinetic model. Specifically, the model must reflect the cometabolic nature of the oxidation reactions, the product toxicity resulting from the chlorinated organic oxidations, and the competitive inhibition resulting from the broad specificity of MMO.

Cometabolic oxidation reactions do not support cell growth, hence, the active microbial concentration will not increase due to the degradation reactions. Conversely, the toxicity of the chlorinated oxidation products may be assumed to cause cell inactivation in proportion to the oxidation reactions [10,14], resulting in the following expression for the concentration of active cells [19]:

$$X = X_0 - \frac{1}{T_{ci}} (S_{oi} - S_i) \quad (3)$$

which can be directly substituted into the Monod equation to yield an expression for the rate of cometabolic oxidations in the presence of product toxicity:

$$r_i = \frac{k_i S_i \left[X_o - \frac{1}{T_{ci}}(S_{oi} - S_i) \right]}{K_{si} + S_i} \quad (4)$$

where:

- S_i = concentration of cometabolized contaminant i at time t (mg/l)
- S_{oi} = initial concentration of cometabolized contaminant i (mg/l)
- X_o = initial active microbial concentration (mg/l)
- T_{ci} = transformation capacity for cometabolized contaminant i (mg i/mg cells).

This model which incorporates the effects of product toxicity into modified Monod kinetics has been shown to adequately describe TCE and CF oxidations by a mixed culture of methane oxidizers [14,19].

For describing the MMO catalyzed oxidation of mixtures of competitive substrates, the following kinetic expression which is based upon competitive enzyme kinetics [36] can be used:

$$r_i = \frac{k_i S_i X}{K_{si}\left(1 + \frac{S_j}{K_{sj}}\right) + S_i} \quad (5)$$

where subscript j refers to the competitive inhibitor and may be extended to include multiple inhibiting compounds. This competitive inhibition equation has been successfully applied by Broholm et al. [23] to describe TCE and methane competitive inhibition kinetics for a mixed methane oxidizing culture, and by Strand et al. [21] for methane inhibited degradation of 1,1,1 TCA.

Finally, the concurrent oxidation of a mixture of competitively inhibiting compounds, each of which exerts specific product toxicity, can be effectively modeled by combining the competitive inhibition and product toxicity modifications of Monod kinetics and assuming that each effect is cumulative:

$$r_i = \frac{k_i S_i \left[X_o - \frac{1}{T_{ci}}(S_{oi} - S_i) - \sum_j \frac{1}{T_{cj}}(S_{oj} - S_j) \right]}{K_{si}\left(1 - \sum_j \frac{S_j}{K_{sj}}\right) + S_i} \quad (6)$$

This expression has been successfully applied to predict the concurrent oxidation rates of TCE and CF in batch studies using a mixed methane oxidizing consortium [14].

4.1 Modeling results

In order to illustrate the relative influences of product toxicity and competitive inhibition on degradation rates and overall reactor performances, chlorinated organic removals estimated using typical Monod kinetics are compared with removals estimated using the above described modifications of the model. The degradation kinetics are modeled for the above proposed reactor design in which the cells are grown in a separate reactor before being mixed with the waste stream. The bioremediation reaction occurs as the waste stream and methanotrophic cells flow into a plug flow reactor. This scenario minimizes the confounding influences of methane inhibition and cell growth while allowing direct examination of the effects of cosubstrate inhibition and product toxicity.

The degradation rate constants used in the modeling calculations (Table 2) were derived from literature values for a mixed methane oxidizing culture maintained in a mechanically aerated suspended bioreactor [14]. The cell density (1600 mg/l) was derived from the above given calculations for a suspended growth bioreactor, fed a methane/air mixture, with a one day detention time. The cells are added to the waste stream in a 1:10 flow ratio, resulting in immediate dilution of the waste concentrations to 90%. Equations were solved using a forward time-step analysis with a Δt of 0.05 hours.

Figure 1 shows the expected TCE disappearance within a plug flow reactor for four initial TCE concentrations. A comparison of TCE removals predicted using Monod kinetics (dashed lines, Eqn. 2) with those predicted using the modified model incorporating product toxicity (solid lines, Eqn. 4), suggests that toxicity exerts substantial effects at high initial TCE concentrations, while showing very little influence at low initial TCE concentrations. In fact, if product toxicity is ignored, conventional Monod kinetics predict that for the conditions described, >99% of the TCE will be degraded in three hours at each of the initial TCE concentrations; whereas the modified model predicts that the three hour TCE removals for the 10 mg/l and 7 mg/l initial conditions would be 62% and 80% respectively, with 99% removal unachievable even given an infinite detention time. These results suggest that at high chlorinated organic concentrations, toxicity exerts a significant influence which if neglected, would result in a substantial overestimation of removal efficiencies. However, since product toxicity inactivates cells in proportion to their transformation capacity, the relative toxic effects for each specific application would be directly related to the ratio of chlorinated organic concentration to cell concentration. That is,

Table 2 Parameters used in kinetic modeling (14)

	TCE	CF
k (d^{-1})	0.84	0.34
K_s (mg/L)	1.5	1.3
T_c (mg i/mg cells)	0.042	0.0083

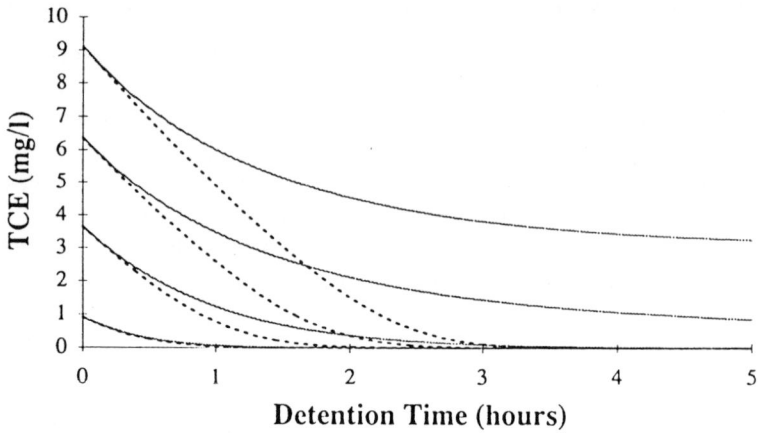

Figure 1 Modeling predictions of TCE degradation within a plug flow reactor using Monod kinetics (dashed lines) and a model incorporating product toxicity (solid lines). Reactor influent is a 1:10 mixture of methane-oxidizing cell suspension and TCE waste at four different initial concentrations.

significant product toxicity would be expected to occur when the TCE to cell ratio approaches the transformation capacity (0.042 mg/mg in this case). Therefore, in order to minimize the effects of product toxicity, it is possible to either increase the cell concentration (which may be limited by methane mass transfer), or decrease the concentration of chlorinated organics.

In order to illustrate the effects of competitive inhibition between chlorinated organics, TCE and CF removals within a plug flow reactor were computed using Eqn. 6 and are shown in Fig. 2a and b respectively. Although both plots represent removals for $20\mu M$ initial concentrations, it is apparent that the overall CF removal is much less efficient than the TCE removal. The lower transformation capacity (T_c) of CF (Table 2) indicates that CF oxidation is significantly more toxic to cells than TCE oxidation. Further, CF exerts a much greater inhibitory effect on TCE oxidation than the reverse, resulting in $>30\%$ reduction in TCE removal for an equivalent ($20\mu M$) CF dose (Fig. 2a), compared to $<14\%$ reduction in CF removal for an equivalent TCE dose (Fig. 2b). However, although product toxicity clearly dominates the kinetics of CF oxidation under these conditions, competitive inhibition significantly effects both the TCE and CF removals, and therefore should be considered when modeling degradation of contaminant mixtures.

5 CONCLUSIONS

Methane oxidizing cultures, which can be maintained in both suspended and biofilm reactors, are capable of effectively degrading a wide range of haloge-

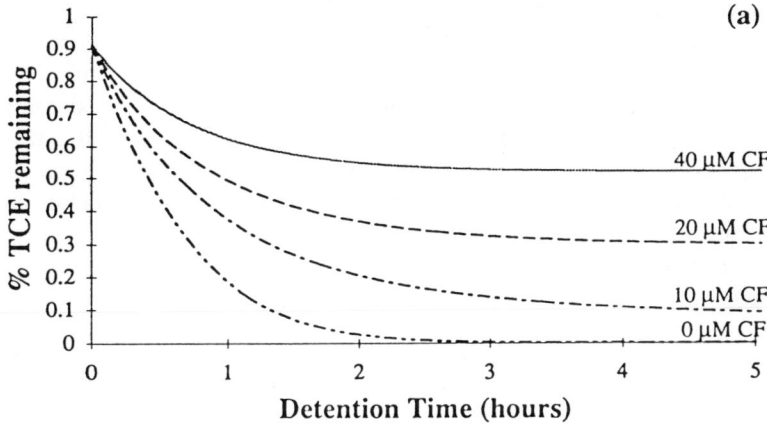

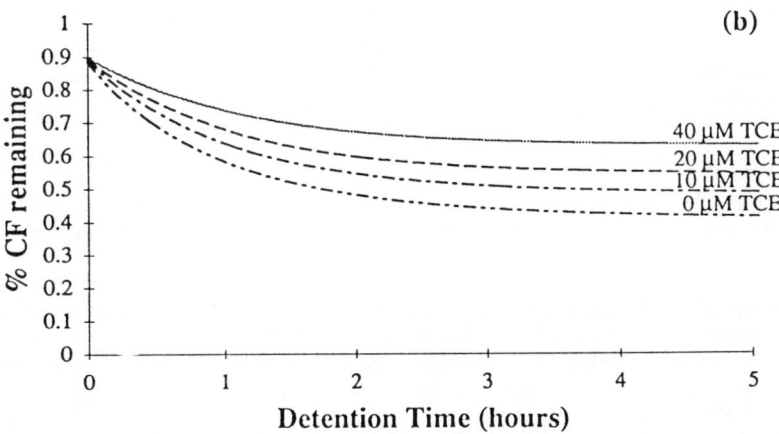

Figure 2 Modeling predictions of TCE (a) and CF (b) degradation within a plug flow reactor using a model incorporating product toxicity and competitive inhibition. Reactor influent is a 1:10 mixture of methane-oxidizing cell suspension and 20μM TCE (a) or CF (b) waste with varying concentrations of competitive inhibitor.

nated organics, indicating their suitability for bioremediation applications. However, the application of methane oxidizers for bioremediation presents a range of unique challenges. Significant among those are the kinetic limitations associated with the toxicity of the chlorinated organic oxidation products, competition for the active sites on the non-specific enzyme, generation of required reducing equivalents, and delivery of the primary substrate to the cells. Kinetic modeling of substrate mass transfer and cell growth illustrates the limitations imposed by low methane solubility. Additionally, a model which incorporates product toxicity and competitive inhibition into degradation kinetics is used along with experimentally derived parameters to illustrate the relative influence of these factors on predicted treatment efficiencies. Results indicate that product toxicity significantly affects degradation kinetics at high ratios of chlorinated organics to cells, and that competitive inhibition substantially influences degradation efficiency.

REFERENCES

[1] McCarty, P.L. (1988) Bioengineering issues related to in-situ remediation of contaminated soils and groundwater. In: Environmental Biotechnology: Reducing Risks from Environmental Chemicals through Biotechnology (Omenn G.S., Ed.), pp.143–162. Plenum Press, New York.

[2] Wilson, J.T. and Wilson, B.H. (1985) Biotransformation of trichloroethylene in soil. *Appl. Environ. Microbiol.* **49**, 242–243.

[3] Little, C.D., Palumbo, A.V., Herbes, S.E., Lidstrom, M.E., Tyndall, R.L. and Gilmer, P.J. (1988) Trichloroethylene biodegradation by a methane-oxidizing bacterium. *Appl. Environ. Microbiol.* **54**, 951–956.

[4] Brusseau G.A., Tsien, H.-C., Hanson, R.S. and Wackett, L.P. (1990) Optimization of trichloroethylene oxidation by methanotrophs and the use of a colorimetric assay to detect soluble methane monooxygenase activity. *Biodegradation*, **1**, 19–29.

[5] Fogel, M.M., Taddeo, A.R. and Fogel, S. (1986) Biodegradation of chlorinated ethenes by a methane-utilizing mixed culture. *Appl. Environ. Microbiol.* **51**, 720–724.

[6] Henson, J.M., Yates, M.V., Cochran, J.W. and Shackleford, D.L. (1988) Microbial removal of halogenated methanes, ethanes, and ethylenes in an aerobic soil exposed to methane. *FEMS Microb. Ecol.*, **53**, 193–201.

[7] Henson, J.M., Yates, M.V. and Cochran, J.W. (1989) Metabolism of chlorinated methanes, ethanes and ethylenes by a mixed baterial culture growing on methane. *J. Indust. Microbiol.* **4**, 29–35.

[8] Oldenhuis, R., Vink, R.L., Janssen, D.B. and Witholt, B. (1989) Degradation of chlorinated aliphatic hydrocarbons by *Methylosinus trichosporium* OB3b expressing soluble methane monooxygenase. *Appl. Environ. Microbiol.* **55**, 2819–2826.

[9] Tsien, H.-C., Brusseau, G.A., Hanson, R.S. and Wackett, L. P. (1989) Biodegradation of trichloroethylene by *Methylosinus trichosporium* OB3b. *Appl. Environ. Microbiol.* **55**, 3155–3161.

[10] Oldenhuis, R., Oedzes, J.Y., van der Waarde, J.J. and Janssen, D.B. (1991) Kinetics of chlorinated hydrocarbon degradation by *Methylosinus trichosporium* OB3b and toxicity of trichloroethylene. *Appl. Environ. Microbiol.* **57**, 7–14.

[11] Fox, B.G., Borneman, J.G., Wackett, L.P. and Lipscomb, J.D. (1990) Haloalkene oxidation by the soluble methane monooxygenase from *Methylosinus trichosporium* OB3b: mechanistic and environmental applications. *Biochemistry* **29**, 6419–6427.

[12] Green, J. and Dalton, H. (1989) Substrate specificity of soluble methane monooxygenase - Mechanistic implications. *J. Biol. Chem.* **264**, 17698–17703.

[13] Alvarez-Cohen, L. and McCarty, P.L. (1991) Effects of toxicity, aeration, and reductant supply on trichloroethylene transformation by a mixed methanotrophic culture. *Appl. Environ. Microbiol.* **57**, 228–235.
[14] Alvarez-Cohen, L. and McCarty. P.L. (1991) Product toxicity and cometabolic modeling of chloroform and trichloroethylene transformation by methanotrophic resting cells, *Appl. Environ. Microbiol.* **57**, 1031–1037.
[15] Henry, S.M. and Grbic-Galic, D. (1991) Influence of endogenous and exogenous electron donors and trichloroethylene oxidation toxicity on trichloroethylene oxidation by methanotrophic cultures from a groundwater aquifer. *Appl. Environ. Microbiol.* **57**, 236–244.
[16] Speitel, G., and McLay, D.S. (1990) Biofilm reactors for treatment of gas streams containing chlorinated solvents. In: Proceedings of Environmental Engineering 1990 Specialty Conference Arlington, VA (O'Melia, R., Ed.), pp. 366–373. American Society of Civil Engineers, New York.
[17] Phelps, T.J., Niedzielski, J.J., Malachowsky, K.J., Schram, R.M., Herbes, R.M. and White, D.C. (1991) Biodegradation of mixed-organic wastes by microbial consortia in continuous-recycle expanded-bed bioreactors. *Environ. Sci. Technol.* **25**, 1461–1465.
[18] McFarland, M.J., Vogel, C.M. and Spain, J.C. (1992) Methanotrophic cometabolism of trichloroethylene (TCE) in a two stage bioreactor system. *Water. Res.* **26**, 259–265.
[19] Alvarez-Cohen, L. and McCarty, P.L. (1991) A cometabolic biotransformation model for halogenated aliphatic compounds exhibiting product toxicity. *Environ. Sci. Technol.* **25**, 1381–1387.
[20] Lanzarone, N.A. and McCarty, P.L. (1990) Column studies on methanotrophic degradation of trichloroethylene and 1,2-dichloroethylene. *Groundwater* **28**, 910–919.
[21] Strand, S.E., Bjelland, M.D. and Stensel, H.D. (1990) Kinetics of chlorinated hydrocarbon degradation by suspended cultures of methane-oxidizing bacteria. *Res. J. Water Pollut. Control Fed.* **62**, 124–129.
[22] Leahy M.C., Findley, M. and Fogel S. (1989) Biodegradation of chlorinated aliphatics by a methanotrophic consortium in a biological reactor. Presented at the Second National Conference on Biotreatment, November 27–29, Washington, DC.
[23] Broholm, K., Jensen, B.K., Christensen, T.H. and Olsen, L. (1990) Toxicity of 1,1,1-trichloroethane and trichloroethene on a mixed culture of methane-oxidizing bacteria. *Appl. Environ. Microbiol.* **56**, 2488–2493.
[24] Palumbo, A.V., Eng, W., Boerman, P.A., Strandberg, G.W., Donaldson, T.L. and Herbes, S.E. (1991). Effects of diverse organic contaminants on trichloroethylene degradation by methanotrophic bacteria and methane-utilizing consortia. In: On-Site Bioreclamation (Hinchee, R.E. and Olfenbuttel, R. F., Eds), pp. 77–91. Butterworth-Heinemann, Boston.
[25] Stirling D.I. and Dalton, H. (1979) The fortuitous oxidation and cometabolism of various carbon compounds by whole-cell suspensions of *Methylococcus capsulatus* (Bath). *FEMS Microbiol. Lett.* **5**, 315–318.
[26] Dalton, H. (1980) Oxidation of hydrocarbons by methane monooxygenases from a variety of microbes. *Adv. Appl. Microbiol.* **26**, 71–87.
[27] Hou, C.T., Patel, R.N., Laskin, A.I. and Barnabe, N. (1980) Microbial oxidation of gaseous hydrocarbons: oxidation of lower n-alkenes and n-alkanes by resting cell suspensions of various methylotrophic bacteria, and the effect of methane metabolites. *FEMS Microbiol. Lett.* **9**, 267–270.
[28] Patel, R.N., Hou, C.T., Laskin, A.I. and Felix, A. (1982) Microbial oxidation of hydrocarbons: properties of a soluble methane monooxygenase from a facultative methane-utilizing organism, *Methylobacterium* sp. Strain CRL-26. *Appl. Environ. Microbiol.* **44**, 1130–1137.

[29] Palumbo, A.V., Boerman, P.A., Herbes, S.E., White, D.C., Strandberg, G.W., Donaldson, T.L., Lucero, A.J., Jennings, H.L. and Phelps, T.J. (1991) A cometabolic approach to groundwater remediation. In: Proceedings from the U. S. Department of Energy Environmental Remediation Conference, pp. 95–100. September, 1991, Pasco Washington.

[30] Alvarez-Cohen, L., and McCarty, P.L. (1991) Two-stage dispersed-growth treatment of halogenated aliphatic compounds by cometabolism. *Environ. Sci. Technol.* **25**, 1387–1393.

[31] Mackay, D. and Shiu, W.Y. (1981) A critical review of Henry's Law constants for chemicals of environmental interest. *J. Phys. Chem. Ref. Data* **10**(4), 1175–1199.

[32] Verschueren, K. (1983) Handbook of Environmental Data on Organic Chemicals, Van Nostrand Reinhold, New York.

[33] Dean, J.A. (1985) Lange's Handbook of Chemistry, Thirteenth Edition, McGraw-Hill, New York.

[34] Perry, R.H. and Green, D. (1984) Perry's Chemical Engineer's Handbook, Sixth Edition, McGraw-Hill, New York.

[35] Gossett, J.M. (1987) Measurement of Henry's Law constants for C_1 and C_2 chlorinated hydrocarbons. *Environ. Sci. Technol.* **21**, 202–208.

[36] Bailey, J.E. and Ollis, D.F. (1986) Biochemical Engineering Fundamentals, McGraw-Hill, New York.

[37] Roberts, P.V. (1983) Dependence of oxygen transfer rate on energy dissipation during surface aeration and in stream flow. International Symposium on Gas Transfer at Water Interfaces, June 13-15, 1983. Cornell University, Ithaca, NY.

[38] Leak, D.J., Stanley, S.H. and Dalton, H. (1985) Implications of the nature of methane monooxygenase on carbon assimilation in methantrophs. In: Microbial Gas Metabolism (Poole, R.K. and Crawford, C.S., Eds) pp. 201–208. Academic Press, London.

[39] Strandberg, G.W., Donaldson, T.L. and Farr, L.L. (1989) Degradation of trichloroethylene and trans-1-2-dichloroethylene by a methanotrophic consortium in a fixed-film, packed-bed bioreactor. *Environ. Sci. Technol.* **23**, 1422–1425.

[40] Strand, S.E., Wodrich, J.V. and Stensel, H.D. (1991) Biodegradation of chlorinated solvents in a sparged, methanotrophic biofilm reactor. *Res. J. Water Pollut. Control Fed.* **62**, 124–129.

28

Chlorinated Methanes as Carbon Sources for Aerobic and Anaerobic Bacteria

T. Leisinger, S. La Roche[1], R. Bader, M. Schmid-Appert, S. Braus-Stromeyer and A.M. Cook

Mikrobiologisches Institut ETH, ETH-Zentrum, CH-8092 Zürich, Switzerland

1 SUMMARY

Chlorinated methanes are significant environmental pollutants which are subject to a variety of microbially catalyzed substitutive, oxygenative and reductive dehalogenations. Dehalogenation reactions providing the degrading bacteria with a source of carbon and energy have been observed for chloromethane and dichloromethane only. A *Hyphomicrobium* sp. grows aerobically in chloromethane-salts medium, and a strictly anaerobic homoacetogenic bacterium grows fermentatively with this substrate. Aerobic utilization of dichloromethane by methylotrophic bacteria depends on the synthesis of dichloromethane dehalogenase, which catalyzes a substitutive reaction yielding formaldehyde and inorganic chloride. Characterization of this enzyme, of its structural gene and of the regulatory system governing its expression, have led to a considerable understanding of aerobic dichloromethane metabolism. In contrast, growth of an acetogenic mixed culture in dichloromethane-salts medium under anaerobic conditions is little understood. Isolation and characterization of the organism(s) capable of anaerobic utilization of dichloromethane, as well as establishing the biochemistry of dehalogenation, represent new challenges.

---

[1] Present address: Frederick Cancer Research & Development Center, Frederick, MD 21701, USA

2 INTRODUCTION

The first reports on microbial degradation of chlorinated methanes appeared a little more than a decade ago, when it was demonstrated that some of these compounds are fortuitously oxygenated [1] or utilized as growth substrates [2,3] by specific bacteria. Since then, several laboratories have tested the capacity of micro-organisms to metabolize chlorinated methanes. As a result of these efforts, a picture begins to emerge of the various microbial interactions with chlorinated methanes, as well as of the potential and limitations of microbes for reintroducing these compounds into the carbon cycle. In this review we focus on the utilization of chloromethane and dichloromethane as carbon and energy sources by aerobic and anaerobic bacteria. These are the only chlorinated methanes that support bacterial growth, and emphasis will be placed on the microbiological aspects of their metabolism. For a competent discussion of the mechanistic features of bacterial chlorinated methane metabolism the reader is referred to a recent review by Wackett et al. [4]. The broader field of bacterial metabolism of halogenated compounds is treated in two excellent recent review articles [5,6].

The obvious incentive for exploring the microbial metabolism of chlorinated methanes is their significance as environmental pollutants. CH_3Cl, CH_2Cl_2, $CHCl_3$ and CCl_4, the four compounds comprising this class, are used as intermediates in chemical syntheses and as industrial solvents. Their physical properties, namely low boiling-points, low water solubility and, with the exception of CH_3Cl, non-flammability, are advantageous for commercial use but at the same time facilitate their escape into soil, water and atmosphere. Chlorinated methanes in the environment represent a public health concern since they have been shown to be toxic and/or carcinogenic. With the exception of the strong methylating agent CH_3Cl, they require metabolic activation by hydroxylation to produce their toxic effects [7]. Chlorinated methanes are industrially produced on a scale of approximately 2 million tons per year. As with many other organochlorine compounds, some chlorinated methanes are also natural products. Natural formation of trace amounts of $CHCl_3$ and CCl_4 by marine macroalgae has been observed [8] whereas the release of CH_2Cl_2 as a natural metabolite has not been reported so far. The most prominent example of a natural organohalogen compound produced in large quantities is CH_3Cl which is formed by *Phellinus pomaceus* and other wood-rotting fungi as well as by a variety of marine algae at an estimated rate of 5 million tons per year (10 times the annual industrial production) [8,9].

3 MICROBIAL METABOLISM OF CHLORINATED METHANES

3.1 *Dehalogenative reactions*

Dehalogenation reactions are key steps in the metabolism of chlorinated methanes. They can be roughly categorized in the three following major groups:

Nucleophilic Substitution

$$-\overset{|}{\underset{|}{C}}-X + \underset{(HO^-)}{RS^-} \longrightarrow -\overset{|}{\underset{|}{C}}-OH + X^-$$

Oxygenation

$$-\overset{H}{\underset{|}{\overset{|}{C}}}-X + [O] \longrightarrow \left[-\overset{OH}{\underset{|}{\overset{|}{C}}}-X \right] \longrightarrow -\overset{O}{\underset{|}{\overset{\|}{C}}} + HX$$

Reduction

$$-\overset{|}{\underset{|}{C}}-X + 2e^- + 2H^+ \longrightarrow -\overset{|}{\underset{|}{C}}-H + HX$$

Nucleophilic substitutions of chlorine substituents are catalyzed by specific enzymes, the dehalogenases. Depending on the nature of the nucleophile, these reactions are hydrolytic (OH$^-$ as nucleophile) or thiolytic (RS$^-$ as nucleophile). So far only one dehalogenase, dichloromethane dehalogenase (DCM dehalogenase), acting on a chlorinated methane has been described, and this enzyme will be discussed below.

Oxygenative dehalogenation of chlorinated methanes is catalyzed by methane monooxygenase of methanotrophic bacteria. CH_3Cl, CH_2Cl_2 and $CHCl_3$ (but not CCl_4) are fortuitously oxygenated by the soluble form of this enzyme to yield formaldehyde, formate and CO_2, respectively, as final transformation products [1,10].

A variety of strictly anaerobic bacteria in pure culture reductively dehalogenate chlorinated methanes either as growing cells or as resting cells kept at low reduction potential [4,6]. The higher its degree of halogenation, the more susceptible is a chlorinated methane to reductive dehalogenation. Thus the dehalogenation of CCl_4 proceeds in a stepwise manner to $CHCl_3$, CH_2Cl_2 and CH_3Cl with transient accumulation of each of the less halogenated transformation products. Reductive dehalogenation of chlorinated methanes is catalyzed by whole cells and cell extracts, and by autoclaved cells of an anaerobic bacterium [e.g. 11] as well as *in vitro* by transition-metal coenzymes such as vitamin B_{12}, coenzyme F_{430} and hematin [12,13]. The similarity in product pattern and reaction kinetics between systems *in vivo* and the model systems employing pure coenzymes suggests that Co-corrinoids and/or the Ni-porphinoid F_{430} are the predominant catalysts of reductive dehalogenation in some anaerobes. In strictly anaerobic and facultatively anaerobic bacteria not possessing these coenzymes, reductive dehalogenation of chlorinated methanes may result from the activity of iron-centered redox-active cofactors.

CCl_4 metabolism by strict anaerobes, concomitantly with reductive dehalogenation, comprises also a pathway leading to CO_2. The reactions involved in this conversion are thought to involve also vitamin B_{12} and coenzyme F_{430}. Possible pathways for CO_2 formation from CCl_4 are discussed elsewhere [4,6].

3.2 Relevance of dehalogenation to the cell

Chlorinated methanes assume different roles in microbial metabolism. They may serve as substrates of broad-specificity enzymes, thereby being fortuitously transformed to non-halogenated products which, however, are not further metabolized. Such cometabolic transformations thus remain without direct benefit to the dehalogenative organisms, and a prime example for this metabolic pattern is the fortuitous oxidation of chlorinated methanes by the soluble methane monooxygenase of methanotrophic bacteria.

Reductive dehalogenation of chlorinated methanes represents cometabolism under anoxic conditions. In this process CCl_4 and $CHCl_3$ serve as electron acceptors without any demonstrable advantage to the organisms. In this context it is important to note the recent discovery of *Dehalobacter restrictus*, a novel anaerobic bacterium in which dehalogenation of tetrachloroethene via trichloroethene to *cis*-1,2-dichloroethene is coupled to growth [14]. In this bacterium the conservation of the free energy liberated in the dehalogenative transformation of tetrachloroethene makes it seem possible that organisms exist, in which an analogous integration into energy metabolism of the reductive dehalogenation of chlorinated methanes has evolved.

Productive metabolism, that is utilization as sole carbon and energy source has been observed for CH_3Cl and CH_2Cl_2. With a view of applying chlorinated methane-degrading bacteria in the treatment of gaseous [15,16,30] and liquid wastes [17], this is the preferred type of metabolism. It confers a selective advantage to the degradative organisms and usually enables higher degradation rates than dehalogenation processes based on cometabolic transformations.

4 GROWTH WITH CHLOROMETHANE

Under aerobic conditions the naturally occurring compound chloromethane is utilized as a growth substrate by a *Hyphomicrobium* that was isolated from industrial sewage. This organism grows with a specific growth rate of 0.09 h^{-1} in minimal medium with 1% (v/v) CH_3Cl added to the gas phase, and its growth rate decreases progressively when the substrate concentration is increased to 10% [18]. The dehalogenation mechanism has not been elucidated. Washed cell suspensions of CH_3Cl-grown cells dechlorinated CH_3Cl in an oxygen-dependent reaction. Methane was not oxidized. This suggests that CH_3Cl is transformed by a monooxygenase, which is not methane monooxygenase, to formaldehyde according to the following reaction:

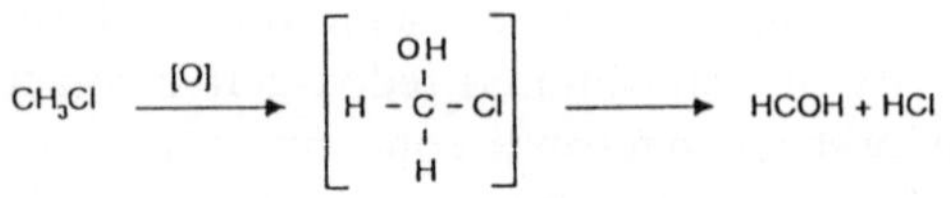

$$CH_3Cl \xrightarrow{[O]} \begin{bmatrix} & OH & \\ & | & \\ H & -C- & Cl \\ & | & \\ & H & \end{bmatrix} \longrightarrow HCOH + HCl$$

The utilization of CH_3Cl as a bacterial growth substrate under anaerobic conditions has recently been discovered by Traunecker *et al.* [19]. They have isolated a strictly anaerobic, Gram-positive, coccoid homoacetogenic bacterium that grows with CH_3Cl and converts this substrate to acetate. The doubling time of this isolate at its optimum growth temperature of 25°C on 2% (v/v) CH_3Cl in the gas phase is 30 h, and faster growth is observed on a mixture of CH_3Cl plus CO. Utilization of CH_3Cl appears to be inducible. CH_2Cl_2 is not utilized by the organism, and $CHCl_3$ and CCl_4 are toxic. The mechanism of dechlorination by this new organism is still unknown. Tetrahydrofolate and Co-corrinoids are discussed as possible dehalogenation catalysts [19]. The fermentation of CH_3Cl and CO_2 to acetate in the anaerobic bacterium, which apparently represents a new species of the homoacetogens, has the following stoichiometry:

$$4\ CH_3Cl + 2\ CO_2 + 2\ H_2O \longrightarrow 3\ CH_3COO^- + 7\ H^+ + 4\ Cl^-$$

5 AEROBIC GROWTH WITH DICHLOROMETHANE

5.1 *Dichloromethane-utilizing methylotrophs*

Methylotrophic bacteria growing in CH_2Cl_2-salts medium are readily isolated from sewage sludge or from soil that has been exposed to this solvent [17,20,21]. In contrast, a large number of methylotrophs from culture collections was unable to grow on CH_2Cl_2 (O. Ghisalba, personal communication). This indicates that acquisition and presumably also maintenance of this property need selective pressure. CH_2Cl_2-utilizing bacteria that have been characterized to a certain extent are listed in Table 1. The majority of these organisms are unidentified Gram-negative facultative methylotrophs. One facultative methylotroph is a *Hyphomicrobium*. The pink- pigmented strain DM4 was assigned to the genus *Methylobacterium* [26]. More recently a restricted facultative methylotroph growing on CH_2Cl_2 was isolated and identified as *Methylophilus* sp. [24,25]. The latter organism grows significantly faster on CH_2Cl_2 than the other facultative methylotrophs (Table 1).

All CH_2Cl_2-utilizers examined so far possess the enzyme dichloromethane (DCM) dehalogenase. This strongly inducible enzyme enables the organisms to transform CH_2Cl_2 to inorganic chloride and formaldehyde, a central metabolite of methylotrophic growth. The reaction catalyzed by DCM dehalogenase is thought to proceed through the following intermediates, with GSH and GS representing free and bound glutathione, respectively:

$$CH_2Cl_2 \xrightarrow[HCl]{GSH} \left[GSCH_2Cl \xrightarrow[HCl]{H_2O} GSCH_2OH \right] \xrightarrow{GSH} HCOH$$

In contrast to other haloalkane dehalogenases [5], nucleophilic displacement of chloride by DCM dehalogenase is not based on the direct attack by hydroxide from water, but on the thiol group of GSH as the initial nucleophile.

Table 1 Dichloromethane-utilizing methylotrophic bacteria

Strain	Organism	Growth rate with 10 mM CH_2Cl_2 (h^{-1})	Plasmid(s)[1]	Reference
DM1	unidentified	0.08	120 kb	[2, 17]
DM3	Gram-negative	0.11	120 kb	[17, 22]
DM5	facultative	0.06	120 kb	[17, 22]
DM6	methylotrophs	0.05	120 kb	[17, 22]
DM10		0.08	120 kb	[17, 22]
DM2	*Hyphomicrobium* sp.	0.07	none detected	[20]
DM4	*Methylobacterium* sp.	0.09	120 kb, 40 kb, 8 kb	[17, 22, 23]
DM11	*Methylophilus* sp.	0.22	70 kb[2]	[21, 24, 25]

[1] Largely unpublished data (R. Gälli).
[2] R. Bader, unpublished.

This proposed mechanism [20] is analogous to the GSH-dependent metabolism of CH_2Cl_2 in rat liver cytosol observed by Ahmed and Anders [27]. The postulated intermediates of the bacterial DCM dehalogenase reaction have not been isolated but, consistent with nucleophilic displacement of chloride, CD_2Cl_2 was shown to yield dideutero-formaldehyde [28].

DCM dehalogenase was first purified from *Hyphomicrobium* sp. DM2 [29]. It is a homo-hexamer (monomer, 33 kDa) with an absolute requirement for GSH as cofactor. CH_2Cl_2, CH_2Br_2, CH_2I_2 and CH_2BrCl are substrates, and among a number of chlorinated ethanes tested only 1,1-dichloroethane is dehalogenated, though the rate is approx. 1000-fold lower than the reaction rate with CH_2Cl_2. DCM dehalogenases with properties closely similar to those of the strain DM2 enzyme have been purified from the facultative methylotrophic strain DM1, from *Methylobacterium* sp. DM4 (Table 1) and from *Hyphomicrobium* sp. GJ21. The four enzymes from facultative methylotrophs form the group A dehalogenase type, whose representatives exhibit identical electrophoretic mobility, close immunological relatedness and identity of the N-terminal amino acid sequences [31]. DCM dehalogenase from the restricted facultative methylotroph *Methylophilus* sp. DM11 differs from the group A enzymes and forms the separate group B dehalogenase type [21]. Its N-terminus is different from the N-terminus of group A, the immunological relatedness to group A is weak, and differences in the kinetic properties of the two enzyme classes were observed. Despite these distinctions, similarity in subunit molecular mass and the common requirement of glutathione in catalysis indicate evolutionary relatedness between group A and group B enzymes.

5.2 The dichloromethane-utilization genes of Methylobacterium sp. DM4

Studies on the genetics of CH_2Cl_2-utilization were initiated with *Methylobacterium* sp. DM4 because in this organism, stable CH_2Cl_2-non-utilizing

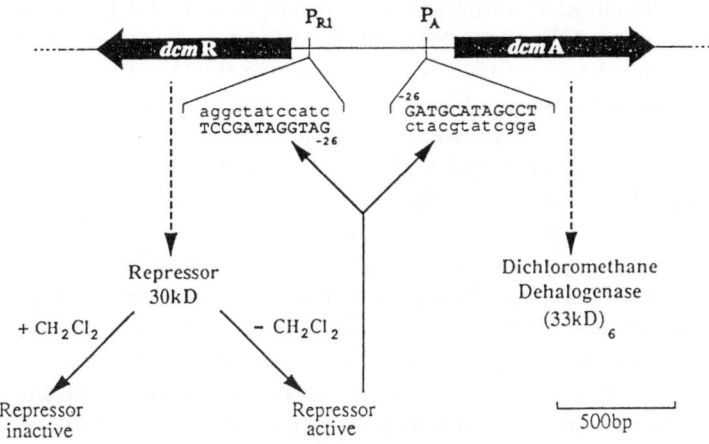

Figure 1 Model for the regulated expression of DCM dehalogenase and the putative repressor protein in *Methylobacterium* sp. DM4. P_A and P_{R1} represent the promoter regions of *dcmA* and *dcmR*, respectively. For explanations see text.

(Dcm⁻) mutants were obtained at high rates after growth at the supra-optimal temperature of 32°C. Dcm⁻ mutants were then used as recipients for cloning the *dcm* (CH_2Cl_2-utilization) genes by complementation [23]. Using the cloned *dcm* genes from strain DM4 as a hybridization probe, their location on the genomes of the facultative methylotrophic CH_2Cl_2-utilizers was explored. The *dcm* genes were shown to be encoded on the 120 kb plasmids detected in strains DM1, DM3, DM5, DM6 and DM10 (Table 1) (R. Gälli, unpublished data) whereas the three cryptic plasmids carried by strain DM4 were clearly unrelated to CH_2Cl_2 metabolism [23].

The genes for CH_2Cl_2-utilization by *Methylobacterium* sp. DM4 are thus located on the chromosome (or on an undetected megaplasmid). They are encoded by a 2.8 kb DNA fragment whose nucleotide sequence has been determined [32]. As shown in Fig. 1, the *dcm* region contains *dcmA*, the structural gene of DCM dehalogenase, and *dcmR*, the regulatory gene responsible for inducibility of DCM dehalogenase by CH_2Cl_2. *dcmA* and *dcmR* are organized in two divergent transcription units. Their translational start points are separated by a 619 bp intergenic region which contains the transcriptional start sites. We have shown that *dcmR* encodes a *trans*-acting factor which negatively controls DCM dehalogenase formation, as well as its own synthesis, at the transcriptional level [33]. Such a function of *dcmR* is supported by the presence of a good helix-turn-helix motif in the N-terminal region of its deduced amino acid sequence. Furthermore both promoter P_A and promoter P_{R1} contain almost identical 12 bp-sequences at identical positions (-15 to -26) relative to the mRNA start sites [33]. These observations have led to the current model for DcmR-mediated regulation of DCM dehalogenase expression shown in Fig. 1. It assumes that *dcmR* encodes a repressor which, in the absence of CH_2Cl_2, binds to the operator sites within the promoters P_A and P_{R1} and thereby sterically hinders initiation of transcription. CH_2Cl_2 would abolish repressor binding and thereby relieve inhibition of transcription initiation. Our present efforts are directed towards overexpressing the DcmR protein in *E. coli* and demonstrating its specific binding to the putative operator regions as well as towards establishing the postulated binding of CH_2Cl_2 to DcmR.

In *Methylobacterium* sp. DM4 and in four other CH_2Cl_2-utilizers examined so far, the regulatory system under investigation provides for 50- to 80-fold induction of DCM dehalogenase by CH_2Cl_2. There seem to be slight differences between the regulatory proteins from different strains. In two *Hyphomicrobium* spp. 1,1-dichloroethane and 1,2-dichloroethane are efficient gratuitous inducers of DCM dehalogenase whereas their inducing effect is marginal in strains DM4 and DM1 [31]. From a physiological point of view efficient regulation of DCM dehalogenase is particularly important since the intracellular concentration of this enzyme in induced cells amounts to 8% of the total soluble protein in *Methylophilus* sp. DM11 and to about 16% in the other methylotrophic CH_2Cl_2-utilizers [21,29].

5.3 Dichloromethane dehalogenase, a bacterial glutathione S-transferase

DCM dehalogenase is the key determinant for the metabolism of this xenobiotic substrate, presumably novel to terrestrial environments. This invites speculations on the evolutionary origin and on the potential for further evolution of this enzyme. Analysis of the amino acid sequence deduced from *dcmA* shows that DCM dehalogenase fits into the glutathione S-transferase (GST) enzyme family. The enzyme contains three regions of amino acid residues that are highly conserved among eukaryotic GSTs. On an overall amino acid similarity basis it is equally related to eukaryotic GSTs (38–51% similarity) as are all the eukaryotic GSTs from the alpha, mu and pi classes among themselves (32–63% similarity) [32].

Recently a fourth class, theta, of eukaryotic GSTs has been defined by Meyer et al. [34]. Unlike other eukaryotic GSTs, representatives of this group lack activity towards the 'universal' substrate for GST, 1-chloro-2,4-dinitrobenzene, exhibit high activity with CH_2Cl_2, and are not retained by GSH affinity matrices. These distinctive properties are shared by DCM dehalogenase. In addition, the N-terminus of DCM dehalogenase from *Methylobacterium* sp. DM4 (and from *Methylophilus* sp. DM11) shows greater sequence similarity to the theta class than to other classes of eukaryotic GSTs (Fig. 2). Functional as well as structural criteria thus suggest an association of bacterial DCM dehalogenases with the theta family.

An even closer structural relationship might be expected between the DCM dehalogenases from methylotrophs and eubacterial GSTs. However, so far there is no evidence for a uniform GST gene family in representatives of the cyanobacteria and the purple bacteria, the only prokaryotic phyla which are reported to produce GSH [35]. GSTs active towards 1-chloro-2,4-dinitro-

```
DM4-DCMD      MSPNPTNIHT GKTLRLLYHP ASQPCRSAHQ FMYEIDVPFE EEVVDISTDI...
DM11-DCMD                STKLRYLHHP ASQPCRAVHQ FMLENNIEFQ EEIVDITTDI...
THETA-GST                GLELYLDL LSQPCRAVYI FAKKNGIPFQ LRTVDLRCGQ...

PROMI-GST                    MKLYYT PGSCSLSPHI VLRQTGLDFS IERIDLRRLL..

ALPHA-GST               PGKPVLHYFD GRGRMEPIRW LLAAAGVEFE EQFLKTRDDL...
MU-GST                  PMTLGYWD IRGLAHAIRL LLEYTDTSYE DKKYTMGDAP...
PI-GST                  PPYTITYFP VRGRCEAMRM LLADQDQSWK EEVVTMETWP...
```

Figure 2 Alignment of the N-terminal amino acid sequences of eubacterial and eukaryotic glutathione S-transferases (GSTs). Identical amino acids within the theta class GSTs are boxed as are identical amino acids among the alpha, mu and pi classes of eukaryotic GSTs. Bold print indicates amino acids occurring three or more times at identical positions among all sequences shown. DM4-DCMD, DCM dehalogenase of *Methylobacterium* sp. DM4 [32] I; DM11-DCMD, DCM dehalogenase of *Methylophilus* sp. DM11 (R. Bader, unpublished); THETA-GST, GST subunit 12 from rat liver [34]; PROMI-GST, GST from *Proteus mirabilis* [36]; ALPHA-GST, GST from rat liver [41]; MU-GST, GST from murine fibroblasts [42]; PI-GST, GST from pig lung [43].

benzene have been characterized from several nonphotosynthetic purple bacteria [35]. Sequence information on these enzymes is not yet available. The N-terminus of GST from *Proteus mirabilis*, the only bacterial GST whose N-terminal sequence has been determined [36], indicates little homology to the established eukaryotic GST classes. No homology with GSTs was observed for the deduced amino acid sequence of a GST from *Serratia marcescens* [37] that catalyzes the formation of a stable conjugate between fosfomycin and GSH and thereby confers resistance against this antibiotic to the host (Fig. 2).

The group B DCM dehalogenase from *Methylophilus* sp. DM11 (see Section 5.1) catalyzes CH_2Cl_2 dehalogenation at an approximately 5-fold greater rate (k_{cat} 3.28 s^{-1}) than the group A dehalogenases (k_{cat} 0.55 s^{-1}) [21]. This has prompted us to clone and sequence the *dcmA* gene of strain DM11 (R. Bader, unpublished). Its deduced amino acid sequence exhibits 56% sequence identity and 75% similarity with the DCM dehalogenase from strain DM4, thus supporting the evolutionary relatedness of these two enzymes from methylotrophs (Fig. 2). Sequence comparison between the two enzyme types and experiments to uncover active site residues participating in catalysis [4] should lead to a better understanding of the primary structure elements determining catalytic activity.

6 ANAEROBIC UTILIZATION OF DICHLOROMETHANE BY AN ACETOGENIC MIXED CULTURE

Recently, anaerobic enrichment cultures capable of growth with CH_2Cl_2 as the sole carbon and energy source have been obtained in two laboratories [38,39]. These anoxic systems are of interest for the treatment of contaminated groundwater, and they call for microbiological and biochemical studies to elucidate the pathway of anaerobic CH_2Cl_2 degradation. The as yet undefined mixed cultures convert CH_2Cl_2 (but not CH_3Cl) to carbon dioxide, methane and acetate, suggesting interactions between fermentative, methanogenic and acetogenic organisms. We now have simplified the system by serial transfer of the enrichment culture in a mineral salts-CH_2Cl_2 medium which does not contain yeast extract or other undefined organic material. This led to an acetogenic mixed culture which, under an atmosphere of N_2 plus CO_2, formed 0.84 mol of acetate and 4 mol of chloride per 2.0 mol of CH_2Cl_2 utilized (S. Braus-Stromeyer, unpublished). This conversion suggests the following theoretical fermentation balance:

$$2\ CH_2Cl_2 + 2\ H_2O \rightarrow CH_3COO^- + 4\ Cl^- + 5\ H^+$$
$$(\Delta G^{\circ\prime} = -492.7\ kJ/reaction)$$

During the conversion of CH_2Cl_2 to acetate we measured a transient accumulation of formate which disappeared toward the end of the reaction. This observation and the fact that $^{14}CH_2Cl_2$ gave rise to acetate preferentially labeled in the methyl group (88%) and only to a minor extent (12%) in the carboxyl group, leads to the proposal for the acetogenic fermentation of CH_2Cl_2 shown in Fig. 3. It suggests that 2 mol of CH_2Cl_2 are dehalogenated to the oxidation state of formaldehyde and oxidized to 2 mol of formate. One formate

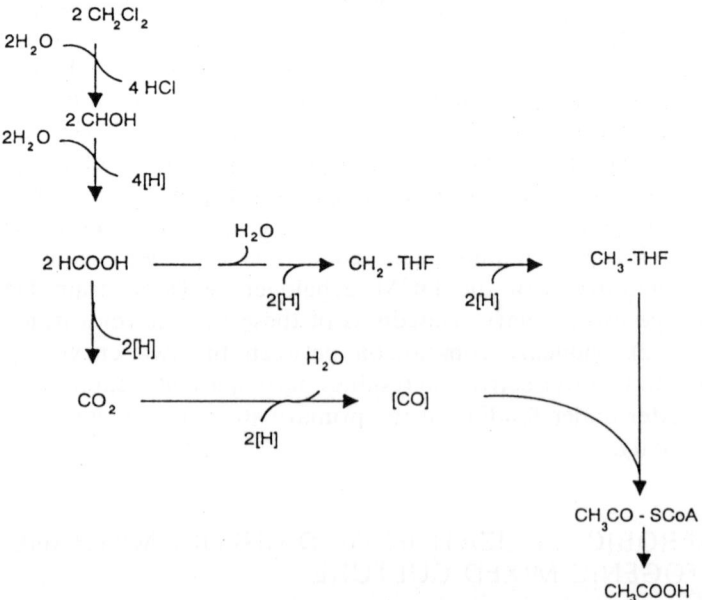

Figure 3 Hypothetical scheme for the fermentative degradation of dichloromethane to acetate by an anaerobic enrichment culture. THF, tetrahydrofolate; [CO], enzyme-bound carbonyl.

would be oxidized further to carbon dioxide while the other could serve for acetogenesis via the acetyl CoA pathway [40], either by the dehalogenative organism itself or by a syntrophic partner. The CH_2Cl_2-fermenting mixed culture at present is composed of three morphologically different bacteria, whose roles in the process are unknown. Equally unknown is the mechanism of the substitutive dehalogenation of CH_2Cl_2 implied in this process. Cell extracts of the mixed culture, like extracts from all anaerobes that have been analyzed to date [35], do not contain GSH, and CH_2Cl_2 dehalogenation by our cell-free extracts is not stimulated by GSH (S. Braus-Stromeyer, unpublished). This indicates that CH_2Cl_2 dehalogenation in the anaerobic system is presumably mechanistically different from the DCM dehalogenase reaction in the aerobic methylotrophs.

ACKNOWLEDGEMENT

Work in the authors' laboratory was supported by grants from the Swiss Federal Institute of Technology, Zürich.

REFERENCES

[1] Stirling, D.I. and Dalton, H. (1979) The fortuitous oxidation and cometabolism of various carbon compounds by whole-cell suspensions of *Methylococcus capsulatus* (Bath). *FEMS Microbiol. Lett.* **5**, 315–318.
[2] Brunner, W., Staub, D. and Leisinger, T. (1980) Bacterial degradation of dichloromethane. *Appl. Environ. Microbiol.* **40**, 950–958.
[3] Rittmann, B.E. and McCarty, P.L. (1980) Utilization of dichloromethane by suspended and fixed-film bacteria. *Appl. Environ. Microbiol.* **39**, 1225–1226.
[4] Wackett, L.P., Logan, M.S.P. and Blocki, F.A. (1992) A mechanistic perspective on bacterial metabolism of chlorinated methanes. *Biodegradation*, in press.
[5] Hardman, D.J. (1991) Biotransformation of halogenated compounds. *Crit. Rev. Biotechnol.* **11**, 1–40.
[6] Mohn, W.W. and Tiedje, J.M. (1992) Microbial reductive dehalogenation. *Microbiol. Revs.* **56**, 482–507.
[7] Anders, M.W. and Pohl, L.R. (1985) Halogenated alkanes. In: Bioactivation of Foreign Compounds (Anders, M.E., Ed.), pp. 283–315. Academic Press, New York.
[8] Gschwend, P.M., McFarlane, J.K. and Newman, K.A. (1985) Volatile halogenated organic compounds released to seawater from temperate marine macroalgae. *Science* **227**, 1033–1035.
[9] Wuosmaa, A.M. and Hager, L.P. (1990) Methyl chloride transferase: a carbocation route for biosynthesis of halometabolites. *Science* **249**, 160–162.
[10] Oldenhuis, R., Oedzes, J.Y., Van der Waarde, J.J. and Janssen, D.B. (1991) Kinetics of chlorinated hydrocarbon degradation by *Methylosinus trichosporium* OB3b and toxicity of trichloroethylene. *Appl. Environ. Microbiol.* **57**, 7–14.
[11] Egli, C., Stromeyer, S., Cook, A.M. and Leisinger, T. (1990) Transformation of tetra- and trichloromethane to CO_2 by anaerobic bacteria is a non-enzymic process. *FEMS Microbiol. Lett.* **68**, 207–212.
[12] Gantzer, C.J. and Wackett, L.P. (1991) Reductive dechlorination catalyzed by

bacterial transition-metal coenzymes. *Environ. Sci. Technol.* **25**, 715–721.
[13] Krone, U.E., Laufer, K. and Thauer, R.K. (1989) Coenzyme F_{430} as a possible catalyst for the reductive dehalogenation of chlorinated C_1 hydrocarbons in methanogenic bacteria. *Biochemistry* **28**, 10061–10065.
[14] Holliger, C. (1992) Reductive dehalogenation by anaerobic bacteria. Ph.D. Dissertation, Agricultural University of Wageningen, The Netherlands.
[15] Hartmans, S. and Tramper, J. (1991) Dichloromethane removal from waste gases with a trickle-bed bioreactor. *Bioprocess Eng.* **6**, 83–92.
[16] Diks, R.M. and Ottengraf, P. (1991) Verification studies of a simplified model for the removal of dichloromethane from waste gases using a biological trickling filter (part II). *Bioprocess Eng.* **6**, 131–140.
[17] Gälli, R. and Leisinger, T. (1985) Specialized bacterial strains for the removal of dichloromethane from industrial waste. *Conserv. Recycl.* **8**, 91–100.
[18] Hartmans, S., Schmuckle, A., Cook, A.M. and Leisinger, T. (1986) Methyl chloride: naturally occurring toxicant and C_1 growth substrate. *J. Gen. Microbiol.* **132**, 1139–1142.
[19] Traunecker, J., Preuss, A. and Diekert, G. (1991) Isolation and characterization of a methyl chloride utilizing, strictly anaerobic bacterium. *Arch. Microbiol.* **156**, 416–421.
[20] Stucki, G., Gälli, R., Ebersold, H.-R. and Leisinger, T. (1981) Dehalogenation of dichloromethane by cell extracts of *Hyphomicrobium* DM2. *Arch. Microbiol.* **130**, 366–371.
[21] Scholtz, R., Wackett, L.P., Egli, C., Cook, A.M. and Leisinger, T. (1988) Dichloromethane dehalogenase with improved catalytic activity isolated from a fast-growing dichloromethane-utilizing bacterium. *J. Bacteriol.* **170**, 5698–5704.
[22] Gälli, R. (1986) Optimierung des mikrobiellen Abbaus von Dichlormethan in einem Wirbelschicht-Bioreaktor. Dissertation ETH Nr. 7994, Eidg. Techn. Hochschule, Zürich, Switzerland.
[23] Gälli, R. and Leisinger, T. (1988) Plasmid analysis and cloning of the dichloromethane-utilization genes of *Methylobacterium* sp. DM4. *J. Gen. Microbiol.* **134**, 943–952.
[24] Tsuji, K., Tsien, H.C., Hanson, R.S., De Palma, S.R., Scholtz, R. and La Roche, S. (1990) 16S ribosomal RNA sequence analysis for determination of phylogenetic relationship among methylotrophs. *J. Gen. Microbiol.* **136**, 1–10.
[25] Doronina, N. and Trotsenko, Y. (1992) Personal communication.
[26] Green, P.N. and Bousfield, I.J. (1982) A taxonomic study of some Gram-negative facultatively methylotrophic bacteria. *J. Gen. Microbiol.* **128**, 623–638.
[27] Ahmed, A.E. and Anders, M.W. (1978) Metabolism of dihalomethanes to formaldehyde and inorganic halide. II. Studies on the mechanism of reaction. *Biochem. Pharmacol.* **27**, 2021–2025.
[28] Gälli, R., Stucki, G. and Leisinger, T. (1982) Mechanism of dehalogenation of dichloromethane by cell extract of *Hyphomicrobium* DM2. *Experientia* **38**, 1378.
[29] Kohler-Staub, D. and Leisinger, T. (1985) Dichloromethane dehalogenase of *Hyphomicrobium* sp. strain DM2. *J. Bacteriol.* **162**, 676–681.
[30] Ottengraf, S.P.P., Meeters, J.J.P., Van den Oever, A.H.C. and Rozema, H.R. (1986) Biological elimination of volatile xenobiotic compounds in biofilters. *Bioprocess Eng.* **1**, 61–69.
[31] Kohler-Staub, D., Hartmans, S., Gälli, R., Suter, F. and Leisinger, T. (1986) Evidence for identical dichloromethane dehalogenases in different methylotrophic bacteria. *J. Gen. Microbiol.* **132**, 2837–2843.
[32] La Roche, S. and Leisinger, T. (1990) Sequence analysis and expression of the bacterial dichloromethane dehalogenase structural gene, a member of the glutathione S-transferase supergene family. *J. Bacteriol.* **172**, 164–171.
[33] La Roche, S. and Leisinger, T. (1991) Identification of *dcmR*, the regulatory gene

governing expression of dichloromethane dehalogenase in *Methylobacterium* sp. strain DM4. *J. Bacteriol.* **173**, 6714–6721.

[34] Meyer, D.J., Coles, B., Pemble, E., Gilmore, K.S., Fraser, G.M. and Ketterer, B. (1991) Theta, a new class of glutathione transferases purified from rat and man. *Biochem. J.* **274**, 409–414.

[35] Fahey, R.C. and Sundquist, A.R. (1991) Evolution of glutathione metabolism. *Adv. Enzymol. Relat. Areas Mol. Biol.* **64**, 1–53.

[36] Di Ilio, C., Aceto, A., Piccolomini, R., Allocati, N., Caccuri, A.M., Barra, D. and Federici, G. (1989) N-terminal region of *Proteus mirabilis* glutathione transferase is not homologous to mammalian and plant glutathione transferases. *FEBS Lett.* **250**, 57–59.

[37] Suarez, J.E., Arca, P., Villar, C.J. and Hardisson, C. (1989) Evolutionary origin, genetics, and biochemistry of clinical fosfomycin resistance. In: Genetics and Molecular Biology of Industrial Microorganisms (Hershberger, C.L., Queener, S.W. and Hegeman, G., Eds), pp. 93–98, American Society for Microbiology, Washington, DC.

[38] Freedman, D.L. and Gossett, J.M. (1991) Biodegradation of dichloromethane and its utilization as a growth substrate under methanogenic conditions. *Appl. Environ. Microbiol.* **57**, 2847–2857.

[39] Stromeyer, S.A., Winkelbauer, W., Kohler, H., Cook, A.M. and Leisinger, T. (1991) Dichloromethane utilized by an anaerobic mixed culture: acetogenesis and methanogenesis. *Biodegradation* **2**, 129–137.

[40] Fuchs, G. (1986) CO_2 fixation in acetogenic bacteria: variations on a theme. *FEMS Microbiol. Rev.* **39**, 181–213.

[41] Telakowski-Hopkins, C.A., Rodkey, J.A., Bennett, C.D., Lu, A.Y.H. and Pickett, C.B. (1985) Rat liver glutathione S-transferases. *J. Biol. Chem.* **260**, 5820–5825.

[42] Townsend, A.J., Goldsmith, M.E., Pickett, C.B. and Cowan, K.H. (1989) Isolation, characterization and expression in *Escherichia coli* of two murine mu class glutathione S-transferase cDNAs homologous to the rat subunits 3 (Yb1) and 4 (Yb2). *J. Biol. Chem.* **264**, 21582–21590.

[43] Dirr, H.W., Mann, K., Huber, R., Ladenstein, R and Reinemer, P. (1991) Class π glutathione S-transferase from pig lung. *Eur. J. Biochem.* **196**, 693–698.

29

Production and Applications of Amidase from *Methylophilus methylotrophus*

M.A. Carver[1] and C.W. Jones[2]

[1]*ICI Bio Products, Billingham, Cleveland TS23 1LB, UK and* [2]*Department of Biochemistry, University of Leicester, Leicester LE1 7RH, UK*

1 SUMMARY

Methylophilus methylotrophus hydrolyses short-chain aliphatic amides to ammonia and the corresponding aliphatic acid using a cytoplasmic amidase. Cellular amidase activity is regulated by changes both in the concentration of the enzyme (induction by acetamide, repression by ammonia) and also in its specific activity (indirect inactivation by ammonia – reversed non-physiologically by heat). Analysis of purified high-activity and low-activity forms of the enzyme indicate that they exhibit significant physico-chemical differences. Novel strains of *M. methylotrophus* in which the amidase is overexpressed and/or exhibits improved kinetic properties can be selected using continuous culture. Methods have been devised for the large-scale preparation of crude enzyme and for its subsequent commercial use in the removal of residual acrylamide from polyacrylamide flocculants.

2 INTRODUCTION

Acrylamide is widely used in the manufacture of polyelectrolyte flocculants of the type used in potable water production, sewage sludge conditioning, papermaking, mining and secondary oil recovery [1,2]. Unfortunately, it is a potent cumulative neurotoxin of high chronic toxicity and has also been classified as a category A2 suspect human carcinogen by the regulatory

---

Microbial Growth on C_1 Compounds
© Intercept Ltd, PO Box 716, Andover, Hampshire SP10 1YG, UK

authorities in the USA and in Germany. Consequently the level of residual acrylamide in these polymers is subject to strict legislative control, particularly in the USA where currently accepted levels of residual monomer in flocculants used in potable water production and various industrial processes are 0.05% (w/w) and 0.1–5% (w/w) respectively; these levels are subject to increasing scrutiny and downward pressure as the toxicological profile of acrylamide is raised.

Current 'best-available' technology for polymer production relies either on a 'post-heating' process which drives residual monomer into the polymer, or on the use of a redox reagent such as sulphite to convert acrylamide to less toxic products. These approaches can produce polymers of low residual acrylamide content, but have associated problems such as increased reactor occupancy time and polymer damage.

In response to a perceived and growing need from both industry and the regulatory authorities, ICI Bio Products has developed a new enzyme-based approach to the problem of residual acrylamide toxicity, based on the use of a bacterial amidase which hydrolyses acrylamide to ammonia and the non-neurotoxic acrylic acid [3,4].

Polyacrylamide flocculants (both homo- and hetero-polymers of acrylamide) are manufactured as high-viscosity aqueous-solution polymers, as dry powder products (usually made by drying solution polymers) and as inverse-emulsion polymers. The latter afford a difficult environment for enzymes to work in, since they are made by the redox- or heat-activated, radical-initiated polymerization of dispersed aqueous monomer emulsions in a non-aqueous phase such as paraffin or xylene, stabilized by combinations of detergents. Consequently a hydrolytic enzyme acting in such an environment needs to be functionally resistant to inactivation by the high temperatures generated by the exothermic polymerization process, by the reactive process constituents and by the resultant flocculants (many of which will react with charged proteinaceous materials), as well as having appropriate kinetic properties.

Preliminary screening and assessment of potential sources of such an enzyme indicated that the amidase (aliphatic acylamide hydrolase; EC 3.5.1.4) from *Methylophilus methylotrophus* is particularly suitable since it combines most of the desired functional properties [3,4] with potential for low development costs (reflecting ICI's experience of growing the organism in large-scale continuous culture and the extensive toxicological testing already carried out [5,6]).

Methylophilus methylotrophus is a restricted facultative methylotroph. Its preferred carbon sources are methanol and various methylated amines, all of which are oxidized to formaldehyde (and ammonia) and thence metabolized via the assimilatory (2-keto 3-deoxy 6-phosphogluconate (KDPG) aldolase/transaldolase variant) and dissimilatory ribulose monophosphate pathways [7–10]; it also grows poorly on glucose and fructose [11; D. Byrom, personal communication]. Ammonia, methylated amines, various short-chain aliphatic amides, urea and formamide can be used as sources of nitrogen; the latter three substrates are hydrolysed to ammonia by the action of amidase, urease and formamidase respectively [3,4; M.A. Carver, J. Hinton and J. Ashworth, unpublished; D. Scherr and C.W. Jones, unpublished], and the ammonia is assimilated via the glutamine synthetase (GS)/glutamate synthase (GOGAT; glutamine 2-oxoglutamate amino transferase) system [12].

3 REGULATION OF AMIDASE SYNTHESIS

Amidase synthesis by *M. methylotrophus* is induced by short-chain aliphatic amides and repressed by ammonia [13], such that the highest amidase concentrations are found during growth in continuous culture at low dilution rate under either acetamide or dual methanol-acetamide limitation (where the ammonia concentration resulting from amide hydrolysis is undetectably low) and maximum repression occurs under methanol limitation (when the ammonia concentration exceeds approximately 0.1 mM). When expressed maximally, amidase comprises up to approximately 5% of the total cell protein [4,13,14; J. Mills and C.W. Jones, unpublished], although this can be substantially increased by directed evolution.

4 AMIDASE PROPERTIES

M. methylotrophus amidase is easily purified to homogeneity using FPLC. It is an acidic homotetramer (pI 4.1, M_r 4 × 38,000) which hydrolyses a wide range of aliphatic amides such as acetamide, propionamide and acrylamide [4,13,14], and exhibits a number of properties in common with the amidases from *Pseudomonas aeruginosa* [15], *Brevibacterium* sp. R312 [16] and *Arthrobacter* J1 [23] (Table 1).

5 LARGE-SCALE CULTURE AND THE PROBLEM OF ACETATE TOXICITY

Amidase is produced commercially from cells grown in large volume at high cell density (> 10 g dry wt/l). The metabolic products of acetamide hydrolysis are ammonia and acetic acid; ammonia is readily assimilated by the GS-GOGAT system, but the metabolic fate of acetate in *M. methylotrophus* is obscure. The organism cannot grow on acetate as a sole source of carbon (D. Byrom, personal communication), since although it can convert acetate into acetyl CoA using acetyl CoA synthetase (the level of which increases in cells grown on acetamide or acetate (A. Lloyd 1990, Ph.D. Thesis, University of Bath) it has an incomplete TCA cycle, has only very low levels of isocitrate lyase and lacks malate synthase. The fate of the acetyl CoA in unclear, but during growth at low cell densities it is presumably metabolized quantitatively into some as yet unidentified end-product since no acetate is spilled under these

Table 1 Comparative properties of various bacterial amidases

Property	Methylophilus methylotrophus [4,13,14]	Brevibacterium R312 [16]	Pseudomonas aeruginosa [15]	Arthrobacter J1 [23]
M_r Subunit (SDS-PAGE)	38,000	43,000	38–40,000	39,000
Native (GF)	155,000		200,000	300,000
(ND-PAGE)	155,000	180,000		320,000
Subunits	4	4	6	8
Isoelectric point	4.1	3.5		3.8
Relative V_{max}				
Acetamide	100	100	100	100
Propionamide	180	1220	274	216
Acrylamide	151	227	189	330
Induction by amides	Yes	Yes	Yes	Yes
Repression by ammonia acetate	Yes No	No Yes	No (Yes)	(No) (Yes)
Acetyltransferase actiity	Yes	Yes	Yes	Yes
Inhibition by thiol agents	Yes	Yes	Yes	Yes
Thiol groups per subunit	7		9	

GF, gel filtration; ND-PAGE, non-dissociating polyacrylamide gel electrophoresis.

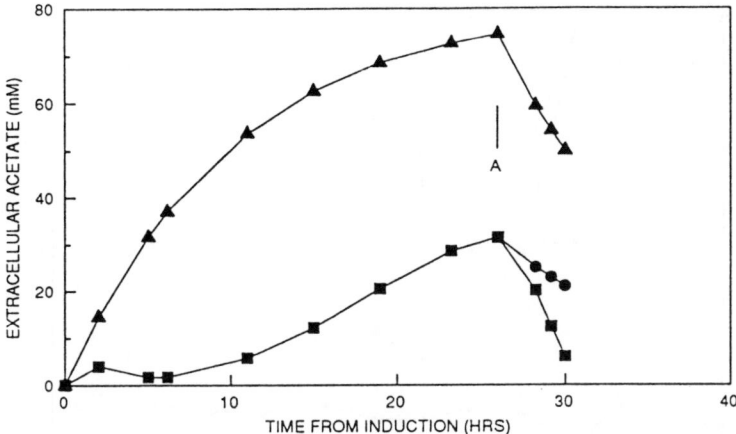

Figure 1 Acetate metabolism in *Methylophilus methylotrophus*. *M. methylotrophus* was grown in continuous culture (12 g dry wt cells l^{-1}) at a dilution rate (D) of 0.1 h^{-1} under dual methanol-acetamide limitation. Acetamide was replaced as the sole nitrogen source by ammonia at point A such that the amount of nitrogen fed to the culture remained constant. Extracellular acetate produced from acetamide was measured by the acetate kinase method [24]. ■, measured acetate concentration; ▲ and ●, theoretical acetate concentration (assuming no metabolism of acetate by the culture).

conditions. In contrast, during growth at high cell density, this limited ability to metabolize acetate is apparently saturated and the organism readily spills acetate until it reaches toxic levels. The mechanism of acetate toxicity is unclear, but growth studies show that acetate has a deleterious effect on growth rate such that in the presence of 30–40 mM acetate μ_{max} is less than the imposed dilution rate and the culture washes out [13; N.J. Silman, 1991, Ph.D. Thesis, University of Leicester].

The limited ability of *M. methylotrophus* to metabolize acetate has been exploited to allow large-scale growth at high cell density under dual C and N limitation by supplying acetamide and ammonia (or urea) alternately to the culture as the source of nitrogen. When the acetamide feed is replaced by ammonia (or urea), the concentration of acetate in the culture supernatant decreases (presumably as a result of metabolism since it falls at a faster rate than would be expected simply by dilution) and when it reaches an acceptably low level the acetamide feed is reinstated (Fig. 1). Provided that the period of ammonia (or urea) addition is kept below 2 h, this regime does not significantly repress amidase synthesis and yields high and stable levels of both biomass and amidase. Interestingly, however, the amidase activity of such cells is considerably lower than that exhibited by cells grown at low cell density under acetamide or dual methanol-acetamide limitation, in spite of the similar concentrations of amidase in all three types of culture, but can be substantially increased by a period of heating (Fig. 2) [3].

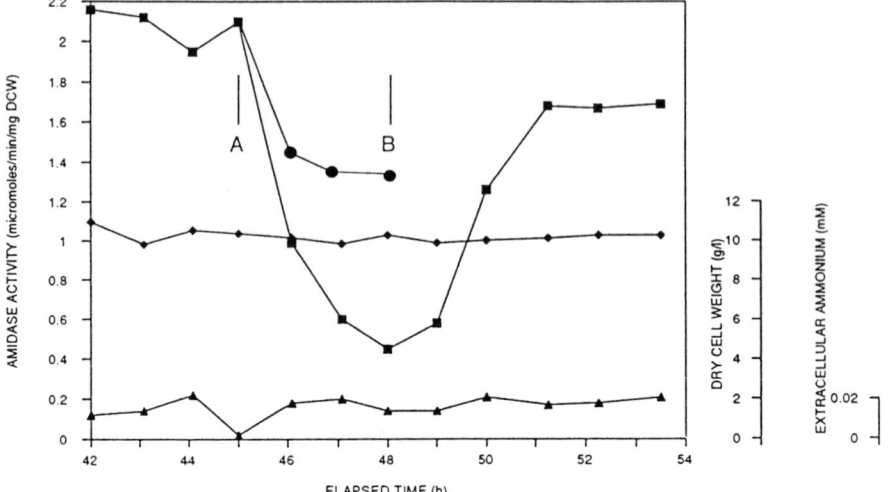

Figure 2 'Switch-off' of amidase activity in *M. methylotrophus*. *M. methylotrophus* was grown in continuous culture (12 g dry wt cells l^{-1}; D 0.1 h^{-1}) under dual methanol-acetamide limitation. Acetamide was replaced as the sole nitrogen source by ammonia at point A such that the amount of nitrogen fed to the culture remained constant; this procedure was reversed at point B. Whole cell amidase activity was measured at 30°C in the presence of 0.1 M acrylamide essentially as described in [13]. Heat-stimulation was carried out by heating whole cells at 60°C for 2 h in the absence of added amide, followed by cooling to 4°C prior to assaying as described above. Ammonium was assayed by the sodium phenate method [13], and cell dry wt. by drying followed by gravimetric analysis. ■, amidase activity (before heat stimulation); ●, amidase activity (after heat stimulation); ▲, ammonium concentration; ◆, cell dry wt.

6 LARGE-SCALE PRODUCTION AND APPLICATIONS OF AMIDASE

Large-scale production of amidase for commercial purposes is carried out *via* an eight-step process (Fig. 3). Cells grown as described above are harvested by disc-stack centrifugation and disrupted by passage through a bead mill or an adiabatic expansion homogeniser. The resultant broken-cell suspension is then subjected to a heat-treatment step which has the three-fold effect of coagulating insoluble material, precipitating many soluble proteins and activating the amidase. The latter is then recovered from the solids using a combined centrifugation/washing procedure before being concentrated by ultrafiltration. A solid product can also be made by spray- or freeze-drying of this concentrated preparation.

The production of inverse emulsion polymers typically takes place in stirred, steel reaction vessels in which the redox environment is carefully controlled by

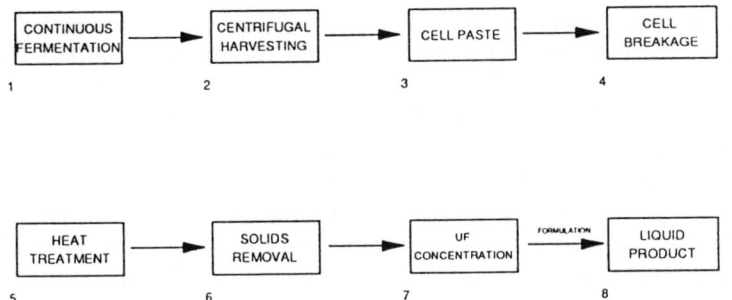

Figure 3 Outline of amidase production process.

sparging with nitrogen and air. The process takes place in several temperature-controlled stages and consists, at its simplest, of an initial polymerization at 50°C in which the concentration of the residual monomer falls to 0.1–1.0% (w/w) followed by a second step in which the temperature is raised to 60–65°C to drive residual monomer into polymer, thus decreasing its concentration below 0.05–0.1% (w/w).

The resultant product is subsequently treated with amidase in one of two ways. The enzyme is either added to the reaction vessel immediately after the heating phase and allowed to react *in situ* (Fig. 4), typically taking 0.5–2.0 h to

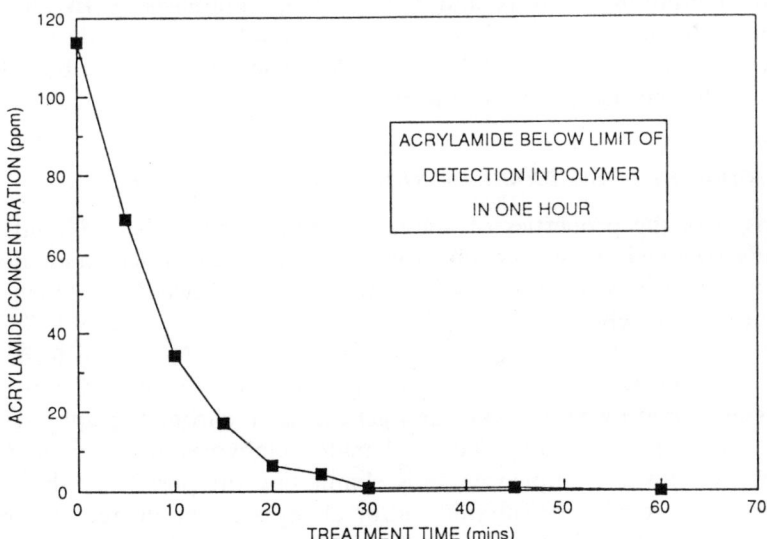

Figure 4 Amidase treatment of an anionic polyacrylamide emulsion. An anionic polyacrylamide emulsion was treated with 1000U of cell-free amidase formulation (produced as described in Fig. 3) at 60°C, pH 6.0. Residual acrylamide concentration was determined using HPLC essentially as described previously [3,4].

reduce the monomer concentration below the limit of detection (0.1–10 ppm depending on the polymer), or a lower amount of enzyme is added and the polymer plus enzyme is pumped into storage tanks where the monomer concentration decays to the desired level more slowly. The method used depends on the exact configuration of the plant and local regulatory requirements.

One of the major advantages of this enzymic method over existing chemical procedures is that it does not adversely affect the functional properties of the polymer. Polyacrylamides of all the major anionic, cationic and nonionic classes can be treated in this way. Commercial quantities of amidase-treated polymer have been manufactured and sold in the USA.

7 REGULATION OF AMIDASE ACTIVITY

7.1 Switch-off

The amidase activity of whole cells, broken cells and cell-free extracts prepared from *M. methylotrophus* grown at high cell density under dual C and N limitation as described above is generally lower than expected from the measured amidase concentration in the cells. The extent of this 'switch-off' of amidase activity is rather variable, but can often be increased by storing a sample of culture anaerobically. Switch-off can also be brought about by adding excess ammonia to low cell density cultures growing under dual methanol-acetamide limitation (J. Mills and C.W. Jones, unpublished). In all cases switch-off is characterized by a large decrease in total cellular amidase activity compared with any change in cellular amidase concentration, i.e. by a large decrease in the specific activity of the enzyme.

7.2 Properties of switched-off (low activity) amidase

Comparison of the properties of amidase purified from cells containing the apparently switched-off enzyme and from cells grown under acetamide limitation at low cell density shows that the enzyme can exist in two distinct high- and low-activity forms which exhibit very different k_{cat} values [4,14; N.J. Silman, N.R. Wyborn and C.W. Jones, unpublished] (Table 2). These two forms are identical in many respects (e.g. substrate specificities, subunit and native M_r values as measured by SDS-PAGE and gel exclusion chromatography respectively, k_m values for acetamide and acrylamide, sulphydryl group content and N-terminal amino acid sequences, thus eliminating the possibility that (dis)-association of a regulatory subunit and/or changes in the degree of subunit association are responsible for the different k_{cat} values. However, the low-activity form exhibits a lower pI (4.0 c.f. 4.1–4.2), an increased sensitivity to the sulphydryl group reagent p-chloromercuribenzoate, and a higher thermostability than the high-activity form. Chemical and spectrophotometric analyses show that neither form of the amidase is modified by the addition of phosphate,

Table 2 Comparison of the high-activity and low-activity forms of the amidase from *M. methylotrophus*

Property	High-activity amidase [4,14]	Low-activity amidase
M_r Subunit (SDS-PAGE)	38,000	38,000
Native (GF)	155,000	155,000
(ND-PAGE)	155,000	155,000
Subunits	4	4
Isoelectric point	4.1	4.0
k_{cat} (s^{-1}) Acetamide	128	4–45
k_m (mM) Acetamide	1	1
Acrylamide	12	12
Kinetics	Linear	Hysteretic
Thiol groups per subunit	7	7
I_{50} pCMB (μM)	300	50
N-terminal sequence	MIH	MIH

Neither form of the enzyme showed any evidence for the presence of bound nucleoside, phosphate (serine, threonine), glucose, mannose or N-terminal acylation. The data on the low-activity enzyme are taken from J. Hinton and M.A. Carver (unpublished) and from N.J. Silman, N. R. Wyborn and C.W. Jones (unpublished).

nucleoside or various sugars, thus eliminating modifications of the type exhibited by other nitrogen-metabolizing enzymes such as nitrogenase (ADP-ribosylation [17,18] or glutamine synthetase (adenylylation [20]). More detailed investigations using electrospray mass spectroscopy show that the high-activity and low-activity forms have very similar M_r values, indicating that they differ from each other by virtue of either a very small (probably post-translational) modification or some sort of locked conformational change.

7.3 Heat-reactivation

Switched-off cellular amidase activity can generally be substantially increased by heating at 55–65°C for up to 6 h, prior to cooling and assaying under standard conditions at 30 or 37°C. The extent of this heat reactivation depends on the exact temperature, pH and protein concentration employed, and can be as much as 50-fold in cells grown at high cell density with acetamide/ammonia or acetamide/urea as the nitrogen source (Fig. 5) (c.f. 0.2 to 0.5-fold heat-reactivation in cells grown at high cell density under acetamide-limitation, and significant heat-inactivation in cells grown under various conditions at low cell density). It is important to note that in all cases heat reactivation is seen as an

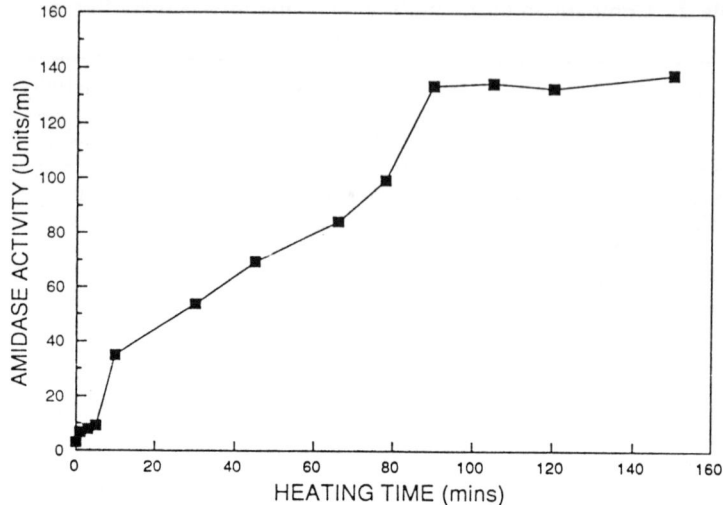

Figure 5 Heat reactivation of 'switched-off' amidase. *M. methylotrophus* was grown in continuous culture (12 g dry wt cell l^{-1}; $D = 0.1\ h^{-1}$) under dual methanol-acetamide/ammonium hydroxide (1:3 N content) limitation. Cells were suspended in 50 mM sodium phosphate buffer pH 7.0, then broken by passage through a French pressure cell at 15,000 psi, centrifuged at 100,000g for 1 h to remove cell debris, and finally filtered through a 0.45 μm filter. The resultant cell-free extract was heat-treated at 60°C and assayed for amidase activity as described in Fig. 2.

increase in total cellular amidase activity, not simply as a spurious increase in specific activity due for example to precipitation of non-amidase protein.

In contrast to whole cells, broken cells or cell-free extracts, purified low-activity amidase cannot be reactivated by heat, indicating that other cellular components are required for this process; furthermore, heat reactivation of cell-free extracts is inhibited by dialysis and EDTA. Fractionation studies of cell-free extracts using anion-exchange chromatography, gel-filtration chromatography and ultrafiltration show that heat reactivation depends on the presence of a constituent protein fraction, probably in association with a low M_r component (possibly a cation). In order to reactivate the low-activity amidase, this fraction can either be heated with the amidase or heated alone, then cooled and added to the amidase; in both cases reactivation occurs stoichiometrically rather than catalytically. It is tentatively concluded that the effect of heat on this protein fraction is to expose a domain which is capable of accepting the modifying group from the low-activity form of the amidase. However, the extent to which this heat-reactivation process reflects the physiological switch-on mechanism (e.g. by stimulating enzymes involved in this mechanism) or occurs *via* solely adventitious chemical effects, remains to be determined.

7.4 Biochemical mechanism and physiological role of switch-off and switch-on

Current evidence indicates that *M. methylotrophus* amidase can exist in two quite different forms (*viz.* a high-activity form and a low-activity form which has probably undergone a post-translational modification) depending on the exact growth conditions. It is likely, therefore, that these two forms comprise part of a reversible, fine-control mechanism (switch-off/switch-on) for the regulation of amidase activity which operates in addition to the coarse-control mechanism (induction/repression) for the regulation of amidase synthesis. Although the enzymology of the switch-off/switch-on process has not been investigated, it is likely, by analogy with the regulation of nitrogenase and glutamine synthetase [17–19], to require discrete enzymes which catalyse the interconversion of the high-activity and low-activity forms of the enzyme, probably in response to the availability of ammonia and/or acetate in the growth environment.

The physiological basis of switch-on/switch-off is probably to conserve resources and prevent acetate toxicity. Following the sudden availability of exogenous ammonia (the preferred nitrogen source for growth) to a culture already hydrolysing acetamide, amidase activity would no longer be required and would be positively deleterious due to its continued ability to produce potentially-toxic acetate. Repression of amidase synthesis would be of little benefit in these circumstances since, in the observed absence of rapid degradation of the enzyme, loss of cellular amidase activity by dilution would be very slow. The ability of the organism to switch off amidase activity rapidly, particularly under conditions of methanol or oxygen shortage and in the presence of high concentrations of acetate, would be of crucial importance to its continued survival. Extending this hypothesis, switch-on would occur in response to ammonia and/or acetate disappearance, thus allowing the organism to scavenge available acetamide efficiently as a source of nitrogen.

8 EVOLUTION OF AMIDASE IN CONTINUOUS CULTURE

As short-chain aliphatic amides enter cells by passive diffusion, amidase is the first enzyme of amide metabolism. This has enabled prolonged growth of *M. methylotrophus* in amide-limited continuous culture to be used to select new strains of this organism with increased 'biological fitness' [20–22] with respect to acrylamide hydrolysis (i.e. which overproduce the amidase and/or synthesize novel forms of the enzyme with improved kinetic properties) such that they outgrow the parent strain and eventually take over the culture.

A sequential selection strategy using acetamide- or acrylamide-limited continuous cultures of *M. methylotrophus* at low dilution rate has been used successfully to isolate three new strains [MM6, MM8 and MM15) [14]. The concentration of amidase in each of these strains following growth under acetamide limitation is 4- to 6-fold higher than that of the wild-type organism, reaching up to 25% of the total cell protein (Table 3). In contrast, the amidase activities of strains MM6 and MM8 are respectively 4-fold and 12-fold that of the wild-type organism, whereas the amidase activity of MM15 is essentially unchanged but its affinity for acrylamide is substantially increased.

Comparison of the kinetic properties of the amidase purified from wild-type organism and from strains MM6, MM8 and MM15 confirm that the wild-type and MM6 enzymes are essentially identical, whereas the MM8 enzyme has a 3-fold higher k_{cat} than the wild-type/MM6 enzymes, and the MM15 enzyme has a similar k_{cat} to the wild-type enzyme but an 8-fold lower k_m for acrylamide (Table 3). Physico-chemical analysis of the MM15 enzyme shows that it has a substantially more open structure than any of the other enzymes (as evidenced by its lower sedimentation coefficient), a property which is presumably responsible for its decreased k_{cat} compared with the MM8 enzyme. The MM15 enzyme is also much more thermolabile than any of the other enzymes, such that it is approximately 80% dissociated into inactive monomers and dimers even in cells grown at the normal growth temperature of 37°C (thus explaining the very low whole cell activity of MM15 and the unexpectedly high purification factor for the amidase). However, when strain MM15 is grown in amide-limited continuous culture at 25°C, the amidase regains its more compact structure and is not significantly dissociated into inactive monomers and dimers; furthermore, the resultant whole cell amidase activity is almost 15-fold higher than during growth at 37°C, and is similar to that of strain MM8.

It is concluded that all four amidase phenotypes that might be expected to be selected by this approach have been isolated, viz: (i) overproduction of wild-type enzyme (MM6), (ii) overproduction of an altered enzyme with a higher k_{cat} (MM8), (iii) overproduction of an altered enzyme with a lower k_m (MM15) and (iv) overproduction of an altered enzyme with both a higher k_{cat} and a lower k_m (MM15 grown at 25°C) [14]. Unfortunately, the increased thermolability of the MM8 and MM15 enzymes preclude their use under process conditions. It is likely, however, that in general terms this approach could be used to improve the production and/or the kinetic properties of certain other microbial enzymes with potential uses in pollution control and detoxification.

9 CONCLUSIONS

The robust nature and acceptable kinetic properties of *M. methylotrophus* amidase have allowed the successful development of a large-scale process in which the crude enzyme is used to remove toxic acrylamide from polyacrylamide flocculants. The regulation of amidase synthesis and the biochemical properties of the enzyme have been investigated, and methods for isolating novel strains in which the amidase is overexpressed and/or exhibits improved

Table 3 Properties of amidases from wild-type and mutant strains of *M. methylotrophus*

Strain	Whole cell amidase Concentration (%)	Whole cell amidase Activity (U mg cells^{-1})	Purification factor Actual	Purification factor Expected	Kinetic properties of pure amidase Acetamide k_{cat} (s^{-1})	Acetamide k_m (mM)	Acrylamide k_{cat} (s^{-1})	Acrylamide k_m (mM)
WT	4.3	1.7	20.5	23.2	128	1.1	151	16.1
MM6	19.8	7.3	4.5	5.1	100	2.2	131	19.0
MM8	24.6	20.0	4.1	4.1	310	1.7	419	12.1
MM15	25.3	2.1	21.2	4.0	96	1.3	127	2.1
MM15 (25°C)	22.4	27.9	4.5	4.5	310	1.2	329	2.5

kinetic properties have been reported. However, much remains to be done in terms of understanding the regulation of amidase activity in this organism. Future work will be directed towards establishing the nature of the low-activity form of the enzyme, and to investigating the enzymology of the switch-on/switch-off process.

REFERENCES

[1] Hubermann, C.E. (1991) Acrylamide. In: Kirk Othmer Encyclopedia of Chemical Technology, 4th edn. (Eckroth, D., Ed.), pp. 251–266, Wiley, New York.
[2] Lipp, D. and Kazakiewicz, J. (1991) Acrylamide polymers. In: Kirk Othmer Encyclopedia of Chemical Technology, 4th edn. (Eckroth, D., Ed.), pp. 260–287, Wiley, New York.
[3] Carver, M.A. and Hinton, J. (1987) Process for the decomposition of acrylamide. European Patent EP 272,026.
[4] Byrom, D. and Laver, M.A. (1987) Process for the decomposition of acrylamide. European Patent EP 272,025.
[5] Senior, P.J. and Windass, J. (1980) The ICI single cell protein process. *Biotechnol. Letts.* **2**, 205–210.
[6] Vasey, R.B. and Powell, K.A. (1984) Single-cell protein. *Biotechnol. Genetic. Eng. Revs.* **2**, 285–310.
[7] Large, P.J. (1981) Microbial growth on methylated amines. In: Microbial Growth on C_1 Compounds (Dalton, H., Ed.), pp. 55–69, Heyden, London.
[8] Zatman, L.J. (1981) A search for patterns in methylotrophic pathways. In: Microbial Growth on C_1 Compounds (Dalton, H., Ed.), pp. 42–54, Heyden, London.
[9] Beardsmore, A.J., Aphergis, P.N.G. and Quayle, J.R. (1982) Characterisation of the assimilatory and dissimilatory pathways of carbon metabolism during growth of *Methylophilus methylotrophus* on methanol. *J. Gen. Microbiol.* **128**, 1423–1429.
[10] Burton, S.M., Byrom, D., Jones, G.D.D. and Jones, C.W. (1983) The oxidation of methylated amines by the methylotrophic bacterium *Methylophilus methylotrophus*. *FEMS Microbiol. Letts.* **17**, 185–190.
[11] Stirling, D.I. and Summit, N.J. (1988) Process for culturing *Methylophilus methylotrophus*. US Patent 4 652 527.
[12] Windass, J.D., Worsey, M.J., Pioli, E.M., Barth, P.J., Atherton, K.T., Dart, E.C., Byrom, D., Powell, K. and Senior, P.J. (1980) Improved conversions of methanol to single-cell protein by *Methylophilus methylotrophus*. *Nature* **237**, 396–401.
[13] Silman, N.J., Carver, M.A. and Jones, C.W. (1989) Physiology of amidase production by *Methylophilus methylotrophus*: isolation of hyperactive strains using continuous culture. *J. Gen. Microbiol.* **135**, 3153–3164.
[14] Silman, N.J., Carver, M.A. and Jones, C.W. (1991) Directed evolution of amidase in *Methylophilus methylotrophus*; purification and properties of amidases from wild-type and mutant strains. *J. Gen. Microbiol.* **137**, 169–178.
[15] Clarke, P.H. (1970) The aliphatic amidases of *Pseudomonas aeruginosa*. *Adv. Microbial Physiol.* **4**, 179–222.
[16] Thiery, A., Maestracchi, M., Arnaud, A., Galzy, P. and Nicholas, M. (1986) Purification and properties of an acylamide aminohydrolase [EC 3.5.1.4] with a wide activity spectrum from *Brevibacterium* sp. R312. *J. Basic Microbiol.* **5**, 299–311.
[17] Pope, M.R., Murrell, S.A. and Ludden, P.W. (1985) Covalent modification of the iron protein of nitrogenase from *Rhodospirillum rubrum* by adenosine diphosphorylation of a specific arginine residue. *Proc. Natl. Acad. Sci. USA* **82**, 3173–3177.

[18] Ludden, P.W. and Roberts, G.P. (1989) Regulation of nitrogenase activity by reversible ADP-ribosylation. *Curr. Top. Cell Reg.* **30**, 23–56.
[19] Ginsburg, A. and Stadtman, E.R. (1973) Regulation of glutamine synthetase in *Escherichia coli*. In: Enzymes of Glutamine Metabolism (Prusiner, S. and Stadtman, E.R., Eds), pp. 9–44, New York, Academic Press.
[20] Dykhuizen, D.E. and Hartl, D.L. (1983) Selection in chemostats. *Microbiol. Revs.* **47**, 150–168.
[21] Dykhuizen, D.E., Dean, A.M. and Hartl, D.L. (1987) Metabolic flux and fitness. *Genetics* **115**, 25–31.
[22] Mortlock, R.P. (1982) Metabolic acquisitions through laboratory selection. *Ann. Rev. Microbiol.* **36**, 259–284.
[23] Asano, Y., Tachibana, M., Tani, Y. and Yamada, H. (1982) Purification and characterisation of an amidase which participates in nitrile degradation. *Agric. Biol. Chem.* **46**, 1175–1181.
[24] Bergmeyer, H.V. and Mollering, H. (1974) Acetate analysis. In: Methods of Enzymatic Analysis. 2nd edn, Vol. 3 (Bermeyer, H. Ed.), pp. 1521–1532, Academic Press, New York.

30

Molecular Biology and Genetics of Methylamine Dehydrogenases

Mary E. Lidstrom and Andrei Y. Chistoserdov

Environmental Engineering Science, W.M. Keck Laboratories 138-78, California Institute of Technology, Pasadena, CA 91125, USA

1 SUMMARY

Methylamine dehydrogenase (MADH) is a periplasmic quinoprotein having a unique cofactor tryptophan tryptophylquinone (TTQ). MADHs from facultative autotrophic (*Paracoccus denitrificans* and *Thiobacillus versutus*), obligate (*Methylobacillus flagellatum* KT) and facultative (*Methylobacterium extorquens* AM1) methylotrophic bacteria use amicyanin as their electron acceptor, whereas MADHs from restricted facultative methylotrophs (*Methylophilus* spp.) apparently use a *c*-type cytochrome as an electron acceptor. Gene clusters coding MADH (*mau*) from *P. denitrificans*, *M. extorquens* AM1 and *Methylophilus* sp. W3A1 were cloned and sequenced. The *mau* gene cluster from *M. extorquens* AM1 encodes eleven polypeptides (*mau*FBEDACJGLMN) possibly involved in methylamine metabolism. The genes for the large (*mau*B) and the small (*mau*A) subunits of MADH and for amicyanin (*mau*C) were identified. Insertion mutations in each gene were introduced into the chromosome of *M. extorquens* AM1 and three phenotypically different groups of mutants were obtained, those unable to grow on alkylamines as a sole source of carbon (*mau*F, *mau*B, *mau*E, *mau*D, *mau*A, *mau*G, and *mau*L), one able to utilize ethylamine but not methylamine as a source of carbon (*mau*C) and those which utilize alkylamines at the level of the wild type (*mau*J, *mau*M and *mau*N). The role of the MauJ, MauM, and MauN proteins in methylamine metabolism is uncertain. None of the 11 mutants is impaired in the ability to utilize alkylamines as a source of nitrogen. Analysis of MADHs from the mutants showed that the *mau*F, *mau*E and *mau*D gene products are likely involved in the small subunit

---

Microbial Growth on C_1 Compounds
© Intercept Ltd, PO Box 716, Andover, Hampshire SP10 1YG, UK

maturation, whereas the *mau*G and *mau*L gene products may be involved in the cross-linking step of cofactor synthesis. Sequence analysis of the *mau* cluster from *P. denitrificans* showed that it is similar to the *mau* cluster of *M. extorquens* AM1. The DNA fragment with the *mau* cluster from *P. denitrificans* was able to complement *mau*E, *mau*D, and *mau*A mutations but not the *mau*F, *mau*B, *mau*G, *mau*C or *mau*L mutations of *M. extorquens* AM1. The *mau* cluster from *Methylophilus* sp. W3A1 differs from those of *M. extorquens* AM1 and *P. denitrificans* by the absence of *mau*C (amicyanin) and *mau*J. It is able to complement *mau*D and *mau*G mutations of *M. extorquens* AM1. Surprisingly the *mau* cluster from *M. flagellatum* KT also does not have *mau*C or *mau*J.

2 INTRODUCTION

Oxidation of amines is a widespread process in nature, found in both eukaryotic and prokaryotic organisms [1,2]. Amine oxidation by microorganisms plays an important role in biogeochemical cycling of C_1 compounds in the biosphere. Three enzymatic systems are used by methylotrophic bacteria for utilization of primary amines. Gram-positive bacteria oxidize primary amines *via* a quinoprotein amine oxidase [3]. Gram-negative methylotrophic bacteria which can grow on primary amines as sole sources of carbon, energy and nitrogen employ two other systems for amine oxidation, either a flavin-containing N-methylglutamate dehydrogenase, a system that is specific for oxidation of methylamine [4–6] or a quinoprotein methylamine dehydrogenase (MADH), which has a broader substrate range [7]. Several physiological and taxonomic groups of Gram-negative bacteria are able to oxidize methylamine via MADH [8–15]. MADH catalyzes the oxidation of methylamine, producing formaldehyde and NH_4^+. Other primary alkyl amines but not diamines or aromatic amines may serve as substrates for MADH [7]. MADH enzymes from different sources consist of two small and two large subunits with MWs ranging from 13–15 kD and 40–45 kD, respectively, and in all known cases this enzyme is periplasmic [8–13,15]. Depending on the type of electron acceptor used MADHs can be divided into two subclasses (Table 1). The majority of MADHs studied use the copper-containing protein amicyanin as an electron acceptor. Such MADHs have been found in facultative autotrophic [16,17] and facultative [18] and obligate [18–20] methylotrophic bacteria. MADHs belonging to the second subclass apparently use a *c*-type cytochrome as an electron acceptor [21] and their distribution is limited to the restricted facultative methylotrophs (*Methylophilus* spp. [21,22]). For this paper, we will use the designations $MADH_a$ and $MADH_c$ for MADHs from these two subclasses, respectively. The MADH small subunits of both subclasses have the same immunological properties and isoelectric points, whereas the large subunits of the MADH subclasses are immunologically distinct [23].

Amino acid sequencing of the small subunit of MADH from *M. extorquens* AM1 (formerly *Pseudomonas* sp. AM1) has shown that it has a quinone cofactor

Table 1 Comparison of properties of MADHs from different methylotrophs [1–24,27–32]

	P_I of the large subunit	The MADH electron acceptor	Preferred substrates	Number of disulfide bonds in the small subunit polypeptide
Facultative autotrophs: *Thiobacillus versutus* *Paracoccus denitrificans*	<7	amicyanin	C1–C3 alkylamines	6
Facultative methylotrophs: *Methylobacterium extorquens* AM1	<7	amicyanin	C1–C4 alkylamines	6
Obligate methylotrophs: *Methylobacillus flagellatum* KT *Methylomonas* sp. J Organism 4025	>7	amicyanin	methylamine	7
Restricted facultative methylotrophs: *Methylophilus methylotrophus* AS1 *Methylophilus* sp. W3A1	>7	c-type cytochrome	methylamine	7

covalently bound to the polypeptide chain [24]. This cofactor has been shown to be a unique quinone cofactor synthesized from two tryptophan residues of the small subunit polypeptide [25], and has been called tryptophan tryptophylquinone (TTQ).

Although a great deal of information has been collected concerning the biochemical properties of MADH [1–25], the genetic studies of this interesting system have begun only recently. This article will review information about the genetics of MADH existing in the literature and include new unpublished data from our laboratory as well. We have proposed to call genes pertaining to methylamine metabolism *via* MADH *mau* (*m*ethylamine *u*tilization) genes [26] and we are using this designation in this paper.

3 ORGANIZATION OF *mau* GENE CLUSTERS

Cloning, identification and sequencing of the following *mau* genes have been reported in the literature: the MADH large [27] and small [26] subunit and amicyanin [28] genes from *M. extorquens* AM1, the MADH large [29] and small [29,30] subunit and amicyanin [30] genes from *P. denitrificans* and the MADH small subunit and amicyanin genes [31] from *Thiobacillus versutus*. In this article, we report the cloning and sequencing of the MADH large and small subunit genes from *Methylophilus* sp. W3A1. E.R. Gak (Institute of Genetics and Selection of Industrial Microorganisms, Moscow) has also provided us with information about the sequence of the MADH small subunit gene from *Methylobacillus flagellatum* KT [32].

The organization of the *mau* gene clusters from different bacteria studied so far is similar (Fig. 1) and corresponding genes show substantial similarity. The *mau* gene cluster from *M. extorquens* AM1 is comprised of eleven open reading frames, which are located on the chromosome in the order *mau* FBEDACJGLMN. A twelfth open reading frame has been found upstream of the *mau* gene cluster, which is transcribed in the orientation opposite to that of the *mau* cluster (denoted *orf*-1 in Fig. 1). The *mau*B, *mau*A and *mau*C open reading frames were identified as the genes for the MADH large subunit polypeptide, the MADH small subunit polypeptide and amicyanin respectively [27–29]. Open reading frames corresponding to *mau*FBEDACJ have been found also in the genome of *P. denitrificans* [29,30]. Less sequence data are available for the *T. versutus mau* region, but genes corresponding to *mau*DACJ have been found so far [31]. The order of genes in the *mau* gene cluster from *Methylophilus* sp. W3A1 (containing $MADH_c$) is similar, although it is missing the gene for amicyanin (*mau*C) and the open reading frame (*mau*J) following the *mau*C open reading frame. Surprisingly, the *mau* cluster from *M. flagellatum* KT, which has a MADH belonging to the $MADH_a$ subclass is also missing *mau*C and *mau*J.

Analysis of the DNA sequences of the *mau* clusters from *M. extorquens* AM1, *P. denitrificans*, *T. versutus* and *Methylophilus* sp. W3A1 revealed the presence

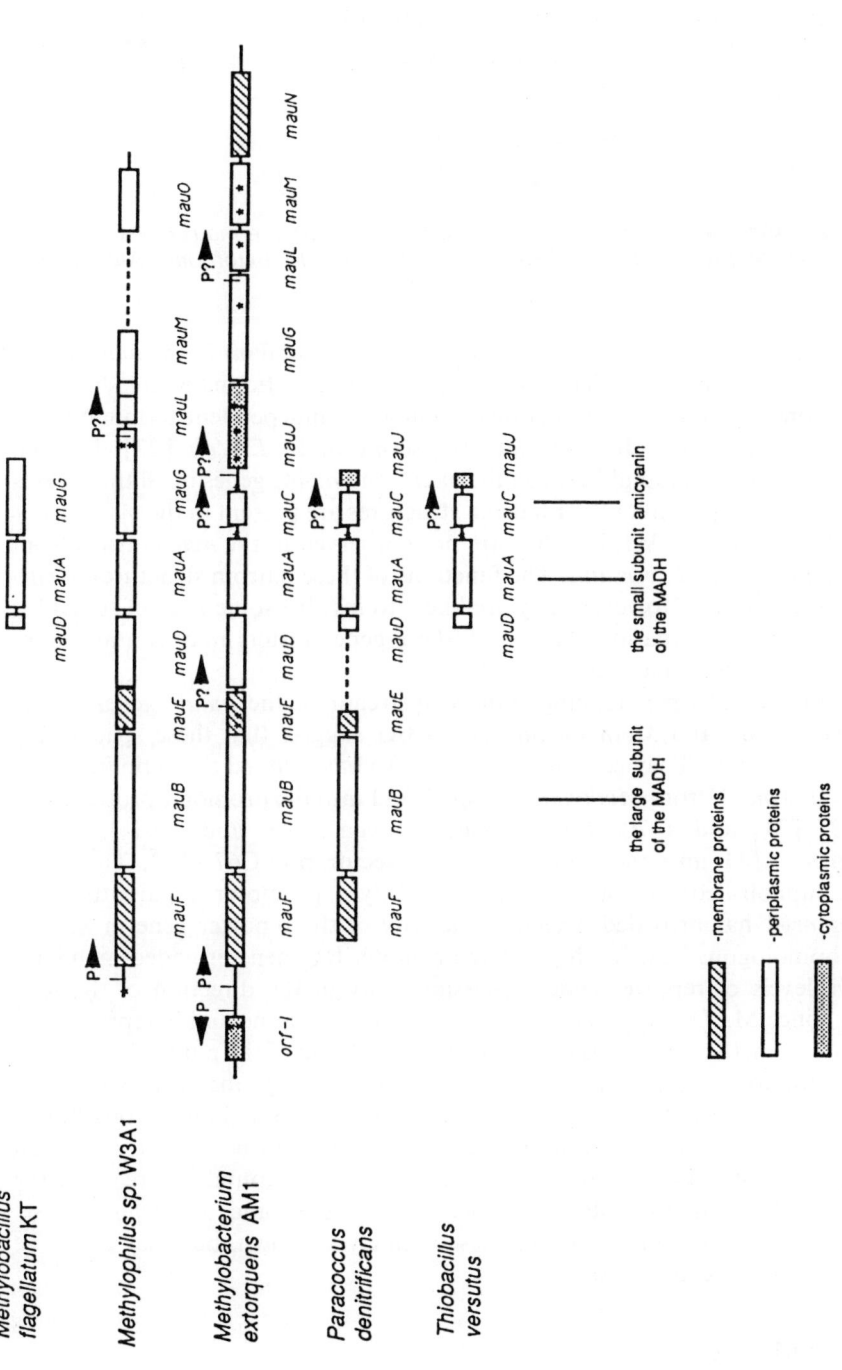

Figure 1 Genetic maps of the *mau* clusters from *M. flagellatum* KT [32], *Methylophilus* sp. W3A1 [48], *M. extorquens* AM1 [27,39], *P. denitrificans* [29,30] and *T. versutus* [31]. P and P? denote proven and suggestive promoters, arrows show direction of transcription from them. Asterisks denote hairpin structures with free energies 10 kcal/mol or more, some of which were shown to function as transcriptional terminators in *E. coli* [27].

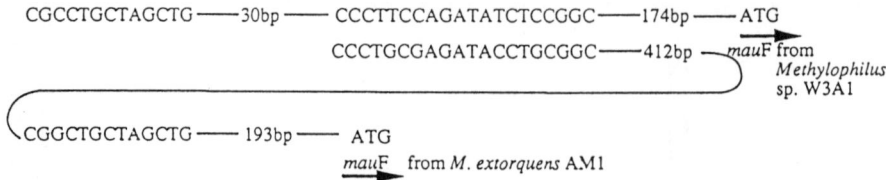

Figure 2 Alignment of the homologous sequences from the *mau* cluster promoter regions from *Methylophilus* sp. W3A1 (upper line) and *M. extorquens* AM1 (lower line).

of putative hairpin structures, which are especially abundant in the *mau* cluster from *M. extorquens* AM1. Their locations on the genetic maps are shown in Fig. 1. None of the hairpin structures resembles ρ-independent terminators of *E. coli* [33] but they do terminate transcription in *E. coli* [27]. Hairpin structures have been found between the *mau*A and *mau*C genes in all three cases in which *mau*C is present. Other hairpin structures are present in the *mau* cluster from *M. extorquens* AM1, but they are not conserved in the *mau* clusters from *P. denitrificans* and *T. versutus*. The functions of these hairpin structures *in vivo* (if any) is unknown. However, they are likely to involve some kind of signal for transcriptional punctuation. Therefore, these gene clusters may contain more than one transcriptional unit.

The absence of open reading frames upstream of the *mau*F genes in *M. extorquens* AM1 and *Methylophilus* sp. W3A1 suggest that these spaces may contain promoters. To check this hypothesis we have cloned a 330 bp fragment upstream of *mau*F from *Methylophilus* sp. W3A1 into the promoter probe vector pAYC37 [34] and a 700 bp fragment between *mau*F and *orf*-1 from *M. extorquens* AM1 into the promoter probe vector pAYC67 (Ap^R, Tc^R, Cm^S, IncP1, unpublished) in both orientations. Only a promoter apparently transcribing *ori*-1 has provided detectable activity of the reporter gene in *E. coli*. In the homologous hosts each putative promoter fragment provided high constitutive levels of reporter gene expression, only in the direction of the *mau* cluster. Since MADH is generally induced by methylamine and is repressed by succinate or methanol [35], the constitutive expression of the putative promoter may be due to the high copy number of the promoter probe vectors used (10 to 12 copies for pAYC37 and 7 or 8 copies for pAYC67). Similar constitutive expression of the *mox*F promoter in a promoter probe vector has been documented [36]. Two regions of high similarity were found upstream of the *mau*F translational start sites in *M. extorquens* AM1 and *Methylophilus* sp. W3A1 (Fig. 2). The significance of these is not yet known, but they could be sequences involved in transcription.

4 *mau* MUTANTS

Preliminary characteristics of sixty-two *M. flagellatum* KT mutants unable to grow on methylamine but capable of growth on methanol [37] indicated the presence of three classes with lesions in the small subunit, both subunits simultaneously, and both subunits and a natural electron acceptor simultaneously. The majority of the mutants synthesized functional MADH although they had lost the ability to utilize methylamine as a carbon source, and all grew normally with methylamine as a source of nitrogen.

An insertion mutation in the amicyanin gene from *P. denitrificans* was obtained by an *in vitro* insertion of a Km^R cassette and the subsequent introduction of the insertion into the chromosome [31]. This mutant was unable to grow on methylamine as a source of carbon, indicating that amicyanin is an obligatory component of electron transfer from MADH to oxygen in *P. denitrificans*. We have used the same approach for the construction of insertion mutants of *M. extorquens* AM1 [38]. Insertional mutations were generated in each open reading frame identified in the sequence. To prove that the Km^R cassette from pUC4K used in these experiments does not have a polar effect on downstream genes, insertions between *mau*D and *mau*A and *mau*A and *mau*C were also generated. The strains with the insertions between genes have a Mau$^+$ phenotype, although the amounts of MADH in these strains were slightly lower compared to the wild type. Only mutants with insertions of the Km^R cassette in the orientation coinciding with that of the transcription of mutagenized genes were taken for further analyses.

All of these mutants fall into three distinct phenotypic classes (Table 2). Mutants in *mau*F, *mau*B, *mau*E, *mau*D, *mau*A, *mau*G and *mau*L were not able to utilize any primary amine as a source of carbon, showing that these are all required for growth on methylamine, and also indicating that the *mau* system is responsible for utilization of primary amines with a carbon chain longer than 1. MADH activities were undetectable in all these mutants. Western blot experiments have shown that the small subunit polypeptide was absent in extracts of the *mau*F, *mau*B, *mau*E, *mau*D, *mau*A mutants, whereas the large subunit was present but at decreased levels compared to the wild type, except in the *mau*B mutant, where it was missing. The small subunit was also not detected in these mutants using a quinone-specific staining technique [39]. The *mau*G and *mau*L mutants have a similar amount of the large and small subunit polypeptides as the wild type.

The amicyanin mutant (*mau*C) was impaired in the ability to utilize methylamine as a source of carbon but still able to utilize ethyl-, propyl- and butylamines as carbon sources with growth rates approximately three times slower than those of the wild type. MADH activity and the large and the small subunit polypeptide were present in extracts of the *mau*C mutant, although in amounts lower than those of the wild type. Other electron acceptors (cytochrome c_H and cytochrome c_L) can participate *in vitro* in electron transfer from *M. extorquens* AM1 MADH to a terminal oxidase [40]. Our experiments with the *mau*C mutant indicate that electron acceptors other than amicyanin are able to discharge reduced MADH. However, either those acceptors do not function with methylamine as a substrate or they do not participate in energy metabolism. Bacteria growing methylotrophically are strongly dependent on obtaining energy from the first step of methanol or methylamine oxidation [41].

Table 2 Properties of the insertion *mau* mutants of *M. extorquens* AM1

Mutant	Mutation point	Ability to utilize methylamine as carbon	as nitrogen	Ability to utilize ethylamine as carbon	as nitrogen	MADH activity	Large subunit	Small subunit
wild type	none	+++	+++	+++	+++	+++	+++	+++
PvuIIKm	inside *mauA*	–	+	–	+	–	+	–
2BcIIKm	between *mauA* and *mauC*	+++	+++	+++	+++	+++	+++	+++
BssHIIKm	inside *mauC*	–	+	+	++	++	++	–
194a-5	inside *mauB*	–	+	–	+	–	–	–
195c-7	inside *mauD*	–	+	–	+	–	++	–
257c-10	inside *mauE*	–	+	–	+	–	++	–
260c-9	inside *mauF*	–	+	–	+	–	+	–
196c-7	between *mauD* and *mauA*	+++	+++	+++	+++	+++	+++	+++
RsrIIKm-4	inside *mauJ*	+++	+++	+++	+++	+++	+++	+++
261c-14	inside *mauJ*	+++	+++	+++	+++	+++	+++	+++
263c-12	inside *mauG*	–	+	–	+	–	+++	+++
264c-1	inside *mauL*	–	+	+++	+++	–	+++	+++
265a-2	inside *mauM*	+++	+++	+++	+++	+++	+++	+++
262c-6	inside *mauN*	+++	+++	+++	+++	+++	+++	+++

+++, a property is the same as wild type; ++, slight decrease, compared to the wild type; +, considerable decrease; –, absence of a property.

The third phenotypic group consists of *mau*J, *mau*M and *mau*N mutants. They can grow on C1-C4 primary amines at the level of the wild type indicating that the *mau*J, *mau*M and *mau*N gene products are not obligatory for growth on methylamine as a carbon source. MADH activity and the large and the small subunit polypeptides in extracts of these mutants are at the level of the wild type. However, the location of *mau*J, *mau*M and *mau*N genes within the *mau* cluster is suggestive of some role in methylamine utilization.

All mutant strains unable to utilize methylamine as a source of carbon are still able to utilize methylamine as a source of nitrogen, however, with growth rates four to five times slower than wild type. This suggests the presence of another (methyl)amine oxidation system in *M. extorquens* AM1. The N-methylglutamate oxidation pathway has not been detected in *M. extorquens* AM1 [42; A.Y. Chistoserdov, unpublished]. Our attempts to find copper-containing amine oxidase activity using phenylmethylamine [44] and phenylethylamine [45] as substrates were negative, and neither substrate is able to serve as a source of nitrogen for *M. extorquens* AM1. In addition the activities of NAD(P)-dependent and dye-linked formaldehyde dehydrogenases and formate dehydrogenases were at wild type levels in all mutants, indicating that none of the *mau* cluster genes are involved in formaldehyde or formate oxidation.

All attempts to generate an insertion of the Km^R cassette into *orf*-1 were futile. Analysis of 1000 Km^R colonies selected on a succinate-containing medium and 500 Km_R colonies selected on a methanol-containing medium has shown that they all retained Tc^R as well as the ability to grow on methylamine as a carbon source. The most likely explanation of this result is that the *orf*-1 gene encodes a product important for cell growth no matter what carbon source is used. The Km^R colonies in this experiment appear as a result of whole vector insertions, which do not lead to a mutation of the *orf*-1 gene.

In order to test complementation by heterologous DNA, plasmids carrying the *mau* genes of *M. extorquens* AM1, *P. denitrificans*, (both orientations) and *Methylophilus* sp. W3A1 (orientation opposite to the *lac* promoter of the vector) respectively were conjugated into all *mau*⁻ mutants of *M. extorquens* AM1. As expected the *mau* cluster from M. extorquens AM1 complemented all the *mau*⁻ mutants. The cloned DNA from *P. denitrificans* does not contain *mau*G and *mau*L, so as expected the *mau*G and *mau*L mutants were not complemented. Mutants in *mau*E, *mau*D, and *mau*A were complemented, but not *mau*B, *mau*C and *mau*F mutations. The *Methylophilus* sp. W3A1 *mau* cluster complemented only the *mau*D and *mau*G mutations. In all cases, the lack of complementation could be due to problems in expression or functional incompatibility, but MauD function is clearly conserved.

5 COMPARISON OF Mau POLYPEPTIDE PROPERTIES

5.1 *MauF*

The entire sequences of MauF from *M. extorquens* AM1 (MauF$_a$, MW 29,464) and *Methylophilus* sp. W3A1 (MauF$_w$, MW 30,429) and a partial sequence of

MauF from *P. denitrificans* (MauF$_p$) were deduced from the nucleotide sequences of the corresponding genes. Similarities are: MauF$_a$/MauF$_w$, 49%; MauF$_a$/MauF$_p$, 65%; MauF$_w$/MauF$_p$, 52%. The MauF polypeptides are predicted (using algorithms [45,46]) to be membrane proteins having four or five transmembrane helices. We did not observe the synthesis of MauF in a T7 expression system [27] probably due to its instability in *E. coli* cells. MauF probably functions in transport and/or maturation of the small subunit polypeptide, since the *mau*F mutant produced no detectable small subunit.

5.2 *MauB*

*mau*B genes encode the large subunit polypeptide [27] and complete sequences are available from *M. extorquens* AM1 (MauB$_a$, MW 44,635), *P. denitrificans* (MauB$_p$, MW 45,440) and *Methylophilus* sp. W3A1 (MauB$_w$, MW 45,243). Similarities are: MauB$_a$/MauB$_p$, 64%; MauB$_a$/MauB$_w$, 45%; MauB$_p$/MauB$_w$, 44%. Such differences are in agreement with the existence of two subclasses of MADH. The MauB polypeptides have normal leader sequences, as expected for periplasmic proteins.

5.3 *MauE*

The entire sequences of MauE from *M. extorquens* AM1 (MauE$_a$; MW$_a$ 19,521) and *Methylophilus* sp. W3A1 (MauE$_w$, MW 20,159) and a partial sequence of MauE from *P. denitrificans* (MauE$_p$) are available. Similarities are: MauE$_a$/MauE$_w$, 39%; MauE$_a$/MauE$_p$, 50%; MauE$_w$/MauE$_p$, 37%. The greatest similarity between MauE$_a$ and MauE$_w$ exists in hydrophilic domains, the lowest in hydrophobic domains. The *mau*E genes were predicted to encode membrane-bound polypeptides having four transmembrane helices. A translational fusion between *mau*E from *M. extorquens* AM1 and *pho*A from *E. coli* was generated in the hydrophilic domain of MauE between the first two hydrophobic helices. This fused protein showed phosphatase activity in *E. coli*, indicating that this domain is exposed to the periplasm [47].

MauE is toxic for *E. coli*. All plasmid constructions in which the *mau*E gene from *M. extorquens* AM1 can be transcribed from an *E. coli* promoter cause retardation of the host growth. We were not able to obtain expression of the *mau* cluster from *Methylophilus* sp. W3A1 in *E. coli* under the *lac* promoter, probably because of the toxicity of MauE. MauE is likely to be involved in transport and/or maturation of the small subunit polypeptide, since the *mau*E mutant has a phenotype similar to the *mau*F mutant.

5.4 *MauD*

The entire sequences of MauD from *M. extorquens* AM1 (MauD$_a$, MW 22,233) and *Methylophilus* sp. W3A1 (MauD$_w$, MW 23,083) and C-terminal sequences of the MauD polypeptides from *P. denitrificans* (MauD$_p$), *T. versutus* (MauD$_t$)

Molecular biology and genetics of MADH 391

```
                                          +         0 0
a.                                       MRALAFAAALAAFSATAALA

                                    0 0+ +0              0
b.                                 MISATKIRSCLAACVLAAFGATGALA

       +00  --   -+ 0++    000++     +   0           *  -++  + 0+  0
c.   MLGKSQFDDLFEKMSRKVAGHTSRRGFIGRVGTAVAGVALVPLLPVDRRGRVSRANA

        0  +  --   -+ 0++   000+++0   +   0          *  -++  + 0+  0
d.    MLGNFRFDDMVEKLSRRVAGQTSRRSVIGKLGTAMLGIGLVPLLPVDRRGRVSRANA

          ++-0  -0+ -+  +00 0+0 ++     +       0       -++0+     -
e.       MKKDTGPDSKIEKLARTTASKTGRRGFIGRLGGFLVGSALLPLLPVDRRSRLGGEVQA

                          +  +0 0++-  00   +      0       0   -   +
f.                      MAKPKSPSRRELLTNGVKAAGVTCLAGLALTAYVESASKA

                                   00    +  +0       0 0
g.                                MMTHAYTKVRQALCWGSATLGAAALA

                                         0      +  0         0
h.                                      MALPPNFMPLFRASLIGLGLGCSALALA

                                              ++0           0  0
i.                                           MKKTAIAIAVALAGFATVAQA

                                           +  0+      00     -1
j.                                        MKATKLVLGAVILGSTLLAGC

                                         0 0      00         -1
k.                                      MTMQFLIASNVLLWLALIGC
```

Figure 3 Leader sequences of Mau polypeptides: a. MauC (amicyanin) from *M. extorquens* AM1; b. MauC (amicyanin) from *P. denitrificans*; c. MauA (small subunit of MADH) from *M. extorquens* AM1; d. MauA (small subunit of MADH) from *P. denitrificans*; e. MauA (small subunit of MADH) from *Methylophilus* sp. W3A1; f. MauM from *M. extorquens* AM1; g. MauB (large subunit of MADH) from *M. extorquens* AM1; h. MauB (large subunit of MADH) from *P. denitrificans*; i. leader sequence of OmpA protein from *E. coli*, shown for comparison; j. leader sequence of the major lipoprotein from *E. coli*, shown for comparison; k. MauD from *M. extorquens* AM1. Positively charged N-termini of sequences interacting with the phospholipid layer are underlined; -1 above lipoprotein leader sequences denotes the last amino acid of the leader, for regular leader sequences the last written amino acid is a -1 amino acid. Symbols $+$, $-$, 0 above amino acids designate positively charged, negatively charged and polar amino acids.

and *M. flagellatum* KT (MauD$_k$) are available. Similarities between MauD$_a$ and MauD$_w$, MauD$_t$, MauD$_k$ and MauD$_p$ are 52%, 82%, 20% and 60%, respectively. Similarities between MauD$_w$ and MauD$_t$, MauD$_k$ and MauD$_p$ are 43%, 56% and 26% respectively. Similarities between MauD$_t$ and MauD$_k$ and MauD$_p$ are 32% and 86%, respectively. Similarity between MauD$_k$ and MauD$_p$ is 26%. Variations in the levels of similarity are connected with the fact that the C-terminal sequences of MauD are the least conserved. MauD is predicted to be periplasmic, but the predicted MauD leader sequences, which are known only for MauD$_w$ and MauD$_a$, do not have positive charges on the N-termini (Fig. 3). It is known that substitutions of the positively charged amino acids in the normal leader sequences of *E. coli* periplasmic proteins leads to considerable slowing of transport into the periplasm [48]. Fused MauD$_a$-PhoA proteins were still able to be transported into the periplasm of *E. coli*, but activities were one-tenth those of a β-lactamase-PhoA (Bla-PhoA) fused protein [47]. The activities of the MauD$_a$-PhoA proteins were higher in the periplasm of *M. extorquens* AM1 and comparable with the activities of the MauA$_a$-PhoA fusion [49]. MauD$_a$ has a leader sequence similar of those of lipoproteins, whereas MauD$_w$ has a regular periplasmic leader sequence. However, the latter can complement the *mauD* mutant of *M. extorquens* AM1. The *mauD* mutant has a phenotype similar to mutants in *mauE* and *mauF*.

5.5 MauA

MauA is the small subunit of MADH. The entire sequences of the MauA polypeptides from *M. extorquens* AM1 (MauA$_a$, MW 20,084), *P. denitrificans* (MauA$_p$, MW 20,393), *Methylophilus* sp. W3A1 (MauA$_w$, MW 20,237), *M. flagellatum* KT (MauA$_k$, MW 20,174) and *T. versutus* (MauA$_t$, MW 20,358) are available. The similarities between MauA$_a$ and MauA$_p$, MauA$_w$, MauA$_k$ and MauA$_t$ are 81%, 64%, 66% and 84%, respectively. The similarities between MauA$_p$ and MauA$_w$, MauA$_k$ and MauA$_t$ are 62%, 66% and 92% respectively. The similarities between MauA$_w$ and MauA$_k$ and MauA$_t$ are 85% and 64%, respectively. Similarity between MauA$_k$ and MauA$_t$ is 66%. Alignment of the amino acid sequences of MauA$_w$, MauA$_a$, MauA$_p$ and MauA$_t$ is shown in Fig. 4, and the similarities are plotted in Fig. 5.

All of these MauA polypeptides share two unusual properties. They contain the TTQ cofactor synthesized from two tryptophans of the polypeptide chain and they have unusual leader sequences (Fig. 3). A leader sequence having positive charges near a signal peptidase recognition site greatly slows translocation into the periplasm of *E. coli* [48]. A fused MauA$_a$-PhoA protein was able to be transported into the periplasm of *M. extorquens* AM1 but not *E. coli* [49], indicating the presence of a specific translocation system in the methylotroph. The leader sequences of MauA are also considerably longer than regular leader sequences and are similar to each other especially in the positively charged area near the signal peptidase cleavage sites. The role of this unusual leader sequence is not yet known, but it might serve as an intramolecular chaperone, or participate in cofactor synthesis.

Although all MauA polypeptides are highly similar, the amino acid sequence of MauA$_w$ belonging to MADH$_c$ differs distinctly from the amino acid sequences of MauA$_a$, MauA$_p$ and MauA$_t$ belonging to MADH$_a$. The greatest difference is in the leader sequences, and, in addition, MauA$_w$ has seven disulfide bonds instead of six disulfide bonds typical for MauAs belonging to MADH$_a$. Once again MauA$_k$ is a surprising exception. It also has seven disulfide bonds and the leader sequence resembles that of MauA$_w$ more than the other MauA polypeptides.

```
a.  MLGKSQFDDLFEKMSRKVAGHTSRRGFIGRVGTAVAGVALVPLLPVDRRGRVS-RANA*
    :::    :::.::: ::::    ::.::::   ...:::::::::::::: ::::
b.  MLGNFRFDDMVEKLSRRVAGQTSRRSVIGKLGTAMLGIGLVPLLPVDRRGRVS-RANA*
    :::::::::::::::::::::: ::::: ::..:::  . :  :::::::::::::: ::::
c.  MLGNFRFDDMVEKLSRRVAGRTSRRGAIGRLGTVLAGAALVPLLPVDRRGRVS-RANA*
     :  ::  .::  :. :  :: ::..:: : : :::.::::::::.    : :
d.  MKKNTGFDSGIEKLARKTASKTGRRSFIGKLGGFLVGSALLPLLPVDRRGRMN-EAHA*
    :::  :::::  ::::::  ::::::::: :::::::::::::::::::. :   :
e.  MKKDTGFDSKIEKLARTTASKTGRRGFIGRLGGFLVGSALLPLLPVDRRSRLGGEVQA*


a.  AE-SAG-DPRGKWKPQDNDVQSCDYWRHCSIDGNICDCSGGSLTSCPPGTKLASSS*WV*
    :.   :.  :::.::  ::::::  :::::::::::::::::::: ::::::::: :::
b.  ADAPAGTDPRAKWVPQDNDIQACDYWRHCSIDGNICDCSGGSLTNCPPGTKLATAS*WV*
    :      :::::::::::::::::::::::::::::::::::::::::::::::::
c.  AGPAEGVDPRAKWQPQDNDIQACDYWRHCSIDGNICDCSGGSLTNCPPGTKLATAS*WV*
    ::   :  . :::  ::::   :::::::::::::  :::::  ::::: :    ::
d.  --ETKGVLGREGYKPQDKDPKSCDYWRHCSIDGNLCDCCGGSLTSCPPGTELSPSS*WV*
    :   : : :  :.::::::: ::::::.::::::::.:::::::. ::::::: 
e.  AT-T-GNLTRSGFKPQDKDPKACDYWRHCTIDGNLCDCCGGTLTSCPPGSSLSPSS*WV*


a.  ASCYNPTDKQSYLISYRDCCGANVSGRCACLNTEGELPVYRPEFGNDIIWCFGAEDDA
    :::::::::  ::::: :::::: :::::: ::::::::::::::::.:::::::::::
b.  ASCYNPTDGQSYLIAYRDCCGYNVSGRCPCLNTEGELPVYRPEFANDIIWCFGAEDDA
    :::::::::::::::::::::::::::::::::::::::::::::::::::::::::
c.  ASCYNPTDGQSYLIAYRDCCGYNVSGRCPCLNTEGELPVYRPEFANDIIWCFGAEDDA
    :::::.:::.:::::::::   :::  :.: :::::::::: :::.::::::.  ::
d.  ASCFNPGDGQTYLIAYRDCCGKQTCGRCNCVNQGELPVYRPEFNNDIVWCFGADNDA
    :::.::::  :::::::::::::::::::::::::: ::::::::::::::::::::
e.  ASCYNPGDQQTYLIAYRDCCGKQTCGRCNCVNTQGELPVYRPEFNNDIVWCFGADNDA


a.  MTYHCTISPIVGKAS
    :::::::::::::::
b.  MTYHCTISPIVGKAS
    :::::::::::::::
c.  MTYHCTISPIVGKAS
    :::::::::::::::
d.  MTYHCTVSPIVGKAS
    :::::::::::::::
e.  MTYHCTISPIVGKAS
```

Figure 4 Alignment of the amino acid sequences for MauA from: a. *M. extorquens* AM1. b. *P. denitrificans* c. *T. versutus* d. *M. flagellatum* KT e. *Methylophilus* sp. W3A1. Asterisks indicate signal peptidase recognition site. Tryptophans involved in TTQ assembly are shown in italics. Pairs of cysteines in the sequences from *M. flagellatum* and *Methylophilus* sp. W3A1 involved in the seventh disulfide bond are shown in bold.

5.6 MauC and MauO

The *mauC* genes encode amicyanins. The amino acid sequences of amicyanins from *M. extorquens* AM1 ($MauC_a$, MW 12,606), *P. denitrificans* ($MauC_p$, MW 13,983) and *T. versutus* ($MauC_t$, MW 14,258) are available. Similarities are: $MauC_a/MauC_p$, 48%; $MauC_a/MauC_t$, 44%; $MauC_p/MauC_t$, 61%.

It has been shown that *M. flagellatum* KT has amicyanin and MADH may use it as an electron acceptor [19], but the amicyanin gene was not found in

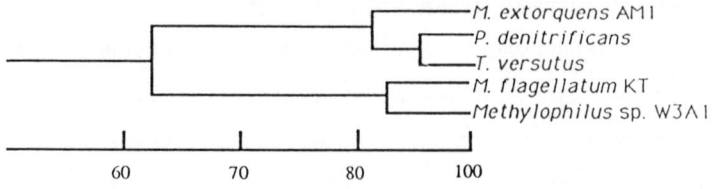

Figure 5 Similarity (%) between the small subunit polypeptides from different methylotrophic bacteria.

the *mau* cluster so far. The amicyanin gene may be located elsewhere in the chromosome. In *Methylophilus* sp. W3A1 amicyanin has not been detected, and *mauC* is not present in the *mau* gene cluster. It probably uses a *c*-type cytochrome as an electron acceptor. The only candidate found in the *mau* cluster which might be such a cytochrome is a product of the *mauO* gene. The sequencing of the *mauO* gene is not completed, but the last 148 amino acids are known, and these include a *c*-type cytochrome heme binding site.

5.7 MauJ

The entire amino acid sequence of the MauJ polypeptide from *M. extorquens* AM1 (MauJ$_a$, MW 32,682) and partial sequences of MauJ from *P. denitrificans* (MauJ$_p$) and *T. versutus* (MauJ$_t$) are available. Similarities are: MauJ$_a$/MauJ$_p$, 53%; MauJ$_a$/MauJ$_t$, 45%; MauJ$_p$/MauJ$_t$, 62%. These are the only polypeptides encoded by the *mau* cluster that are predicted to be cytoplasmic. The *mauJ* gene is absent in the *mau* clusters of *Methylophilus* sp. W3A1 and *M. flagellatum* KT. Two *mauJ* mutants of *M. extorquens* AM1 were generated, and they were not impaired in their ability to grow on methylamine as a source of carbon.

5.8 MauG

The entire amino acid sequences of MauG from *M. extorquens* AM1 (MauG$_a$, MW 38,145) and *Methylophilus* sp. W3A1 (MauG$_w$, MW 37,265) and a partial amino acid sequence of MauG from *M. flagellatum* KT are available. Similarities are: MauG$_a$/MauG$_w$, 51%; MauG$_a$/MauG$_k$, 55%; MauG$_w$/MauG$_k$, 65%. MauG$_w$ can function normally in *M. extorquens* AM1, complementing the corresponding mutation.

Analysis of the amino acid sequences of MauG suggested that they are periplasmic diheme cytochromes. They show similarity with cytochrome c peroxidase from *Pseudomonas aeruginosa*, which is also a periplasmic diheme cytochrome [50]. The *mauG* mutant of *M. extorquens* AM1 produces the large and small subunit polypeptides of the mature sizes, but MADH is not functional. It is possible that MauG is involved in synthesizing the Trp-Trp bond of TTQ, which should occur in the periplasm.

5.9 MauL

The amino acid sequences of MauL from *M. extorquens* AM1 (MauL$_a$, MW 18,771) and *Methylophilus* sp. W3A1 (MauL$_w$, MW 19,085) are available. They show 31% similarity. These two polypeptides are predicted to be localized in the periplasm. MauL$_a$ has a regular leader sequence whereas MauL$_w$ is predicted to be a lipoprotein (Fig. 3). The properties of the *mauL* mutants resemble those of the *mauG* mutant, implying their participation in the same stage of TTQ synthesis. MauL might provide spatial orientation and recognition of the small subunit and the peroxidase.

5.10 MauM and MauN

The amino acid sequences of the entire MauM polypeptides from *M. extorquens* AM1 (MauM$_a$, MW 23,308) and *Methylophilus* sp. W3A1 (MauM$_w$, MW 23,868) and the sequence of the MauN polypeptide (MW 30,683) from *M. extorquens* AM1 are available. The similarity between MauM$_a$ and MauM$_w$ is 55%.

The products of *mauM* and *mauN* are not obligatory for methylamine utilization. The MauM polypeptides from *M. extorquens* AM1 and *Methylophilus* sp. W3A1 are predicted to be periplasmic proteins. The MauN polypeptide is predicted to be membrane bound. Homology searches in the GenBank library reveal substantial homology of MauM and the C-terminal part of MauN with ferredoxins, indicating the presence of FeS clusters in these proteins. One of the possible functions of MauM and MauN could be to produce hydrogen peroxide for a peroxidase reaction.

6   A MODEL OF MADH ASSEMBLY

A working model of MADH assembly which takes into account all available data concerning mutant phenotypes and protein sequences is shown in Fig. 6. In this model, the small subunit polypeptide attaches to the inner membrane as it is synthesized. Due to the presence of positively charged amino acids in the leader sequence, the polypeptide is not transported into the periplasm co-translationally. This allows a modifying enzyme (oxygenase or hydroxylase) to convert the first tryptophan into tryptophylquinone. It does not seem likely that any of the known Mau proteins can be responsible for this Trp modification. It is possible that the long leader sequence of MauA participates in this event, for example, by supporting necessary conformation of the polypeptide (a kind of intramolecular chaperone). MauF and MauE may also participate at this stage. It is also possible that the unfolded conformation of the maturing small subunit polypeptide may be maintained by the *mauD* gene product, which continues to provide its chaperone function in the periplasm. Once the small subunit is modified it could then be transported into the periplasm. We assume

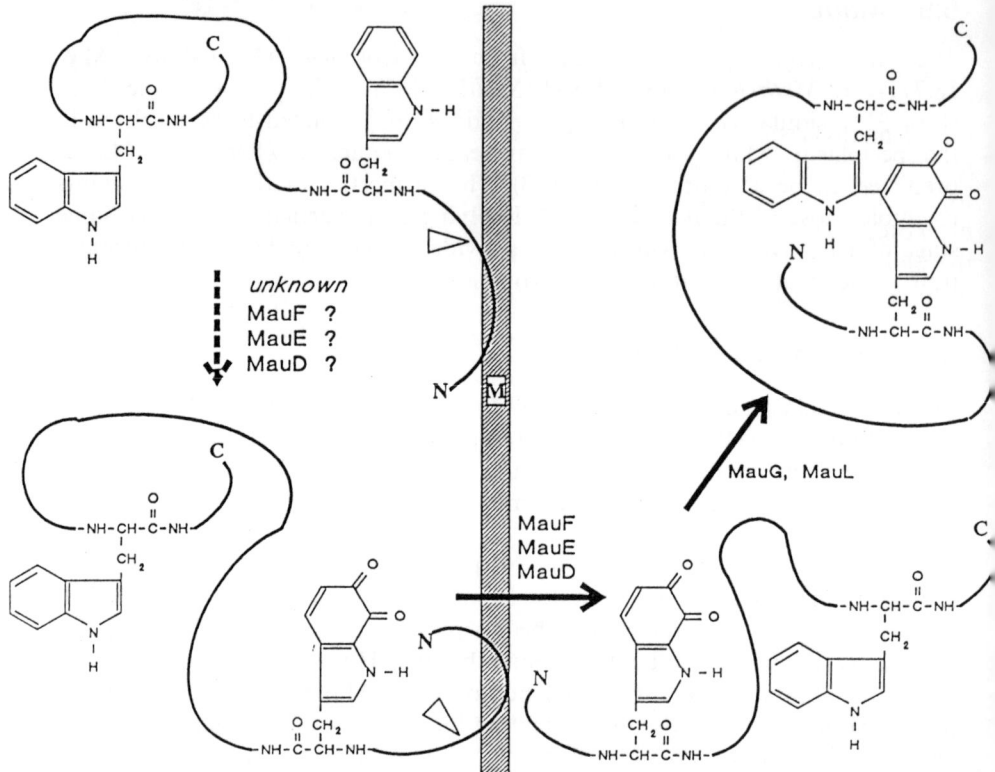

Figure 6 A model for maturation of the MADH small subunit polypeptide. 'M' denotes the inner membrane. The cytoplasm is on the left of the figure, the periplasm is on the right. Triangle denotes the signal peptidase recognition site. Synthesized TTQ is shown in the right upper corner of the figure.

that normal transport machinery functions at this stage, probably assisted by the *mau*F and *mau*E gene products. The transport of the large subunit, the amicyanin (or other acceptor), MauG and MauL into the periplasm should occur independently. In the periplasm a MauA-MauD complex could attach to the inner membrane through the lipid part of MauD (in the case of $MADH_a$) or through the lipid part of MauL (in the case of $MADH_c$). The function of MauL in this model is to provide the recognition of the tryptophylquinone and the second tryptophan by the MauG peroxidase. In the end the MauD-MauA-MauL complex can be attacked by the MauG peroxidase, completing the maturation of the small subunit polypeptide. Hydrogen peroxide required for this reaction can be provided by the *mau*M and *mau*N gene products or an amine oxidase. It is not clear at which stage of the small subunit maturation the large subunit polypeptide becomes attached to form a holoenzyme. It seems likely that this would occur at the later stages of the process or after it is complete.

We have attempted to determine whether all known products of the *mau* gene cluster are sufficient to allow a methanol-utilizing bacterium to use methylamine. The plasmid carrying the *mau* cluster from *M. extorquens* AM1 was conjugated into three different pink-pigmented methylotrophs unable to utilize methylamine as isolated (*Methylobacterium* sp. 317, *Methylobacterium* sp. D12 and *Methylobacterium* R14 [51]) . These bacteria are closely related to *M. extorquens* AM1, therefore, any genes from *M. extorquens* AM1 should be expressed in them without problems. The *mau* gene cluster does not provide the ability to grow on methylamine for these bacteria suggesting the necessity of some other gene products outside of the *mau* cluster, possibly those responsible for quinone cofactor synthesis.

7 CONCLUDING REMARKS: THE EVOLUTION OF THE *mau* SYSTEM

All MADHs can be divided into two groups based on the type of electron acceptor they use: $MADH_c$ and $MADH_a$. Data shown here indicate that this division does not reflect taxonomic relationships between bacteria harboring *mau* genes. MADH and the *mau* cluster from *M. flagellatum* KT share some properties with those of *Methylophilus* sp. W3A1 (alkaline isoelectrofocusing point of the large subunit, seven disulfide bonds in the small subunit, the absence of the amicyanin and *mau*J genes in the cluster, higher similarity of all *mau* genes products to those from *Methylophilus* sp. W3A1 rather than *M. extorquens* AM1) and some with *M. extorquens* AM1 (antibodies raised against the large subunit from *M. flagellatum* KT bind the large subunit from *M. extorquens* AM1 well, whereas antibodies raised against the large subunit from *Methylophilus sp.* W3A1 do not bind the large subunit from *M. extorquens* AM1; both MADH use amicyanin as an electron acceptor). *M. flagellatum* KT and *Methylophilus* sp. W3A1 are presumably phylogenetically related, since all known strains of these groups classify in the beta-subgroup of the proteobacteria [52]. *M. extorquens* AM1, *P. denitrificans* and *T. versutus* on the other hand belong to the alpha-subgroup of proteobacteria [53]. This is in keeping with the higher relatedness of the *mau* cluster from *M. flagellatum* KT and *Methylophilus* sp. W3A1, as compared to the strains from the alpha proteobacteria. An interesting question now is whether *Methylophilus* spp. really do not have amicyanin or whether amicyanin in *Methylobacillus* spp. is an obligatory component of the electron transfer chain during growth on methylamine.

REFERENCES

[1] Knowles, P.F. and Yadav., K.D.S. (1984) Amino oxidases. In: Copper Proteins and Copper Enzymes (Lontie, R., Ed.), pp. 103–129. CRC Press, Boca Raton, FL.
[2] Large, P.J. and Green, J. (1984) Oxidation of mono-, di-, and trimethylamine by methazotrophic yeasts: properties of the microsomal and peroxisomal enzymes

involved and comparison with bacterial enzyme systems. In: Microbial Growth on C_1 Compounds (R.L. Crawford and R.S. Hanson, Eds), pp. 155–164. American Society for Microbiology, Washington, DC.

[3] Levering, P.R., Van Dijken, J.P., Veenhuis, M. and Harder, W. (1981) *Arthrobacter* P1, a fast growing versatile methylotroph with amine oxidase. *Arch. Microbiol.* **129**, 72–80.

[4] Boulton, C.A., Haywood, G.W. and Large, P.J. (1980) N-methylglutamate dehydrogenase, a flavohaemoprotein purified from a new pink trimethylamine-utilising bacterium. *J. Gen. Microbiol.* **117**, 293–304.

[5] Hersh, L.B., Peterson, J.A. and Tompson, A.A. (1971) An N-methylglutamate dehydrogenase from *Pseudomonas* MA. *Arch. Biochem. Biophys.* **145**, 115–120.

[6] Loginova, N.V., Shishkina, V.N. and Trotsenko, Yu.A. (1976) Primary metabolic pathways of methylated amines in *Hyphomicrobium vulgare*. *Microbiology (USSR)* **45**, 34–40.

[7] Eady, R.R. and Large, P.J. (1971) Microbial oxidation of amines. Spectral and kinetic properties of the primary amine dehydrogenase of *Pseudomonas* AM1. *Biochem. J.* **123**, 757–771.

[8] Haywood, G.W., Janschke, N.S., Large, P.J. and Wallis, J.M. (1982) Properties and subunit structure of methylamine dehydrogenase from *Thiobacillus* A2 and *Methylophilus methylotrophus*. *FEMS Microbiol. Lett.* **15**, 79–82.

[9] Husain, M. and Davidson, V.L. (1987) Purification and properties of methylamine dehydrogenase from *Paracoccus denitrificans*. *J. Bacteriol.* **169**, 1712–1717.

[10] Kenny, W.C. and McIntire, W. (1983) Characterization of methylamine dehydrogenase from bacterium W3A1: interaction with reductant and amino-containing compounds. *Biochemistry* **22**, 3858–3868.

[11] Matsumoto, T. (1978) Methylamine dehydrogenase of *Pseudomonas* sp. J: purification and properties of the subunits. *Biochim. Biophys. Acta* **522**, 291–302.

[12] Kirukhin, M.Y., Chistoserdov, A.Y. and Tsygankov, Y.D. (1990) Methylamine dehydrogenase from *Methylobacillus flagellatum*. In: Methods in Enzymology (M.E. Lidstrom, Ed.), Vol. 188, pp. 247–256. Academic Press, San Diego.

[13] Mehta, R.J. (1977) Methylamine dehydrogenase from the obligate methylotroph *Methylomonas methylovora*. *Can. J. Microbiol.* **23**, 402–406.

[14] Eady, R.R. and Large, P.J. (1968) Purification and properties of an amine dehydrogenase from *Pseudomonas* AM1 and its role in growth on methylamine. *Biochem. J.* **106**, 245–255.

[15] Lawton, S.A. and Anthony, C. (1985) The role of copper proteins in the oxidation of methylamine by an obligate methylotroph. *Biochem. J.* **228**, 719–725.

[16] Husain, M. and Davidson, V.L. (1985) An inducible periplasmic blue copper protein from *Paracoccus denitrificans*: purification, properties, and physiological role. *J. Biol. Chem.* **260**, 14626–14629.

[17] Houwelingen, T. van, Canters, G.W., Stobbelaar, G., Duine, J.A., Frank, Jzn. J. and Tsugita, A. (1985) Isolation and characterization of a blue copper protein from *Thiobacillus versutus*. *Eur. J. Biochem.* **153**, 75–80.

[18] Tobari, J. (1984) Blue copper proteins in electron transport in methylotrophic bacteria. In: Microbial Growth on C_1 Compounds (R.L. Crawford and R.S. Hanson, Eds), pp. 106–112, American Society for Microbiology, Washington, DC.

[19] Dinarieva T. and Netrusov, A. (1989) Cupredoxines of obligate methylotroph. *FEBS Lett.* **259**, 47–49.

[20] Auton, K.A. and Anthony, C. (1989) The role of cytochromes and blue copper proteins in growth of an obligate methylotroph on methanol and methylamine. *J. Gen. Microbiol.* **135**, 1923–1931.

[21] Chandrasekar, R. and Klapper, M.H. (1986) Methylamine dehydrogenase and cytochrome c_{552} from the bacterium W3A1. *J. Biol. Chem.* **261**, 3616–3619.

[22] Burton, S.M, Byrom, D., Carver, M., Jones, G.D.D. and Jones, C.W. (1983) The oxidation of methylated amines by methylotrophic bacterium *Methylophillus methylotrophus. FEMS Microbiol. Lett.* **17**, 185–190.
[23] Davidson, V.L. and Neher, J.W. (1987) Evidence for two subclasses of methylamine dehydrogenases with distinct large subunits and conserved PQQ-bearing small subunits. *FEMS Microbiol. Lett.* **44**, 121–124.
[24] Ishii, Y., Hase, T., Fukumori, Y., Matsubara, H. and Tobari, J. (1983) Amino acid sequence studies of the light subunit of methylamine dehydrogenase from *Pseudomonas* AM1: existence of two residues binding the prosthetic group. *J. Biochem.(Tokyo)* **93**, 107–119.
[25] McIntire, W.S., Wemmer, D.E., Chistoserdov, A.Y. and Lidstrom, M.E. (1991) A new cofactor in a procaryotic enzyme: tryptophan tryptophylquinone as the redox prosthetic group in methylamine dehydrogenase. *Science* **252**, 817–824.
[26] Chistoserdov, A.Y., Tsygankov, Y.D. and Lidstrom, M.E. (1990) Cloning and sequencing of the structural gene for the small subunit of methylamine dehydrogenase from *Methylobacterium extorquens* AM1: Evidence for two tryptophan residues involved in the active center. *Biochem. Biophys. Res. Comm.* **172**, 211–216.
[27] Chistoserdov, A.Y., Tsygankov, Y.D. and Lidstrom, M.E. (1991) Genetic organization of methylamine utilization genes from *Methylobacterium extorquens AM1*. *J. Bacteriol.* **173**, 5901–5908.
[28] Chistoserdov, A.Y., Tsygankov, Y.D. and Lidstrom, M.E. (1991) Nucleotide sequence of the amicyanin gene from *Methylobacterium extorquens* AM1. *DNA Sequence* **2**, 53–55.
[29] Chistoserdov, A.Y., Boyd, J., Mathews, F.S. and Lidstrom, M.E. (1992) The genetic organization of the *mau* gene cluster of the facultative autotroph *Paracoccus denitrificans. Biochem. Biophys. Res. Comm.*, **184**, 1226–1234.
[30] Spanning, van, R.J.M., Wansell, C.W., Reijnders, W.N.M., Oltman, L.F. and Stouthamer A.H. (1990) Mutagenesis of the gene encoding amicyanin of *Paracoccus denitrificans* and the resultant effect on methylamine oxidation. *FEBS Lett.* **275**, 217–220.
[31] Ubbink, M., van Kleef, M.A.G., Kleinjan, D.-J., Hoitink, C.W.G., Huitema, F., Beintema, J.J., Duine J.A. and Canters G.W. (1991) Cloning, sequencing and expression studies of the genes encoding amicyanin and the β-subunit of methylamine dehydrogenase from *Thiobacillus versutus. Eur. J. Biochem.* **202**, 1003–1012.
[32] Gak, E.R., personal communication.
[33] Auberton Carafa, Y. d', Brody, E. and Thermes, C. (1990) Prediction of Rho-independent *Escherichia coli* transcription terminators. A statistical analysis of their RNA stem-loop structures. *J. Mol. Biol.* **216**, 835–858.
[34] Tsygankov, Y.D. and Chistoserdov, A.Y. (1985) Specific-purpose broad-host-range vectors. *Plasmid* **14**, 118–125
[35] Laufer, K. and Lidstrom, M.E. (1992) Regulation and expression of bacterial quinoproteins. In: Quinoproteins (Davidson, V., Ed.), Marcel-Dekker, New York.
[36] Morris, C.J. and Lidstrom, M.E. (1992) Cloning of a methanol-inducible *mox*F promoter and its analysis in *mox*B mutants of *Methylobacterium extorquens* AM1. *J. Bacteriol.* **174**, 4444–4449.
[37] Gak, E.R., Chistoserdov, A.Y. and Tsygankov, Y.D. (1989) Mutants of *Methylobacillus flagellatum* defective in catabolism of methylamine. In: Abstracts of 6th Int. Symp. Microbial Growth on C_1 Compounds, p. 417. Göttingen, Germany.
[38] Chistoserdov, A.Y., Chistoserdova, L.V., McIntire, W. and Lidstrom, M.E., unpublished.
[39] Paz, M.A., Fluckiger, R., Boak, A., Kagan, H.M. and Gallop, P.M. (1991) Specific detection of quinoproteins by redox-cycling staining. *J. Biol. Chem.* **266**, 689–692.
[40] Fukumori, Y. and Yamanaka, T. (1987) The methylamine oxidizing system of

Pseudomonas AM1 reconstituted with purified components. *J. Biochem.* **101**, 441–445.
[41] Anthony, C. 1982. The Biochemistry of Methylotrophs. Academic Press, London.
[42] Wagner, C. and Quayle, J.R. (1972) Carbon assimilation pathways during growth of *Pseudomonas* AM1 on methylamine and *Pseudomonas* MS on methylamine and trimethylsulphonium salts. *J. Gen. Microbiol.* **72**, 485–491.
[43] Dooly, D.M., McIntire, W.S., McGuirl, M.A., Cote, C.E. and Bates J.L. (1990) Characterization of the active site of *Arthrobacter* P1 methylamine oxidase: evidence for copper-quinone interactions. *J. Am. Chem. Soc.* **112**, 2782–2789.
[44] Parrott, S., Jones, S. and Cooper, R.A. (1987) 2-phenylalanine catabolism by *Escherichia coli* K12. *J. Gen. Microbiol.* **133**, 347–351.
[45] Rao, J.K.M. and Argos, P. (1986) A conformational preference parameter to predict helices in integral membrane proteins. *Biochim. Biophys. Acta.* **869**, 197–214.
[46] Eisenberg D., Schwarz, E., Komaromy, M. and Wall, R. (1984) Analysis of membrane and surface protein sequences with the hydrophobic moment plot. *J. Mol. Biol.* **179**, 125–142.
[47] Chistoserdov A. Y. and Lidstrom, M. E., unpublished observations.
[48] Gennity, J., Goldstain, J. and Inouye M. (1990) Signal peptide mutants of *Escherichia coli.* *J. Bioenerg. Biomembr.* **22**, 233–269.
[49] Chistoserdov A.Y., and Lidstrom, M.E. (1991) The small subunit polypeptide of methylamine dehydrogenase from *Methylobacterium extorquens* AM1 has an unusual leader sequence. *J. Bacteriol.* **173**, 5909–5913.
[50] Ronnberg, M., Kalkkinen, N. and Ellfolk, N. (1989) The primary structure of *Pseudomonas* cytochrome *c* peroxidase. *FEBS Lett.* **250**, 175–178.
[51] Green, P.N. (1991) The genus *Methylobacterium*. In: The Prokaryotes II, (Balows, A. et al., Eds), Vol. 1, pp. 2342–2349. Springer-Verlag, New York.
[52] Tsuji, K., Tsien, H.C., Hanson, R.S., Depalma, S.R., Scholtz, R. and Laroche, S. (1990) 16S ribosomal RNA sequence analysis for determination of phylogenetic relationship among methylotrophs. *J. Gen. Microbiol.* **136**, 1–10.
[53] Stouthamer, A.H. (1992) Metabolic pathways in *Paracoccus denitrificans* and closely related bacteria in relation to the phylogeny of prokaryotes. *Antonie v.Leeuwenhock J. Microbiol. Serol.* **61**, 1–34.

31

New Developments in the Biochemistry of Quinoproteins in Methylotrophs

J.A. Duine

Department of Microbiology and Enzymology, Delft University of Technology, Julianalaan 67, 2628 BC Delft, The Netherlands

1 SUMMARY

Microbial oxidation of C_1 compounds is catalyzed by a number of oxidoreductases having a quinone as cofactor. These so-called quinoproteins oxidize methanol (the PQQ-containing methanol dehydrogenase) and methylamine the tryptophanyl tryptophanquinone (TTQ)-containing methylamine dehydrogenase (MADH) and the topaquinone (TPQ)-containing methylamine oxidase (MeAO)). MeAO is widespread, occurring in yeasts as well as in Gram-negative and Gram-positive bacteria. TTQ and TPQ are protein-chain integrated quinone-cofactors derived from 2 tryptophans and tyrosine, respectively. It is highly likely that the formation of these cofactors occurs by post-translational modification, just as has been found for enzymes using the free radical forms of tryptophan or tyrosine as cofactor. Evidence is now emerging that biosynthesis of PQQ may proceed in a similar way, a peptide being the building block out of which a number of enzymes transform tyrosine and glutamic acid into PQQ, after which a metalloprotease excises PQQ.

Formaldehyde oxidation in methylamine-grown *Hyphomicrobium zavarzinii* occurs by a dye-linked enzyme containing a quinone cofactor of unknown structure. However, as with methanol oxidation, formaldehyde oxidation is catalyzed by a large variety of enzymes having different coenzyme/cofactor requirements, other examples being: (1) NAD-linked enzymes, some of these requiring the participation of a co-substrate (either GSH or Factor); (2)

Microbial Growth on C_1 Compounds
© Intercept Ltd, PO Box 716, Andover, Hampshire SP10 1YG, UK

nicotinoproteins, enzymes containing NAD(P) as cofactor, and catalyzing dismutation of formaldehyde into methanol and formic acid or oxidation of formaldehyde with a nitroso-compound (the *in vitro* assay); (3) dye-linked molybdoprotein formaldehyde dehydrogenase containing 5 different cofactors (Fe, S, Mo, FAD, pterine). The latter enzyme appears to be very interesting as it also catalyzes the oxidation of formyl esters. Indications exist that this reaction occurs in Gram-positive bacteria (oxidation of Factor-formyl ester to Factor-carbonate ester, decomposing into Factor and CO_2) and may form an oxidative alternative in Gram-negative bacteria for the well-known hydrolytic pathway in which S-formylglutathione is transformed into GSH and formic acid.

2 INTRODUCTION

Although quinoproteins (enzymes with a quinone cofactor) were discovered 13 years ago, new developments have taken place in this field during the past few years. One of these was the determination of the structures of the protein-chain integrated cofactors (Fig. 1), topaquinone (TPQ) [1] and tryptophanyl tryptophanquinone (TTQ) [2]. These cofactors are involved in catalysis by quinoprotein amine oxidoreductases, the enzymes relevant for methylotrophs indicated as methylamine dehydrogenase (MADH, EC 1.4.99.3) and methylamine oxidase (MeAO) (the latter belonging to the large group of copper-containing amine oxidases, EC 1.4.3.6). Cloning and sequencing of the genes encoding these enzymes have helped tremendously in the structure assignment and in stressing that TPQ and TTQ are aromatic amino acid-derived cofactors, and not covalently-bound PQQ [3]. Elucidation of the 3-dimensional structure of MADH [4] and its complex with the natural electron acceptor, the blue copper protein amicyanin [15], has now provided the tools to allow rational modification of the mechanism of electron transfer in the soluble segment of a microbial respiratory chain.

Several C_1 compound-converting oxidoreductases have been described of which the quinoprotein nature has been presumed but has not definitely been proven. In the search for the PQQ-containing methanol-oxidizing multi-enzyme

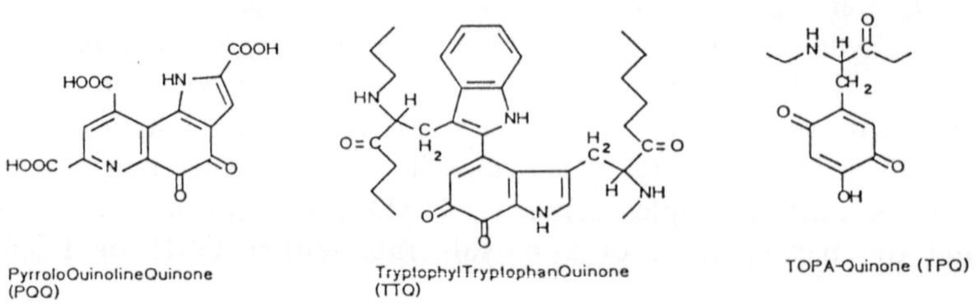

Figure 1 Structures of pyrroloquinoline quinone (PQQ), tryptophanyl tryptophanquinon (TTQ), and topaquinone (TPQ).

complex of *Amycolatopsis methanolica* [6], enzymes have been isolated which are called 'nicotinoproteins', i.e. dehydrogenases containing NAD(P) as cofactor [7,8]. Evidence has been obtained for a quinone cofactor (which is not PQQ) in formaldehyde dehydrogenase of methylamine-grown *Hyphomicrobium zavarzinii* [9]. Several novel formaldehyde dehydrogenases have been characterized: NAD/Factor-dependent formaldehyde dehydrogenase [10], present in Gram-positive bacteria, which is in fact an alcohol dehydrogenase class III and similar to NAD/GSH-dependent formaldehyde dehydrogenase [11]; molybdoprotein aldehyde dehydrogenase [12], also able to oxidize formate esters, a property used for oxidation of Factor-formyl ester in Gram-positive bacteria, and perhaps forming an oxidative alternative for the well-known hydrolytic route of S-formylglutathione conversion in other organisms.

At least 6 genes are involved in biosynthesis of PQQ in methylotrophs [13] as well as in the non-methylotroph, *Klebsiella aerogenes* [14]. Although many papers and patents have appeared on the beneficial effects of PQQ on mammals, because of the absence of an established physiological role for PQQ in eukaryotes, the potentials of Gram-negative methanol-utilizers for production of PQQ have not been exploited.

3 METHYLAMINE DEHYDROGENASE

Certain Gram-negative methylamine-utilizers contain a dye-linked dehydrogenase for the direct oxidation of methylamine. The enzyme, indicated here as quinoprotein methylamine dehydrogenase (MADH), has been well studied for *Pseudomonas* sp., *Paracoccus denitrificans*, *Thiobacillus versutus*, *Methylobacterium extorquens* AM1 and Bacterium W3A1.

The primary structures of a number of MADHs have been elucidated (partly) by amino acid sequencing [15–17], the others derived from the nucleotide sequence of their genes [16–18]. The most striking features are the diversity in structures of the α-subunit and the consistency of that of the β-subunit. The latter has a very compact protein structure, as found in denaturation studies [15] and predicted by the high number of cross-links originating from S-S bridges, observable in the 3-dimensional structure, [4] and that of TTQ [19], having a covalent bond between two tryptophan moieties forming part of the protein chain. Folding of the subunits and association to the tetrameric structure seems to present no problem since the denatured subunits easily combine into active holo-enzyme *in vitro* [15]. The secondary structure is rather peculiar since seven 'W-motifs' are found. This motif also occurs in viral neuraminidase [4]. The active site is located between the boundaries of the α and β subunit and has a rather hydrophobic character, although Resonance Raman spectroscopy and differences in the spectral behavior upon reaction with ammonium salts indicate some variation in the topology [20,21]. MADH forms a complex with (one) of its natural electron acceptors, the blue copper protein amicyanin, as demonstrated by a shift in its redox potential. The complex has even be crystallized and its 3-dimensional structure derived [5].

As expected from the o-quinone moeity present in TTQ, the enzyme is inhibited by carbonyl-group reagents such as semicarbazide and hydrazines. The kinetic parameters of MADH have been determined for artificial dyes as well as for amicyanin (see [23] for a compilation). For the latter situation, the k_{cat} is hardly affected by pH or ionic strength and the reaction has an activation energy of 42 kJ/mol. The methylamine-oxidation step is strongly affected by the pH and salt composition of the buffer used. Ammonium salts exert a special effect on the system, most probably related to the fact that TTQ can exist in the 'amino-phenol' and 'iminoquinone' forms. The latter play a role in the catalytic cycle as it has been found that the spectral intermediate formed upon reduction of oxidized MADH with methylamine is different from that obtained by dithionite-reduction [23].

Blocking of the amicyanin-gene in *P. denitrificans* has proven that the protein is the sole natural electron acceptor for this organism [24]. However, indications exist that alternatives, such as a cytochrome, are is possible in other organisms [25]. The three-dimensional structure of amicyanin has been elucidated and its gene has been cloned [26] so that a model system is now available to study the structure-function relationships of a simple, soluble redox protein chain. Eventually, this may lead to an explanation of the events taking place in the periplasmic space, enabling differentiation between a mechanism in which electron transfer takes place in ordered suprastructural complexes and one based on simple diffusion-controlled collisions [27].

4 METHYLAMINE OXIDASE

The first prokaryotic methylamine oxidase (MeAO) was characterized from the Gram-positive bacterium, *Arthrobacter* P1 [28]. Enzymes which seem similar with respect to cofactor identity have been discovered in yeasts which used methylamine used as a N-source [29]. Since aromatic amine oxidases have been observed in *Escherichia coli* [30], *Salmonella typhimurium* [31] and in *Klebsiella aerogenes* [32] and it is highly likely that these enzymes contain the cofactor TPQ, there seems no barrier with respect to the cofactor so that MeAO may also occur in Gram-negatives. In this respect it is also interesting to mention that mammalian semicarbazide-sensitive amine oxidase is an excellent catalyst for methylamine oxidation [32a]. If further evidence can be provided that this has physiological significance, the implication is that these organisms behave as methylotrophs.

The primary structure of MeAO of *Hansenula polymorpha* has been reported [33], and although the enzyme has not yet been characterized, the presence of the consensus sequence containing the tyrosine from which TPQ is derived [1], proves that it belongs to the group of the quinoprotein, copper-containing amine oxidases (EC 1.4.3.6). The formation of TPQ from the tyrosine residue is still unknown but two mechanisms can be envisaged: (1) a 'spontaneous process' in which the specific tyrosyl residue is oxidized to TPQ by the Cu^{2+} in the enzyme; (2) a 'biosynthesis route' in which one or more enzymes catalyze the transformation.

Unfortunately, no three-dimensional structure is available for one of those enzymes. Consequently, the topology of the active site and the positioning of the Cu^{2+} and TPQ towards each other, are largely unknown. A wealth of spectroscopic data is available, some of these pointing to a rather large distance between the two cofactors [34]. A disturbing factor in estimating the relevance of the data is caused by the uncertainty about the number of TPQs per enzyme molecule, the reports pointing to either one, two or one reactive and the other less reactive [35,36]. Since other structural properties are in accordance with a symmetric structure (a dimeric enzyme with subunits of equal size and containing two copper ions), the disagreement on the number and uniformity of the organic cofactor hampers interpretation of structural data in a mechanistic sense.

In a search for the intermediates in the catalytic cycle, it has been shown that electrons donated from the substrate to TPQ can eventually reduce Cu^{2+} to Cu^+ in the enzyme [37]. Based on the existence of the $TPQH^{\cdot}/Cu^+$ intermediate, it has been postulated that O_2 reacts with the Cu^+ after which the second electron is transfered and H_2O_2 is formed [37] (Fig. 2). Since quite different constraints are imposed on organisms using either MADH or MeAO (removal of toxic formaldehyde and H_2O_2 for MADH and cytosolic- or peroxisome-located MeAO, respectively; regulation and transport of formaldehyde to assimilation and dissimilation for periplasm-located MADH; with or without generation of useful energy from the oxidation step) one wonders what could be the benefits of each enzyme for the methylotrophic organism concerned. Finally, it should be mentioned that PQQ as such is an excellent catalyst for amine oxidation and decarboxylation of amino acids. However, no PQQ-

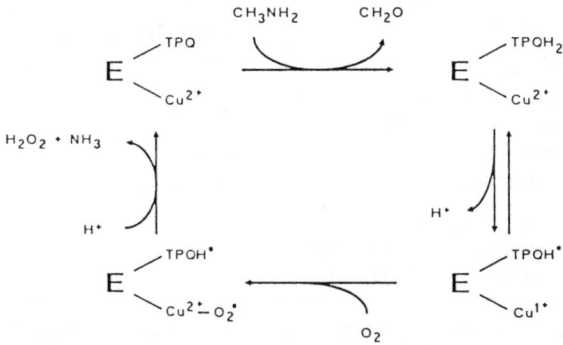

Figure 2 Steps in the catalytic cycle of quinoprotein methylamine oxidase (adapted from ref. [37]).

containing enzyme has been found that catalyzes such a reaction. Therefore, for unknown reasons, biological oxidation of amines is catalyzed by enzymes containing either TTQ or TPQ.

5 FORMALDEHYDE DEHYDROGENASES

Oxidation of formaldehyde occurs with specific enzymes (Table 1) such as the NAD-dependent dehydrogenase found in *Pseudomonas putida* [38], with cosubstrate-requiring enzymes like the widely distributed GSH-requiring enzyme and the factor-requiring enzyme observed in *Amycolatopsis methanolica* and *Rhodococcus erythropolis* [10]; with the nicotinoprotein formaldehyde dismutase found in *Pseudomonas aeruginosa* [39]; or with dye-linked dehydrogenase induced in *Hyphomicrobium zavarzinii* grown on methylamine. The latter has been investigated with respect to the nature of its cofactor. ESR spectroscopy as well as an assay for o-quinones revealed that it is an o-quinone, but not PQQ as such [9].

Oxidation (*in vitro*) of formaldehyde can also occur with aspecific enzymes, usually named aldehyde dehydrogenases as their role cannnot be derived from their substrate specificity nor from their induction pattern. Also for this case, NAD(P)-dependent as well as dye-linked enzymes have been found. The latter group (Table 1) consists of several enzymes with quite different cofactor identities. Haem-containing aldehyde dehydrogenase has been found in *Pseudomonas* species [40] and the EC number 1.2.99.3 was officially assigned to it. Later, an aldehyde dehydrogenase was purified from *Acetobacter aceti* and the

Table 1 Formaldehyde dehydrogenases

Enzyme	Coenzyme	Cofactor	EC number
Aspecific enzymes, formate-producing			
Aldehyde dehydrogenase[a]	NAD(P)		1.2.1.x
Aldehyde dehydrogenase[b]		Fe/S/FAD/Mo/Pter.	
Aldehyde dehydrogenase		haem	1.2.99.3
Specific enzymes, formate producing			
Formaldehyde dehydrogenase	NAD		1.2.1.46
Formaldehyde dismutase		NAD	1.2.99.4
Formaldehyde dehydrogenase[c]		quinone	
Specific enzymes, formate ester producing			
Formaldehyde dehydrogenase	NAD (GSH)		1.2.1.1
Formaldehyde dehydrogenase	NAD (Factor)		

[a] A recent example [59] shows that such a general aldehyde dehydrogenase can be transformed into 'formaldehyde-specific' by a modifying protein.
[b] This type of enzyme also oxidizes formate esters [12].
[c] From methylamine-grown *Hyphomicrobium zavarzinii* [9].

enzyme was reported to contain PQQ [41]. So far this has not been confirmed by others and the amino acid sequence deduced from the DNA sequence of the gene for this enzyme [42] does not show similarity with that of PQQ-containing dehydrogenases [43]. Unfortunately, in a later reevaluation [44] the EC number refers now to the putative quinoprotein, although it is structurally quite different from the haemoprotein.

Molybdoprotein aldehyde dehydrogenases have been purified and characterized from *Comamonas testosteroni* and *A. methanolica* (the latter organism also containing a quite different dye-linked aldehyde dehydrogenase which has not been further investigated [12]). The enzymes do not oxidize heterocyclic compounds like xanthine (as is the case for the mammalian molybdoprotein, aldehyde oxidase). As for other molybdoproteins, methanol is an inhibitor, although the concentration required presumably implies that this property has no significance *in vivo*. Interestingly, the enzymes are able to oxidize formate esters to the corresponding carbonate esters, the latter spontaneously decomposing into the alcohol moiety and CO_2. Evidence has been provided to show that this capability is required in *A. methanolica* to oxidize the Factor-formyl ester since no hydrolase could be detected and formate dehydrogenase activity is very low [12]. Although a specific hydrolase exists for the cleavage of S-formylglutathione in Gram-negative bacteria, the occurrence of formate ester dehydrogenase in *C. testosteroni* (this organism is able to demethoxylate aromatic compounds, so that a formaldehyde-converting system is required) and *P. denitrificans* (unpublished results) could indicate that oxidative conversion of S-formylglutathione might be an alternative for the hydrolytic pathway (in which S-formylglutathione is converted into GSH and formic acid) in Gram-negative methylotrophs.

6 PQQ-BIOSYNTHESIS

Despite extensive attempts to detect intermediates in cross-feeding experiments with different classes of PQQ-mutants, positive results were never obtained. This formed one of the main reasons to postulate [45] a mechanism based on a protein matrix as template (Fig. 3). Some evidence is emerging for this hypothesis: the first gene in a fragment of DNA encoding the whole PQQ-biosynthesis gene cluster of *K. aerogenes*, has a size corresponding to a peptide of 21 amino acids; the last gene codes for a protein which shows structural similarity with a novel class of metallo-protein proteases [14] (the other genes code for proteins with no structural resemblance to those present in the current data banks). Combining these results with the steps postulated, the peptide could form the matrix for a process in which a tyrosine is transformed into a precursor-like structure to which a glutamyl residue is attached and the final step consisting of excision of PQQ by a protease. (In this connection it should be noted that a tyrosine is conserved in the 'peptide-gene' of *Acinetobacter calcoaceticus* [46] and *K. aerogenes* [14] and that replacement with phenyl-

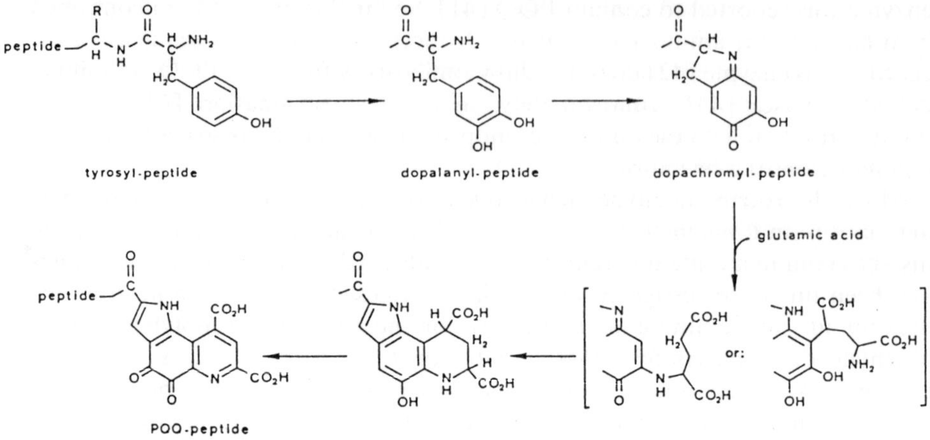

Figure 3 Proposed route of PQQ biosynthesis (from ref. [45]).

alanine leads to inability of PQQ production. If the hypothesis apears to be correct, this would mean that all quinone cofactors are originating from a specific aromatic amino acid in a peptide/protein (comparable to tyr˙ and trp˙ and tyr-cys˙ cofactors, recently detected in a number of oxidoreductases [47,48]), PQQ being exceptional as it is excised and used as a free cofactor. In this view it could be imagined that other combinations of amino acids still await detection as quinone cofactor. Recently, evidence has been provided showing that the non-aromatic amino acid, glycine, can act as a cofactor in its free radical form [49].

Although PQQ-biosynthesis was originally thought to be mainly restricted to methylotrophic bacteria, recent findings cast some doubt on that. One of the mysteries in quinoprotein research is the fact that many bacteria produce the protein part of the quinoprotein but not PQQ, a striking example being glucose dehydrogenase, present in many Gram-negative bacteria but nearly always occurring in the apo-enzyme form [50]. Although *E. coli* is an example of the inability of Enterobacteriacea to produce PQQ (see [51]), a recent report [52] gives evidence for the presence of silent genes for PQQ-biosynthesis. However, it has also been reported [53] that a DNA fragment much smaller than the one encoding the six genes of PQQ biosynthesis was able to transform an *E. coli* strain into PQQ-production. Hopefully, ongoing studies on PQQ-biosynthesis in methylotrophs may shed light on these controversial points.

7 CONCLUSIONS AND PROSPECTS

Oxidation of methanol is possible with the well known NAD-dependent alcohol dehydrogenases. However, due to the hydrophilic properties of methanol and a bad fit in the active site, and an unfavorable pK_a, the catalytic performance of these enzymes for this substrate is rather bad [54]. Cofactor-containing methanol oxidoreductases (either flavoprotein [55], quinoprotein, or nicotinoprotein [56]) appear to be much better catalysts for this reaction. However, the configuration of the active site with respect to amino acid residues involved is another decisive factor since flavoproteins, quinoproteins, and nicotinoproteins exist which are able to oxidize primary alcohols except methanol [57]. Therefore, elucidation of the 3-dimensional structure of the variants could provide insight into the factors responsible.

Oxidation of methylamine is catalyzed by TTQ-containing dehydrogenase (present in certain Gram-negative bacteria) and by TPQ-containing methylamine oxidase (present from man to bacteria). The amine oxidoreductases form part of a novel class of enzymes in which a specific aromatic amino acid residue is transformed by a post-translational modification process into an oxidized form acting as an organic cofactor (either a free radical or quinone). It seems that even PQQ, which is used as a free cofactor, is produced in a similar way. Although being an excellent catalyst in biomimetic studies on amine oxidation, PQQ has not been found to act as cofactor in amine oxidoreductases.

Microbial oxidation of formaldehyde is catalyzed by a multitude of oxidoreductases having different coenzyme/cofactor requirements, with some of the organisms harboring a number of these enzymes. Among these are specific formaldehyde dehydrogenases and general aldehyde dehydrogenases able to oxidize formaldehyde (at least *in vitro*). Specific enzymes include NAD/GSH- and NAD/Factor-dependent formaldehyde dehydrogenases, both giving a formate ester as a product. Although a hydrolytic pathway is known for splitting the S-formylglutathione ester, conversion of Factor-formyl ester (produced in Gram-positive bacteria, lacking GSH) seems to occur in an oxidative step, producing Factor-carbonate ester, decomposing into Factor and CO_2. The enzyme responsible for this, formate ester dehydrogenase, may also play a role in conversion of methylformate, observed in Gram-positives incubated with formaldehyde [58] and most probably formed from oxidation of the hemiketal adduct of methanol and formaldehyde catalyzed by formaldehyde dismutase [53] and/or methanol/formaldehyde oxidoreductase [57]). Formate ester dehydrogenase not only occurs in Gram-positives but also in Gram-negatives, perhaps providing an oxidative alternative for hydrolytic conversion of S-formylglutathione in these organisms.

The variety in cofactors/coenzymes of enzymes catalyzing one and the same reaction reflects the versatility of microbes, including methylotrophs, in catalytic capabilities. Most probably, this is related to constraints put on a certain organism with respect to rapidity of substrate conversion, economics in obtaining useful energy from the oxidation steps and providing flexibility in maintaining the internal redox balance and proton motive force. Present insight into the properties of the diverse categories of enzymes is insufficient to correlate these with physiological roles. It is likely that comparative studies on quinoproteins and other enzymes will eventually provide some answers.

REFERENCES

[1] Janes, S.M., Mu, D., Wemmer, D., Smith, A.J., Kaur, S., Maltby, D., Burlingname, A.L. and Klinman, J.P. (1990) A new redox cofactor in eukaryotic enzymes: 6-hydroxydopa at the active site of serum amine oxidase. Science 248, 981–987.

[2] McIntire, W.S., Wemmer, D.E., Chistoserdov, A. and Lidstrom, M.E. (1991) A new cofactor in a prokaryotic enzyme: tryptophan tryptophylquinone as the redox prosthetic group in methylamine dehydrogenase. Science 252, 817–824.

[3] Duine, J.A. (1991) Quinoproteins: enzymes containing the quinoid cofactor pyrroloquinoline quinone (PQQ), topaquinone (TPQ) or tryptophanyl tryptophanquinone (TTQ). Eur. J. Biochem. 200, 271–284.

[4] Vellieux, F.M.D., Huitema, F., Groendijk, H., Kalk, K.H., Frank, J., Jongejan, J.A., Duine, J.A., Petratos, K., Drenth, J. and Hol, W.G.J. (1989) Structure of quinoprotein methylamine dehydrogenase at 2.25 Å resolution. EMBO J. 8, 2171–2178.

[5] Chen, L., Durley, R., Poliks, B.J., Hamada, K., Chen, Z., Mathews, F.S., Davidson, V.L., Satow, Y., Huizinga, E., Vellieux, F.M.D. and Hol, W.G.J. (1992) Crystal structure of an electron-transfer complex between methylamine dehydrogenase and amicyanin. Biochemistry 31, 4959–4964.

[6] Duine, J.A., Frank, J. and Berkhout, M.P.J. (1984) NAD-dependent, PQQ-containing methanol dehydrogenase: a bacterial dehydrogenase in a multienzyme complex. FEBS Lett. 168, 217–221.

[7] Bystrykh, L.V., Govorukhina, N.I., van Ophen, P.W., Hektor, H., Dijkhuizen, L. and Duine, J.A. Methanol/formaldehyde oxidoreductase and formaldehyde dismutase activities in Gram-positive bacteria oxidizing methanol. Submitted.

[8] van Ophen, P.W., van Beeumen, J. and Duine, J.A. (1993) Nicotinoprotein (NAD(P)-containing) alcohol dehydrogenases. Purification and characterization of a novel type from Amycolatopsis methanolica. Eur. J. Biochem., in press.

[9] Klein, C., Frank, J., Duine, J.A. and Schwartz, A.C. (1993) Evidence for the presence of a quinoid cofactor in dye-linked formaldehyde dehydrogenase from methylamine-grown Hyphomicrobium zavarzinii. Eur. J. Biochem., in press.

[10] van Ophen, P.W., van Beeumen, J. and Duine, J.A. (1992) NAD-linked, factor-dependent formaldehyde dehydrogenaseor trimeric, zinc-containing, long-chain alcohol dehydrogenase from Amycolatopsis methanolica. Eur. J. Biochem. 206, 511–518.

[11] Koivusalo, M. Baumann, M. and Uotila, L. (1989) Evidence for the identity of glutathione-dependent formaldehyde dehydrogenase and class III alcohol dehydrogenase. FEBS Lett. 257, 105–109.

[12] van Ophem, P.W., Bystrykh, L.V. and Duine, J.A. (1992) Dehydrogenase activities (dye-linked) for formate and formate esters in Amycolatopsis methanolica. Eur. J. Biochem. 206, 519–525.

[13] Biville, F., Turlin, E. and Gasser, F. (1989) Cloning and genetic analysis of six pyrroloquinoline quinone biosynthesis genes in Methylobacterium organophilum DSM 760 J. Gen. Microbiol. 135, 2917–2929.

[14] Meulenberg, J.J.M., Sellink, E., Riegman, N.H., Loenen, W.A.M. and Postma, P.W. (1992) Nucleotide sequence and structure of the Klebsiella pneumoniae pqq operon. Mol. Gen. Genet. 232, 284–294.

[15] Ishii, Y., Hase, T., Fukumori, Y., Matsubara, H. and Tobari, J. (1983) Amino acid sequence studies of the light subunit of methylamine dehydrogenase from Pseudomonas AM1: existence of two residues binding the prosthetic group. J. Biochem. 93, 107–119.

[16] Ubbink, M., van Kleef, M.A.G., Kleinjan, D.-J., Hoitink, C.W.G., Huitema, F., Beintema, J.J., Duine, J.A. and Canters, G.W. (1991) Cloning, sequencing and

expression studies of the genes encoding amicyanin and the β-subunit of methylamine dehydrogenase from *Thiobacillus versutus*. *Eur. J. Biochem.* **202**, 1003–1012.

[17] Huitema, F., Canters, G.W., van Beeumen, J. and Duine, J.A. (1993) Cloning and sequencing the gene coding for the large subunit of methylamine dehydrogenase from *Thiobacillus versutus*. Submitted.

[18] Chistoserdov, A.Y., Tsygankov, Y.D. and Lidstrom, M.E. (1990) Cloning and sequencing of the structural gene for the small subunit of methylamine dehydrogenase from *Methylobacterium extorquens* AM1. *Biochem. Biophys. Res. Commun.* **172**, 211–216.

[19] Chen, L., Matthews, F.S., Davidson, V.L., Huizinga, E.G., Vellieux, F.M.D., Duine, J.A. and Hol, W.G.J. (1991) Crystallographic investigations of the tryptophan-derived cofactor in the quinoprotein methylamine dehydrogenase. *FEBS Lett.* **287**, 163–166.

[20] McIntire, W.S., Bates, J.L., Brown, D.E. and Dooley, D.M. (1991) Resonance Raman spectroscopy of methylamine dehydrogenase from Bacterium W3A1. *Biochemistry* **30**, 125–133.

[21] Backes, G., Davidson, V.L., Huitema, F., Duine, J.A. and Sanders-Loehr, J. (1991) Characterization of the tryptophan-derived quinone cofactor of methylamine dehydrogenase by resonance Raman spectroscopy. *Biochemistry* **30**, 9201–9210.

[22] Gray, K.A., Davidson, V.L. and Knaff, D.B. (1988) Complex formation between methylamine dehydrogenase and amicyanin from *Paracoccus denitrificans*. *J. Biol. Chem.* **263**, 13987–13990.

[23] Gorren, A.C.F. and Duine, J.A. (1993) Kinetic studies on methylamine dehydrogenase and amicyanin from *Thiobacillus versutus*. Submitted.

[24] van Spanning, R.J.M., Wansell, C.W., Reijnders, W.N.M., Oltmann, L.F. and Stouthamer, A.H. (1990) Mutagenesis of the gene encoding amicyanin of *Paracoccus denitrificans* and the resultant effect on methylamine oxidation. *FEBS Lett.* **275**, 217–220.

[25] Auton, K.A. and Anthony, C. (1989) the role of cytochromes and blue copper proteins in growth of an obligate methylotroph on methanol and methylamine. *J. Gen. Microbiol.* **135**, 1923–1931.

[26] Lommen, A., Canters, G.W. and van Beeumen, J. (1988) A ^{1}H-NMR study on the blue copper protein amicyanin from *Thiobacillus versutus*. *Eur. J. Biochem.* **176**, 213–223.

[27] van Wielink, J.E. and Duine, J.A. (1990) How big is the periplasmic space? *Trends Biochem. Sci.* **15**, 136–137.

[28] van Iersel, J., van der Meer, R.A. and Duine, J.A. (1986) Methylamine oxidase from *Arthrobacter* P1: a bacterial copper-quinoprotein amine oxidase. *Eur. J. Biochem.* **161**, 415–419.

[29] Large, P.J. and Haywood, G.W. (1990) Amine oxidases from methylotrophic yeasts. *Methods in Enzymolology* **1881**, 427–435.

[30] Parrott, S., Jones, S. and Cooper, R.A. (1987) 2-Phenylethylamine catabolism by *Escherichia coli* K12. *J. Gen. Microbiol.* **133**, 347–351.

[31] Loretti, P., Groen, B.W., Mondovi, B. and Duine, J.A. (1993) Copper-quinoprotein amine oxidase from *Salmonella typhimurium*. Submitted.

[32] Sugino, H., Sasaki, M., Azakami, H., Yamashita, M. and Murooka, Y. (1992) A monoamine-regulated *Klebsiella aerogenes* operon containing the monoamine oxidase structural gene and the *maoC* gene. *J. Bacteriol.* **174**, 2485–2492.

[32a] Precious, E., Gunn, C.E. and Lyles, G.A. (1988) Deamination of methylamine by semicarbazide-sensitive amine oxidase in human umbilical artery and rat aorta. *Biochem. Pharmacol.* **37**, 707–713.

[33] Bruinenberg, P.G., Evers, M., Waterham, H.R., Kuipers, J., Arnberg, A.C. and Ab, G. (1989) Cloning and sequencing of the peroxisomal amine oxidase from *Hansenula polymorpha*. *Biochim. Biophys. Acta* **1008**, 157–167.

[34] Lamkin, M.S., Williams, T.J. and Falk, M.C. (1988) Excitation energy transfer study of the spatial relationship between the carbonyl and metal cofactors in pig plasma amine oxidase. *Arch. Biochem. Biophys.* **262**, 72–79.
[35] Janes, S.M. and Klinman, J.P. (1991) An investigation of bovine serum amine oxidase active site stoichiometry. *Biochemistry* **30**, 4599–4605.
[36] Morpugo, L., Agostinelli, E., Mondovi, B., Avigliano, L., Silvestri, R., Stefanich, G. and Artico, M. (1992) Bovine serum amine oxidase: half-site reactivity with phenyldrazine, semicarbazide and aromatic hydrazides. *Biochemistry* **31**, 2615–2621.
[37] Dooley, D.M., McGuirl, M.A., Brown, D.E., Turowski, P.N., McIntire, W.S. and Knowles, P.F. (1991) A Cu(I)-semiquinone state in substrate-reduced amine oxidases. *Nature* **349**, 262–264.
[38] Ogushi, S., Ando, M. and Tsuru, D. (1984) Substrate specificity of formaldehyde dehydrogenase from *Pseudomonas putida*. *Agr. Biol. Chem.* **48**, 597–601.
[39] Kato, N., Yamagami, T., Shimao, M. and Sakazawa, C. (1986) Formaldehyde dismutase, a novel NAD-binding oxidoreductase from *Pseudomonas putida* F61. *Eur. J. Biochem.* **156**, 59–64.
[40] Patel, R.N., Hou, C.T., Derelanko, P. and Felix, A. (1980) Purification and properties of a heme-containing aldehyde dehydrogenase from *Methylosinus trichosporium*. *Arch. Biochem. Biophys.* **203**, 654–662.
[41] Ameyama, M. and Adachi, O. (1982) Aldehyde dehydrogenase from acetic acid bacteria, membrane-bound. *Methods in Enzymology* **89**, 491–497.
[42] Tamaki, T., Horinouchi, S., Fukaya, M., Okumura, H., Kawamura, Y. and Beppu, T. (1989) Nucleotide sequence of the membrane-bound aldehyde dehydrogenase gene from *Acetobacter polyoxogenes*. *J. Biochem.* **106**, 541–544.
[43] Cleton-Jansen, A.M., Goosen, N., Fayet, O. and van de Putte, P. (1990) Cloning, mapping and sequencing of the gene encoding *Escherichia coli* quinoprotein glucose dehydrogenase. *J. Bacteriol.* **172**, 6308–6315.
[44] Enzyme Nomenclature (1990), as published in *Eur. J. Biochem.* **187**, supplement.
[45] Duine, J.A. and Jongejan, J.A. (1989) Pyrroloquinoline quinone: a novel cofactor. *Vitamin Horm.* **45**, 233–262.
[46] Goosen, N., Horsman, H.P.A., Huinen, R.G.M. and van de Putte, P. (1989) The *Acinetobacter calcoaceticus* genes involved in the biosynthesis of the coenzyme pyrrolo-quinoline-quinone: nucleotide sequence and expression in *Escherichia coli* K-12. *J. Bacteriol.* **171**, 447–455.
[47] Prince, R.C. (1988) Tyrosine radicals. *Trends Biochem. Sci.* **13**, 286–288.
[48] Prince, R.C. and George, G.N. (1990) Tryptophan radicals. *Trends Biochem. Sci.* **15**, 170–172.
[49] Wagner, A.F.V., Frey, M., Neugebauer, F.A., Schäfer, W. and Knappe, J. (1992) The free radical in pyruvate formate-lyase is located on glycine-734. *Proc. Natl. Acad. Sci. USA* **89**, 996–1000.
[50] Duine, J.A. (1991) Energy generation and the glucose dehydrogenase pathway in *Acinetobacter*. In: The Biology of *Acinetobacter* (Towner, K.J. et al., Eds), pp. 295–312, Plenum Press, New York.
[51] Bouvet, O.M.M., Lenormand, P. and Grimont, P.A.D. (1989) Taxonomic diversity of the D-glucose oxidation pathway in the *Enterobacteriaceae*. *Int. J. Syst. Bacteriol.* **39**, 61–67.
[52] Biville, F., Turlin, E. and Gasser, F. (1991) Mutants of *Escherichia coli* producing pyrroloquinoline quinone. *J. Gen. Microbiol.* **137**, 1775–1782.
[53] Goldstein, A.H. and Liu, S.T. (1987) Molecular cloning and regulation of a mineral phosphate solubilizing gene from *Erwinia herbicola*. *Biotechnol.* **5**, 72–74.
[54] Pockert, Y. and Page, J.D. (1990) Zinc-activated alcohols in ternary complexes of liver alcohol dehydrogenase. *J. Biol. Chem.* **265**, 22101–22108.

[55] van der Klei, I.J., Bystrykh, L.V. and Harder, W. (1990) Alcohol oxidase from *Hansenula polymorpha*. *Methods in Enzymolology* **188**, 420–427.
[56] Arfman, A., van Beeumen, J., de Vries, G.E., Harder, W. and Dijkhuizen, L. (1991) Purification and characterization of an activator protein for methanol dehydrogenase from thermotolerant *Bacillus* spp. *J. Biol. Chem.* **266**, 3955–3960.
[57] van Ophem, P.W. and Duine, J.A. (1993) Microbial alcohol, aldehyde and formate ester oxidoreductases. In: Carbonyl Metabolism 4 (Weiner, H., ed.), Plenum Press, New York, in press.
[58] Mason, R.P. and Sanders, J.K.M. (1989) *In vivo* enzymology: a deuterium NMR study of formaldehyde dismutase in *Pseudomonas putida* F61a and *Staphylococcus aureus*. *Biochemistry* **28**, 2160–2168.
[59] Tate, S., Millet, J., Green, J. and Dalton, H. (1993) The modifier protein of NAD linked formaldehyde dehydrogenase from *Methylococcus capsulatus* (Bath), in preparation.

32
Methylamine Utilization in Yeast and Bacteria: Studies Using *in vivo* NMR

Edward Bellion and John G. Jones[1]

Department of Chemistry and Biochemistry, The University of Texas at Arlington, Arlington, TX 76019-0066, USA

1 SUMMARY

The metabolism of [^{13}C] methylamine in the methylotrophic yeast *Hansenula polymorpha* and the methylotrophic bacteria *Pseudomonas* MA and *Methylobacterium extorquens* AMI was studied both *in vivo* and in perchlorate cell extracts using NMR methods. Methylamine metabolism was also followed with nitrogen-15 NMR in *Pseudomonas* MA. The results with *Pseudomonas* MA demonstrate that N-methylglutamate is formed *in vivo* from methylamine by replacement of the α-amino group of glutamate by methylamine, whereas γ-glutamylmethylamide was observed only under conditions of low oxygenation, and was ascribed a role in methylamine sequestration rather than oxidation or assimilation. No evidence was found for the *in vivo* formation of 5-hydroxy-*N*-methylpyroglutamate. In *M. extorquens*, no *N*-methylated derivatives were observed. Carbon from methylamine was rapidly converted in this organism to trehalose, the labeling pattern of which was consistent with its formation via [3-^{13}C]serine-derived trioses. Methylamine cannot support the growth of *H. polymorpha* as a carbon source but nevertheless carbon from methylamine was incorporated into trehalose and glycerol, rather than being oxidized. The labeling pattern in these derivatives indicated that methylamine-

---

[1] Present address: Department of Radiology, The University of Texas Southwestern Medical Center at Dallas, Dallas, TX 75235, USA

Microbial Growth on C₁ Compounds
© Intercept Ltd, PO Box 716, Andover, Hampshire SP10 1YG, UK

derived formaldehyde was not incorporated via dihydroxyacetone synthase (DHAS); this was supported by the lack of cross-reactivity during immunoblotting of SDS-PAGE-resolved cell extract with antiserum against DHAS.

2 INTRODUCTION

Methylamine can be utilized as both carbon and nitrogen source by bacteria, which may be grouped into two types depending on the means of methylamine cleavage. Members of one group such as *Methylobacterium extorquens* AM1 possess an enzyme that directly cleaves methylamine into formaldehyde and ammonia [1]. Members of the other group including *Pseudomonas* MA do not possess such a direct cleavage enzyme and have a complex pathway of methylamine utilization, involving formation of N-methylated derivatives of glutamate [2]. N-methylglutamate is oxidized, resulting in the cleavage of the original methylamine carbon-nitrogen bond, regenerating glutamate and producing formaldehyde [3]. The metabolism of γ-glutamylmethylamide [4] within this organism is less clearly understood. Also 5-hydroxy-N-methylpyroglutamate is formed from α-ketoglutarate *via* N-methyl-α-ketoglutaramate [5]. Whereas the role and function of N-methylglutamate is clearly defined, no function has yet been ascribed to these other N-methylated compounds in the metabolism or assimilation of methylamine.

Methylotrophic yeast cannot utilize methylamine as a carbon source, but a wide range of amines, including methylamine, can be readily utilized by yeasts as sources of metabolic nitrogen [6]. These compounds are generally oxidized by FAD-linked oxidases, which cleave the carbon-nitrogen bond and result in the formation of ammonia and an aldehyde residue [7]. Since the carbon moiety cannot be used for growth, an alternative readily metabolized carbon source such as glucose is required. The reason for the inability to exploit the formaldehyde in the case of the methylotrophs remains a puzzle, since it is readily assimilated during methanol growth. It has been suggested that perhaps the excess ammonia that would be generated causes repression of some metabolic enzymes [8].

This article describes our recent studies on methylamine metabolism by these organisms. To follow the fate of methylamine inside the cell, which is tedious with radioisotope methodology, we used *in vivo* NMR to observe both ^{15}N- and ^{13}C-labeled methylamine. This method monitors the fate of both the carbon and nitrogen and, simultaneously, reliably quantifies the rate of endogenous methylamine oxidation. ^{13}C-NMR has become a powerful and dependable tool for obtaining *in vivo* rates of metabolism for many metabolites in a wide range of organisms and conditions [9–13]. We also used this technique to detect changes in glucose metabolism when methylamine was used as the nitrogen source.

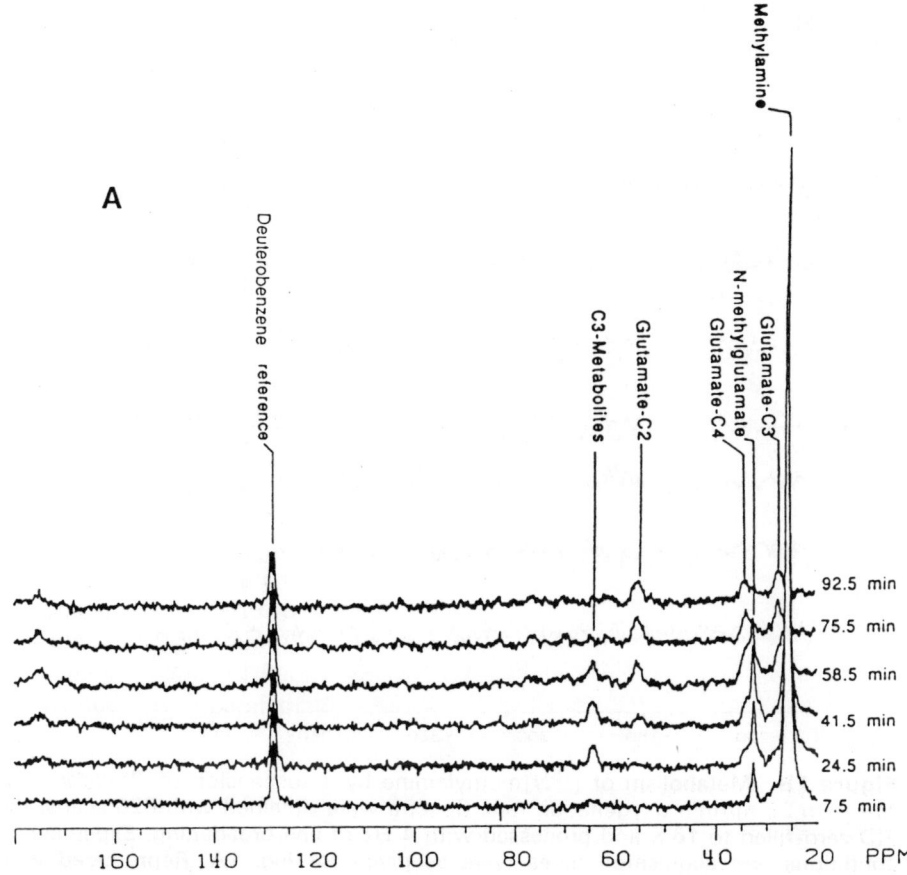

Figure 1A 50.33 MHz ^{13}C-NMR spectra of [^{13}C]methylamine metabolism by a suspension of methylamine-grown *Pseudomonas* MA cells at a rate of oxygenation of 25 ml/min. Spectra were obtained with 2000 scans, the FID zero-filled to 16 K and processed with 6Hz of line broadening. The probe temperature was maintained at 30°C throughout the experiment. The times shown represent the mid-point of the interval for each set of accumulations. Each spectrum took 13 minutes to collect; the interval between sets was 2 minutes. Reproduced with permission of the American Society for Biochemistry and Molecular Biology.

3 METHYLAMINE METABOLISM IN BACTERIA

3.1 Pseudomonas MA

Figure 1A shows a time-course of [^{13}C]methylamine metabolism by a fully oxygenated suspension of cells. Spectra taken over a period of 1.5–2 hr showed the appearance of the N-methylglutamate methyl peak at 32.7 ppm (see Table 1) along with a broad set of resonances at 63–64 ppm. These chemical shifts are characteristic of hydroxymethyl carbons of hydroxypyruvate, glycerate,

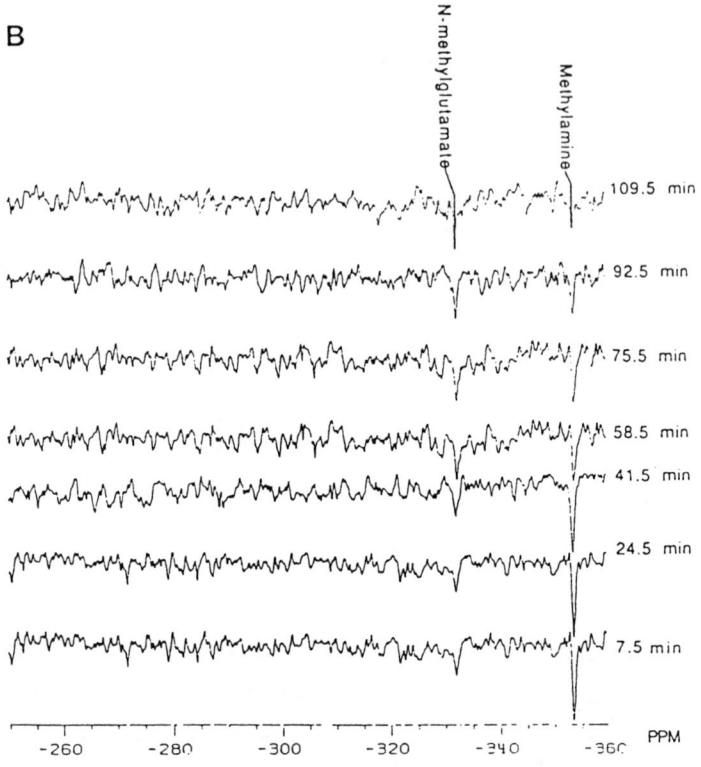

Figure 1B Metabolism of [^{15}N]methylamine by a suspension of *Pseudomonas* MA with 25 ml/min oxygenation rate. Spectra were obtained with 820 scans, the FID zero-filled to 16 K and processed with 4 Hz of line broadening. Experimental conditions and acquisition times were as given for Fig. 1A. Reproduced with permission of the American Society for Biochemistry and Molecular Biology.

2-phosphoglycerate and 3-phosphoglycerate which would arise from C3-labeled serine formed by the condensation of ^{13}C-formaldehyde (as N^5,N^{10}-methylene-FH$_4$) and endogenous glycine. No resonances corresponding to the C3 carbons of pyruvate or alanine were detected, supporting the view that the route of labeled carbon into the C2 of acetate is *via* the phosphoenolpyruvate carboxylase and malyl-CoA lyase steps [14]. When the oxygenation rate was reduced from 25 ml/min to 12 ml/min (Fig. 2A), the rate of conversion of N-methylglutamate to products was reduced, and the intensities of the TCA cycle-derived metabolites were also lower. An additional resonance slightly downfield of methylamine was also detected, which could not be assigned on the basis of its chemical shift because of coinciding resonances from other compounds, (see Table 2) but was postulated to be that of γ-glutamylmethylamide.

Under both these sets of conditions, there was no significant production of formaldehyde, formate or bicarbonate from the oxidation of N-methylglutamate. This indicates that this step is tightly coupled to subsequent steps that

Table 1 Carbon-13 chemical shifts of various metabolic intermediates of methylamine metabolism in *Pseudomonas* MA

Compound	Chemical shift (ppm from TMS)						
	C1	C2	C3	C4	C5	C2-N-methyl	C5-N-methyl
Glutamic acid	175.8	55.8	28.3	34.6	182.2		
N-Methylglutamate[a]	180.0	64.5	26.3	33.3	174.6	32.7	
Glutamine	175.0	55.4	27.5	32.0	178.5		
γ-Glutamylmethylamide[a]	173.4	54.0	26.1	31.2	174.6		25.6
Glycine	173.6	40.1					
Serine	173.1	57.4	61.3				
Formaldehyde[b]	83.2						
Bicarbonate	161.0						
Formate	169–172						
Methylamine	25.4						

[a] 1.0 M solution pH 4.2.
[b] as the hydrated form.
Reproduced with permission of the American Society for Biochemistry and Molecular Biology.

Table 2 Nitrogen-15 chemical shifts for various metabolic intermediates of methylamine metabolism in *Pseudomonas* MA

Compound	Chemical shift (ppm)[a]
Glutamic acid	−334.44
Glutamine:	
Amide nitrogen	−264.00
Amine nitrogen	−334.25
N-methylglutamate	−332.90[b]
γ-glutamylmethylamide:	
Amide nitrogen	−261.42[b]
Amino nitogen	−333.90[b]
Methylamine	−353.45[c]
Ammonium chloride	−353.00[c]

[a] relative to 1 M $^{15}NHNO_3$ in 50% D_2O.
[b] 1 M solution in 50% D_2O pH 4.0-4.2.
[c] 0.1 M ^{15}N-enriched samples in 50% D_2O.
Reproduced with permission of the American Society for Biochemistry and Molecular Biology.

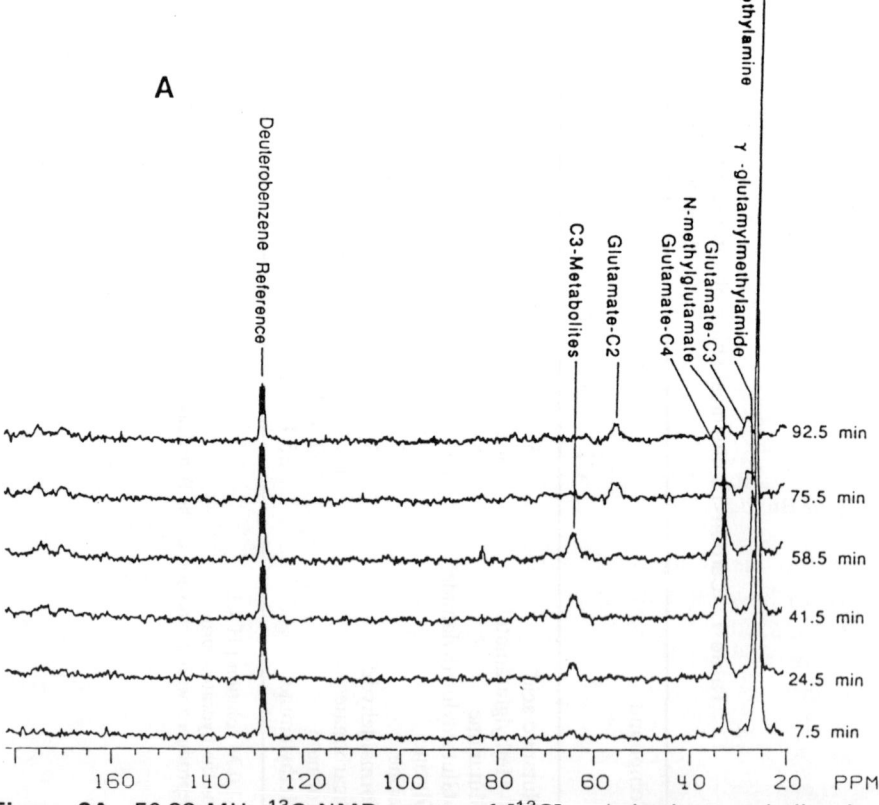

Figure 2A 50.33 MHz ^{13}C-NMR spectra of [^{13}C]methylamine metabolism by a suspension methylamine-grown cells of *Pseudomonas* MA with an oxygenation rate of 12 ml/min. Experimental conditions, spectral parameters and acquisition times were as given for Fig. 1A. Reproduced with permission of the American Society for Biochemistry and Molecular Biology.

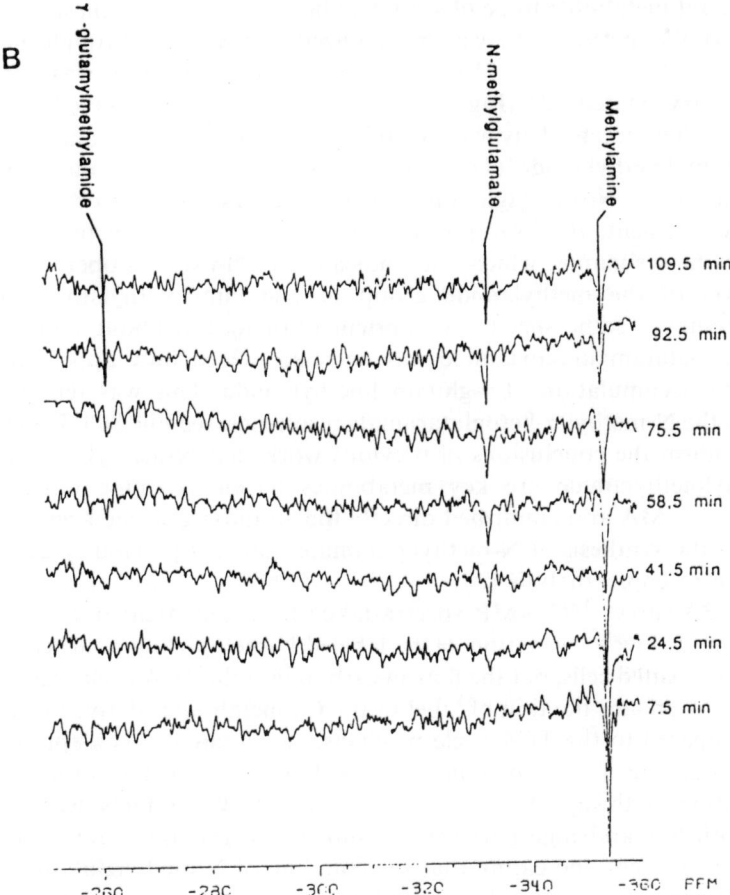

Figure 2B Metabolism of [^{15}N]methylamine by a suspension of *Pseudomonas* MA with an oxygenation rate of 12 ml/min. Spectral parameters were as given for Fig. 1B. Reproduced with permission of the American Society for Biochemistry and Molecular Biology.

involve the metabolism of formaldehyde. Because of the high toxicity of formaldehyde, such a mechanism is obligatory for the organism under these growth conditions to prevent any accumulation of this product.

The flow of label from methylamine into γ-glutamylmethylamide could not be properly monitored with ^{13}C-NMR, since the ^{13}C chemical shift of the N-methyl group lies very close to that of methylamine and several other ^{13}C resonances of TCA cycle as well as the N-methyl amino acids. The ^{15}N shifts however are readily separable (see Table 2) and by this method, the methylamine moiety can be unambiguously traced into both of the N-methylated amino acids. Figure 1B shows the metabolism of ^{15}N-labeled methylamine under highly-oxygenated conditions. The N-methylglutamate resonance was the only predominant metabolite to be observed, indicating that under these conditions, the bulk of the methylamine was metabolized immediately through the oxidative pathway and into the TCA cycle. When [^{15}N]methylamine was introduced to poorly-oxygenated cells (Fig. 2B), a new resonance was observed in addition to N-methylglutamate. This was identified as belonging to the amido nitrogen of γ-glutamylmethylamide (Table 1). It was observed only after significant accumulation of N-methylglutamate had occurred, and its occurrence was relatively transient. Its disappearance coincided with a brief increase in the methylamine resonance, which may indicate a pathway was operating for the conversion of the methylamide group of γ-glutamylmethylamide back to methylamine. In some spectra, a transient resonance at about 1 ppm upfield from the γ-glutamylmethylamide was detected, which only appeared after a significant accumulation of γ-glutamylmethylamide. This was tentatively assigned as the N-methyl-α-ketoglutaramate resonance (see Fig. 3C). These experiments confirm the conclusions of previous work that N-methylglutamate and γ-glutamylmethylamide are key metabolites in methylamine utilization by *Pseudomonas* MA and confirmed directly the amino-group replacement mechanism for the synthesis of N-methylglutamate that was previously deduced by mass spectroscopic analysis [2].

Figure 3A shows ^{13}C-NMR spectra taken from cells treated with 50 μg/ml tetracycline at low oxygenation. Here, labeled N-methylglutamate accumulated as in the untreated cells, but the flow of carbon into the TCA cycle was impaired as seen by a greater fraction of label in the C_3-metabolites derived from serine when compared to the TCA cycle products, i.e. glutamate. The appearance of free formaldehyde and formate indicated a shunting of carbon through a linear C_1-oxidation pathway. The C_1-oxidation pathway products were detected under both low and high (spectra not shown) oxygenation rates when tetracycline was present, indicating that the coupling of N-methylglutamate oxidation to formaldehyde assimilation had been disrupted. The oxygenation rate also had a marked effect on the flow of carbon through the C_1-oxidation pathway. When low rates were used, the principal products detected were bicarbonate and formate, with only minor amounts of formaldehyde. However, during high oxygenation rates, high levels of formaldehyde were detected with little of the other products observed. Absolute amounts of formate and bicarbonate are undoubtedly higher than they appear in these spectra due to pulse saturation and low NOE. Also, during low oxygenation in the presence of tetracycline, a strong peak downfield of the methylamine resonance was observed. This resonance did not originate from any TCA cycle reactions since, under these conditions, very little labeling of any TCA cycle-derived metabolites occurred. On the basis of these observations, we believe that this resonance belonged to the methyl group of γ-glutamylmethylamide.

^{15}N-NMR spectra of cells exposed to tetracycline (Figs. 3B and 3C) showed prolonged accumulation of N-methylglutamate compared to the untreated cells, which is in good agreement with the ^{13}C-NMR data. During low oxygenation rates, the γ-glutamylmethylamide resonance accumulated over time, and did not disappear in contrast to the result with the untreated cells.

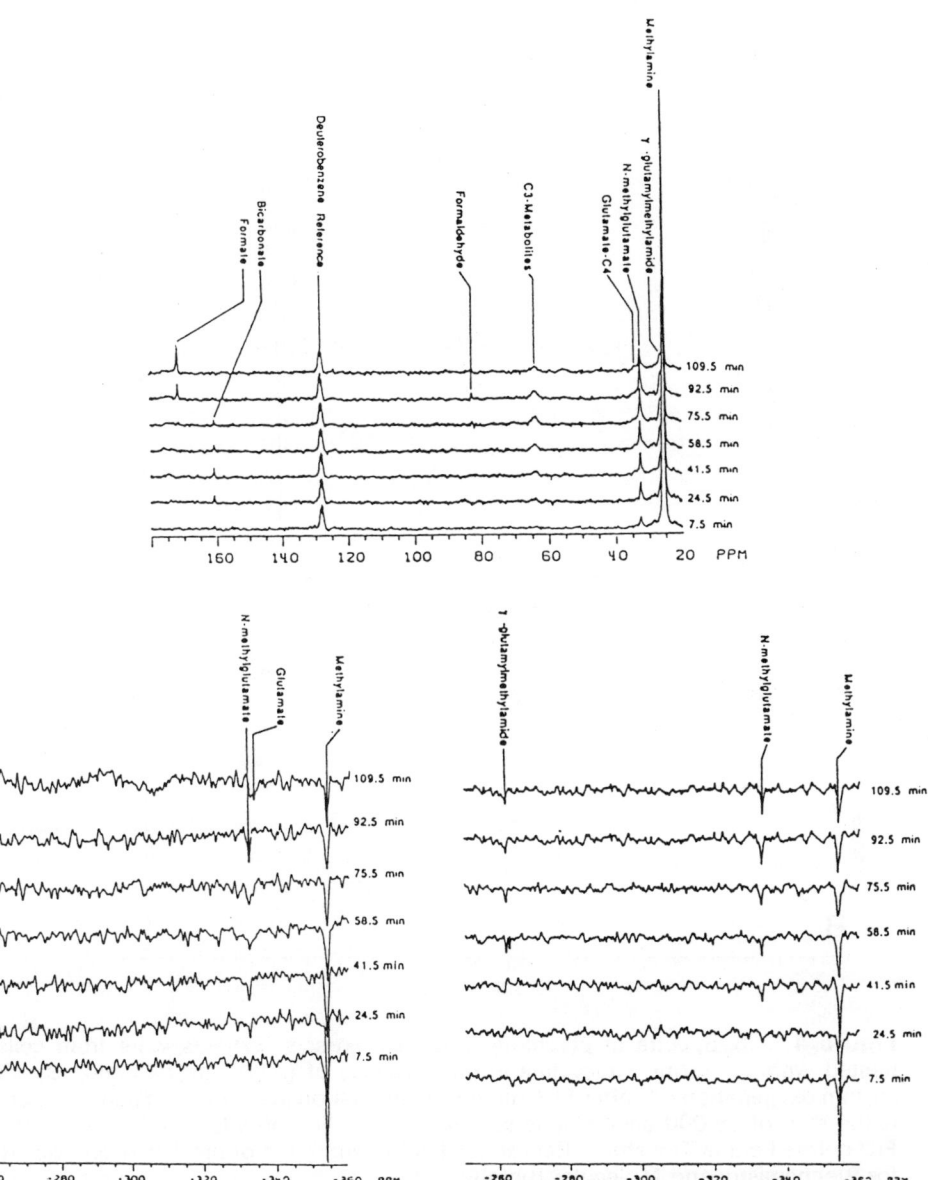

Figure 3 A. 50.33 MHz ^{13}C-NMR spectra of [^{13}C]methylamine metabolism by a suspension of methylamine-grown *Pseudomonas* MA cells in the presence of 50 μg/ml tetracycline and an oxygenation rate of 12 ml/min. Spectral parameters were as given for Fig. 1A. B. Metabolism of [^{15}N]methylamine by a suspension of *Pseudomonas* MA in the presence of 50 μg/ml tetracycline and 25 ml/min oxygenation rate. Experimental conditions, spectral parameters and acquisition times were as given for Fig. 1B. C. Metabolism of [^{15}N]methylamine by a suspension of *Pseudomonas* MA in the presence of 50 μg/ml tetracycline and 12 ml/min oxygenation rate. Spectral parameters were as given for Fig. 1B. Reproduced with permission of the American Society for Biochemistry and Molecular Biology.

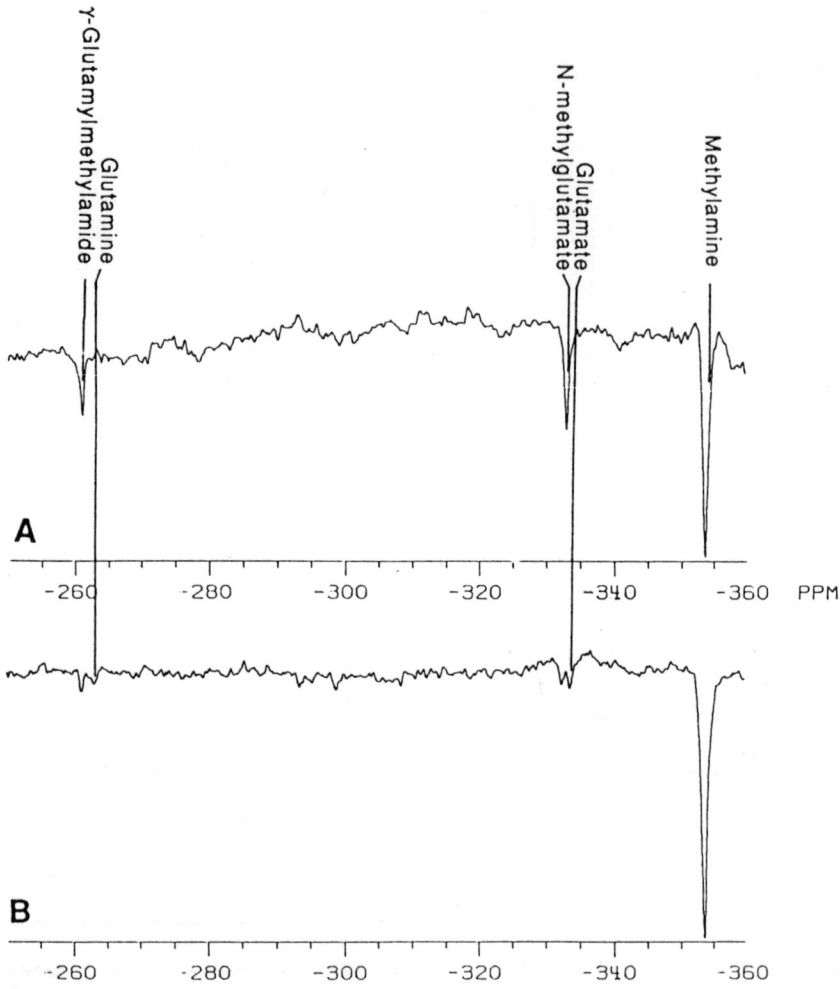

Figure 4 ^{15}N-Spectra of perchloric acid cell extracts. Extracts were from cells treated with 50 μg/ml tetracycline in the presence of (A) 12 ml/min and (B) 25 ml/min oxygenation rate after 117 minutes of incubation in the probe. Each spectrum is the sum of 15,000 accumulations with 2 Hz of line broadening applied to the FID before Fourier Transform. Reproduced with permission of the American Society for Biochemistry and Molecular Biology.

When the oxygenation rate was increased, this metabolite was not detected *in vivo* and was only a minor feature in the cell-extract spectrum (Fig. 4). Overall these results indicate that the carbon was also diverted from potential pathways of amino-acid synthesis to a C_1-oxidation pathway involving formate and bicarbonate (carbon dioxide), neither of which has a strong role in amino-acid metabolism. A linear pathway of formaldehyde oxidation has not previously been reported for this organism but is common in many other methylotrophs where it is the major energy-producing pathway [15]. Energy production in *Pseudomonas* MA is via a cyclic pathway involving enzymes of the assimilation pathway and the TCA cycle [14]. However, substantial levels of separate NAD-linked dehydrogenases specific for both formaldehyde and formate are induced in methylamine-grown *Pseudomonas* MA compared to succinate-grown cells (E. Bellion and D. Tan, unpublished data). This could point to an alternative energy-producing pathway or the pathway may simply be present as a means of detoxification as has recently been postulated for methylotrophic yeast [16]. The lack of substantial amounts of formate occurring under normal conditions tends to support the latter hypothesis. The appearance of glutamate labeled in its nitrogen was detected when the tetracycline-treated samples were highly oxygenated, even though the glutamate pool was low as seen by ^{13}C-NMR. This suggests that nitrogen incorporation into proteins was hindered, resulting in the accumulation of the label in amino acids.

The physiological role of γ-glutamylmethylamide in this organism is not clear; γ-glutamylmethylamide was observed only during low oxygenation rates, indicating that its synthesis was probably not related to the levels of the other observable metabolites. It could be made in response to a decrease in methylamine metabolism through the N-methylglutamate dehydrogenase pathway, providing the methylamine with another metabolic sink. The oxidation of N-methylglutamate generates free ammonia; since the levels of N-methylglutamate are high, this step is probably a major contributor to the endogenous ammonia pool. In contrast, further metabolism of γ-glutamylmethylamide could be by a mechanism that may not result in the liberation of ammonia. The spectra show that loss of γ-glutamylmethylamide is accompanied by a transient reappearance of methylamine, which could occur by direct hydrolysis. It could also occur by the hydrolysis of N-methyl-α-ketoglutaramate to methylamine and α-ketoglutarate, a known reaction in this organism catalyzed by 5-hydroxy-N-methylpyrpglutamate (HMPG) synthase, the equilibrium for which greatly favors hydrolysis [5]. N-methyl-α-ketoglutaramate could be formed from γ-glutamylmethylamide by transamination. In this way the excess nitrogen would be held within the cell as organic nitrogen. Thus it is possible that both γ-glutamylmethylamide synthetase and HMPG synthase are involved in the temporary removal of excess methylamine-derived nitrogen. The lack of accumulation of HMPG rules out any direct role of this compound in either methylamine metabolism or storage. Its observed formation in cell extracts from methylamine and α-ketoglutarate is simply a result of its chemical equilibrium with N-methyl-α-ketoglutaramate, formed by reversal of the reaction intended for its hydrolysis. It is also possible that this pathway is involved in the import of methylamine into the cells.

3.2 Methylobacterium extorquens AM1

The time course of [^{13}C]methylamine metabolism by methylamine-grown cells is depicted in Fig. 5. There was no signal observed that corresponded to

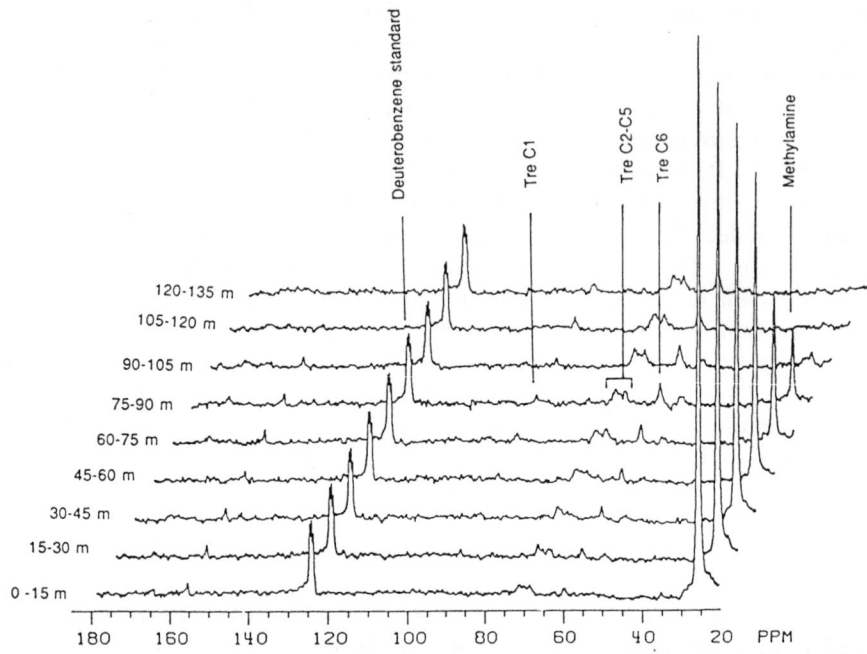

Figure 5 50.33 MHz ^{13}C-NMR spectra of [^{13}C]methylamine metabolism by methylamine-grown *Methylobacterium extorquens* AMI. Spectral parameters were as given for Fig. 1A. For clarity the labels for the metabolites above the spectra coincide with the 75–90 minute spectrum. The chemical shift scale corresponds to the lowermost (0–15 min) spectrum.

N-methylated intermediates in concurrence with the known direct PQQ-type pathway for methylamine oxidation in this organism. Also, in contrast to *Pseudomonas* MA, we observed the formation of substantial quantities of trehalose. This sugar has been previously observed in this organism following growth on methanol by both ^{14}C pulse chase experiments [17] and by ^{13}C NMR work [18]. The distribution of label in the various carbons of the disaccharide may be more clearly seen in the spectrum of the perchloric acid extract, Fig. 6. Carbon 6 shows the greatest enrichment, with carbons 2 and 5 having the weakest enrichment. Although trehalose C1 should be as enriched as C6 (because both C1 and C6 derive from C3 of serine), the relatively lower enrichment of trehalose C1 is indicative of substantial activity of the pentose phosphate cycle. These findings are therefore in accord with hexose formation from trioses derived directly from serine, the initial C_1 fixation compound, and not from TCA cycle activity, from which C2 and C5 should be more enriched than C3 and C4. Thus the labeling pattern of the carbons of trehalose derived from [^{13}C]methylamine is unable to resolve the question of how glyoxylate is regenerated from acetate in this organism and cannot definitely confirm or rule out any participation of an unknown type of isocitrate cleaving enzyme. The

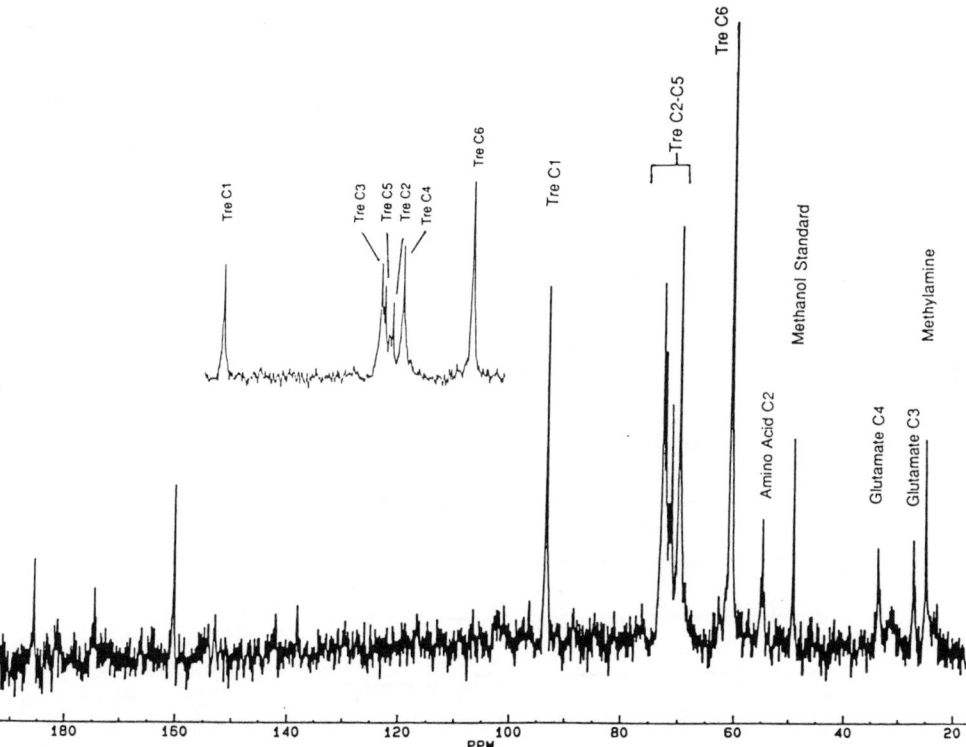

Figure 6 75.5 MHz ^{13}C-NMR spectrum of a perchloric acid extract obtained from the cell suspension utilised in Fig. 5 after 135 min incubation. Methanol was used as an internal standard. The inset shows the 60–100 ppm region of the spectrum expanded to facilitate analysis of the trehalose C2–C5 multiplet. The spectrum was obtained with 6300 accumulations with 4 Hz of line broadening applied to the FID before Fourier Transform. The unlabeled resonances above 160 ppm are attributable to various carboxylates.

presence of such an activity would compensate for the lack of the normal isocitrate lyase, the absence of which has been determined with direct enzyme analyses [19,20] and Western immunoblots using antiserum to isocitrate lyase from *Pseudomonas* MA [21; E. Bellion and H.S. Dodson, unpublished data].

4 METHYLAMINE METABOLISM IN YEAST

4.1 ^{13}C-Labeled methylamine metabolism in Hansenula polymorpha

Figure 7 shows a set of spectra obtained from methylamine/glucose-grown cells fed with ^{13}C-labeled methylamine. The label was found exclusively in metabolites of the assimilation pathway with no label detected in either formate or

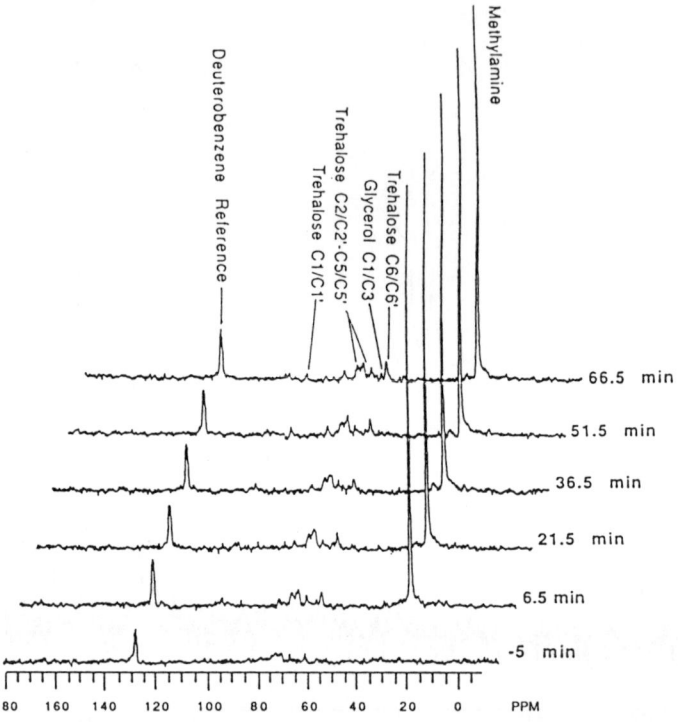

Figure 7 50.33 MHz ^{13}C-NMR spectra of [^{13}C]methylamine metabolism by glucose/methylamine-grown *Hansenula polymorpha*. [^{13}C]Methylamine was added at t = 0 to a concentration of 72 mM. The cells were oxygenated at a rate of 15 ml/min with a cell density of 65 mg dry wt/ml and the probe temperature was maintained at 37°C. 1800 accumulations were taken per spectrum (13 min) with 6 Hz line broadening applied to the FID before Fourier Transform. Times represent the mid-point of each set of accumulations. The chemical shift scale corresponds to the lowermost (−5 m) spectrum. Reproduced with permission of the American Society for Microbiology.

bicarbonate. Trehalose was one of the main labeled products accounting for all the peaks in the cell extract spectrum except for a singlet at 63 ppm and another at 72 ppm, indicating that a significant fraction of the formaldehyde generated by methylamine oxidation had been assimilated into carbohydrates, rather than oxidized to formate and bicarbonate. The two other peaks were assigned as the C1/C3 and C2 resonances of glycerol based on a previous ^{13}C-NMR study of glycolysis in *Saccharomyces cerevisiae* [10]. The even distribution of the label in all three carbons of glycerol suggests that this product was not directly formed from newly-synthesized dihydroxyacetone. During the assimilation of formaldehyde in *methanol-adapted* cells, there is high activity of dihydroxyacetone synthase, DHAS, a transketolase that donates a 2-carbon fragment from xylulose-5-phosphate to formaldehyde resulting in the assimilation of the single carbon of formaldehyde into triose. The first labeled product

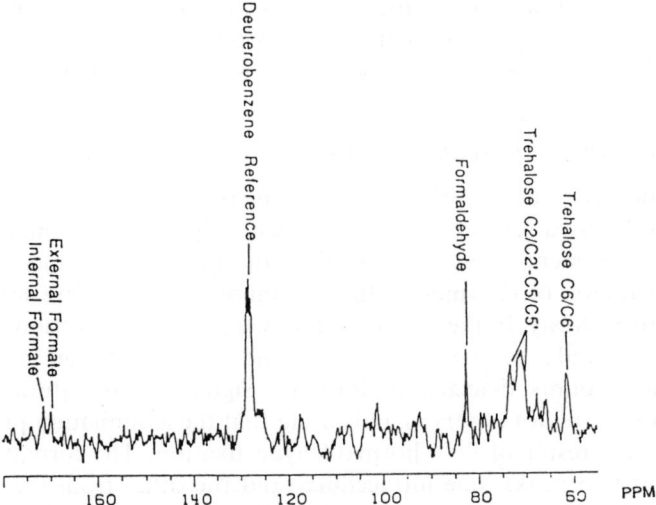

Figure 8 50.33 MHz ^{13}C-NMR spectra of [^{13}C]formaldehyde metabolism in glucose/methylamine-grown *Hansenula polymorpha*. Labeled formaldehyde was pre-prepared from [^{13}C]methanol and purified alcohol oxidase in the NMR resuspension buffer and the pH adjusted to 5.5. The formaldehyde + paraformaldehyde concentration was calculated to be 80 mM in formaldehyde equivalents from the change in intensity of the methanol peak before and after the alcohol oxidase addition. The cells were resuspended in 2.4 ml of buffer to a density of approx. 78 mg dry wt/ml, glucose was added (110 mM final concentration) and data were collected immediately after the addition of 600 μl of the formaldehyde solution. The spectrum consisted of 1000 scans (8 min) and was processed in the same way as the [^{13}C]methylamine spectra (Fig. 7). The methanol peak did not change in intensity during incubation with the cells. Reproduced with permission of the American Society for Microbiology.

formed is dihydroxyacetone that would be exclusively labeled at the C1 and C3 carbons [22,23]. A reductive pathway exists for the conversion of this metabolite to glycerol [11], with the result that label would be exclusively found in the C1 and C3 carbons of glycerol, with none in the C2 carbon. Thus this pathway was not being used to incorporate formaldehyde. Additionally a Western-blot analysis of extracts from glucose/methylamine-grown cells using antiserum obtained against purified DHAS from methanol-grown cells gave a negative result (data not shown). These facts indicate that a different transketolase was responsible for formaldehyde assimilation in glucose/methylamine-grown cells. This enzyme was most likely the normal or classical transketolase acting non-specifically as suggested by Waites and Quayle [24]. This enzyme has been recently confirmed to be able to perform this function in a DHAS-negative mutant of *Hansenula polymorpha* growing on xylose/methanol mixtures in which substantial methanol carbon was assimilated [25]. Thus a possible reason for the failure of this organism to grow on methylamine as a sole carbon source could be due to its inability to induce DHAS in sufficient quantities under these conditions. The spectrum showed a slower evolution of the trehalose C1/C1' peak compared to the C6/C6' peak, indicating that a loss of hexose C1 by the pentose phosphate pathway was occurring [12].

4.2 [^{13}C]Formaldehyde utilization

The assimilatory route of carbon flow from methylamine-derived formaldehyde was supported by a series of experiments with ^{13}C-labeled formaldehyde (Fig. 8). These experiments were conducted in the presence of glucose to repress possible synthesis of enzymes of the methanol (and therefore formaldehyde) assimilation pathway. In these spectra, following a rapid depletion of the added substrate, the bulk of the label was found in assimilation products with only small amounts detected in formate. Signals from natural abundance [^{13}C]glucose were not observed, indicating that the assimilation products were produced as a result of [^{13}C]formaldehyde fixation. The formate resonance appeared as two peaks, one intracellular and the other from the more acidic external medium. This indicates that the cells actively secreted this metabolite as it was being produced. The lack of carbon flow through the oxidative pathway except when the cells were challenged with formaldehyde was surprising considering the high activities of formaldehyde dehydrogenase in these cells [26]. This suggests that the role of the oxidative pathway is to scavenge excess formaldehyde that is not assimilated into carbohydrates thus protecting the cells from possible deleterious effects of this compound. Similar conclusions have been reached from studies of methanol utilization also [16,23].

5 CONCLUSIONS

This work has shown that use of *in vivo* NMR analysis of actively metabolizing cells and NMR analysis of perchlorate extracts can provide valuable information on methylamine assimilation by micro-organisms. Differences are readily seen between both bacteria and yeast and also between different strains of methylotrophic bacteria. Thus methylamine is converted to trehalose in the yeast *Hansenula polymorpha* and in *M. extorquens* AM1 and both of these organisms utilize a single reaction to oxidize methylamine initially. In contrast, *Pseudomonas* MA oxidizes methylamine via the N-methylglutamate pathway and does not produce trehalose. Both bacterial strains assimilate formaldehyde via serine.

ACKNOWLEDGEMENTS

The research work described in this paper was supported by research grants from The National Institute of General Medical Sciences (GM 38571) and The Texas Higher Education Coordinating Board Advanced Research Program (grant number 00356-118).

REFERENCES

[1] Eady, R.R. and Large, P.J. (1968) Purification and properties of an amine dehydrogenase from *Pseudomonas* AM1 and its role in growth on methylamine. *Biochem. J.* **168**, 245–255.

[2] Shaw, W.V., Tsai, L. and Stadtman, E.R. (1966) The enzymatic synthesis of N-methylglutamic acid. *J. Biol. Chem.* **241**, 935–944.

[3] Hersh, L.B., Peterson, J.A. and Thompson, A.W. (1971) An *N*-methylglutamate dehydrogenase from *Pseudomonas* MA. *Arch. Biochem. Biophys.* **145**, 115–120.

[4] Kung, H. and Wagner, C. (1969) γ-Glutamylmethylamide. A new intermediate in the metabolism of methylamine. *J. Biol.Chem.* **244**, 4136–4140.

[5] Hersh, L.B. (1970) 5-Hydroxy-*N*-methylpyroglutamate synthase-purification and mechanism of action. *J. Biol. Chem.* **245**, 3526–3535.

[6] van Dijken, J.P. and Bos, P. (1981) Utilization of amines by yeasts. *Arch. Microbiol.* **128**, 320–324.

[7] Green, G., Haywood, G.W. and Large, P.J. (1982) More than one amine oxidase is involved in the metabolism by yeasts of primary amines supplied as nitrogen source. *J. Gen. Microbiol.* **128**, 991–996.

[8] Zwart, K. and Harder, W. (1983) Regulation of the metabolism of some alkylated amines in the yeasts *Candida utilis* and *Hansenula polymorpha*. *J. Gen. Microbiol.* **129**, 3157–3169.

[9] den Hollander, J.A., Behar, K.L. and Shulman, R.G. (1981) ^{13}C-NMR study of transamination during acetate utilization by *Saccharomyces cerevisiae*. *Proc. Nat. Acad. Sci. USA* **78**, 2693–2697.

[10] den Hollander, J.A., Brown, T.R., Ugurbil, K. and Shulman, R.G. (1979) ^{13}C nuclear magnetic resonance studies of anaerobic glycolysis in suspensions of yeast cells. *Proc. Nat. Acad. Sci. USA* **76**, 6096–6100.

[11] Dickinson, J.R., Dawes, I.W., Boyd, A.S.F. and Baxter, R.L. (1983) ^{13}C-NMR studies of acetate metabolism during sporulation of *Saccharomyces cerevisiae*. *Proc. Nat. Acad. Sci. USA* **80**, 5847–5851.

[12] Ugurbil, K., Brown, T.R., Glynn, P. and Shulman, R.G. (1978) High-resolution ^{13}C nuclear magnetic resonance studies of glucose metabolism in *E. coli*. *Proc. Nat. Acad. Sci. USA* **75**, 3742–3746.

[13] Walter, T.E., Han, C.H., Kollman, V.H., London, R.E. and Matwyioff, N.A. (1982) ^{13}C-Nuclear magnetic resonance studies of the biosynthesis by *Microbacterium ammoniaphilum* of L-glutamate selectively enriched with Carbon-13. *J. Biol. Chem.* **257**, 1189–1195.

[14] Newaz, S.S. and Hersh, L.B. (1975) Reduced nicotinamide adenine dinucleotide-activated phosphoenolpyruvate carboxylase in *Pseudomonas* MA: potential regulation between carbon assimilation and energy production. *J. Bacteriol.* **124**, 825–833.

[15] Anthony, C. (1982) The Biochemistry of Methylotrophs, Academic Press, London and New York.

[16] Sibirny, A.A., Ubiyvok, V.M., Gonchar, M.V., Titorenko, V.I., Voronovsky, A.Y., Kapultsevich, Y.G. and Bliznik, K.M. (1990) Reactions of direct formaldehyde oxidation to CO_2 are non-essential for energy supply of yeast methylotrophic growth. *Arch. Microbiol.* **154**, 566–575.

[17] Large, P.J., Peel, D. and Quayle, J.R. (1961) Microbial growth on C_1 compounds. 2. Synthesis of cell constituents by methanol-and formate-grown *Pseudomonas* AM1 and methanol-grown *Hyphomicrobium vulgare*. *Biochem. J.* **81**, 470–480.

[18] Narbad, A., Hewlins, M.J.E. and Callely, A.G. (1989) ^{13}C-NMR studies of acetate and methanol metabolism by methylotrophic *Pseudomonas* strains. *J. Gen Microbiol.* **135**, 1469–1477.

[19] Large, P.J. and Quayle, J.R. (1963) Microbial growth on C_1 compounds. 5. Enzyme activities in extracts of *Pseudomonas* AM1. *Biochem. J.* **87**, 386–396.

[20] Bellion, E., Bolbot, J.A. and Lash, T.D. (1981) Generation of glyoxylate in methylotrophic bacteria. *Curr. Microbiol.* **6**, 367–372.
[21] Dodson, H.S. (1992) Chemical and immunological studies with isocitrate lyase. M.S. Thesis, The University of Texas at Arlington.
[22] Waites, M.J., Lindley, M.D. and Quayle, J.R. (1981) Determination of the labelling pattern of dihydroxyacetone and hexose phosphate following a brief incubation of methanol-grown *Hansenula polymorpha* with [^{14}C]methanol. *J. Gen. Microbiol.* **122**, 193–199.
[23] Jones, J.G. and Bellion, E. (1991) Methanol oxidation and assimilation in *Hansenula polymorpha*. An analysis by ^{13}C n.m.r. *in vivo*. *Biochem. J.* **280**, 475–481.
[24] Waites, M.J. and Quayle, J.R. (1980) The interrelationship between transketolase and dihydroxyacetone synthase activities in the methylotrophic yeast *Candida boidinii*. *J. Gen. Microbiol.* **124**, 309–316.
[25] de Koning, W., Bonting, K., Harder, W. and Dijkhuizen, L. (1990) Classical transketolase functions as the formaldehyde-assimilating enzyme during growth of a dihydroxyacetone synthase-negative mutant of the methylotrophic yeast *Hansenula polymorpha*. *Yeast* **6**, 117–125.
[26] Zwart, K. and Harder, W. (1983) Regulation of the metabolism of some alkylated amines in the yeasts *Candida utilis* and *Hansenula polymorpha*. *J. Gen. Microbiol.* **129**, 3157–3169.

33

Biochemistry of the Aerobic Utilization of Carbon Monoxide

Ortwin Meyer, Kurt Frunzke and Gerhard Mörsdorf

Lehrstuhl für Mikrobiologie, Universität Bayreuth, Universitätsstraße 30, D-8580 Bayreuth, Germany

1 SUMMARY

The aerobic chemolithoautotrophic utilization of CO or H_2 by *Pseudomonas carboxydovorans* and other carboxidotrophic bacteria requires the molybdenum iron-sulfur flavoprotein CO dehydrogenase for the catabolic oxidation of CO, a membrane-bound nickel-containing uptake hydrogenase, for the utilization of H_2 as energy source, a cytochrome b_{561} serving as membrane anchor and electron acceptor of CO dehydrogenase, a cytochrome b_{563} serving as a CO-insensitive alternative terminal oxidase, ribulosebisphosphate carboxylase (Rubisco) and phosphoribulokinase (PRK) as key enzymes for the fixation of CO_2 in the reductive pentose phosphate cycle.

This review discusses the biochemical properties of the above components, the reactions they catalyze and how they work together. In *P. carboxydovorans* the structural genes coding for the CO dehydrogenase subunits (*cox L, M* and *S*) and the uptake hydrogenase (*hox*) are carried on the plasmid pHCG3, those encoding Rubisco (*cfxL*) and PRK (*cfxP*) are carried on the plasmid as well as on the chromosome and those, coding for the cytochromes b_{561} (*pac*) and b_{563} (*tox*) are chromosomal. *Cox L, M* and *S* are clustered in *P. carboxydovorans* and arranged in the order *coxM-coxS-coxL*. All CO dehydrogenases examined in that respect contain molybdopterin cytosine dinucleotide (MCD) as the organic portion of the molybdenum cofactor. Several carboxidotrophic bacteria were found to excrete urothione, and there is a metabolic relationship between MCD and urothione.

Microbial Growth on C_1 Compounds
© Intercept Ltd, PO Box 716, Andover, Hampshire SP10 1YG, UK

2 INTRODUCTION

Carboxidotrophic bacteria are characterized by the chemolithoautotrophic utilization of carbon monoxide (CO) as a sole source of carbon and energy under aerobic or denitrifying conditions [1]. Their habitat is soil or water, and natural enrichments are found in the covering soil of burning charcoal piles [2,3]. Carboxidotrophic bacteria represent an assemblage of about 20 phylogenetically distinct bacteria. Their metabolism is facultative, i.e. besides CO, most strains also use H_2 plus CO_2 or organic substrates, although in most instances only a limited range. A noteworthy exception is the obligately chemolithoautotrophic thermophile *Streptomyces thermoautotrophicus*, which feeds on CO or H_2 plus CO_2, exclusively [2] and can fix dinitrogen [4].

The chemolithoautotrophic utilization of CO or H_2 by carboxidotrophic bacteria requires several biochemical components which are not present in other micro-organisms or at least not in that combination. In *P. carboxydovorans* these components are the molybdenum iron-sulfur flavoprotein CO dehydrogenase for the catabolic oxidation of CO, a membrane-bound uptake hydrogenase for the catabolic utilization of H_2, a cytochrome b_{561} serving as membrane anchor and electron acceptor of CO dehydrogenase, a cytochrome b_{563} functioning as a CO-insensitive alternative terminal oxidase, ribulosebisphosphate carboxylase (Rubisco) and phosphoribulokinase (PRK) as key enzymes for the fixation of CO_2 in the reductive pentosephosphate cycle.

Several carboxidotrophic bacteria harbor one to three relatively large plasmids [5,6]. *Pseudomonas carboxydovorans* carries the 128 kb plasmid pHCG3 (Fig. 1). In this bacterium the structural genes encoding CO dehydrogenase (*coxL, M* and *S*) and hydrogenase (*hox*) are plasmid encoded [1,7,8]. The genes are carried on the chromosome in plasmid-free carboxidotrophic bacteria. In *P. carboxydovorans* the genes *cfx*L and *cfx*P, encoding Rubisco and PRK, respectively, are duplicated and reside on pHCG3 and also on the chromosome [9]. The genes encoding the cytochromes b_{561} (*pac*) and b_{563} (*tox*) are chromosomal.

The study of carboxidotrophic CO dehydrogenases has led to the identification of molybdopterin dinucleotides as the organic component of the bacterial molybdenum cofactor and, specifically, molybdopterin cytosine dinucleotide (MCD) in *P. carboxydovorans* CO dehydrogenase [10].

This review will discuss the recent advances made in the understanding of the biochemistry of the special components in carboxidotrophic bacteria.

3 CARBON MONOXIDE DEHYDROGENASES

CO dehydrogenase is the key enzyme in the utilization of CO as a growth substrate by carboxidotrophic bacteria. The enzyme is best understood in *P. carboxydovorans* [1,11,12] but has also been purified from a wide range of taxonomically distinct carboxidotrophic bacteria, including mesophilic and

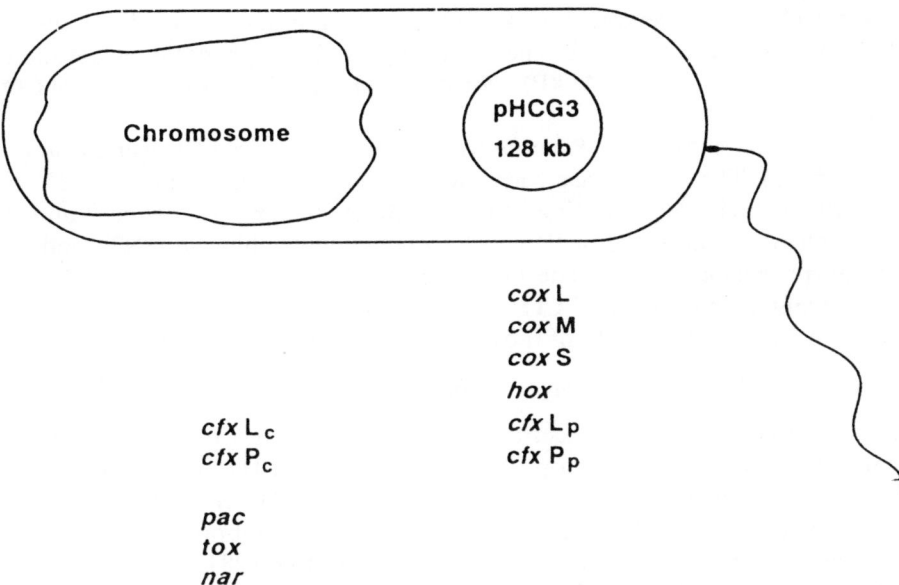

Figure 1 Location of carboxidotrophy genes on the genome of *Pseudomonas carboxydovorans*.

thermophilic species [2,13–17]. Except with *Streptomyces thermoautotrophicus* [2,16,17] and *Streptomyces* G26 [15] the basic biochemical features of CO dehydrogenases are remarkably similar in most carboxidotrophic bacteria studied [18–21].

3.1 Reactions catalyzed

CO dehydrogenase is the common expression for CO: acceptor oxidoreductase (EC 1.2.2.4). For a while the enzyme has been termed a 'CO oxidase' in analogy to xanthine oxidase (a molybdoenzyme which oxidizes xanthine with O_2 and water to uric acid and superoxide). However, the analogy is not quite correct since CO dehydrogenase does not act on oxygen, and usage of the term CO oxidase thus should be discouraged. CO dehydrogenase catalyzes the oxidation of CO to gaseous CO_2, using water as the oxidant [11,12,22,23]. There is evidence that in *P. carboxydovorans* the two electrons produced are transferred to a membrane-bound cytochrome b_{561} [1,24,25]:

$$CO + H_2O + 2\, b_{561}(ox) \rightarrow CO_2 + 2\, b_{561}(red) + 2\, H^+.$$

In cell-free systems or in the purified state all CO dehydrogenases examined in that respect use artificial oxidized electron acceptors (A) which are thereby reduced (for a review see [1,15,19]):

$$CO + H_2O + A(ox) \rightarrow CO_2 + AH_2(red).$$

Suitable electron acceptors for all CO dehydrogenases are oxidized methylene blue, thionine, toluylene blue, dichlorophenol indophenol or pyocyanine [11–13,15]. FAD, FMN, NAD(P)$^+$ could not serve as electron acceptors of CO dehydrogenase.

CO dehydrogenase from *P. carboxydovorans* also carries hydrogenase activity [1,26–28]. The enzyme has been shown to catalyze the oxidation of H_2 *in vitro* with a specific H_2-oxidizing activity of about 200 nmol H_2 oxidized min^{-1} (mg protein)$^{-1}$ (measured with iodonitrotetrazolium chloride (INT) and 1-methoxyphenazine methosulfate (MPMS) as electron acceptors). In different independent preparations of CO dehydrogenase, the H_2-oxidizing activities were consistently 10 to 16% of those of the CO-oxidizing activities [26,27]:

$$H_2 + A(ox) \rightarrow AH_2.$$

Range and type of artificial electron acceptors used for H_2 oxidation (A) were similar to those in CO oxidation [26]. The oxidation of H_2 by CO dehydrogenase *in vivo* seems not to be growth supporting [24]. The utilization of H_2 as a growth substrate by *P. carboxydovorans* requires a membrane-bound uptake hydrogenase which has been purified and characterized [28].

Under appropriate conditions, CO dehydrogenase as well as intact cells of *P. carboxydovorans* catalyze the formation of H_2 [26,27]:

$$DH_2(red) \rightarrow H_2 + D(ox).$$

The best electron donor (DH_2) for the evolution of H_2 by CO dehydrogenase was reduced methyl viologen [27]. Specific activities for H_2 evolution by CO dehydrogenase were 366 nmol of H_2 formed h^{-1} (mg protein)$^{-1}$. This is about 0.4% of the CO-oxidizing activity of the enzyme. Other suitable electron donors were reduced benzyl viologen, iodonitrotetrazolium chloride (INT), methylene blue, thionine, dichlorophenole indophenol (DCPIP), $NADH + H^+$, $FADH_2$, $FMNH_2$ and reduced riboflavin. $NADPH + H^+$ was ineffective. The CO- and H_2-oxidizing activities of CO dehydrogenase were inactive with oxidized soluble physiological electron acceptors. In contrast, the evolution of H_2 by CO dehydrogenase could be linked to reduced soluble physiological electron donors (e.g. $NADH + H^+$ or flavins).

It has been found recently that the CO dehydrogenases purified from *P. carboxydovorans*, *P. carboxydoflava* [29; Z. Alikulov, K. Frunzke, and O. Meyer, unpublished] and *S. thermoautotrophicus* [17] carry nitrate reductase activity:

$$Nitrate + DH_2(red) \rightarrow nitrite + D(ox) + H_2O.$$

Suitable electron donors were reduced methyl- or benzylviologen (DH_2) at pH 6.5. Similar to respiratory nitrate reductases (for a review refer to [30] and [31]), nitrate reduction by CO dehydrogenase was not linked to $NAD(P)H + H^+$. The specific nitrate reductase activities of CO dehydrogenases from different sources were low (Table 1) and did not exceed about 2.5% of the CO-oxidizing activity. The molybdenum-containing enzymes mammalian xanthine oxidase

Table 1 Nitrate reductase activity of different purified CO dehydrogenases

Source	Nitrate reductase activity[a]	Reference
Pseudomonas carboxydovorans grown aerobically with CO	2 (30°C)	[17]; Alikulov, Frunzke and Meyer, unpublished
Pseudomonas carboxydoflava grown aerobically with CO	2 (25°C)	[29]
Pseudomonas carboxydoflava grown under denitrifying conditions in the presence of CO	164 (25°C)	[29]
	8 (60°C)	
	11 (50°C)	
	760 (50°C)	
Streptomyces thermoautotrophicus grown aerobically with CO	0.6 (30°C)	[17]

[a] Activities were measured with reduced methyl viologen as electron donor and are in nmol min^{-1} (mg protein)$^{-1}$ at the temperatures given in brackets.

and chicken liver xanthine oxidase exhibited reduced methylviologen-nitrate reductase and to a lesser extent $FADH_2$-nitrate reductase activities, which amounted, however, to only about 0.05% of their xanthine-hydroxylating activity [32]. $NAD(P)H + H^+$ or $FADH_2$ could not serve as electron donors for nitrate reduction by CO dehydrogenase. That the molybdenum center of CO dehydrogenase is functionally involved in nitrate reduction is indicated by the susceptibility of the reaction to inhibition by methanol or cyanide. With CO as electron donor under anaerobic conditions, CO dehydrogenase displayed very low nitrate reductase activities [about 0.3 nmol nitrite formed min^{-1} (mg protein)$^{-1}$] (Z. Alikulov, K. Frunzke and O. Meyer, unpublished):

$$\text{Nitrate} + CO \rightarrow \text{nitrite} + CO_2.$$

That the electrons employed for nitrate reduction are indeed generated through the oxidation of CO has been concluded from simultaneous inactivation of the oxidation of CO with INT or nitrate by guanidine hydrochloride (0–250 mM) whereas the benzylviologen-nitrate reductase activity remained unaffected under these conditions.

Substrates not utilized by CO dehydrogenase were xanthine, hypoxanthine, purine, adenine, allopurinol, formaldehyde and salicylaldehyde [12], and side activities other than those mentioned above have not been reported.

3.2 Intracellular location and membrane anchor

After ultracentrifugation of crude extracts of *P. carboxydovorans*, more than 50% of CO dehydrogenase appeared in the soluble fraction and the remaining portion in the membrane fraction [23,33]. Using modified immunoferritin and protein A-gold techniques, it was demonstrated that in the exponential growth phase, 87% of CO dehydrogenase was associated with the inner aspect of the cytoplasmic membrane and 13% occurred in the cytoplasm [33–35]. In bacteria from the stationary growth phase, only about half of the total amount of the enzyme remained membrane-bound, and the dissociation of the enzyme from the cytoplasmic membrane coincided with a decrease of the CO-oxidizing activity with O_2. The CO-oxidizing activity with the unphysiological electron acceptor methylene blue, which does not require contact of CO dehydrogenase with the membrane, always exceeded that with O_2 [34]. Measurements of respiration rates of extracts with different electron donors in addition to CO have shown that the electron transport chain is not rate-limiting in intact cells of *P. carboxydovorans* and the electron flow from CO to O_2 is controlled by the amount of CO dehydrogenase attached to the membrane-bound electron acceptor.

In *P. carboxydovorans*, CO dehydrogenase and hydrogenase were found in association with the cytoplasmic membrane in a weakly bound and a tightly bound pool [25]. The pools could be experimentally distinguished on the basis of resistance to removal from membranes by washes in low-ionic-strength buffer. The tightly bound pool of the enzyme could be differentially solubilized under conditions leaving the electron transport system intact, and with the nondenaturing zwitterionic detergent 3-(3-cholamidopropyl) dimethylammonio-1-propanesulfonic acid (CHAPS) and the nonionic detergent dodecyl β-D-maltoside. In vitro reconstitution of depleted membranes with the corresponding supernatants containing CO dehydrogenase led to binding of the enzyme and to reactivation of respiratory activities with CO. The reconstitution reaction required cations which increased in effectiveness with increasing ionic charge: monovalent (Li^+), divalent (Mg^{2+}, Mn^{2+}), or trivalent (Cr^{3+}, La^{3+}). Reconstitution of depleted membranes with CO dehydrogenase was specific for CO-grown bacteria. Cytoplasmic membranes from H_2 or heterotrophically grown *P. carboxydovorans* had no affinity for CO dehydrogenase at all, indicating the absence of cytochrome b_{561} which is the physiological electron acceptor of the enzyme.

3.3 Physical and chemical properties

In contrast to the Ni-containing CO dehydrogenases from anaerobic bacteria which are oxygen-labile [19,36–38], the Mo-containing CO dehydrogenases from carboxidotrophic aerobes are perfectly stable when exposed to air. They have been purified from crude bacterial extracts employing ammonium sulfate precipitation, FPLC on the anion exchanger Accell QMA, sucrose density gradient centrifugation, chromatography on hydroxylapatite, anion exchange chromatography on DEAE-Sepharose and eventually preparative PAGE. Hydrophobic interaction chromatography on butyl-Sepharose was found to be especially effective in the purification of CO dehydrogenase from *Streptomyces thermoautotrophicus* [16,17]. The properties of CO dehydrogenases have been extensively reviewed [1,15,18,19,20,39].

CO dehydrogenases have been and still are routinely purified from various bacterial sources, and in many instances (at least in our own laboratory) the results have not been published in form of regular papers (Table 2). The preparations of CO dehydrogenases obtained in practice often vary considerably with respect to specific activity, fold purification and recovery (Table 2). About 1% or less of the total cell protein is usually represented by CO dehydrogenase. Fully active CO dehydrogenases with a complete cofactor composition should have the following properties (Table 3): A molecular weight of 266,800 (230,000 to 310,000, depending on the method employed) [11,13,15, 16–18,40]. They are composed of large (L), medium (M) and small (S) subunits with mean molecular weights of 84,400, 30,200 and 16,800, respectively (Table 3). The subunit structure is $(LMS)_2$. One enzyme molecule contains 2 Mo, 2 MCD, 8 Fe, 8 'S' and 2 FAD. The pH optimum is around 7 to 7.5. The temperature for maximum activity is around 65°C (e.g. *P. carboxydovorans, S. thermoautotrophicus*), 80°C (*P. thermocarboxydovorans*) or greater than 95°C (*Bacillus schlegelii*). The apparent K_M values of purified CO dehydrogenases for CO (μM) were 53 (*P. carboxydovorans* [11]), 63 (*P. carboxydohydrogena* [13]),

Table 2 Purification of CO dehydrogenases from various bacterial sources

Source	Sp act[a]	Fold purification	Yield	Reference
Pseudomonas carboxydovorans	1.94	35.5	26	[11]
	3.91	64.1	54	[12]
	1.65	12.7	49	[67]
	0.602	9.5	35	[24]
Pseudomonas carboxydohydrogena	180	35	30	[13]
	1.2	9.2	12.5	[67]
Pseudomonas carboxydoflava	0.64	6.7	8.7	[67]
	0.53	3.8	11	[68]
	6.9	15	27	[29]
Pseudomonas thermocarboxydovorans	11.6[b]	14.5	31	[15]
Streptomyces thermoautotrophicus	16.1	34.9	26	[17]
	6.1	7.4	8.9	[16]

[a] μmol CO oxidized min^{-1} (mg protein)$^{-1}$ at 30°C.
[b] At 50°C.

0.5 (*P. thermocarboxydovorans* [15]), 1.3 (*Streptomyces* G26 [15]) and 0.35 (*S. thermoautotrophicus* [17]).

The molybdenum content in preparations of CO dehydrogenases is frequently considerably below 2 Mo per enzyme molecule, referring to contamination with demolybdo-CO dehydrogenase and strengthening the necessity to characterize preparations of the enzyme also with respect to the molybdenum content. In contrast, purified CO dehydrogenases usually show a full complement of iron, acid-labile sulfur and FAD. Methanol (but not ethanol) is a specific inhibitor of CO dehydrogenases since it binds to Mo (+V) in a turnover-dependent, time increasing reaction [12].

The CO dehydrogenase purified from *P. thermocarboxydovorans* showed the following unique properties [15]: (i) its optimum temperature in the usual assay

Table 3 Molecular weights of CO dehydrogenase subunits

Source	$M_r (\times 10^3)$				Reference
	L	M	S	HE[a]	
Pseudomonas carboxydovorans	86	34	17	274	[35]
	87	30	17	286	[7]
	85	25	14	248	[40,69]
Pseudomonas carboxydohydrogena	85	28	14	254	[13]
Pseudomonas carboxydoflava	70	33	17	240	[18]
Pseudomonas thermocarboxydovorans	87	29	21	274	[70]
Streptomyces thermoautotrophicus	87	30	17	268	[17]
	88	32.5	17.5	276	[16]

[a] HE, holoenzyme; calculated from the molecular weights of subunits assuming a subunit structure of (LMS)$_2$.

Table 4 Amino terminal amino acid sequences of CO dehydrogenase subunits

L-Subunit[a]

```
                                                     5              10
Pseudomonas carboxydovorans [7]           M N I Q T - V E P T A G E
Pseudomonas carboxydoflava [7]            M N A P V Q D A E
Pseudomonas carboxydohydrogena [7]        M G H P
Pseudomonas thermocarboxydovorans [66]    M N A P L S D R E K
Streptomyces thermoautotrophicus [17]    (M) A V K E E R P I G F G
```

M-Subunit[b]

```
                                                    5              10            15
Pseudomonas carboxydovorans [7]           M M I P G S F D Y H R P K S I
Pseudomonas carboxydoflava [7]            M M I P G - F E Y H A P K H V
Pseudomonas carboxydohydrogena [7]        M M I P G H F D Y H R P K S V
Streptomyces thermoautotrophicus [17]     M Q V P A P F E Y Q R A K S V ... R
                                                                              30
```

S-Subunit[c]

```
                                                    5              10            15            20
Pseudomonas carboxydovorans [7]           M A K A H I - E L T I N G H P V E A L V E P
Pseudomonas carboxydoflava [7]            M A K K I I - T V N V - G K A Q E K A V E P
Pseudomonas carboxydohydrogena [7]        M A K A
Pseudomonas thermocarboxydovorans [66]    M S K - H I V S M T V N G R K V E E A V E A
Streptomyces thermoautotrophicus [17]     M Q - - - I - T I N V N G E D Y T R E I E P ... H
                                                                                            36
```

[a] Highest homologies: *P. carboxydoflava* and *P. thermocarboxydovorans* with 66.7% identity (6 amino acids out of 9) and 77.8% similarity (7 amino acids out of 9).

[b] Highest homologies: *P. carboxydovorans* and *P. carboxydohydrogena*, 86% identity (13 amino acids out of 15) and 93.3% similarity (14 amino acids out of 15).

[c] Highest homologies: *P. carboxydovorans* and *P. thermocarboxydovorans*, 52.4% identity (11 amino acids out of 21) and 76.2% similarity (16 amino acids out of 21).

was 80°C, (ii) its affinity for CO was very high (see above) and the enzyme used both horse heart cytochrome c and potassium ferricyanide as electron acceptors *in vitro*, in addition to other artificial electron acceptors such as methylene blue and phenazinemethosulfate (PMS) used by the *P. carboxydovorans* and *P. carboxydohydrogena* enzymes. The CO dehydrogenase from *S. thermoautotrophicus* revealed the following unique properties [16,17]: (i) it showed the highest affinity for CO of any carboxidotrophic bacterium examined (see above), (ii) it could reduce low-potential electron acceptors such as methyl- or benzyl viologen, (iii) its N-terminal amino acid sequences showed distinct differences to those of the other CO dehydrogenases (Table 4), and (iv) it occurred in two different catalytically active forms, a LMS heteromonomer and a $(LMS)_2$ heterodimer.

The CO dehydrogenases from *P. carboxydovorans* and *P. carboxydohydrogena* are still the only ones that have been studied by EPR [41]. For a recent treatment of enzymes depending on the pterin molybdenum cofactor including CO dehydrogenases, refer to [42].

3.4 *Molybdenum cofactor, molybdopterin dinucleotides and urothione*

CO dehydrogenases contain the pterin molybdenum cofactor which is part of the active centre of CO dehydrogenase (Fig. 2). The cofactor is composed of molybdopterin cytosine dinucleotide (MCD), molybdenum and an oxygen- or sulfur-substituent attached to the metal. There are indications that the cofactor is released from CO dehydrogenase during post-exponential dissociation of the enzyme from the cytoplasmic membrane and disintegrates into its components, whereupon MCD is converted to urothione and excreted into the culture medium.

3.4.1 Molybdopterin dinucleotides in the molybdenum cofactor The pterin cofactor (bactopterin) in CO dehydrogenase isolated from *P. carboxydoflava* has been shown to differ from molybdopterin in molecular mass, phosphate content, stability, and other properties, implying a novel structure [43–45]. Subsequently, the novel pterin was also resolved in other bacterial molybdoenzymes [44]. The structure of the CO dehydrogenase pterin has been investigated by alkylation and isolation of the carboxamidomethyl derivative [10]. The alkylated pterin was identified as [di(carboxamidomethyl)]molybdopterin cytosine dinucleotide (Fig. 2) on the basis of its absorption properties and by degradation with nucleotide pyrophosphatase yielding dicarboxamidomethyl molybdopterin and CMP. Further treatments of these products with alkaline phosphatase produced species with absorption and chromatographic properties identical to those of the corresponding dephospho compounds. The exact chemical structure of [di(carboxamidomethyl)]molybdopterin cytosine dinucleotide has recently been further substantiated by electron impact mass spectroscopy, ^1H-^1H COSY, ^1H-and ^{13}C-NMR [46, J. Tachil, personal com-

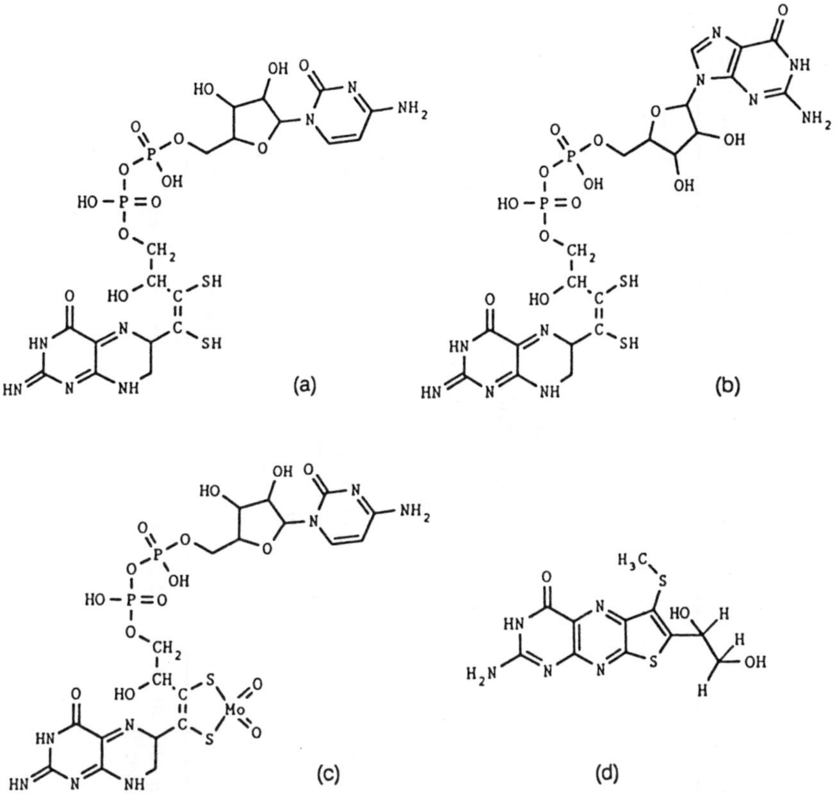

Figure 2 Proposed chemical structures of (a) molybdopterin cytosine dinucleotide (MCD), (b) molybdopterin guanine dinucleotide (MGD), (c) the native molybdenum cofactor (MoCo), composed of MCD, molybdenum and an O-substituent attached to the metal and (d) urothione, an excretion product of carboxidotrophic bacteria. The pterins are shown in their dihydro (a–c) or oxidized (d) state.

munication]. So far, MCD has uniquely been identified in carboxidotrophic dehydrogenases and, recently, in a bacterial quinoline oxidoreductase, whereas different pterins or pterin dinucleotides have been resolved in other molybdoenzymes (Table 5).

3.4.2 *Excretion of urothione* Urothione [(-)-2-amino-7-[1,2-dihydroxy-ethyl] -6-methylmercapto-3H-thieno[3,2-g]pteridin-4-on] (Fig. 2) is excreted into human urine [47,48], presumably as a degradation product of molybdopterin released from the molybdenum cofactor upon the turnover of the liver enzymes xanthine dehydrogenase, aldehyde oxidase and sulfite oxidase [49,50]. The excretion of urothione into the medium by carboxidotrophic bacteria has recently been reported [51–53]. The excreted material was isolated from culture supernatants of CO-autotrophically grown *Pseudomonas carboxydoflava* and

Table 5 Occurrence of molybdopterin and molybdopterin dinucleotides in molybdoenzymes from various sources

Type of pterin[a]	Enzyme	Source	Reference
MPT	Sulfite oxidase	Chicken liver	[71]
	Xanthine oxidase	Bovine milk	[71]
	Xanthine dehydrogenase	*Pseudomonas aeruginosa*	[72]
MCD	CO dehydrogenase	*Pseudomonas carboxydoflava*	[10,73]
		Pseudomonas carboxydohydrogena	[73]
		Pseudomonas carboxydovorans	[73]
		Streptomyces thermoautotrophicus	[16,17]
	Quinoline oxidoreductase	*Pseudomonas putida*	[74]
		Rhodococcus sp.	[74]
MGD	Dimethyl sulfoxide reductase	*Rhodobacter sphaeroides*	[75]
	Formylmethanofuran dehydrogenase	*Methanosarcina barkeri*	[76]
	Formate dehydrogenase	*Methanobacterium formicium*	[77]
	Nitrate reductase	*Pseudomonas carboxydoflava*	[55]
		Pseudomonas stutzeri	[78]
		Escherichia coli	[79]
MAD, MHD	Formylmethanofuran dehydrogenase	*Methanobacterium thermoautotrophicum*	[80]

[a] MPT, molydopterin; MCD, molybdopterin cytosine dinucleotide; MGD, molybdopterin guanine dinucleotide; MAD, molybdopterin adenine dinucleotide; MHD, molybdopterin hypoxanthine dinucleotide.

its chemical structure identified as urothione (Fig. 2) on the basis of co-chromatography on HPLC with authentic urothione, the characteristic pH-dependence of its UV/VIS spectra, derivatization to pterin-6-carboxylic-7-sulfonic acid, a molecular mass of 451 of its triacetyl-derivative analyzed by electron impact mass spectroscopy, ^{1}H-NMR and ^{1}H-^{1}H COSY [51–53]. Cells of *P. carboxydoflava* growing chemolithoautotrophically with CO revealed maximum production of urothione [34 µg urothione (g d.wt.)$^{-1}$]. Excretion of the compound was also observed with other carboxidotrophic bacteria (e.g. *P. carboxydovorans* or *S. thermoautotrophicus*) and always coincided with conditions where CO dehydrogenase was expressed [51,53].

Pseudomonas carboxydoflava has been shown to grow with organic substrates under denitrifying conditions forming N_2 [1,54]. Nitrate reductase from the bacterium has been purified 36-fold with a yield of 13% and a specific activity of 19.6 µmol nitrite formed min^{-1} (mg protein)$^{-1}$ [29,55]. The enzyme had a molecular weight of 200,000 and contained molybdopterin guanine dinucleotide (MGD) in its molybdenum cofactor (Fig. 2, Table 5). Under denitrifying growth conditions and in the absence of CO, *P. carboxydoflava* was devoid of MCD and did not excrete urothione [51, 53].

A metabolic relationship between MCD and urothione is suggested by the coincidence of inactivation of CO dehydrogenase in the post-exponential growth phase and excretion of urothione. This is further substantiated by the finding that MCD in solutions of unfolded CO dehydrogenase was converted to carboxamidomethyl-norurothione through the action of nucleotide pyrophosphatase, alkaline phosphatase and iodoacetamide. The presence of active nucleotide pyrophosphatase [6.7 nmol of di(cam)MCD cleaved h^{-1} (mg protein)$^{-1}$] and alkaline phosphatase [250 nmol of p-nitrophenol formed from p-nitrophenylphosphate h^{-1} (mg protein)$^{-1}$] in extracts of *P. carboxydoflava* has also been demonstrated.

3.5 Activation by selenium

The CO-oxidizing activities of CO dehydrogenase with methylene blue or dichlorophenol indophenol were specifically activated upon aerobic incubation with selenite [45,56]. A similar activation of CO dehydrogenase could be achieved upon anaerobic incubation with sulfide plus dithionite. The extent of activation achieved depended on the enzyme sample used but was practically the same with selenite or sulfide plus dithionite. Fully active Se-CO dehydrogenase contained selenium, molybdenum, and flavin adenine dinucleotide in a 1:1:1 molar ratio. The incorporated selenium was shown to be a covalently bound constituent of the protein of CO dehydrogenase and not part of the noncovalently bound molybdenum cofactor. It has been suggested that upon activation of CO dehydrogenase, selenium became bound between the sulfurs of half-cystine residues forming seleniumtrisulfides:

$$4 \text{ RSH} + H_2SeO_3 \rightarrow \text{RSSeSR} + \text{RSSR} + 3 H_2O.$$

Selenite-activated CO dehydrogenase displayed EPR spectra of the Mo(V) resting I type that were practically identical to those obtained with the native enzyme, indicating that selenite treatment does not alter the redox state of the molybdenum. The signal was not evident in the EPR spectrum of the enzyme activated with sulfide plus dithionite and reoxidized under appropriate conditions, and it was thus concluded that S-CO dehydrogenase and Se-CO dehydrogenase are different enzyme species. This was further substantiated by differences in the susceptibility to inactivation upon exposure to air. Se-CO dehydrogenase was stable, whereas S-CO dehydrogenase inactivated with a half-life of approximately 2 days.

The action of selenite was directed to the cytoplasmic species of CO dehydrogenase exclusively, whereas the CO → methylene blue activity of the membrane-bound enzyme remained unaffected.

4 CYTOCHROMES

The metabolism of carboxidotrophic bacteria is strictly respiratory, employing O_2 as a terminal electron acceptor [57], for a recent review see [1,18,20]. In the absence of O_2, *P. carboxydoflava* and few other carboxidobacterial strains could grow with organic substrates and nitrate serving as terminal electron acceptor [1,54]. Carboxidotrophic bacteria contain cytochromes of the *b*-, *c*- and *a*-type [58] and are not inhibited by up to 90% (by volume) CO in the gas phase [57]. Electron transport is best understood in *P. carboxydovorans* [59,60]. Experiments employing electron transport inhibitors, room- and low-temperature spectroscopy, and photochemical action spectra have led to a model for the respiratory chain of *P. carboxydovorans* [59] (Fig. 3). The chain is branched at the level of *b*-type cytochromes or ubiquinone-10. In the absence of CO, electrons from the oxidation of organic substrates were preferentially channelled into the CO-sensitive branch which contains cytochromes b_{558}, c, and a_1. The CO-insensitive branch allowed growth in the presence of CO and contained cytochromes b_{561} and b_{563}. Because of its insensitivity towards CO, cytochrome b_{563} revealed no photochemical action spectrum. However, reoxidation experiments established the functioning of cytochrome b_{563} ($= o$) as an alternative CO-insensitive terminal oxidase of the autotrophic branch. It was the least reducible cytochrome, and it was the first to react with O_2 during reoxidation of reduced extracts. Cytochrome *d* and the salicylhydroxamic acid-sensitive alternative terminal oxidase of many eukaryotes were absent. Hydrogen peroxide was not formed by the CO-insensitive branch and, in CO difference spectra, a maximum at 416 nm and troughs at 433 and 563 nm were indicative of cytochrome *o*. The functioning of cytochrome a_1 as a terminal oxidase was established by photochemical action spectra. Tetramethyl-*p*-phenylenediamine was oxidized via cytochromes *c* and *a* exclusively. The CO-sensitive branch was also sensitive to antimycin A and micromolar concentrations of cyanide. The CO-insensitive branch was sensitive to 2-*n*-heptyl-4-hydroxyquinoline-*N*-oxide and to millimolar concentrations of cyanide.

Aerobic utilization of carbon monoxide 447

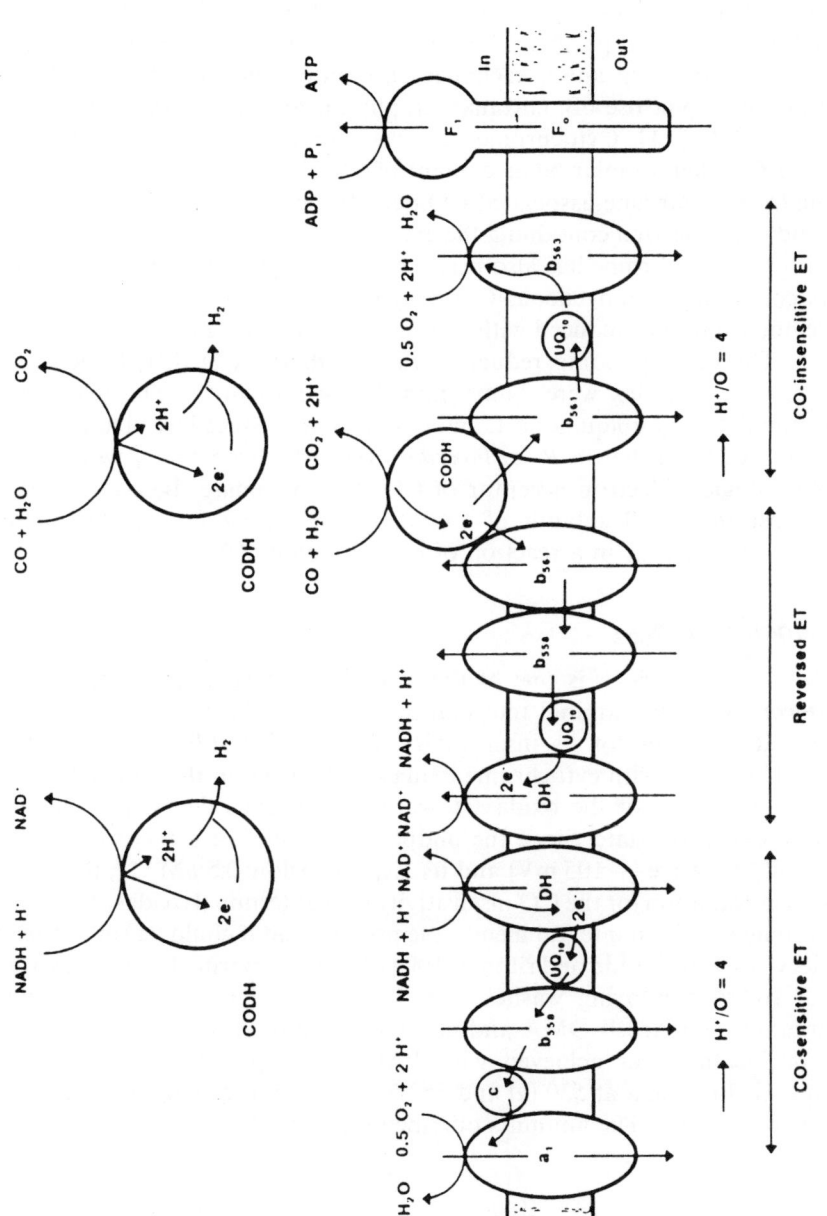

Figure 3 Hypothetical scheme for the CO-sensitive, insensitive and reversed (NADH+H⁺yielding) electron transport in *Pseudomonas carboxydovorans*, interaction of CO dehydrogenase with electron transport and functioning of cytoplasmic CO dehydrogenase. Abbreviations: CODH, CO dehydrogenase; UQ_{10}, ubiquinone-10; ET, electron transport; DH, NADH-dehydrogenase.

4.1 Cytochrome b_{561}

Cytochrome b_{561} in *P. carboxydovorans* is an integral membrane protein, and dodecyl β-D-maltoside has been found to be suitable for solubilization [24]. The cytochrome has been purified 21.5-fold from cytoplasmic membranes with a yield of 6% employing anion exchange chromatography on Accell QMA, gel filtration on Sepharose 6B, chromatography on hydroxylapatite and nondenaturing PAGE [24]. Cytochrome b_{561} was composed of a single 55,000 subunit and revealed a molar heme content of 0.8.

Starting from membrane-associated CO dehydrogenase, a complex could be isolated and characterized containing the enzyme and cytochrome b_{561} in a 1:2 molar ratio, the cytochrome has also been obtained from such complexes [24]. In assays containing homogeneous CO dehydrogenase, cytochrome b_{561} and CO, the cytochrome was reduced with a specific activity of 1.2 nmol min^{-1} (mg protein)$^{-1}$. The corresponding reduction rates with the CO dehydrogenase-cytochrome b_{561} complex were 2 nmol min^{-1} (mg protein)$^{-1}$. The reactions were not stimulated by ubiquinone-1, and b_{561} was not reduced by ubiquinol-1. It was thus concluded that in *P. carboxydovorans* cytochrome b_{561} would act as the physiological electron acceptor of CO dehydrogenase. IgG-antibodies raised against the small subunit of CO dehydrogenase cross reacted with purified cytochrome b_{561} in a reaction of partial identity [24].

4.2 Cytochrome b_{563}

Although cytochrome b_{563} is one of the most interesting components in *P. carboxydovorans*, only marginal information on its biochemical properties is available and the basis for its insensitivity towards CO still remains to be elucidated. CO reacts with cytochrome oxidase only when in the reduced form (Fe^{2+}). The sensitivity of the oxidase towards CO is thus dependent on the steady-state oxidation state. Since the midpoint potential of cytochrome b_{563} was unusually negative (-105 mV) and its K_M was below 0.5 μM O_2, this was taken as an explanation for the CO insensitivity of that terminal oxidase [21].

Cytochrome b_{563} is an integral membrane protein, and it could be solubilized with dodecyl-β-D-maltoside or Triton X-100 [61]. It was purified 26.5-fold with a yield of 16.6% employing washes with Triton plus KCl, anion exchange chromatography on Accell QMA and chromatography on phenyl-Sepharose. Further purification was achieved with PAGE. The pyridine hemochrome spectra revealed maxima at 530 (α) and 557.5 (β), which are characteristic of a protoheme compound. The amounts obtained did not allow further characterization.

5 REVERSE ELECTRON TRANSFER

None of the CO dehydrogenases known so far is capable of linking the CO oxidation reaction to the formation of NAD(P)H + H$^+$. This is surprising since the metabolism of carboxidotrophic bacteria is chemolithoautotrophic and requires reduced pyridine nucleotides for the assimilation of CO_2 in the reductive pentose phosphate cycle. It has been shown in *P. carboxydovorans*, growing with CO or H_2, that electrons and a proton motive force, generated by respiration, are required to drive a reverse electron transfer for the formation of reduced pyridine nucleotides [60]. In cell suspensions of *P. carboxydovorans* pulsed with lithotrophic substrates (CO or H_2) in the presence of oxygen, formation of reduced pyridine nucleotides and of ATP could be demonstrated using the bioluminescent assay. Experiments employing base-acid transition, an uncoupler and inhibitors of ATPase or electron transport have led to a model for the formation of $NAD(P)H + H^+$ in chemolithotrophically growing *P. carboxydovorans* [60] which has been adopted in Fig. 3.

The protonophore FCCP (carbonyl-p-trifluoromethoxy-phenylhydrazone) inhibited both formation of $NAD(P)H + H^+$ and ATP. In the absence of oxygen, a chemical potential imposed by base-acid transition resulted in the formation of $NAD(P)H + H^+$ and ATP when electrogenic substrates (CO or H_2) were present. This indicated proton motive force-driven $NAD(P)H + H^+$ formation. The proton motive force was generated by oxidation of the substrate and not by ATP hydrolysis, as is obvious from $NAD(P)H + H^+$ formation during inhibition of ATP synthesis by oligomycin and N,N'-dicyclohexyl-carbodiimide.

The CO-borne electrons are transferred *via* the ubiquinone-10-cytochrome *b* region to NADH dehydrogenase functioning in the reverse direction, as indicated by inhibition of $NAD(P)H + H^+$ formation by HQNO (2-*n*-heptyl-4-hydroxyquinoline-N-oxide) and rotenone, and by resistance to antimycin A.

6 GROWTH YIELDS AND PROTON TRANSLOCATION

The molar growth yields of different carboxidotrophic bacteria in energy source-limited cultures were similar (Table 6). The yields obtained under chemolithoautotrophic conditions with CO or H_2 were only one-tenth those with pyruvate. That CO does not inhibit heterotrophic growth is evident from the same yields with pyruvate, irrespective of whether CO is present or absent (Table 6). The molar growth yields of carboxidotrophic bacteria on CO or H_2 under chemolithoautotrophic conditions (Table 6) compare to the Y_m of 2.9 (g biomass per mole) of *Paracoccus denitrificans* during autotrophic growth on formate [62]. The respiration-driven proton translocation has been measured with the O_2 pulse method and intact cells of *P. carboxydovorans* [63,64], *P. carboxydoflava* [64] and *Pseudomonas carboxydohydrogena* [64]. Proton translocation rates ($\rightarrow H^+/O$ ratios) were independent on the growth phase. Ratios of $\rightarrow H^+/O$ measured in *P. carboxydovorans* grown with pyruvate, CO, pyruvate plus CO, or H_2 yielded respective average values of 3.5, 3.2, 3.5 and 5.8 for endogenous substrates [64]. With pyruvate, CO or H_2 as electron donors, the corresponding average $\rightarrow H^+/O$ ratios were 3.8, 0.6 and 0.6

Table 6 Molar growth yields of carboxydotrophic bacteria with different substrates [64]

	Molar growth yields $(Y_m)^a$		
Substrate	P. carboxydovorans	P. carboxydoflava	P. carboxydohydrogena
Pyruvate	23.4	23.8	21.0
Pyruvate + CO	23.9	24.6	22.3
Pyruvate + H_2	23.2	22.8	18.3
CO	1.9	2.0	2.2
H_2	2.5	1.7	2.2

[a] Molar growth yields are in g bacterial d.wt. per mole of substrate; with the combinations of different substrates they are given per mole of pyruvate.

(pyruvate-grown bacteria), 3.6, 0.5 and 5.8 (bacteria grown on pyruvate plus CO), 3.5, 3.9 and 6.2 (CO-grown bacteria) and 3.8, 4.4 and 5.6 (H_2-grown bacteria). Additional evidence for the insensitivity of pyruvate oxidation towards CO was obtained from average $\rightarrow H^+/O$ ratios with pyruvate or H_2, which were the same irrespective of whether CO was present or absent [64]. The data suggest that the CO-sensitive and -insensitive branch of the respiratory chain of *P. carboxydovorans* are equally efficient with respect to proton translocation (Fig. 3). Compared to pyruvate and CO, an additional proton translocation site seems to be available in the oxidation of H_2.

7 HYDROGENASE

Cells of *P. carboxydovorans* growing with H_2 plus CO_2 contain an inducible uptake hydrogenase. The enzyme is an integral membrane protein and has been isolated and characterized [28]. Hydrogenase has been solubilized employing dodecyl-β-D-maltoside and purified 25-fold with a yield of 8% and a specific activity of 155 μmol of H_2 oxidized min^{-1} (mg protein)$^{-1}$. The purified enzyme revealed Michaelis-Menten kinetics and a K_M of 77 μM H_2. The molecular weight of the holoenzyme was 105,000. It was a heterodimer with a subunit structure of LS (L = 67,000; S = 30,000). One enzyme molecule contained 0.7 to 0.75 Ni, 7.25 Fe and 7.2 to 7.5 acid labile sulfurs, suggesting the presence of iron-sulfur centers. Maximum reaction rates occurred at pH 7 (membrane-bound enzyme) and pH 5.8 (purified enzyme). The enzyme behaved as a weak anion (pI = 6.0). Hydrogenase was not able to act on cytochrome b_{561} [24]. The two subunits of *P. carboxydovorans* hydrogenase have been sequenced from the amino-terminal end. The sequences obtained, and especially those of the small subunit, are very homologous to those of hydrogenase subunits from other bacterial sources (Table 7). The high degree of similarity between the

Table 7 Amino terminal amino acid sequences of membrane bound hydrogenase subunits

		5					10					15					20						
L-Subunit																							
Pseudomonas carboxydovorans[a]	S	V	I	Q	T	P	N	G	Y	K	L	D	N	S	G	R	R	V	V	V	D	P	V	T	R	I	E
Bradyrhizobium japonicum [81]	M	G	I	Q	T	P	N	G	F	N	L	D	N	S	G	K	R	I	V	V	D	P	V	T	R	I	E (76% Identity)
Rhizobium leguminosarum [82]	M	T	I	Q	T	P	N	H	F	T	L	D	N	S	G	K	R	I	V	V	D	P	V	T	R	I	E (76% Identity)
Alcaligenes eutrophus [83]	–	S	A	Y	A	T	Q	G	F	N	L	D	D	R	G	R	R	I	V	V	D	P	V	T	R	I	E (56% Identity)
S-Subunit																							
Pseudomonas carboxydovorans[a]	M	E	T	K	P	R	T	P	V	L	W	L	H	G	L	E	E	T									
Rhodocyclus gelatinosus [84]	M	E	T	K	P	R	T	P	V	L	W	L	H	G	L	E	C	T (94% Identity)									
Azotobacter vinelandii [85]	M	E	T	K	P	R	T	P	V	L	W	L	H	G	L	E	C	T (94% Identity)									
Alcaligenes eutrophus [83]	M	E	T	K	P	R	T	P	V	L	W	L	H	G	L	E		(100% Identity)									

[a] B. Santiago, personal communication.

uptake hydrogenase from *P. carboxydovorans* and the corresponding enzymes from other bacterial sources is also evident from subunit structure, molecular weight, cofactor composition and the presence of Ni.

8 CARBOXIDOTROPHY GENES

8.1 cox genes

Employing deoxyoligonucleotide probes and Southern hybridizations, the structural genes encoding the large, medium and small subunits of CO dehydrogenase (*coxL*, *M* and *S*, respectively) were found to be conserved in carboxidotrophic bacteria of distinct taxonomic position and can be carried either on a plasmid or the chromosome [7,8]. It was, therefore, speculated that the conserved *cox* genes have arisen from horizontal gene transfer of a common ancestor, e.g. through a CAR-plasmid (from CARbon monoxide) [7]. It seems, however, that the N-terminal amino acid sequences of the *Streptomyces thermoautotrophicus* CO dehydrogenase subunits do not quite fit into this picture (Table 4). *Streptomyces thermoautotrophicus* is a noteworthy exception since none of the three *cox* genes could be detected [8]. In *P. carboxydovorans*, *coxL*, *M* and *S* were carried on the plasmid pHCG3 and were absent on the chromosome (Fig. 1). *coxL*, *M* and *S* from *P. carboxydovorans* were localized on restriction fragments of pHCG3 and cloned into a T7 RNA polymerase-dependent expression vector. After induction of transcription at the T7 promoter, *cut*L and S were heterologously expressed in *Escherichia coli*, as was evident from crossreactivity of the gene products with specific antibodies raised against CO dehydrogenase subunits and indistinguishable electrophoretic mobility [65]. The 5'-coding regions of the cox genes were sequenced, and the amino-terminal amino acid sequences deduced were identical with the amino acid sequences of the corresponding CO dehydrogenase subunits. *coxM* was not expressed in *E. coli*, but its coding region could be identified upstream of *coxS* from the amino acid sequence deduced from the nucleotide sequence. The *cox* genes were found to be clustered in *P. carboxydovorans* and arranged in the order *coxM-coxS-coxL* [65]. The first 148 amino acids from the N-terminal end of *coxL* known so far were highly homologous with the N-terminus of the molybdenum cofactor-containing domain of xanthine dehydrogenases from eukaryotic sources [65]. Similar results have been obtained for the large subunit of CO dehydrogenase from *P. thermocarboxydovorans*. For a review on the molecular genetics of CO dehydrogenase the reader is referred to O'Reilly *et al.*, this volume.

8.2 cfx genes

Heterologous gene probes derived from *cfx*Lp and *cfx*Pp genes of *Alcaligenes eutrophus* H16 revealed the presence of structural genes encoding ribulosebisphosphate carboxylase (Rubisco) and phosphoribulokinase (PRK) on the genome of carboxidotrophic bacteria [9]. The two genes were found to be conserved. In *P. carboxydovorans cfx* genes are duplicated and reside on the plasmid pHCG3 and the chromosome as well (Fig. 1). Also, in all plasmid-harboring carboxidotrophic bacteria, *cfx*L and *cfx*P structural genes were found to be plasmid coded. A *cfx*L gene probe from *Rhodospirillum rubrum* did not detectably hybridize with DNA from a wide range of carboxidotrophic bacteria [9].

8.3 Other genes

The location on the genome of genes of *P. carboxydovorans* involved in the expression of hydrogenase (*hox*), the physiological electron acceptor of CO dehydrogenase cytochrome b_{561} (*pac*), the CO-insensitive terminal oxidase cytochrome b_{563} (*tox*) and nitrate reductase *(nar)* have been deduced from experiments with the pHCG3 cured mutant strain OM5-12 (Fig. 1).

9 CONCLUSIONS

It is plain from the foregoing findings and considerations that much more information is required to substantiate preliminary conclusions about the functions of CO dehydrogenase in a carboxidobacterial cell, in addition to oxidizing CO for the production of energy and assimilable CO_2. Furthermore, the electron transport system of *P. carboxydovorans* depicted in Fig. 3 is far from being complete or fully understood and none of the components has been shown individually to function as a proton pump. The structural basis for the insensitivity of cytochrome b_{563} towards CO remains particularly enigmatic. We also do not know how the molybdopterin cytosine dinucleotides, flavin adenine dinucleotides and iron-sulfur centers are distributed on the CO dehydrogenase subunits, how they are precisely linked to the protein backbone and in which way they interact with CO, water, the electron acceptor and each other to establish an appropriate electron transport. Moreover, the factor(s) determining association of CO dehydrogenase with and dissociation from the cytoplasmic membrane *in vivo* are not understood, and it is still not trivial to obtain preparations of CO dehydrogenases which are fully active and not contaminated with cofactor-deficient enzyme molecules, i.e. demolybdo-CO dehydrogenase. The enzyme is too large for protein-NMR, and crystals have not yet been obtained. Nevertheless, since the last review on CO dehydrogenases [1], considerable progress has been made in some areas. The chemical structure of the organic portion of the CO dehydrogenase molybdenum cofactor has been recognized as molybdopterin cytosine dinucleotide. In addition, CO dehydrogenase structural genes have been cloned and expressed in *E. coli*. It will probably not take very long to determine the complete nucleotide sequences of the CO dehydrogenase structural genes. In *P. carboxydovorans* all *cox* genes are linked, as are the genes for the small and large subunits of CO dehydrogenase in *P. thermocarboxydovorans*, and further work will show whether *cox* genes are part of a larger transcription unit.

It seems that *E. coli* cannot synthesize molybdopterin cytosine dinucleotide. Consequently, heterologous expression of *cox* structural genes alone in *E. coli* is not likely to yield active CO dehydrogenase. This points to the requirement of a homologous expression system. On the other hand, homologous expression of the small CO dehydrogenase subunit or of partial proteins of all subunits most probably will furnish proteins amenable to structure analysis by protein NMR.

ACKNOWLEDGEMENTS

We appreciate stimulating discussion on carbon monoxide research with our collaborators at the University of Bayreuth, in particular Maria Kraut, Beatrix Santiago, Jörg Tachil, Matthias Volk and Dr Dilip Gadkari. We also thank Bernd Köppel, Klaus-Peter Fuchs, Dieter Willbold and Dr. Zerek Alikulov for making their unpublished material available to us. Finally we thank Waltraud Meyer who helped to make the drawings and last but not least, Molly Daniel and Sigrid Glas who typed the draft and made helpful comments in improving the manuscript. Experimental work carried out in our laboratory was financially supported by the Deutsche Forschungsgemeinschaft (grant no. Me 732/5-1) and the Fonds der Chemischen Industrie.

REFERENCES

[1] Meyer, O., Frunzke, K., Gadkari, D., Jacobitz, S., Hugendieck, I. and Kraut, M. (1990) Utilization of carbon monoxide by aerobes–recent advances. *FEMS Microbiol. Rev.* **87**, 253–260.

[2] Gadkari, D., Schricker, K., Acker, G., Kroppenstedt, R.M. and Meyer, O. (1990) *Streptomyces thermoautotrophicus* sp. nov., a thermophilic CO and H_2 oxidizing obligately chemolithoautotroph. *Appl. Environ. Microbiol.* **56**, 3727–3734.

[3] Meyer, O., Meyer, W., Gadkari, D., Zellmann, H., Schricker, K. and Schmitt, M. (1991) Microorganisms and their activities in the covering soil of a burning charcoal pile. In: Proceedings of the Third Symposium on Biotechnology of Coal and Coal-Derived Substances (Rehm, H.J., Fakoussa, R.M., Schacht, S. and Klein, J., Eds), pp. 111-121. DMT-Institute for Applied Environmental Chemistry, Essen.

[4] Gadkari, D., Mörsdorf, G. and Meyer, O. (1992) Chemolithoautotrophic assimilation of dinitrogen by *Streptomyces thermoautotrophicus* UBT1: Identification of an unusual N_2-fixing system. *J. Bacteriol.* **174**, 6840–6843.

[5] Kraut, M. and Meyer, O. (1988) Plasmids in carboxydotrophic bacteria: physical and restriction analysis. *Arch. Microbiol.* **149**, 540–546.

[6] Gerstenberg, C., Friedrich, B. and Schlegel, H.G. (1982). Physical evidence for plasmids in autotrophic, especially hydrogen-oxidizing bacteria. *Arch. Microbiol.* **133**, 90–96.

[7] Kraut, M., Hugendieck, I., Herwig, S. and Meyer, O. (1989) Homology and distribution of CO dehydrogenase structural genes in carboxydotrophic bacteria. *Arch. Microbiol.* **152**, 335–341.
[8] Hugendieck, I. and Meyer, O. (1992) The structural genes encoding CO dehydrogenase subunits (*coxL*, *M* and *S*) in *Pseudomonas carboxydovorans* OM5 reside on plasmid pHCG3 and are, with the exception of *Streptomyces thermoautotrophicus*, conserved in carboxydotrophic bacteria. *Arch. Microbiol.* **157**, 301–304.
[9] Hugendieck, I. and Meyer, O. (1991) Genes encoding ribulosebisphosphate carboxylase and phosphoribulokinase are duplicated in *Pseudomonas carboxydovorans* OM5 and conserved in carboxydotrophic bacteria. *Arch. Microbiol.* **157**, 92–96.
[10] Johnson, J.L., Rajagopalan, K.V. and Meyer, O. (1990) Isolation and characterization of a second molybdopterin dinucleotide: Molybdopterin cytosine dinucleotide. *Arch. Biochem. Biophys.* **283**, 542–545.
[11] Meyer, O. and Schlegel, H.G. (1980) Carbon monoxide: methylene blue oxidoreductase from *Pseudomonas carboxydovorans*. *J. Bacteriol.* **141**, 74–80.
[12] Meyer, O. (1982) Chemical and spectral properties of carbon monoxide: methylene blue oxidoreductase. *J. Biol. Chem.* **257**, 1333–1341.
[13] Kim, Y.M. and Hegeman, G.D. (1981) Purification and some properties of carbon monoxide dehydrogenase from *Pseudomonas carboxydohydrogena*. *J. Bacteriol.* **148**, 904–911.
[14] Krüger, B. and Meyer, O. (1984) Thermophilic *Bacilli* growing with carbon monoxide. *Arch Microbiol.* **139**, 402–408.
[15] Bell, J.M., Williams, E. and Colby, J. (1985) Carbon monoxide oxidoreductase from thermophilic carboxydobacteria. In: Microbial Gas Metabolism (Poole, R.K. and Dow, C.S., Eds), pp. 153-160, Academic Press, London.
[16] Willbold, D. (1991) Isolierung und Charakterisierung von Kohlenmonoxid-Dehydrogenase aus dem thermophilen obligat chemolithoautotrophen Bakterium *Streptomyces thermoautotrophicus*. Diplomthesis, University of Bayreuth.
[17] Fuchs, K.P. (1992) Biochemische Charakterisierung der Kohlenmonoxid-Dehydrogenase aus *Streptomyces thermoautotrophicus* und Analyse der N-terminalen Aminosäuresequenzen der drei Untereinheiten. Diplomthesis, University of Bayreuth.
[18] Meyer, O., Jacobitz, S. and Krüger, B. (1986) Biochemistry and physiology of aerobic carbon monoxide-utilizing bacteria. *FEMS Microbiol. Rev.* **39**, 161–179.
[19] Meyer, O. and Fiebig, K. (1985) Enzymes oxidizing carbon monoxide. In: Gas Enzymology (Cox, R.P. and Degn, H., Eds), pp. 147-168, D. Reidel Publishing Comp., Dordrecht, Holland.
[20] Meyer, O. (1988) Biology and biotechnology of aerobic carbon monoxide-oxidising bacteria. In: Biotechnology Focus 1 (Finn, R.K., Präve, P., Schlingmann, M., Crueger, W., Esser, K., Thauer, R. and Wagner, F., Eds), pp. 3–31, Hanser Publishers, Munich.
[21] Meyer, O. (1985) Metabolism of aerobic carbon monoxide-utilizing bacteria. In: Microbial Gas Metabolism (Poole, R.K. and Dow, C.S., Eds), pp. 131–151, Academic Press, London.
[22] Futo, S. and Meyer, O. (1986) CO_2 is the first species formed upon CO oxidation by CO dehydrogenase from *Pseudomonas carboxydovorans*. *Arch. Microbiol.* **145**, 358–360.
[23] Meyer, O. and Schlegel, H.G. (1979) Oxidation of carbon monoxide in cell extracts of *Pseudomonas carboxydovorans*. *J. Bacteriol.* **137**, 811–817.
[24] Jacobitz, S. (1989) Isolierung und Charakterisierung von Cytochrom b_{561} aus *Pseudomonas carboxydovorans* und Identifizierung als physiologischem Elektronenakzeptor von CO-Dehydrogenase. Ph.D. thesis, University of Bayreuth.

[25] Jacobitz, S. and Meyer, O. (1989) Removal of CO dehydrogenase from *Pseudomonas carboxydovorans* cytoplasmic membranes, rebinding of CO dehydrogenase to depleted membranes, and restoration of respiratory activities. *J. Bacteriol.* **171**, 6294–6299.
[26] Santiago, B. (1990) Charakterisierung der Hydrogenaseaktivität der CO-Dehydrogenase aus *Pseudomonas carboxydovorans* OM5. Diplomthesis, University of Bayreuth.
[27] Santiago, B. and Meyer, O. (1991) CO dehydrogenase from *Pseudomonas carboxydovorans* carries the activity of a H_2-evolving hydrogenase. *Bioforum* **14**, 68.
[28] Santiago, B. and Meyer, O. (1992) Hydrogenase activities under different growth conditions in *Pseudomonas carboxydovorans* OM5. *BioEngineering* **8** (Suppl.), 75.
[29] Hoffmüller, P. (1991) Analyse der Molybdän-Kofaktoren in respiratorischer Nitratreduktase und CO-Dehydrogenase aus *Pseudomonas carboxydoflava*. Diplomthesis, University of Bayreuth.
[30] Zumft, W.G. (1992) The denitrifying prokaryotes. In: The Prokaryotes: A Handbook on the Biology of Bacteria: Ecophysiology, Isolation, Identification, Application, 2nd Edn (Balows, A., Trüper, H.G., Dworkin, M., Harder, W. and Schleifer, K.-H., Eds), pp. 554-582, Springer-Verlag, Berlin.
[31] Zumft, W.G., Viebrock, A. and Körner, H. (1987) Biochemical and physiological aspects of denitrification. In: The Nitrogen and Sulphur Cycles (Cole, J.A. and Ferguson, S., Eds) pp. 245-278, Society for General Microbiology, Cambridge.
[32] Ketchum, P., Cambier, H.Y., Frazier, W.A., Madansky C.H. and Nason, A. (1970) In vitro assembly of *Neurospora* assimilatory nitrate reductase from protein subunits of a *Neurospora* mutant and xanthine oxidizing or aldehyde oxidase system of higher animals. *Proc. Natl. Acad. Sci. USA* **66**, 1016–1023.
[33] Rohde, M., Mayer, F. and Meyer, O. (1984) Immunocytochemical localization of carbon monoxide oxidase in *Pseudomonas carboxydovorans*. *J. Biol. Chem.* **259**, 14788–14792.
[34] Rohde, M., Mayer, F., Jacobitz, S. and Meyer, O. (1985) Attachment of CO dehydrogenase to the cytoplasmic membrane is limiting the respiratory rate of *Pseudomonas carboxydovorans*. *FEMS Microbiol. Lett.* **28**, 141–144.
[35] Meyer, O. and Rohde, M. (1984) Enzymology and bioenergetics of carbon monoxide-oxidizing bacteria. In: Microbial Growth on C_1 Compounds, Proceedings of the 4th International Symposium (Crawford, R.L. and Hanson, R.S., Eds), pp. 26-33. American Society of Microbiology, Washington, DC.
[36] Ragsdale, S.W., Clark, J.E., Ljungdahl, L.G., Lundie, L.L. and Drake, H.L. (1983) Properties of purified carbon monoxide dehydrogenase from *Clostridium thermoaceticum*, a nickel, iron sulfur protein. *J. Biol. Chem.* **258**, 2364–2369.
[37] Ragsdale, S.W., Ljungdahl, L.G. and DerVartanian, D.V. (1983) Isolation of carbon monoxide dehydrogenase from *Acetobacterium woodii* and comparison of its properties with those of the *Clostridium thermoaceticum* enzyme. *J. Bacteriol.* **155**, 1224–1237.
[38] Diekert, G., Fuchs, G. and Thauer, R. (1985) Properties and function of carbon monoxide dehydrogenase from anaerobic bacteria. In: Microbial Gas Metabolism (Poole, R.K. and Dow, C.S., Eds), pp. 115-130, Academic Press, London.
[39] Williams, E. and Colby, J. (1986) Biotechnological applications of carboxydotrophic bacteria. *Microbiol. Sci.* **3**, 149–153.
[40] Kim, E.S. and Kim, Y.M. (1984) Subunit structure of carbon monoxide oxidase from *Pseudomonas carboxydovorans*. *Korean Biochem. J.* **17**, 141–147.
[41] Bray, R.C., George, G.N., Lange, R. and Meyer, O. (1983) Studies by e.p.r. spectroscopy of carbon monoxide oxidases from *Pseudomonas carboxydovorans* and *Pseudomonas carboxydohydrogena*. *Biochem. J.* **211**, 687–694.
[42] Wootton, J.C., Nicolson, R.E., Cock, J.M., Walters, D.E., Burke, J.F., Doyle, W.A. and Bray, R.C. (1991) Enzymes depending on the pterin molybdenum cofactor:

sequence families, spectroscopic properties of molybdenum and possible cofactor-binding domains. *Biochim. Biophys. Acta* **1057**, 157–185.
[43] Krüger, B. and Meyer, O. (1987) Structural elements of bactopterin from *Pseudomonas carboxydoflava* carbon monoxide dehydrogenase. *Biochim. Biophys. Acta* **912**, 357–364.
[44] Krüger, B., Meyer, O., Nagel, M., Andreesen, J.R., Meincke, M., Bock, E., Blümle, S. and Zumft, W.G. (1987) Evidence for the presence of bactopterin in the eubacterial molybdoenzymes nicotinic acid dehydrogenase, nitrite oxidoreductase, and respiratory nitrate reductase. *FEMS Microbiol. Lett.* **48**, 225–227.
[45] Meyer, O. and Rajagopalan, K.V. (1984) Molybdopterin in carbon monoxide oxidase from carboxydotrophic bacteria. *J. Bacteriol.* **157**, 643– 648.
[46] Tachil, J., Frunzke, K. and Meyer, O. (1992) The structure of molybdopterin cytosine dinucleotide. *Bio. Engineering* **8** (suppl.), 43.
[47] Koschara, W. (1940) *Hoppe-Seyler's Z. physiol. Chem.* **263**, 78–79.
[48] Goto, M., Sakurai, A., Ohta, K. and Yamakami, H. (1969) Die Struktur des Urothions. *J. Biochem.* **65**, 611–620.
[49] Johnson, J.L. and Rajagopalan, K.V. (1982) Structural and metabolic relationship between the molybdenum cofactor and urothione. *Proc. Natl. Acad. Sci. USA* **79**, 6856–6860.
[50] Bamforth, J.F., Johnson, J.L., Davidson, G. F., Wong, L.T.K., Locktich, G. and Applegarth, D.A. (1990) Biochemical investigation of a child with molybdenum cofactor deficiency. *Clinic. Biochem.* **23**, 537–542.
[51] Volk, M., Frunzke, K. and Meyer, O. (1992) Urothione is the degradation and excretion product of the molybdenum-cofactors from various eubacteria and eucaryotes. *BioEngineering* **8** (suppl.), 43.
[52] Volk, M., Frunzke, K. and Meyer, O. (1992) Urothione is the degradation and excretion product of the molybdenum-cofactors from various eubacteria and eucaryotes. In: 11th Winter Workshop on Biochemical and Clinical Aspects of Pteridines, St. Christoph, Austria.
[53] Volk, M., Frunzke, K. and Meyer, O. (1993) Excretion of urothione by *Pseudomonas carboxydoflava* and *in vitro* degradation of the CO dehydrogenase molybdenum cofactor to norurothione. *Eur. J. Biochem.*, submitted.
[54] Frunzke, K. and Meyer, O. (1990) Nitrate respiration, denitrification, and utilization of nitrogen sources by aerobic carbon monoxide-oxidizing bacteria. *Arch. Microbiol.* **154**, 168–174.
[55] Hoffmüller, P., Frunzke, K. and Meyer, O. (1991) Usage of different nucleotides for the synthesis of molybdenum cofactors in respiratory nitrate reductase and CO dehydrogenase in *Pseudomonas carboxydoflava*. *Bioforum* **14**, 47.
[56] Meyer, O. (1983) Activation by selenite of carbon monoxide oxidase from *Pseudomonas carboxydovorans*. In: Proceedings of the Fourth International Conference on the Organic Chemistry of Selenium and Tellurium (Berry, F.J. and McWhinnie, W.R., Eds), pp. 588-605, The University of Aston in Birmingham.
[57] Cypionka, H. and Meyer, O. (1982) Influence of carbon monoxide on growth and respiration of carboxydobacteria and other aerobic organisms. *FEMS Microbiol. Lett.* **15**, 209–214.
[58] Cypionka, H. and Meyer, O. (1983) The cytochrome composition of carboxydotrophic bacteria. *Arch. Microbiol.* **135**, 293–298.
[59] Cypionka, H. and Meyer, O. (1983) Carbon monoxide-insensitive respiratory chain of *Pseudomonas carboxydovorans*. *J. Bacteriol.* **156**, 1178–1187.
[60] Jacobitz, S. and Meyer, O. (1986) Reduced pyridine nucleotides in *Pseudomonas carboxydovorans* are formed by reverse electron transfer linked to proton motive force. *Arch. Microbiol.* **145**, 372–377.
[61] Fiala, M. (1989) Isolierung der alternativen, CO-insensitiven, terminalen Oxidase aus *Pseudomonas carboxydovorans* OM 5. Diplomthesis, University of Bayreuth.

[62] van Verseveld, H.W. and Stouthamer, A.H. (1978) Growth yields and the efficiency of oxidative phosphorylation during autotrophic growth of *Paracoccus denitrificans* on methanol and formate. *Arch. Microbiol.* **118**, 21–26.
[63] Cypionka, H., van Verseveld, H.W. and Stouthamer, A.H. (1984) Proton translocation coupled to carbon monoxide-insensitive and -sensitive electron transport in *Pseudomonas carboxydovorans*. *FEMS Microbiol. Lett.* **22**, 209–213.
[64] Köppel, B. (1990) Protonentranslokationsmessungen an carboxydotrophen Bakterien. Diplomthesis, University of Bayreuth.
[65] Kraut, M., Schübel, U., Mörsdorf, G. and Meyer, O. (1993) Cloning and heterologous expression of cox-genes encoding carbon monoxide dehydrogenase from *Pseudomonas carboxydovorans*. *Arch. Microbiol.*, submitted.
[66] O'Reilly, C., Colby, J., Pearson, D. M. and Black, G. W., Molecular genetics of carbon monoxide dehydrogenase. This volume.
[67] Herwig, S. (1987) N-terminale Aminosäuresequenzen der Untereinheiten von CO-Dehydrogenase. Diplomthesis, University of Berlin.
[68] Tachil, J. (1988) Charakterisierung von Bactopterin aus Kohlenmonoxid-Dehydrogenase. Diplomthesis, University of Bayreuth.
[69] Kim, Y.M. and Lee, W.H. (1986) Stable subunit pattern of carbon monoxide dehydrogenases from *Pseudomonas carboxydohydrogena* and *Pseudomonas carboxydovorans*. *Korean Biochem. J.* **19**, 75–80.
[70] Black, G.W., Lyons, C.M., Williams, E., Colby, J., Kehoe, M. and O'Reilly, C. (1990) Cloning and expression of the carbon monoxide dehydrogenase genes from *Pseudomonas thermocarboxydovorans* strain C2. *FEMS Microbiol. Lett.* **70**, 249–254.
[71] Kramer, S.P., Johnson, J.L., Ribeiro, A.A., Millington, D.S. and Rajagopalan, K.V. (1987) The structure of the molybdenum cofactor: Characterization of di-(carboxamidomethyl)molybdopterin from sulfite oxidase and xanthine oxidase. *J. Biol. Chem.*, **262**, 16357–16363.
[72] Johnson, J.L., Chaudhury, M. and Rajagopalan, K.V. (1991) Identification of a molybdopterin-containing molybdenum cofactor in xanthine dehydrogenase from *Pseudomonas aeruginosa*. *Biofactors*, **3**, 103–107.
[73] Frunzke, K. and Meyer, O. (1991) Molybdopterin cytosine dinucleotide is the common molybdenum cofactor in CO dehydrogenases of carboxydotrophic *Pseudomonads*. *Bioforum* **14**, 83.
[74] Hettrich, D., Peschke, B., Tshisuaka, B. and Lingens, F. (1991) Microbial metabolism of quinoline and related compounds: X the molybdopterin cofactors of quinoline oxidoreductases from *Pseudomonas putida* 86 and *Rhodococcus spec.* B1 and xanthine dehydrogenase from *Pseudomonas putida* 86. *Biol Chem. Hoppe-Seyler*, **372**, 513–517.
[75] Johnson, J.L., Bastian, N.R., Rajagopalan, K.V. (1990) Molybdopterin guanine dinucleotide: A modified form of molybdopterin identified in the molybdenum cofactor of dimethyl sulfoxide reductase from *Rhodobacter spheroides* forma specialis *denitrificans*. *Proc. Natl. Acad. Sci. USA* **87**, 1390–1394.
[76] Karrasch, M., Börner, G. and Thauer, R.K. (1990) The molybdenum cofactor of formylmethanofuran dehydrogenase from *Methanosarcina barkeri* is a molybdopterin guanine dinucleotide. *FEBS Lett.*, **274**, 48–52.
[77] Johnson, J.L., Bastian, N.R., Schauer, N.L., Ferry, J.G. and Rajagopalan, K.V. (1991) Identification of molybdopterin guanine dinucleotide in formate dehydrogenase from *Methanobacterium formicium*. *FEMS Microbiol. Lett.* **77**, 213–216.
[78] Frunzke, K., Heiß, B., Meyer, O. and Zumft, W.G. (1993) Molybdopterin guanine dinucleotide (MGD) is the organic moiety of the molybdenum co-factor in respiratory nitrate reductase from *Pseudomonas stutzeri FEBS Lett.*, submitted.
[79] Johnson, J.L., Indermaur, L.W. and Rajagopalan, K.V. (1991) Molybdenum cofactor biosynthesis in *E. coli*. *J. Biol. Chem.* **266**, 12140–12145.

[80] Börner, G. Karrasch, M. and Thauer, R.K. (1991) Molybdopterin adenine dinucleotide and molybdopterin hypoxanthine dinucleotide in formylmethanofuran dehydrogenase from *Methanobacterium thermoautotrophicum* (Marburg). *FEBS Lett.* **290**, 10158–10161.

[81] Sayavedra-Soto, L.A., Powell, G.K., Evans, H.J. and Morris, R.O. (1988) Nucleotide sequence of the genetic loci encoding subunits of *Bradyrhizobium japonicum* uptake hydrogenase, *Proc. Natl. Acad. Sci. USA* **85**, 8395–8399.

[82] Schneider, C.G., Schmitt, H.J., Schild, Ch., Tichy, H.V. and Lotz, W. (1990) DNA sequence of the two structural genes for the uptake hydrogenase of *Rhizobium leguminosarum* bv. viciae B10. *Nucleic Acids Res.* **18**, 5285.

[83] Lorenz, B., Schneider, K., Kratzin, H. and Schlegel, H.G. (1989) Immunological comparison of subunits isolated from various hydrogenases of aerobic hydrogen bacteria. *Biophys. Acta* **995**, 1–9.

[84] Uffen, R.L., Colbeau, R., Richaud, P. and Vignais, P.M. (1990) Cloning and sequencing of the genes encoding uptake-hydrogenase subunits of *Rhodocyclus gelatinosus. Mol. Gen. Genet.* **221**, 49–58.

[85] Menon, A.L., Stults, L.W., Robson, R.L. and Mortenson, L.E. (1991) Cloning, sequencing and characterization of the [NiFe] hydrogenase-encoding structural genes (*hoxK* and *hoxG*) from *Azotobacter vinelandii. Gene* **96**, 67–74.

34
Molecular Genetics of Carbon Monoxide Dehydrogenase

Catherine O'Reilly, John Colby, Danita M. Pearson and Gary W. Black[1]

School of Health Sciences, University of Sunderland, Sunderland SR1 3SD, UK

1 SUMMARY

The structural genes encoding carbon monoxide dehydrogenase (CODH) can be carried either on a plasmid or on the bacterial chromosome. Preliminary analysis of the cloned CODH encoding genes from the moderately thermophilic bacterium *Pseudomonas thermocarboxydovorans* C2 suggests that at least the genes for the small and large subunits of CODH are linked in this species and appear to form part of a larger transcriptional unit.

2 INTRODUCTION

While much work has been done on the biochemistry of carbon monoxide (CO) utilization by aerobic carboxidotrophic bacteria, very little is known about the genetics of the system. The key enzyme involved in CO utilization is carbon monoxide dehydrogenase (CODH). The enzyme has been analysed in a range of aerobic CO utilizing bacteria and the structure of the enzyme is remarkably similar in most species studied [1,2]. CODH is a molybdoenzyme, made up of three subunits, molybdenum, complexed by molybdopterin cytosine dinucleotide [1,3], 2 FAD and 2 different iron-sulphur centres [1]. The enzyme activity *in vitro* is increased by the binding of selenium [4]. The complex nature of the

---

[1] Present address: Department of Agricultural Biochemistry, University of Newcastle-upon-Tyne, Newcastle-upon-Tyne SR1 3SD, UK.

Microbial Growth on C_1 Compounds
© Intercept Ltd, PO Box 716, Andover, Hampshire SP10 1YG, UK

enzyme indicates that the production of active enzyme will involve a large number of genes. These will be required to produce not only the basic structural subunits but also the molybdenum cofactor. CODH activity is inducible in most carboxidotrophic bacteria [5,6,7] although it is not yet clear whether this is controlled at the level of transcription or post-translationally. Recent results have shown that heterotrophically grown cells of *Pseudomonas carboxydovorans* contain an inhibitor of CODH activity [8]. However it appears likely, by comparison with other systems, that there will be transcriptional control of expression and therefore regulatory genes will also be required for the production of CODH. In this review, the recent advances which have been made in the understanding of the genetics of CODH production will be discussed. The genes encoding the large, medium and small subunits of CODH will be referred to as *cut*A, *cut*B and *cut*C respectively.

3 LOCALIZATION OF *cut* GENES

The aerobic carboxidotrophic bacteria are taxonomically diverse [5] although the CODH enzyme from most species appears remarkably similar. It is therefore not surprising that in many species the *cut* genes are plasmid borne. This had been shown most clearly for the *cut*B and C genes of *Pseudomonas carboxydovorans* OM5 [9,10]. *Pseudomonas carboxydovorans* OM5 contains a large plasmid, pHCG3, of approximate size 128kb [9]. Fifty-nine mutants of *P. carboxydovorans* OM5 which lack the ability to use CO as a carbon and energy source were isolated by chemical mutagenesis [10]. In eight of these mutants the plasmid was no longer present while one contained a deleted derivative of pHCG3. It has been shown, by oligonucleotide probing with sequences derived from N-terminal sequencing of the medium and small subunits, that at least the *cut*B and C genes are carried on pHCG3 [10]. In general it appears that those carboxidotrophic bacteria that contain plasmids carry the *cut* genes on a plasmid [10,11] although there are a number of strains of CO utilizing bacteria that do not contain plasmids [9,11–13] and in these cases the *cut* genes must be carried on the bacterial chromosome.

4 ISOLATION OF *cut* GENES

The first *cut* genes from aerobic carboxidotrophic bacteria isolated were those from the moderately thermophilic bacterium *Pseudomonas thermocarboxydovorans* C2 [14]. The *cut* genes in this strain are carried on the bacterial chromosome as the strain contains no plasmids [9,12]. The structure of CODH in this strain is very similar to the enzyme from other aerobic carboxidotrophic bacteria. A gene library of *P. thermocarboxydovorans* C2 was generated in the lambda vector L47.1 and was screened immunologically using antiserum raised against purified CODH. A number of clones were isolated which appeared on

Molecular genetics of CODH 463

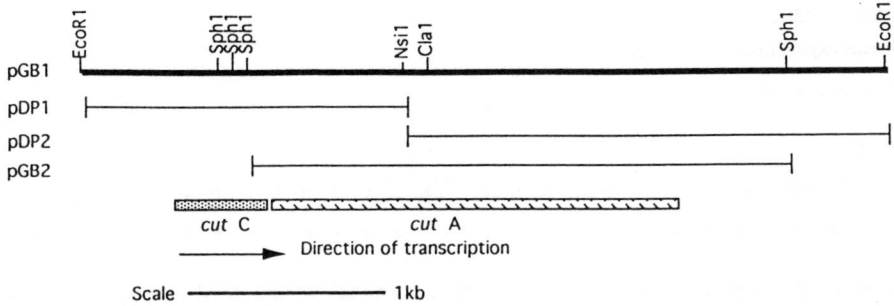

Figure 1 Restriction map of pGB1 indicating the various subclones derived from it. The approximate positions of the *cut*A and C genes are also shown together with the direction of transcription.

Western blotting to be producing at least the large subunit of CODH. A 4kb *Eco*R1 fragment of one of the 'phage clones was subcloned into pUC18 to give pGB1 (Fig. 1). Western blotting of crude cell extracts of *Escherichia coli* strain JM83 containing pGB1 indicated that all three subunits of CODH were being produced by this clone [13,15] although we are now less sure that the medium subunit is encoded by this plasmid (see below). Recently the plasmid borne *cut* genes of *P. carboxydovorans* OM5 have been isolated using oligonucleotide probing and DNA sequencing to identify the genes (O. Meyer, personal communication).

5 SEQUENCE ANALYSIS OF pGB1

The sequence of the entire 4kb region of pGB1 has been determined although at present we are completing the sequence on both strands. However to date the sequencing reveals 5 potential open reading frames. Two of the open reading frames, ORF2 and ORF3, are preceded by an identical 10bp sequence, 5′ CAGAGGAGAC 3″ which appears to be a ribosome binding site with good homology to the consensus sequence [16]. This sequence is found 7bp upstream of ORF2 and 5bp upstream of ORF3. Comparison of the predicted N-terminal sequences of the proteins encoded by these two open reading frames with the limited N-terminal sequence information available on other CODHs [10] reveals that ORF2 has significant homology with the small subunit of CODH from *P. carboxydovorans* OM5 and ORF3 has homology with the large subunit of CODH from *P. carboxydoflava* (Fig. 2). The predicted molecular mass of the protein encoded by ORF2 and 3 are 17 kDa and 72 kDa respectively which is in good agreement with the estimated size of the large and small subunits of CODH. We therefore propose that ORF2 is *cut*C and ORF3 is *cut*A and the position of these genes and the direction of transcription is shown in Fig. 1. The distance between ORF2 and 3 is 25bp and neither ORF2 or 3 are preceded

Large subunit

```
P.carboxydoflava           M N A P V Q D A E
                           | | | | : | | | |
P.thermocarboxydovorans    M N A P L S D R E K
```

Small subunit

```
P.carboxydovorans OM5    M A K A H I   E L T I N G H P V E A L V E P
                         | : | | | |   : | : | | : | | | | : | | | |
P.thermocarboxydovorans  M S K   H I V S M T V N G R K V E E A V E A
```

Figure 2 Comparison of the deduced N-terminal amino acid sequence of the large and small subunits of the CODH of *Pseudomonas thermocarboxydovorans* with the N-terminal sequences of the small subunit of *P. carboxydovorans* OM5 and the large subunit of *P. carboxydoflava* [10].

by an identifiable promoter structure. To date we have not been able to identify clearly the position of the *cut*B gene and none of the open reading frames we have identified has any similarity in the N-terminal region to those published for the medium subunit [10].

6 DELETION ANALYSIS OF pGB1

In order to determine whether the *cut*A and C genes have been correctly assigned, a number of deletion derivatives of pGB1 have been generated. The first of these to be tested for expression of CODH subunits was pGB2 which apparently expresses the large and medium subunits of CODH [13]. We have since made two other subclones from pGB1 and these are shown in Fig. 1. pDP1 extends from the *Eco*R1 site to a *Nsi*1 site which is in the middle of the proposed *cut*A gene and therefore should leave the *cut*C gene intact and truncate the *cut*A gene. *Escherichia coli* containing pDP1 expressed the small subunit and a truncated version of the large subunit of CODH as would be expected from our positioning of the genes in pGB1. pDP2, which extends from the *Nsi*1 site to the righthand *Eco*R1 site apparently expresses neither the large or small subunit of CODH. We are therefore confident that our assignment of the *cut*A and C genes is correct. The gene order for the plasmid borne genes of *P. carboxydovorans* appears to be *cut*B-*cut*C-*cut*A (O. Meyer, personal communication).

7 DO THE *cut* GENES FORM AN OPERON?

Our sequencing data, albeit partially single stranded, indicate that the *cut*C and A genes are separated by only 25bp and while we have identified potential ribosome binding sites adjacent to each gene, we have no evidence for a promoter for either gene. This would appear to indicate that these genes form part of a larger transcribed region and this is supported by the fact that no transcription termination sequence is found in the 4kb region of pGB1 that has been sequenced. Further work needs to be carried out to clarify the situation and at present we are investigating the possibility that the *cut*B gene is 5' to the *cut*C gene and that while our original 'phage clone had this gene intact, the pGB1 derivative has deleted a portion of the *cut*B gene.

8 SEQUENCE SIMILARITY OF THE CODH LARGE SUBUNIT TO OTHER DEHYDROGENASES

The sequence of the predicted protein product of *cut*A has been compared to all sequences in the NBRF protein data base using Microgenie™ software. The best homology observed was with the xanthine dehydrogenase of *Drosophila melanogaster* [17,18]. Homology of approximately 21% was seen over a region of 370 amino acids. Interestingly the homology covered the region thought to be important in the binding of the molybdenum cofactor in XDH [19]. This is in agreement with an earlier finding which suggested that the binding of the Mo-Co in CODH is more similar to that of XDH and xanthine oxidase than to other molybdenum containing hydroxylases [20]. The homology of CODH large subunit with XDH spans a region of 370 amino acids with five regions of clear homology, however homology is also seen over a shorter region of 193 amino acids to the aldehyde dehydrogenase (ALDH) of *Acetobacter polyoxogenes* [21]. This is not likely to be a molybdoenzyme but contains pyrroloquinoline quinone (PQQ) as the prosthetic group [22]. The region showing

```
DMXDH   765       L E L F C S T Q H P S E V Q K L V A H V
CODHLS  227       L T V W I T H Q   A P H V V R T V V S M
AADH    381       Y E I H A G N Q W Q S L I L P T L A K S

DMXDH   785       L T A L P A H R V V C R A K   R L G G G
CODHLS  246       L S G L P E S K V R I I S P   D I G G G
AADH    401       L Q   V P E S K V   I L R S Y L L G G G

DMXDH   803       F G E S R
CODHLS  265       F G N K V   G I Y P
AADH    419       F G R R L N G D Y M
```

Figure 3 Alignment of the homologous regions of the CODH large subunit of *P. thermocarboxydovorans* (CODHLS) with xanthine dehydrogenase from *Drosophila melanogaster* (DMXDH) [17] and aldehyde dehydrogenase from *Acetobacter polyoxogenes* (AADH) [20]. The numbers indicate the amino acid position in the proteins.

homology to ALDH is within the region showing homology to XDH. The greatest homology is seen over a region of 44 amino acids and this is shown in Fig. 3. The homology seen between the three proteins may suggest functional similarities in the enzymes while the homology with XDH may reflect similarities in the binding of the Mo-Co. No similarity is found between the CODH large subunit and other prokaryotic molybdenum containing enzymes or to the two anaerobic CODH enzymes whose sequence has been determined [23,24].

9   EXPRESSION OF CLONED cut GENES

As already described, at least the cutA and cutC genes of P. thermocarboxydovorans C2 are constitutively expressed in E. coli although at present it is unclear whether these genes are being expressed from their own promoter or from a promoter in the cloning vector. It might be expected by analogy to other systems that these genes would be expressed from a σ^{54} type promoter [25,26] and would therefore not be expressed in E. coli under the conditions we have used [13,15]. Further work is required to identify the start of transcription of the cloned genes in E. coli and also to identify the normal start point of transcription in P. thermocarboxydovorans C2.

No CODH enzyme activity is detectable in E. coli containing either the phage clone, P22 or the plasmid subclone pGB1 [13,15]. This is perhaps not unexpected given the complex nature of the enzyme. The cdh genes encoding the anaerobic CODHs of Clostridium thermoaceticum and Methanothrix soehngenii have been cloned into E. coli and also fail to produce active enzyme [24,27]. We have recently used Tn5 transposon mutagenesis to isolate a mutation in the cut A gene of P. thermocarboxydovorans. We plan to introduce into this strain copies of the putative cdh operon cloned on a broad host range plasmid. It should then be possible to analyse the control of gene expression and enzyme production in more detail by site directed mutagenesis.

10   CONCLUSION

The availability of cloned cut genes means that rapid advances should be made in the next few years in our understanding of the genetic control of CO utilization by carboxidotrophic bacteria. The limited data presently available suggest that there are no major differences between the plasmid encoded and chromosomally encoded enzymes and this will probably be reflected in the organization and control of the cut genes.

ACKNOWLEDGEMENTS

The authors wish to thank Professor Ortwin Meyer and coworkers for the communication of unpublished results.

REFERENCES

[1] Meyer, O. (1989) Aerobic carbon monoxide-oxidizing bacteria. In: Autotrophic Bacteria (Schlegel, H.G. and Bowien, B., Eds), pp. 331–350, Springer-Verlag, Berlin.
[2] Kim, Y.M., Kirkconnell, S. and Hegeman, G. (1982) Immunological relationships among carbon monoxide dehydrogenases of carboxydobacteria. FEMS Microbiol. Lett. 13, 219–223.
[3] Johnson, J.L., Rajagopalan, K.V. and Meyer, O. (1990) Isolation and characterisation of a second molybdopterin dinucleotide: molybdopterin cytosine dinucleotide. Arch. Biochem. Biophys. 283, 542–545.
[4] Meyer, O. and Rajagopalan, K.Y. (1984) Selenite-binding to carbon monoxide oxidase from Pseudomonas carboxydovorans. Selenium binds covalently to the protein and activates specifically the CO-methylene blue reaction. J. Biol. Chem. 259, 5612–5617.
[5] Meyer, O. and Schlegel, H.G. (1983) Biology of aerobic carbon monoxide-oxidizing bacteria. Ann. Rev. Microbiol. 37, 277–310.
[6] Lyons, C.M., Justin, P., Colby, J. and Williams, E. (1984) Isolation, characterization and autotrophic metabolism of a moderately thermophilic carboxydobacterium, Pseudomonas thermocarboxydovorans sp. nov. J. Gen. Microbiol. 130, 1097–1105.
[7] Kim, Y.J. and Kim, Y.M. (1989) Induction of carbon monoxide dehydrogenase during heterotrophic growth of Acinetobacter sp. Strain JC! DSM 3803 in the presence of carbon monoxide. FEMS Microbiol. Letts 59, 207–210.
[8] Do, Y.S., Kim, E. and Kim, Y.M. (1990) Carbon monoxide dehydrogenase inhibitor in cell extracts of Pseudomonas carboxydovorans. J. Bacteriol. 172, 1267–1270.
[9] Kraut, M. and Meyer, O. (1988) Plasmids in carboxydotrophic bacteria: physical and restriction analysis. Arch. Microbiol. 149, 540–546.
[10] Kraut, M., Hugendieck, I., Herwig, S. and Meyer, O. (1989) Homology and distribution of CO dehydrogenase structural genes in carboxydotrophic bacteria. Arch. Microbiol. 152, 335–341.
[11] Meyer, O., Frunzke, K., Gadkari, D., Jacobitz, S., Hugendieck, I. and Kraut, M. (1990) Utilization of carbon monoxide by aerobes: recent advances. FEMS Microbiol. Rev. 87, 253–260.
[12] Lyons, C.M. (1987) Physiology and genetics of thermophilic carboxydobacteria. PhD Thesis, University of Newcastle-upon-Tyne, UK.
[13] Black, G.W., Lyons, C.M., Williams, E., Colby, J., Kehoe, M. and O'Reilly, C. (1990) Cloning and expression of the carbon monoxide dehydrogenase genes from Pseudomonas thermocarboxydovorans strain C2. FEMS Microbiol. Letts. 70, 249–254.
[14] Turner, A.P.F., Aston, W.J., Higgins, I.J., Bell, J.M., Colby, J., Davis, G. and Hill, H.A.O. (1984) Carbon monoxide: acceptor oxidoreductase from Pseudomonas thermocarboxydovorans strain C2 and its uses in a carbon monoxide sensor. Anal. Chim. Acta 163, 161–174.
[15] Black, G. (1992) The molecular analysis of the genes encoding the enzyme carbon monoxide dehydrogenase from the carboxydobacterium Pseudomonas thermocarboxydovorans strain C2. PhD Thesis, University of Sunderland, UK.
[16] Gold, L. and Stormo, G. (1987) Translational initiation in Escherichia coli and Salmonella typhimurium. In: Cellular and Molecular Biology Vol. 2 (Neihardt, F.C., Ed.), pp. 1302–1307, American Society of Microbiology, Washington DC.
[17] Keith, T.P., Riley, M.A., Kreitman, M., Lewontin, R.C., Curtis, D. and Chambers, G. (1987) Sequence of the structural gene for xanthine dehydrogenase (rosy locus) in Drosophila melanogaster. Genetics 116, 67–73.

[18] Lee, C.S., Curtis, D., McCarron, M., Love, C., Gray, M., Bender, W. and Chovnick, A. (1987) Mutations affecting expression of the *rosy* locus in *Drosophila melanogaster*. *Genetics* **116**, 55–66.
[19] Wooton, J.C., Nicolson, R.E., Cock, J.M., Walters, D.E., Burke, J.F., Doyle, W.A. and Bray, R.C. (1991) Enzymes depending on the pterin molybdenum cofactor: sequence families, spectroscopic properties of molybdenum and possible cofactor-binding domains. *Biochim. Biophy. Acta* **1057**, 157–185.
[20] Bray, R.C., George, G.N., Lange, R. and Meyer, O. (1983) Studies by E.P.R. spectroscopy of carbon monoxide oxidases from *Pseudomonas carboxydovorans* and *Pseudomonas carboxydohydrogena*. *Biochem. J.* **211**, 687–694.
[21] Tamaki, T., Horinouchi, S., Fukaya, M., Okumura, H., Kawamura, Y. and Beppu, T. (1989) Nucleotide sequence of the membrane-bound aldehyde dehydrogenase gene from *Acetobacter polyoxogenes*. *J. Biochem.* **106**, 541–544.
[22] Ameyama, M., Osada, K., Shinagawa, E., Matsushita, K. and Adachi, O. (1981) Purification and characterisation of aldehyde dehydrogenase of *Acetobacter aceti*. *Agric. Biol. Chem.* **50**, 1889–1890.
[23] Morton, T.A., Runquist, J.A., Ragsdale, S.W., Shanmugasundaram, T., Wood, H.G. and Ljungdahl, L.G. (1991) The primary structure of the subunits of carbon monoxide dehydrogenase/acetyl-CoA synthase from *Clostridium thermoaceticum*. *J. Biol. Chem.* **266**, 23824–23828.
[24] Eggen, R.I.L., Geerling, A.C.M., Jetten, M.S.M. and de Vos, W.M. (1991) Cloning, expression and sequence analysis of the genes for carbon monoxide dehydrogenase of *Methanothrix soehngenii*. *J. Biol. Chem.* **266**, 6883–6887.
[25] Helmann, J.D. and Chamberlain, M.J. (1988) Structure and function of bacterial sigma factors. *Ann. Rev. Biochem.* **57**, 839–872.
[26] Friedrich, B. (1990) The plasmid encoded hydrogenase gene cluster in *Alcaligenes eutrophus*. *FEMS Microbiol. Rev.* **87**, 425–430.
[27] Roberts, D.L., James-Hagstrom, J.E., Garvin, D.K., Gorst, C.M., Runquist, J.A., Baur, J.R., Hasse, F.C. and Ragsdale, S.W. (1989) Cloning and expression of the gene cluster encoding key proteins involved in acetyl-CoA synthesis in *Clostridium thermoaceticum*: CO dehydrogenase, the corrinoid/Fe-S protein, and methyltransferase. *Proc. Natl. Acad. Sci. USA* **86**, 32–36.

35

Current Studies on the Molecular Biology and Biochemistry of CO_2 Fixation in Phototrophic Bacteria

F. Robert Tabita, Janet L. Gibson, Deane L. Falcone, Xing Wang, Lih-Ann Li, Betsy A. Read, Katherine C. Terlesky and George C. Paoli

Department of Microbiology and the Biotechnology Center, The Ohio State University, 484 West 12th Avenue, Columbus, OH 43210-1292, USA

1 SUMMARY

Phototrophic bacteria, particularly purple nonsulfur bacteria and cyanobacteria, have been shown to organize and regulate the expression of Calvin cycle structural gene expression by diverse mechanisms. In *Rhodobacter sphaeroides*, proposed to contain two separate chromosomes, there are separate CO_2 fixation operons on each genetic element which are regulated by the product of a single transcriptional activator gene on the large chromosome. This scenario is somewhat different in other purple nonsulfur photosynthetic bacteria and is totally different from the situation in cyanobacteria. In addition, other genes which indirectly regulate CO_2 fixation have been isolated in *R. sphaeroides* and are presently being characterized. In addition to transcriptional control, there is excellent evidence for post-translational control of RubisCO activity in both *R. sphaeroides* and *Rhodospirillum rubrum*, and much potential for such control in heterocystous cyanobacteria.

2 INTRODUCTION

Photosynthetic bacteria and cyanobacteria are excellent systems for studies of the regulation and biochemistry of CO_2 fixation [1–3]. Genetic manipulation of genes encoding key enzymes and regulatory sequences important for CO_2 metabolism has particularly emphasized the utility of these organisms. In this communication, we summarize recent findings from our laboratory that have directed our recent research efforts.

3 DIFFERENCES IN CALVIN CYCLE GENE ORGANIZATION AND REGULATION

The initial discovery of distinct ribulose bisphosphate carboxylase/oxygenase (RubisCO) enzymes, form I and form II, in *Rhodobacter sphaeroides* [4], has led to the finding that there are two distinct CO_2 fixation structural gene operons in this organism [5]. Work over the last three years has shown that the structural genes of each operon are cotranscribed and regulated by a single promoter 5' of the first gene of the form I and form II operons [6–8]. Genes encoding fructose 1,6-bisphosphatase (*fbp*), phosphoribulokinase (*prk*), and aldolase (*fba*) are found upstream from the *rbcLrbcS* and *rbpL* genes of the form I and form II operons, respectively. In addition, there is an insertion of approximately 3 kb of extra sequence between the *prkB* and *fbaB* genes of the form II or B operon. Two open reading frames were found within this 3 kb insert and were shown to encode genes for transketolase (*tk1B*) and glyceraldehyde-phosphate dehydrogenase (*gapB*) [8]. Interestingly, this represents the first transketolase gene sequenced from any bacterium. All the genes of each operon have been expressed in *Escherichia coli*. Recombinant proteins have been shown to be synthesized to high levels and most of these have been purified to homogeneity and high specific activity [1]; current studies are directed at the enzymology of phosphoribulokinase, fructose bisphosphatase and transketolase, using site-directed mutagenesis and other molecular biological procedures to facilitate examination of structure/function relationships. Recently, a *lysR*-type divergently transcribed transcriptional activator gene (*cfxR*), whose product regulates transcription of both the form I and form II operons, has been found upstream of *fbpA* of the form I operon (F.R. Gibson and J.L. Tabita, unpublished). Since the CO_2 fixation operons are found on separate genetic elements in *R. sphaeroides*, hypothesized to represent two separate chromosomes [9,10], it will be fascinating to explore the mechanism by which the *cfxR* regulatory signal is transduced from operon I on chromosome I to the binding site upstream from the promoter of operon II on chromosome II.

Other regulatory genes have been isolated by complementing mutants that are incapable of derepressing RubisCO synthesis (AUT⁻ strains) [11,12] or are incapable of growing with CO_2 as the electron acceptor (CEA⁻ strains) (W.G. Meijer and J.L. Tabita, unpublished). The former strains are capable of photoheterotrophic growth with CO_2 as electron acceptor but cannot grow under conditions where large amounts of RubisCO are required; e.g. as a chemolithoautotroph, a photolithoautotroph, or photoheterotrophically with butyrate or other reduced organic compounds as electron donors. At least 3

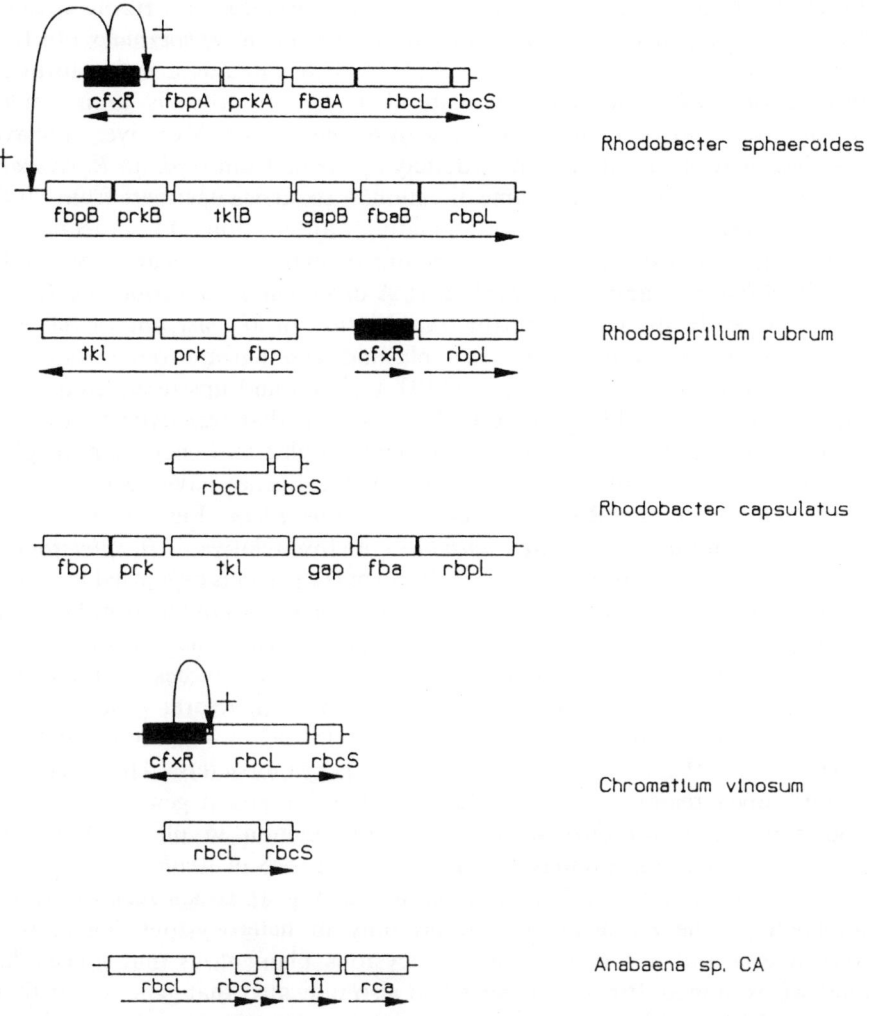

Figure 1 Organization of Calvin cycle structural genes in diverse phototrophic bacteria. Transcriptional activator genes (*cfxR*) have been shown to positively (+) regulate transcription in a number of instances.

separate loci have been isolated by complementation [12]; a 1.5 kb fragment that complements one of these mutants, has been shown to encode a 17,000 molecular weight protein [13]. Thus far, sequencing of the entire 1.5 kb fragment has not resulted in the identification of a protein that would provide a clue as to its role in autotrophic CO_2 fixation. The CEA^- strains and their complementing fragments are also under intense study.

Studies with other phototrophic bacteria and cyanobacteria have shown that there are distinct differences in the organization, and presumably regulation, of the CO_2 fixation structural genes. Certainly, the situation in aerobic chemolithotrophic bacteria is also somewhat different from *R. sphaeroides* [14,15]. Recent studies with *Rhodospirillum rubrum*, which contains a single form II RubisCO or *rbpL* gene [16], indicate that the promoter is immediately upstream from *rbpL* [17], unlike the case in *R. sphaeroides*. Moreover, we have found that there is a *cfxR* gene immediately upstream from *rbpL* in *R. rubrum*, however it is transcribed in the same direction. A sequence upstream from *rbpL* has been shown to be important for regulating transcription (D.L. Falcone and F.R. Tabita, unpublished); upstream from this regulatory region are divergently transcribed *fbp*, *prk*, and *tkl* genes (Fig. 1). A divergently transcribed *cfxR*-like gene was recently shown to control transcription of at least one of the two *rbcLrbcS* pairs in *Chromatium* [18], plasmid and chromosomal *rbcLrbcS* transcription of *Alcaligenes eutrophus* [19], and is found upstream from *rbcL* in *Xanthobacter flavus* [15]. These findings suggest that *lysR*-type genes may be universally required for regulation in phototrophic and chemolithotrophic bacteria. Thus far, no other Calvin cycle structural genes have been found to be associated with the RubisCO genes of *Chromatium* (Fig. 1). A curious situation is found in *Rhodobacter capsulatus*, a close relative of *R. sphaeroides*. In this organism, it has been found that the form II operon is organized basically the same as in *R. sphaeroides*, however, the *rbcL rbcS* genes of the form I operon do not appear to be closely associated with other Calvin cycle structural genes (J.M. Shively, personal communication). Finally, recent studies in our laboratory indicate that the cyanobacterial Calvin cycle structural genes are not associated on the cyanobacterial chromosome, i.e. *prk* is not found near the *rbcLrbcS* genes of unicellular and heterocystous cyanobacteria. However, two unknown open reading frames (ORF1 and ORF2), and a gene (*rca*), which encodes an important enzyme for RubisCO function in plants, RubisCO activase, are found immediately downstream from *rbcS* in *Anabaena* sp. Strain CA [28] (L-A. Li and F.R. Tabita, unpublished; Fig. 1). In a screen of diverse cyanobacteria, the *rca* gene was found only in heterocystous filamentous cyanobacteria and not in the nonheterocystous filamentous and unicellular strains we examined. Previous studies had also indicated that genes other than *rca*, and ORF's 1 and 2, were present downstream from *rbcS* in the unicellular organism *Synechococcus* strain 7942 [20].

4 CATALYSIS AND REGULATION OF RubisCO ACTIVITY

When synthesis is fully derepressed, RubisCO may constitute from 15–50% of the soluble protein of *R. sphaeroides* [21,22] and *R. rubrum* [23]. The substantial expenditure of energy for RubisCO synthesis suggests that the activity of this enzyme might be controlled *in vivo*. In higher plants, there are two well-described mechanisms to regulate RubisCO activity *in vivo*; these involve activation of inactivated RubisCO by the enzyme RubisCO activase [24,25] as well as reversible binding of the intracellular inhibitor 2-carboxyarabinitol-1-

phosphate [26,27], an analog of the transition state formed prior to product formation. The finding of the *rca* gene downstream from the *rbcLrbcS* genes of *Anabaena* [28] suggests that RubisCO activase might play an important role in regulating CO_2 fixation in cyanobacteria as well. In bacteria, there have been several indications that RubisCO activity might be post-translationally regulated (reviewed in 1 and 3). With respect to *R. rubrum* and *R. sphaeroides* there have been two processes described that appear to relate to *in vivo* control of RubisCO activity and post-translational modification. The activity of the type II RubisCO of *R. rubrum* was found to decrease when cells were exposed to oxygen; at the same time the amount of RubisCO protein remained constant, suggesting that the introduction of oxygen resulted in some modification or alteration of the enzyme that affected its activity [29]. At later times of exposure to oxygen, the levels of RubisCO protein rapidly decreased from the culture, suggesting that the modified enzyme is a substrate for some protease, allowing the cell to remove the large amounts of irreversibly inactivated protein. This scenario is reminiscent of the oxidative modification of glutamine synthetase of *Klebsiella* [30], in which the 'marked' or modified enzyme becomes a substrate for a specific protease [31]. An artificial *in vitro* system was subsequently developed to study the oxidative modification of the *R. rubrum* RubisCO [32], however the amino acid(s) targeted for oxidative modification remain to be identified.

The *R. sphaeroides* form I RubisCO, containing both large and small subunits in a L_8S_8 arrangement, becomes inactivated when metabolizable organic acids are added to cells growing with CO_2 as the sole source of carbon [33]. Subsequent studies showed that the inactivation process is reversible in *in vivo* and that the activity recovers after the dissipation of the organic acid from the culture. Interestingly, the inactivation process was found to depend on both the concentration of the organic compound and the nitrogen status of the cells [34]; blockage of ammonia assimilation by inhibition of glutamine synthetase with methionine sulfoximine prevented the recovery of form I RubisCO from pyruvate-mediated inactivation, suggesting the presence of regulatory mechanisms common to both CO_2 fixation and ammonia assimilation [35]. The propensity of the inactivated form I RubisCO to become reactivated during purification [33] greatly slowed our efforts to obtain homogeneous inactivated form I RubisCO for subsequent enzymoligical analyses. However, we recently succeeded in isolating preparations of purified inactivated enzyme which possessed 40–50% of the activity of active form I RubisCO. Despite the fact that these preparations were only 'partially inactive' when purified, the inactivated enzyme exhibited properties that were quite distinct from the active form of the enzyme. On both nondenaturing and SDS gels, the active and inactivated forms of the enzyme exhibited mobility differences [34,36]. The purified inactivated RubisCO could be activated *in vitro* by increasing the temperature, the levels of Mg (II), or by adding ATP, and to some extent, other nucleotides. This activation was accompanied by changes in the electrophoretic mobility of the protein, such that the enzyme was indistinguishable from the

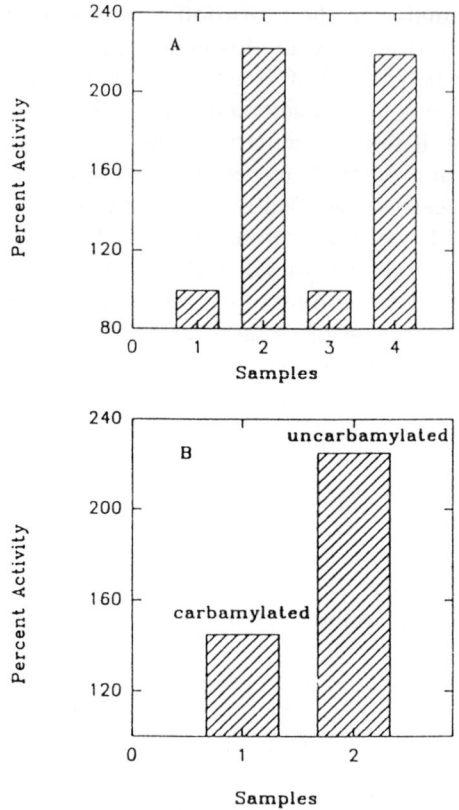

Figure 2 Properties of inactivated form I RubisCO from *R. sphaeroides*. (A) Retention of activity of ATP-modulated reactivated inactivated form I RubisCO after dialysis. Samples of inactivated RubisCO were reactivated in the presence (samples 2 and 4) or absence (samples 1 and 3) of ATP at room temperature. Samples 3 and 4 were not dialyzed. (B) Reactivation of carbamylated and noncarbamylated inactivated form I RubisCO. Inactivated enzyme was carbamylated by incubation of the enzyme with 10 mM Mg^{2+} and 20 mM HCO_3^- for 5 min at room temperature prior to the addition of 20 mM ATP for 30 min at room temperature. A slight increase in the activity of carbamylated enzyme not treated with ATP occurred and was subtracted from the level of activity obtained for the carbamylated enzyme treated with ATP. The untreated enzyme refers to the noncarbamylated sample.

active protein [34,36]. The activity of the ATP-reactivated inactivated enzyme was maintained after dialysis (Fig. 2A), suggesting that the reactivated enzyme had attained a stable conformation favoring maximum activity. The extent to which ATP could reactivate the inactivated enzyme was significantly reduced when the carbamylated ternary complex, or enzyme-Mg^{2+}-CO_2 form of the inactivated enzyme, was prepared (Fig. 2B). In some cases, ATP analogs could mimic the reactivation process and it was found that discrete proportions of

Table 1 Complementation to autotrophic growth and expression of diverse RubisCO genes in the RubisCO deletion strain *Rhodobacter sphaeroides* strain 16

Promoter source	RubisCO genes expressed
Rhodobacter sphaeroides form I operon	*R. sphaeroides* rbcL rbcS
Rhodobacter sphaeroides form II operon	*R. sphaeroides* rbpL
Xanthobacter flavus cfx operon	*X. flavus* rbcL rbcS
Bradyrhizobium japonicum cfx operon	*B. japonicum* rbcL rbcS
Rhodospirillum rubrum pRPS	*R. rubrum* rbpL
R. rubrum pRPS	*Synechococcus* 6301 rbcL rbcS (WT)[a]
R. rubrum pRPS	*Synechococcus* 6301 rbcL rbcS* (G91V mutant)[a]
R. rubrum pRPS	*Synechococcus* 6301 rbcL rbcS* (F92L mutant)
R. rubrum pRPS	*Synechococcus* 6301 rbcL rbcS* (R88K mutant)
R. rubrum pRPS	*Synechococcus* 6301 rbcL *Cylindrotheca* rbcS (hybrid)[b]
R. rubrum pRPS	*R. sphaeroides* rbpL
R. rubrum pRPS	*R. sphaeroides* rbcL rbcS

[a] WT, wildtype; the mutations were in the indicated residues of the recombinant small subunit*.
[b] The recombinant hybrid enzyme contained eukaryotic small subunits from the marine diatom *Cylindrotheca* sp. Strain N1 [39].

active (or inactivated) enzyme and the chaperonin 60 protein of *R. sphaeroides* aggregated in the presence of ATP. Thus, it is apparent that the form I RubisCO might contain a specific ATP-binding site, similar to the plant enzyme [37]; from our recent studies this ATP-binding site might contribute to both the regulation of activity and the assembly of active enzyme. Current studies are directed at further definition of the regulatory mechanism since other type I RubisCO enzymes synthesized in a *R. sphaeroides* background failed to become inactivated [34].

The chaperonin 60 protein of *R. sphaeroides* was purified and shown to catalyze a weak ATPase activity, much like the enzyme from *E. coli*. Like *E. coli*, its synthesis was regulated by heat shock; in addition, the amount of chaperonin 60 increased under cultural conditions which favored enhanced RubisCO synthesis [38]. The *R. sphaeroides* chaperonin 10, also purified to homogeneity, inhibited the ATPase activity of chaperonin 60, much like the situation in *E. coli*, but *E. coli* chaperonin 10 was incapable of inhibiting *R. sphaeroides* chaperonin 60 ATPase activity. These results suggest that studies of chaperonin-mediated RubisCO assembly might be facilitated using proteins from the same source (*R. sphaeroides*); *in vitro* reconstitution of denatured form II RubisCO has recently been accomplished using the *R. sphaeroides* chaperonin 60 and chaperonin 10 proteins (K.C. Terlesky and F.R. Tabita, unpublished).

Finally, we recently constructed an *R. sphaeroides* RubisCO deletion strain, strain 16, that served as a host for the expression of several diverse sources of

RubisCO [22]. Complementation of the RubisCO deletion strain was possible with DNA fragments containing either of the two CO_2 fixation gene clusters (and their promoters) of *R. sphaeroides* or large gene/promoter-containing fragments from the aerobic chemolithoautotrophic bacteria *Xanthobacter flavus* and *Bradyrhizobium japonicum* (D.L. Falcone and F.R. Tabita, unpublished). In addition, a promoter-vector system was developed such that RubisCO genes from any bacterial source, including cyanobacteria, could be expressed to high levels in strain 16, while complementing strain 16 to autotrophic growth. Table 1 summarizes the various wild-type, mutant, and hybrid RubisCO constructs that have been expressed in *R. sphaeroides* strain 16, using either the *R. rubrum* promoter or a promoter sequence from the original RubisCO gene source to drive expression. Of particular interest are the *Synechococcus (Anacystis)* sp. strain 6301 *rbcLrbcS* genes, which encode RubisCO with a relatively high K_{CO_2} [39]. Strain 16, complemented to photosynthetic growth using the cyanobacterial RubisCO to fix CO_2, is thus very dependent on the exogenous concentrations of CO_2 supplied to the cells, a fact which will make selection of mutant RubisCO molecules altered in K_{CO_2} extremely feasible. Moreover, mutant [39] and hybrid [40] prokaryotic-eukaryotic RubisCO molecules of altered specificity factor may also be expressed in strain 16 (Table 1), such that second-site suppressor mutant RubisCO enzymes may also be selected. Strain 16, and a RubisCO-negative mutant of *R. rubrum* (strain I-19), has also proven to be invaluable for studies of the contribution of alternative CO_2 fixation pathways to net carbon assimilation (X. Wang and F.R. Tabita, unpublished).

5 CONCLUSIONS

Studies of the molecular biology and the biochemistry of CO_2 fixation in diverse photosynthetic prokaryotes has the potential, and in some cases has already succeeded, to contribute towards the elucidation of mechanisms which regulate both gene expression and enzyme activity.

ACKNOWLEDGEMENTS

This work was supported by NIH grants GM24497 and GM45404, DOE grant DE-FG02-91ER20033, and USDA grant 91-37306-6325.

REFERENCES

[1] Tabita, F.R. (1988) Molecular and cellular regulation of autotrophic carbon dioxide fixation in microorganisms. *Microbiol. Rev.* **51**, 155–159.
[2] Tabita, F.R. (1987) Carbon dioxide fixation and its regulation in cyanobacteria. In: The Cyanobacteria (Fay, P. and Van Baalen, C., Eds), pp. 96–117, Elsevier, Amsterdam.

[3] Tabita, F.R., Gibson, J.L., Falcone, D.L., Lee, B. and Chen, J.-H. (1990) Recent studies on the molecular biology and biochemistry of CO_2 fixation in phototrophic bacteria. FEMS Microbiol. Rev. **87**, 437–444.
[4] Gibson, J.L. and Tabita, F.R. (1977) Different molecular forms of ribulose 1,5-bisphosphate carboxylase from Rhodopseudomonas sphaeroides. J. Biol. Chem. **252**, 943–949.
[5] Gibson, J.L. and Tabita, F.R. (1988) Localization and mapping of CO_2 fixation genes within two gene clusters in Rhodobacter sphaeroides. J. Bacteriol. **170**, 2153–2158.
[6] Gibson, J.L., Chen, J.-H., Tower, P.A. and Tabita, F.R. (1990) The form II fructose 1,6-bisphosphatase and phosphoribulokinase genes form part of a large operon in Rhodobacter sphaeroides: primary structure and insertional mutagenesis analysis. Biochemistry, **29**, 8085–8093.
[7] Gibson, J.L., Falcone, D.L. and Tabita, F.R. (1991) Nucleotide sequence, transcriptional analysis, and expression of genes encoded within the form I CO_2 fixation operon of Rhodobacter sphaeroides. J. Biol. Chem. **266**, 14646–14653.
[8] Chen, J.-H., Gibson, J.L., McCue, L.A. and Tabita, F.R. (1991) Identification, expression, and deduced primary structure of transketolase and other enzymes encoded within the form II CO_2 fixation operon of Rhodobacter sphaeroides. J. Biol. Chem. **266**, 20447–20452.
[9] Suwanto, A. and Kaplan, S. (1989) Physical and genetic mapping of the Rhodobacter sphaeroides 2.4.1 genome: presence of two unique circular chromosomes. J. Bacteriol. **171**, 5850–5859.
[10] Suwanto, A. and Kaplan, S. (1992) Chromosome transfer in Rhodobacter sphaeroides: Hfr formation and genetic evidence for two unique circular chromosomes. J. Bacteriol. **174**, 1135–1145.
[11] Weaver, K.E. and Tabita, F.R. (1985) Complementation of a Rhodopseudomonas sphaeroides ribulose bisphosphate carboxylase-oxygenase regulatory mutant from a genomic library. J. Bacteriol. **164**, 147–154.
[12] Rainey, A.M. and Tabita, F.R. (1989) Isolation of plasmid DNA sequences that complement Rhodobacter sphaeroides mutants deficient in the capacity for CO_2-dependent growth. J. Gen. Microbiol. **135**, 1699–1713.
[13] Paoli, G.C. and Tabita, F.R. (1992) Analysis of a DNA fragment that complements Aut⁻ Rhodobacter sphaeroides strain KW 25/11 to photolithoautotrophic growth. Abstr. Ann. Mting. Amer. Soc. Microbiol. p. 281, Abstr. K-149.
[14] Husemann, M., Klintworth, R., Buttcher, B., Salnikow, J., Weissenborn, C. and Bowien, B. (1988) Chromosomally and plasmid-encoded gene clusters for CO_2 fixation (cfx genes) in Alcaligenes eutrophus. Mol. Gen. Genet. **214**, 112–120.
[15] Meijer, W.G., Arnberg, A.C., Enequist, H.G., Terpstra, P., Lidstrom, M.E. and Dijkhuizen, L. (1991) Identification and organization of carbon dioxide fixation genes in Xanthobacter flavus H4-14. Mol. Gen. Genet. **225**, 320–330.
[16] Somerville, C.R. and Somerville, S.C. (1984) Cloning and expression of the Rhodospirillum rubrum ribulose bisphosphate carboxylase gene in E. coli. Mol. Gen. Genet. **193**, 214–219.
[17] Leustek, T., Hartwig, R., Weissbach, H. and Brot, N. (1988) Regulation of ribulose bisphosphate carboxylase expression in Rhodospirillum rubrum: characteristics of mRNA synthesized in vivo and in vitro. J. Bacteriol. **170**, 4065–4071.
[18] Viale, A.M., Kobayashi, H., Akazawa, T. and Henikoff, S. (1991) rbcR, a gene coding for a member of the LysR family of transcriptional regulators, is located upstream of the expressed set of ribulose 1,5-bisphosphate carboxylase/oxygenase genes in the photosynthetic bacterium Chromatium vinosum. J. Bacteriol. **172**, 5224–5339.
[19] Windhövel, U. and Bowien, B. (1991) Identification of cfxR, an activator gene of autotrophic CO_2 fixation in Alcaligenes eutrophus. Molec. Mibrobiol. **5**, 2695–2705.

[20] Lieman-Hurwitz, J., Schwarz, R., Martinez, F., Maor, Z., Reinhold, L. and Kaplan, A. (1991) Molecular analysis of high CO_2 requiring mutants: involvement of genes in the region of rbc, including rbcS, in the ability of cyanobacteria to grow under low CO_2. Can. J. Bot. **69**, 945–950.

[21] Jouanneau, Y. and Tabita, F.R. (1986) Independent regulation of synthesis of form I and form II ribulose bisphosphate carboxylase-oxygenase in *Rhodopseudomonas sphaeroides*. J. Bacteriol. **165**, 620–624.

[22] Falcone, D.L. and Tabita, F.R. (1991) Expression of endogenous and foreign ribulose 1,5-bisphosphate carboxylase-oxygenase (RubisCO) genes in a RubisCO deletion mutant of *Rhodobacter sphaeroides*. J. Bacteriol. **173**, 2099–2108.

[23] Sarles, L.S. and Tabita, F.R. (1983) Derepression of the synthesis of D-ribulose, 1,5-bisphosphate carboxylase/oxygenase from *Rhodospirillum rubrum*. J. Bacteriol. **153**, 458–464.

[24] Salvucci, M.E., Portis, A.E., Jr. and Ogren, W.E. (1985) A soluble chloroplast protein catalyzes activation of ribulose bisphosphate carboxylase/oxygenase *in vivo*. Photosynth. Res. **7**, 193–201.

[25] Werneke, J.M., Zielinski, R.E. and Ogren, W.L. (1988) Structure and expression of spinach leaf cDNA encoding ribulose bisphosphate carobxylase/oxygenase activase. Proc. Natl. Acad. Sci. USA **85**, 787–791.

[26] Berry, J.A., Lorimer, G.H., Pierce, J., Seemann, J.R., Meek, J. and Freas, S. (1987) Isolation, identification, and synthesis of a 2-carboxyarabinitol-1-phosphate, a diurnal regulator of ribulose-bisphosphate carboxylase activity. Proc. Natl. Acad. Sci. USA **84**, 734–738.

[27] Gutteridge, S., Parry, M.A.J., Burton, S., Keys, A.J., Mudd, A., Feeney, J., Servaites, J.C. and Pierce, J. (1986) A nocturnal inhibitor of carboxylase in leaves. Nature (Lond.) **324**, 274–276.

[28] Li, L.-A. and Tabita, F.R. (1991) Organization of CO_2 fixation genes in *Anabaena*. Abstr. VII Int. Symp. on Photosynth. Proc., p. 126, Abstr. 126A.

[29] Cook, L.S. and Tabita, F.R. (1988) Oxygen regulation of ribulose 1,5-bisphosphate carboxylase activity in *Rhodospirillum rubrum*. J. Bacteriol. **170**, 5468–5472.

[30] Levine, R.L., Oliver, C.N., Fulks, R.M. and Stadtman, E.R. (1981) Turnover of bacterial glutamine synthetase: oxidative inactivation precedes proteolysis. Proc. Natl. Acad. Sci. USA **78**, 2120–2124.

[31] Roseman, J.E. and Levine, R.L. (1987) Purification of a protease from *Escherichia coli* with specificity for oxidized glutamine synthetase. J. Biol. Chem. **262**, 2101–2110.

[32] Cook, L.S. and Tabita, F.R. (1988) Oxygen-dependent inactivation of ribulose 1,5-bisphosphate carboxylase/oxygenase in crude extracts of *Rhodospirillum rubrum* and establishment of a model inactivation system with purified enzyme. J. Bacteriol. **170**, 5473–5478.

[33] Jouanneau, Y. and Tabita, F.R. (1987) *In vivo* regulation of form I ribulose, 1,5-bisphosphate carboxylase/oxygenase from *Rhodopseudomonas sphaeroides*. Arch. Biochem. Biophys. **254**, 290–303.

[34] Wang, X. and Tabita, F.R. (1992) Reversible inactivation and characterization of purified inactivated form I ribulose 1,5-bisphosphate carboxylase/oxygenase of *Rhodobacter sphaeroides*. J. Bacteriol. **174**, 3593–3600.

[35] Wang, X. and Tabita, F.R. (1992) Interaction between ribulose 1,5-bisphosphate carboxylase/oxygenase activity and the ammonia assimilatory system of *Rhodobacter sphaeroides*. J. Bacteriol. **174**, 3601–3606.

[36] Wang, X. and Tabita, F.R. (1992) Interaction of inactivated and active ribulose 1,5-bisphosphate carboxylase/oxygenase of *Rhodobacter sphaeroides* with nucleotides and the chaperonin 60 (groEL) protein. J. Bacteriol. **174**, 3607–3611.

[37] Salvucci, M.E. and Haley, B.E. (1990) Photoaffinity labeling of ribulose bisphosphate carboxylase/oxygenase with 8-azidoadenosine 5′-triphosphate. Planta **181**, 287–295.

[38] Terlesky, K.C. and Tabita, F.R. (1991) Purification and characterization of the chaperonin 10 and chaperonin 60 proteins from *Rhodobacter sphaeroides*. Biochemistry **30**, 8181–8186.

[39] Read, B.A. and Tabita, F.R. (1992) Amino acid substitutions in the small subunit of ribulose-1,5-bisphosphate carboxylase/oxygenase that influence catalytic activity of the holoenzyme. Biochemistry **31**, 519–525.

[40] Read, B.A. and Tabita, F.R. (1992) A hybrid ribulose bisphosphate carboxylase/oxygenase enzyme exhibiting a substantial increase in substrate specificity factor. Biochemistry **31**, 5553–5560.

36

Genetic Regulation of CO_2 Assimilation in Chemoautotrophs

B. Bowien, R. Bednarski, B. Kusian, U. Windhövel[1], A. Freter, J. Schäferjohann and J.-G. Yoo

Institut für Mikrobiologie, Georg-August-Universität Göttingen, Grisebachstrasse 8, D-W-3400 Göttingen, Germany

1 SUMMARY

Most aerobic chemoautotrophs assimilate CO_2 via the Calvin cycle. Studies intended to disclose the genetics of CO_2 assimilation (Cfx) revealed genomic locations of *cfx* genes on the chromosomes and/or plasmids of these bacteria. The facultative hydrogen autotroph *Alcaligenes eutrophus* H16 contains highly homologous *cfx* gene clusters encoding eight Calvin cycle enzymes. One cluster is located on the chromosome and the other on megaplasmid pHG1. The genes in each cluster form an operon of about 13 kilobase pairs (kb), and both operons are simultaneously expressed. A regulatory gene, *cfxR*, located adjacent to the chromosomal operon and coding for a transcriptional activator, controls the expression of both operons. The back-to-back promoter regions of *cfxR* and the operon overlap. In contrast to the *cfx* operons which are partially or fully repressed in heterotrophic cells, *cfxR* is constitutively expressed at a low level. Since genes homologous to *cfxR* were also found in other facultative autotrophs, the genetic control of Cfx in these bacteria might be based on a common principle.

2 INTRODUCTION

All organic matter of the biosphere is originally being formed by the primary producers which depend on CO_2 assimilation (Cfx) for the synthesis of cell

---

[1] Present address: Lehrstuhl für Physiologie und Biochemie der Pflanzen, Universität Konstanz, Postfach 5560, D-W-7750 Konstanz, Germany.

carbon. Most of these organisms such as the higher plants, algae, cyanobacteria and anaerobic phototrophic bacteria are photosynthetic. They contribute by far the overwhelming portion to global assimilation of inorganic carbon. In addition, many chemosynthetic bacteria are known that either obligately or facultatively perform Cfx by gaining the enormous amounts of metabolic energy and reducing power required for this process through oxidation of inorganic or, in exceptional cases, organic compounds [1]. The Cfx pathways used by different groups of autotrophs are diverse and occasionally not even fully elucidated [2], but the reductive pentose phosphate cycle (Calvin cycle) found in all oxygenic photoautotrophs also operates in most chemoautotrophs except for strictly anaerobic bacterial and archeal species.

At present, only very little is known about the genetics of Cfx in non-Calvin-cycle organisms, mainly because of their difficult genetic accessibility. Some aerobic chemoautotrophs and anoxygenic photoautotrophs fixing CO_2 via the Calvin cycle have, however, received increasing attention in recent years with the aim to unravel their Cfx systems on the genetic level [3–5]. This contribution will discuss relevant information about the genetic structure and regulation of such systems, focussing on aerobic chemoautotrophs.

3 PHYSICAL ORGANIZATION OF *cfx* GENES

The Calvin cycle comprises fourteen reactions catalyzed in bacteria by ten enzymes, two of which, the ribulose-1,5-bisphosphate carboxylase/oxygenase (RuBisCO) and the phosphoribulokinase (PRK), are unique to this primary biosynthetic pathway. Fructose-1,6-bisphosphatase/sedoheptulose-1,7-bisphosphatase (FBP) and fructose-1,6-bisphosphate/sedoheptulose-1,7-bisphosphate aldolase (FBA) are bifunctional enzymes. Like the remaining six enzymes of the cycle, FBP and FBA catalyze reactions also occurring in the central carbon metabolism of heterotrophic organisms. Because of the metabolic key positions of RuBisCO and PRK during Cfx, the search for genes of Calvin cycle enzymes (*cfx* genes) in the genomes of chemoautotrophs was concentrated on the RuBisCO genes (*cfxLS* or *rbcLS*), in some cases also including the PRK gene (*cfxP*).

3.1 Genomic locations

The facultative hydrogen autotroph *Alcaligenes eutrophus* was the first chemoautotrophic bacterium studied for its *cfx* genes. An unexpected finding was the detection of two copies of both the *cfxLS* and *cfxP* genes in the genomes of the strains examined with one copy located on the chromosome and the other on an indigenous megaplasmid [6,7]. The plasmid also carries the gene cluster necessary for the hydrogen oxidation capacity (*hox* genes) of the organism [8]. Although most strains of *A. eutrophus* studied possess chromosomal and plasmid copies of *cfx* genes, the type strain surprisingly lacks the

Table 1 Genomic locations of *cfx* genes in aerobic chemoautotrophic bacteria

Strain		Location of cfx genes		Reference
		Chromosome	Plasmid	
Alcaligenes eutrophus	TF93[a]	+	–[b]	[3]
	H16	+	pHG1	[7,9]
	ATCC17707	+	pAE7	[6]
	H20	+	pHG7	[3]
	CH34	+	–	[3]
A. hydrogenophilus	M50	+	pHG21-a	[3]
Pseudomonas facilis	K	–	pHG22-a	[10]
	J	+	–	[10]
P. carboxydovorans	OM5	+	pHCG3	[11]
Xanthobacter flavus	H4-14	+	–	[12]
Nitrobacter hamburgensis X_{14}		+	pPB13	[13]
Thiobacillus ferrooxidans Fe1		+	–	[14]
Nocardia opaca	MR11	–	pHG201[c]	[15]

[a] Type strain harboring megaplasmid pHG2.
[b] –, no plasmid/chromosomal location of *cfx* genes reported or found.
[c] Linear megaplasmid.

plasmid genes (Table 1), indicating that one copy is sufficient for autotrophic growth. Chromosomal and plasmid locations of *cfx* genes were also observed in some other chemoautotrophs (Table 1). Only a broader survey will show if this type of *cfx* gene location is a widespread feature among these bacteria. However, it could be a strain-specific trait, and a chromosomal location of the *cfx* genes does not appear to be mandatory.

Thiobacillus ferrooxidans Fe1 [14] contains two identical sets of chromosomal *cfxLS* genes, suggesting that the existence of multiple *cfx* gene loci on the same genetic entity could be characteristic of some autotrophic bacteria. The phototrophic purple sulfur bacterium *Chromatium vinosum* was shown to have one functional and one nonexpressed copy of *cfxLS* genes [16]. In the genome of another phototroph, the purple nonsulfur bacterium *Rhodobacter sphaeroides*, two different *cfx* gene clusters are present [5] that are separately encoded on the two different chromosomes of the organism [17].

3.2 The cfx *gene clusters of* A. eutrophus

Strain H16 of *A. eutrophus* harboring megaplasmid pHG1 has been studied in great detail to elucidate the arrangement of its *cfx* genes. The genome of the strain contains two identically organized clusters of twelve structural *cfx* genes, each forming a large *cfx* operon (see 4.2 below) and comprising nearly 13 kb. One cluster is located on the chromosome and the other on the plasmid. Eight of the ten Calvin cycle enzymes are encoded within each operon with the

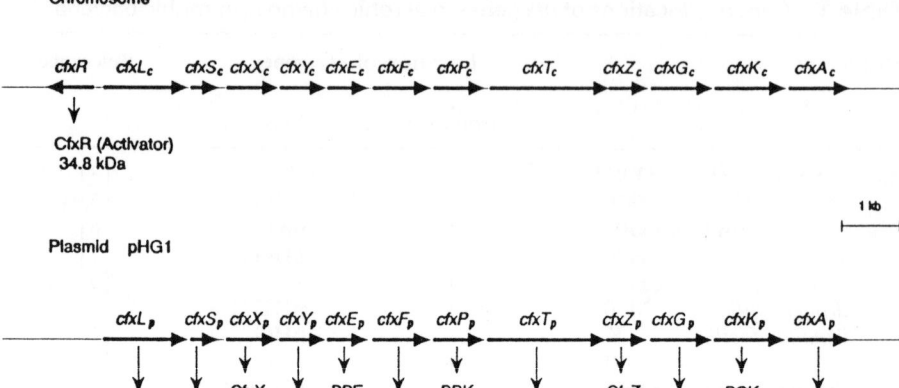

Figure 1 Organization of two *cfx* gene clusters located on the chromosome and the megaplasmid pHG1 of *A. eutrophus* H16. The gene products are shown with their molecular masses in kilodaltons (kDa). Enzyme abbreviations not mentioned in the text: PPE, pentose-5-phosphate 3-epimerase; TK, transketolase; GAP, glyceraldehyde-3-phosphate dehydrogenase; PGK, phosphoglycerate kinase.

RuBisCO genes as initial and the FBA gene as terminal genes (Fig.1). Only the genes for the triose and pentose phosphate isomerases are absent from the operons, whereas three constituent genes, *cfxX*, *cfxY* and *cfxZ*, code for proteins of still unknown functions. It is tempting to speculate that these functions might directly or indirectly be related to Cfx. The nucleotide sequence homology between the operons is very high exceeding 95%, and the amino acid sequence homologies between the deduced corresponding protein products approach the same level. Some type of gene duplication event must have given rise to this peculiar genetic similarity.

The chromosomal *cfx* gene cluster contains one additional gene, *cfxR* (Fig.1), which is part of a separate transcriptional unit and regulates both *cfx* operons (see 4.3 below). There is no functional counterpart of *cfxR* on pHG1, although most of its sequence is conserved on the plasmid [18]. Thus, the duplicated total region within each cluster extends over about 14 kb. The plasmid-encoded cluster is adjacent to the *hox* gene region to form a much larger region of about 100 kb that combines the information for the lithoautotrophic capacity of *A. eutrophus* H16. No *hox* genes were detected on the chromosome of the organism [8], restricting the gene duplication to the *cfx* gene cluster.

The organization of the *cfx* gene region of *Xanthobacter flavus* H4-14 is strikingly similar to that of the chromosomal region of *A. eutrophus*. There is the same order of genes including *cfxP* at one end of the presently known part of the cluster, and only the genes corresponding to *cfxY* and *cfxE* are lacking [12,19].

4 REGULATION OF THE Cfx OPERONS IN *A. eutrophus*

When discussing the regulation of Cfx in autotrophic bacteria, a distinction must be drawn between obligately and facultatively autotrophic organisms. The former do not need to control this process as strictly as the latter, particularly at the genetic level, because of their absolute dependence on one type of carbon metabolism [20]. Versatile autotrophs respond to changing nutritional regimens by switching between autotrophic and heterotrophic modes of carbon assimilation or might even thrive mixotrophically [3]. Such metabolic flexibility is based on the ability to regulate the synthesis of the required enzyme systems primarily by controlling gene transcription. As expected from energetic considerations, the *cfx* genes of these bacteria are generally severely repressed under heterotrophic conditions, whereas derepression or induction takes place during autotrophic and mixotrophic growth [3]. Studies with *A. eutrophus* have provided support for this principle.

4.1 Phenotypic observations

Like many other facultative chemoautotrophs, *A. eutrophus* H16 can grow organoautotrophically with formate as energy source [21]. Mutants cured of megaplasmid pHG1 lost the ability for lithoautotrophic growth on H_2/CO_2 as a result of lacking *hox* genes without being affected in formate utilization [22]. This phenotype is consistent with the chromosomal and plasmid locations of the *cfx* gene clusters in the wild type strain. During heterotrophic growth on fructose, gluconate, glycerol or citrate, synthesis of RuBisCO and PRK is partially derepressed in strain H16 [23,24], in plasmid-cured mutants, however, heterotrophic derepression of the *cfx* genes is drastically reduced [25]. The basis of this regulation could be a gene dosage effect, since a pHG1-free Cfx$^-$ mutant with an inactivated chromosomal *cfx* operon but a functional *cfxR* gene showed very low heterotrophic derepression after conjugational retransfer of pHG1 ([26]; authors' unpublished results), although growth on H_2/CO_2 or formate was regained. Both *cfx* operons are thus functional and equivalent in conferring a Cfx$^+$ phenotype.

The activity levels of most Calvin cycle enzymes are correlated in *A. eutrophus* when grown under autotrophic or heterotrophic conditions, probably reflecting the synthesis rates of the enzymes encoded within a common *cfx* operon [4]. High derepression of the enzymes occurs in lithoautotrophic cells, and CO_2 limitation [27] enhances the derepression even further. H_2 and CO_2 do not appear to be inducers of the *cfx* operons, but we obtained preliminary evidence for an inducing action of formate. Severe repression of the system prevails during growth on organic acids such as pyruvate, succinate and acetate. Therefore, in the absence of formate the system is physiologically controlled by repression/derepression as is characteristic of biosynthetic pathways. This mechanism was also proposed for the Cfx regulation in *Xanthobacter* strain 25a [28], and it might be typical of facultative chemoautotrophs.

Cfx⁻ mutants of *A. eutrophus* deficient in the expression of the *cfx* operons are unaffected in the ability to grow heterotrophically on fructose or organic acids [26], indicating that glycolysis *via* the Entner-Doudoroff pathway and gluconeogenesis are fully functional. This phenotype allows the conclusion that the organism forms isoenzymes, at least for PGK, GAP, FBA, FBP, TK and PPE, which operate in the central heterotrophic carbon metabolism and are encoded by separate non-*cfx* genes. A chromosomal GAP gene (*gap*) has already been detected [29]. Formation of 'autotrophic' and 'heterotrophic' isoenzymes in the central carbon metabolism might not be uncommon in facultative chemoautotrophs. Two FBP isoenzymes with the expected catalytic and regulatory properties were purified from another hydrogen autotroph, *Nocardia opaca* MR11 (formerly strain 1b) [30], and the existence of FBA and GAP isoenzymes must also be postulated by genetic criteria. Like RuBisCO and PRK, the isoenzymes thought to be associated with the Calvin cycle are encoded on the linear megaplasmid pHG201 that also carries the *hox* genes of the organism ([15]; M. Reh, personal communication].

4.2 Operon structure of the cfx clusters

Several findings collectively provide evidence that the genes in each of the two *cfx* clusters of *A. eutrophus* H16, except for *cfxR*, are constituents of a common large operon: (i) The genes have the same relative orientation (see Fig.1); (ii) The properties of transposon Tn5-*mob* insertional Cfx⁻ mutants of a pHG1-cured strain: a) Insertion in the 5' end of the promoter-proximal *cfxL_c* gene resulted in a deficient expression of the *cfx* operon due to the polarly inactivating effect of the transposon on downstream genes [26]; b) insertion within *cfxX_c* left *cfxLS_c* expression intact, whereas the remaining genes were inactivated [26]; c) cointegrational insertion of the transposon, together with its delivery vector, causing an introduction of a weak constitutive promoter located on an IS50 element, inactivated *cfxL_c* but allowed constitutive low-level expression of the other genes [26]; (iii) The heterologous expression in *Escherischia coli* of all operon genes, except for *cfxLS*, depended on the activation by the vector-borne promoter ([9,29,31]; authors' unpublished results). The divergently oriented chromosomal *cfxR* gene belongs to a separate operon.

Despite the existence of a common operon, a primary *cfx* transcript with an expected size of about 13 kilobases could not be detected in *A. eutrophus* cells. The only distinct, abundant mRNA found to be specific for the operons is the *cfxLS* transcript. It is about 2.1 kilobases long and present only when Cfx enzyme synthesis is derepressed, reflecting the tight transcriptional control of the operons [9]. There is a pronounced differential expression of the promoter-proximal *cfxLS* genes and the rest of the genes in the operons. The molar ratio between the RuBisCO and PRK proteins in autotrophic cells amounts to about 5:1 [24]. On the mRNA level, two mechanisms are most likely to cause the differential expression: (i) The 5' segment (*cfxLS*) of the primary transcript might be much more stable than the remaining sections; (ii) Frequent premature transcription termination could occur immediately downstream of *cfxLS*. Measurements of the chemical half lives of individual transcript pools showed that the *cfxLS* mRNA pool is in the same stability range as that of the other *cfx* genes [unpublished results]. Transcription termination between *cfxS* and *cfxX* might thus be the prime mechanism determining differential *cfx* gene expression.

4.3 Function of the cfxR gene

The chromosomally encoded *cfxR* gene of *A. eutrophus* H16 has been shown to be a regulatory gene controlling the expression of both *cfx* operons which thus form a regulon [18]. Mutants defective in this gene fail to grow autotrophically, since their *cfx* operons can no longer be expressed. The product of *cfxR* is an activator protein, CfxR, required for the activity of the operon promoter. Sequence comparisons identified CfxR as a member of the LysR family [32] of bacterial transcriptional regulators, mostly activators [18]. LysR family proteins are characterized by a helix-turn-helix secondary structure motif in their N-terminal parts that is the presumed DNA-binding site. The sequences are most strongly conserved within this region, although the overall similarity between CfxR and LysR is also significant (47% residue identity; Fig. 2). *X. flavus* H4-14 and *C. vinosum* also possess a *cfxR*-homologous gene adjacent

```
                              helix-turn-helix
LysR        MAAVNLRHIEIFHAVMTAGSLTEAAHLLHTSQPTVSRELARFEKVIGLKLFERVRGRLHPTVQGLRLFEEVQRSWY     76
CfxR        MSSFLRALTLRQLQIFVTVARHASFVRAAEELHLTQPAVSMQVKQLESVVGMALFERVKGQLTLTEPGDRLLHHASRILG   80
RbcR        MHVSLRQLRVFEAVARHNSYTRAAEELHLSQPAVSMQVRQLEDEIGLSLFERLGKQVVLTEAGREVFHYSRAIGQ      75
ORFD        MAPHWTLRQLRLVALAAASGSYAKAAQDMGLSPPAVTAQMKALEEDIGVPMFERVDGRLRPTAAGQELLSAQERIAR    77

LysR        GLDRIVSAAESLREFRQGELSIACLPVFSQSFLPQLLQPFLARYPDVSLNIVPQESPLLEEWLSAQRHDLGLTETLHTPA  156
CfxR        EVKDAEEGLQAVKDVEQGSITIG-LISTSKYFAPKLLAGFTALHPGVDLRIAEGNRETLLRLLQDNAIDLALMGRPPREL  159
RbcR        SLREMEEVLESLKGVSRGSLRIA-VASTVNYFAPRLMAIFQQRHSGIGLRLDVTNRESLVQMLDSNSVDLVLMGVPPRNV  156
ORFD        ALSEAERAIAALKSPERGSVVVG-VVSTAKYFAPMALAAFRRRRPEIELRLIIGNREDIIRGIVSLDFDVAIMG...    150

LysR        GTERTELLSLDEVCVLPPGHPLAVKKVLTPDDFQGENYISLSRTDSYRQLLDQLFTEHQVKRRMIVETHSAASVCAMVRA  236
CfxR        DAVSEPIAAHPHVLVASPRHPLHDAKGFDLQELRHETFLLREPGSGTRTVAEYMFRDHLFTPAKVITLGSNETIKQAVMA  239
RbcR        EVEAEAFMDNPLVVIAPPDHPLAGERAISLARLAEETFVMREEGSGTRQAMERFFSERGQTIRHGMQMTRNEAVKQAVRS  236

LysR        GVGISVVNPLT-ALDYAASGLV---VRRFSIAVPFTVSLIRPLH-RPSSALVQAFSGHLQAGLPKLVTSLDAILSSATTA  311
CfxR        GMGISLLSLHTLGLELRTGEIGLLDVAGTPIERIWHVAHMSSKRLSPASESCRAYL--LEHTAEFLGREYGGLMPGRRVA  317
RbcR        GLGLSVVSLHTIELELETRRLVTLDVEGFPDRRQWYLVYRRGKRLSPAAGAFREFV--LSEAARMHCRLG           302
```

Identity/similarity with CfxR: LysR 26.7/46.9 %; RbcR 43.9/64.4 %; ORFD 39.5/63.9 %.

Figure 2 Sequence comparison of the regulatory proteins LysR from *Escherichia coli* [33], CfxR from *Alcaligenes eutrophus* H16, RbcR from *Chromatium vinosum* and incompletely known ORFD from *Xanthobacter flavus* H4-14. The marked region is the presumed DNA-binding helix-turn-helix motif. Identical amino acid residues relative to CfxR are bold. Conservative substitutions of residues, according to the groups ILMV, YFW, PAGST, QNED and HRK, were included into the similarity calculations. Gaps (-) were introduced to optimize the alignment.

and divergently oriented to the *cfxLS* genes [12,34]. The sequences of CfxR, ORFD (product of the partially known *X. flavus* gene) and RbcR (*C. vinosum*) are strikingly similar (Fig. 2), even in their central and C-terminal parts believed to be involved in effector binding. Analogous functions of these proteins in *cfx* gene expression can be assumed.

In *A. eutrophus* an intergenic region of only 167 base pairs separates *cfxR* from *cfxL$_c$* and contains the promoter and control regions of the operons. The region is fully conserved on pHG1 [18], suggesting that the plasmid-borne *cfxR* promoter might principally be functional. Transcript analyses indicated a constitutive low-level expression of *cfxR*, which contrasts with the control of the *cfx* operons (see 4.2 below). The promoter regions of *cfxR* and *cfxL* overlap, and there are indications for an autoregulation of *cfxR* (authors' unpublished results). A permanent synthesis of CfxR could enable the organism to activate *cfx* operon transcription rapidly in response to changing nutritional conditions.

Control regions of genes regulated by LysR-type proteins contain AT-rich segments as potential regulatory sites with the consensus sequence 5'-ATATTGTTT-3' [35]. Similar sequences are present in the *cfx* intergenic regions of *A. eutrophus*, *X. flavus* and *C. vinosum* [18,12,34]. DNA footprinting experiments employing CfxR and an intergenic fragment from *A. eutrophus* have in fact shown binding of CfxR to the presumed site (unpublished results). A specific interaction of CfxR with a signal metabolite as effector is likely to occur *in vivo* to accomplish the activation process leading to derepression of the *cfx* operons. The metabolite is still unknown.

5 CONCLUSIONS

The very few chemoautotrophs so far studied for the genetics of their Cfx system exhibit similarities in location and arrangement of the *cfx* genes. Chromosomal and/or plasmid locations as well as reiterations and duplications might exist in many cases. Clustering of the genes within special *cfx* operons, as in *A. eutrophus*, could be a common feature among these organisms. Investigations of facultative and obligate autotrophs should provide information on possible differences between *cfx* gene organizations in these groups.

As exemplified by *A. eutrophus*, one mechanism of controlling expression of *cfx* genes is (de)repression that might also operate in other facultative chemoautotrophs and anaerobic phototrophic bacteria. The control is achieved by the action of a transcriptional activator protein, CfxR, which belongs to the LysR family of bacterial regulatory proteins. A generally occurring molecular mechanism thus functions in expression control of the *cfx* genes in *A. eutrophus*. The fact that genes, homologous to *cfxR*, were also found in two other facultative autotrophs suggests an involvement of a similar regulatory principle in these organisms. It should be of great interest to see if *cfxR* genes from different bacteria can functionally substitute each other. Knowledge of the nature of the signal metabolite(s) acting as effector(s) of the regulator is of major importance for the understanding of the control. Obligate autotrophs might have evolved a different control strategy for their *cfx* genes which are constitutively expressed.

ACKNOWLEDGEMENT

The work reported from the authors' laboratory was supported by grants from the Deutsche Forschungsgemeinschaft.

REFERENCES

[1] Barton, L.L., Shively, J.M. and Lascelles, J. (1991) Autotrophs: variations and versatilities. In: Variations in Autotrophic Life (Shively, J.M. and Barton, L.L., Eds), pp. 1–23, Academic Press, London.
[2] Fuchs, G. (1989) Alternative pathways of autotrophic CO_2 fixation. In: Autotrophic Bacteria (Schlegel, H.G. and Bowien, B., Eds), pp. 365–382, Science Tech, Madison.
[3] Bowien, B. (1989) Molecular biology of carbon dioxide assimilation in aerobic chemolithotrophs. In: Autotrophic Bacteria (Schlegel, H.G. and Bowien, B., Eds), pp.437–460, Science Tech, Madison.
[4] Bowien, B., Windhövel, U., Yoo, J.-G., Bednarski, R. and Kusian, B. (1990) Genetics of CO_2 assimilation in the chemoautotroph *Alcaligenes eutrophus*. FEMS Microbiol. Rev. 87, 445–450.
[5] Tabita, F.R., Gibson, J.L., Falcone, D.L., Lee, B. and Chen, J.-H. (1990) Recent studies on the molecular biology and biochemistry of CO_2 fixation in phototrophic bacteria. FEMS Microbiol. Rev. 87, 437–443.
[6] Andersen, K. and Wilke-Douglas, M. (1984) Construction and use of a gene bank of *Alcaligenes eutrophus* in the analysis of ribulose bisphosphate carboxylase genes. J. Bacteriol. 159, 973–978.
[7] Klintworth, R., Husemann, M., Salnikow, J. and Bowien, B. (1985) Chromosomal and plasmid locations for phosphoribulokinase genes in *Alcaligenes eutrophus*. J. Bacteriol. 164, 954–956.
[8] Friedrich, B. (1990) The plasmid-encoded hydrogenase gene cluster in *Alcaligenes eutrophus*. FEMS Microbiol. Rev. 87, 425–430.
[9] Husemann, M., Klintworth, R., Büttcher, V., Salnikow, J., Weissenborn, C. and Bowien, B. (1988) Chromosomally and plasmid-encoded gene clusters for CO_2 fixation (*cfx* genes) in *Alcaligenes eutrophus*. Mol. Gen. Genet. 214, 112–120.
[10] Warrelmann, J. and Friedrich, B. (1989) Genetic transfer of lithoautotrophy mediated by a plasmid-cointegrate from *Pseudomonas facilis*. Arch. Microbiol. 151, 359–364.
[11] Hugendieck, I. and Meyer, O. (1991) Genes encoding ribulosebisphosphate carboxylase and phosphoribulokinase are duplicated in *Pseudomonas carboxydovorans* and conserved in carboxydotrophic bacteria. Arch. Microbiol. 157, 92–96.
[12] Meijer, W.G., Arnberg, A.C., Enequist, H.G., Terpstra, P., Lidstrom, M.E. and Dijkhuizen, L. (1991) Identification and organization of carbon dioxide fixation genes in *Xanthobacter flavus* H4-14. Mol. Gen. Genet. 225, 320–330.
[13] Harris, S., Ebert, A., Schütze, E., Diercks, M., Bock, E. and Shively, J.M. (1988) Two different genes and gene products for the large subunit of ribulose-1,5-bisphosphate carboxylase/oxygenase (RuBisCOase) in *Nitrobacter hamburgensis*. FEMS Microbiol. Lett. 49, 267–271.

[14] Kusano, T., Takeshima, T., Inoue, C. and Sugawara, K. (1991) Evidence for two sets of structural genes coding for ribulose bisphosphate carboxylase in *Thiobacillus ferrooxidans. J. Bacteriol.* **173**, 7313–7323.
[15] Kalkus, J., Reh, M. and Schlegel, H.G. (1990) Hydrogen autotrophy of *Nocardia opaca* strains is encoded by linear megaplasmids. *J. Gen. Microbiol.* **136**, 1145–1151.
[16] Kobayashi, H., Viale, A.M., Takabe, T., Akazawa, T., Wada, K., Shinozaki, K., Kobayashi, K. and Sugiura, M. (1991) Sequence and expression of genes encoding the large and small subunits of ribulose 1,5-bisphosphate carboxylase/oxygenase from *Chromatium vinosum. Gene* **97**, 55–62.
[17] Suwanto, A. and Kaplan, S. (1989) Physical and genetic mapping of the *Rhodobacter sphaeroides* 2.4.1 genome: presence of two circular chromosomes. *J. Bacteriol.* **171**, 5850–5859.
[18] Windhövel, U. and Bowien, B. (1991) Identification of *cfxR*, an activator gene of autotrophic CO_2 fixation in *Alcaligenes eutrophus. Mol. Microbiol.* **5**, 2695–2705.
[19] Meijer, W.G., Enequist, H.G., Terpstra, P. and Dijkhuizen, L. (1990) Nucleotide sequences of the genes encoding fructosebisphosphatase and phosphoribulokinase from *Xanthobacter flavus* H4-14. *J. Gen. Microbiol.* **136**, 2225–2230.
[20] Smith, A.J. and Hoare, D.S. (1977) Specialist phototrophs, lithotrophs, and methylotrophs: a unity among a diversity of procaryotes? *Bacteriol. Rev.* **41**, 419–448.
[21] Friedrich, C.G., Bowien, B. and Friedrich, B. (1979) Formate and oxalate metabolism in *Alcaligenes eutrophus. J. Gen. Microbiol.* **115**, 185–192.
[22] Friedrich, B., Hogrefe, C. and Schlegel, H.G. (1981) Naturally occurring genetic transfer of hydrogen-oxidizing ability between strains of *Alcaligenes eutrophus. J. Bacteriol.* **147**, 198–205.
[23] Friedrich, C.G., Friedrich, B. and Bowien, B. (1981) Formation of enzymes of autotrophic metabolism during heterotrophic growth of *Alcaligenes eutrophus. J. Gen. Microbiol.* **122**, 69–78.
[24] Leadbeater, L., Siebert, K., Schobert, P. and Bowien, B. (1982) Relationship between activities and protein levels of ribulosebisphosphate carboxylase and phosphoribulokinase in *Alcaligenes eutrophus. FEMS Microbiol. Lett.* **14**, 263–266.
[25] Bowien, B., Friedrich, B. and Friedrich, C.G. (1984) Involvement of megaplasmids in heterotrophic derepression of the carbon-dioxide assimilating enzyme system in *Alcaligenes* spp. *Arch. Microbiol.* **139**, 305–310.
[26] Windhövel, U. and Bowien, B. (1990) On the operon structure of the *cfx* gene clusters in *Alcaligenes eutrophus. Arch. Microbiol.* **154**, 85–91.
[27] Friedrich, C.G. (1982) Derepression of hydrogenase during limitation of electron donors and derepression of ribulosebisphosphate carboxylase during carbon limitation of *Alcaligenes eutrophus. J. Bacteriol.* **149**, 203–210.
[28] Croes, L.M., Meijer, W.G. and Dijkhuizen, L. (1991) Regulation of methanol oxidation and carbon dioxide fixation in *Xanthobacter* strain 25a grown in continuous culture. *Arch. Microbiol.* **155**, 159–163.
[29] Windhövel, U. and Bowien, B. (1990) Cloning and expression of chromosomally and plasmid-encoded glyceraldehyde-3-phosphate dehydrogenase genes from the chemoautotroph *Alcaligenes eutrophus. FEMS Microbiol. Lett.* **66**, 29–34.
[30] Amachi, T. and Bowien, B. (1979) Characterization of two fructose bisphosphatase isoenzymes from the hydrogen bacterium *Nocardia opaca* 1b. *J. Gen. Microbiol.* **113**, 347–356.
[31] Klintworth, R., Husemann, M., Weissenborn, C. and Bowien, B. (1988) Expression of the plasmid-encoded phosphoribulokinase gene from *Alcaligenes eutrophus. FEMS Microbiol. Lett.* **49**, 1–6.
[32] Henikoff, S., Haughn, G.W., Calvo, J.M. and Wallace, J.C. (1988) A large family of bacterial activator proteins. *Proc. Natl. Acad. Sci. USA* **85**, 6602–6606.

[33] Stragier, P. and Patte, J.-C. (1983) Regulation of diaminopimelate decarboxylase synthesis in *Escherichia coli*. III. Nucleotide sequence and regulation of the *lysR* gene. *J. Mol. Biol.* **168**, 333–350.

[34] Viale, A.M., Kobayashi, H., Akazawa, T. and Henikoff, S. (1991) *rbcR*, a gene coding for a member of the LysR family of transcriptional regulators, is located upstream of the expressed set of ribulose 1,5-bisphosphate carboxylase/oxygenase genes in the photosynthetic bacterium *Chromatium vinosum*. *J. Bacteriol.* **173**, 5224–5229.

[35] Bohannon, D.E. and Sonenshein, A.L. (1989) Positive regulation of glutamate biosynthesis in *Bacillus subtilis*. *J. Bacteriol.* **171**, 4718–4727.

37

CO_2, Reductant, and the Autotrophic Acetyl-CoA Pathway: Alternative Origins and Destinations

Harold L. Drake

Lehrstuhl für Ökologische Mikrobiologie, BITÖK, Universität Bayreuth, Postfach 10 12 51, D-8580 Bayreuth, Germany

1 SUMMARY

Recent investigations demonstrate that diverse oxidizable, C_1-generating compounds and reductant sinks (including certain aromatic substituent groups) can be utilized by acetogens. Some of these substrates compete with or form CO_2, the terminal electron of the acetyl-CoA pathway. The dissimilation or utilization of such compounds and sinks provide a potential basis by which the acetyl-CoA pathway might be regulated by the native environments of acetogens.

2 INTRODUCTION

The study (history) of acetogenic bacteria is intimately tied to the microbial metabolism of C_1 compounds because the terminal electron-accepting acetyl-CoA synthesizing pathway utilized by these organisms is an 'autotrophic' C_1 pathway [1,2]. For this review, acetogenic bacteria are defined as obligately anaerobic bacteria that use the acetyl-coenzyme A (CoA) pathway: (i) for the reductive synthesis of acetate from CO_2; (ii) as a terminal electron-accepting process; (iii) for the conservation of energy; and (iv) for the synthesis of cell carbon [3]. Thus, not all bacteria capable of engaging the acetyl-CoA pathway (e.g. autotrophic methanogens that utilize a similar pathway only for biomass production) or other anaerobic bacteria that synthesize acetate from CO_2 by alternative mechanisms (i.e. those that use the glycine synthase-dependent pathway or the reductive citric acid cycle) are considered in the present statement.

The acetyl-CoA pathway forms the catalytic basis for the synthesis of acetyl-CoA from CO_2 under both autotrophic and heterotrophic conditions. The resolution of the pathway was largely the result of many investigations by Harland G. Wood (deceased) and Lars G. Ljungdahl, and the pathway is often referred to as the Wood or Wood-Ljungdahl pathway. One of the main identifying features of the pathway and acetogenic bacteria is the capacity to utilize chemolithoautotrophic substrates (CO or H_2/CO_2) as sole sources of carbon and energy under strictly anaerobic conditions [1–6].

Acetogens harbor diverse metabolic potentials and are often capable of both chemolithoautotrophic and chemoorganotrophic growth. The purpose of the present statement is to provide a brief overview of (i) the acetyl-CoA pathway, (ii) newly resolved metabolic potentials that form metabolic tangents to the pathway, and (iii) the potential physiological and ecological consequences of these diverse metabolic potentials.

3 OVERVIEW OF THE ACETYL-CoA PATHWAY

The acetyl-CoA pathway serves both catabolic and anabolic functions. Catabolically, it serves as a terminal electron accepting sink. Since CO_2 is the primary electron accepting molecule of the pathway, CO_2 is regarded as the terminal electron acceptor of acetogenesis and acetogenic bacteria. As we will see below, this is not always the case.

Overall in the acetyl-CoA pathway, two molecules of CO_2 collectively accept 8 reducing equivalents and yield acetate according to the following reaction: $2\ CO_2 + 8\ H^+ + 8\ e^- \rightarrow CH_3COOH + 2\ H_2O$. The origin of the reducing equivalents (or reductant) that flow towards CO_2 may be derived from both organotrophic (e.g. glucose) or lithotrophic (e.g. H_2) substrates. Anabolically, the pathway forms the basis for the fixation of carbon (CO_2) for the formation of cell carbon. Figure 1 outlines these basic features of the pathway, and the reader should consult recent reviews for the biochemical and enzymological details of the pathway [1–6].

3.1 *Reductive formation of the methyl and carbonyl carbons of acetyl-CoA*

There are two major reductive portions (or paths) of the pathway, both of which consume reductant and reduce CO_2. These are termed the methyl and carbonyl paths in Fig. 1. The methyl path results in the overall reduction of CO_2 to a methyl-level intermediate that ultimately is fixed into the methyl-carbon position of acetate. The methyl path requires first the reduction of CO_2 to formate and the subsequent reduction of formate to the methyl level via tetrahydrofolate

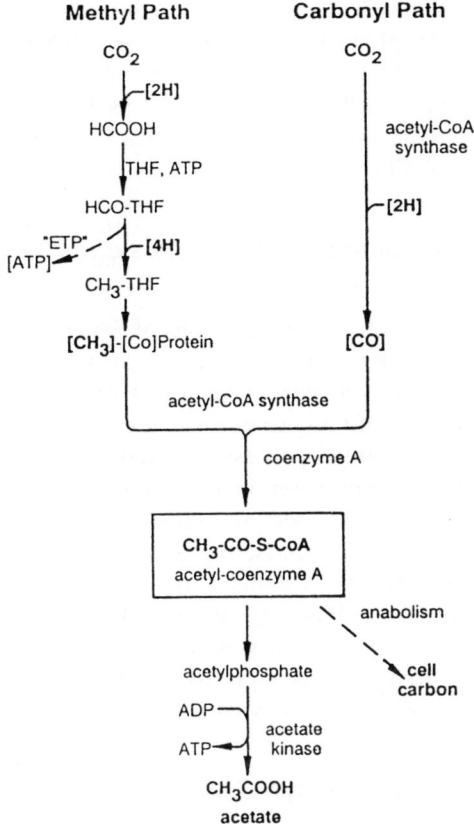

Figure 1 Simplified scheme for the acetyl-CoA pathway. THF, tetrahydrofolate; [Co]protein, corrinoid protein. Broken line and bracketed ATP in the methyl path indicates the conservation of energy via chemiosmotic processes and electron transport phosphorylation (ETP); the exact amount of ATP formed via ETP is not known and likely depends on both growth conditions and acetogen.

(THF)-bound intermediates. ATP is required in the initial activation of formate during the synthesis of formyl-THF by formyl-THF synthetase. On the carbonyl path, CO_2 is reduced to a bound carbonyl ([CO]) group, that ultimately is fixed into the carboxyl-carbon position of acetate. This reaction is thermodynamically unfavorable, and it is estimated that a 1/3 ATP-energy-equivalent is required to drive the reductive formation of [CO] from CO_2 [6].

Acetyl-CoA forms the first two-carbon metabolic intermediate en route to acetate and is synthesized by acetyl-CoA synthase. Acetyl-CoA synthase is also referred to as CO dehydrogenase because of its function in the reversible reaction

$$CO + H_2O \rightleftarrows CO_2 + 2H^+ + 2e^-$$

in the carbonyl path. CO dehydrogenase of acetogens is biochemically distinct from, and serves different physiological roles from, CO dehydrogenase of carboxydotrophic bacteria [7,8]. Anabolically, acetyl-CoA is utilized to synthesize cell carbon, while catabolically, acetyl-CoA proceeds *via* phosphotransacetylase and acetate kinase to acetate and thus yields *via* substrate-level phosphorylation (SLP) one mole of ATP per mole of acetate.

3.2 Energy gain via respiration during the synthesis of acetate from CO_2

Since one mole of ATP is consumed per mole of formate activated, no net gain of ATP is possible via SLP when CO_2 serves as the C_1 substrate for acetate synthesis (Fig. 1). Despite the fact there is no direct net gain of ATP possible via SLP during the reduction of CO_2 to acetate, the pathway yields utilizable energy through respiratory-coupled processes linked to the methyl path. The exact nature of acetogenic respiration is not well established. Respiration involves membranous electron transport and chemiosmotic systems with both proton and sodium ion-coupled circuits, the precise respiratory mechanism employed depending on both the organism and growth conditions [9–14]. Proton- and sodium-driven ATPases have been reported [9,12], and the potential involvement of proton gradients as the driving force for the thermodynamically unfavorable reduction of CO_2 to CO (carbonyl path) has been postulated [6].

3.3 Coupling between the acetyl-CoA pathway and other processes

Recall, however, that the pathway is principally an electron sink, and oxidative processes not directly coupled to the reduction of CO_2 in the acetyl-CoA pathway can yield ATP *via* SLP. Figure 2 illustrates how the acetyl-CoA pathway is integrated to the oxidation (fermentation) of glucose; the cell experiences a substantial energy gain (4 ATP per glucose oxidized) independent of the additional energy conserved *via* respiration and the acetyl-CoA pathway. What is arguably more important in this case is that the pathway provides the basis for regenerating the oxidized forms of catabolic electron carriers (e.g. NAD). In other words, the direct capacity of the acetyl-CoA pathway to consume reductant and recycle electron carriers is, under many conditions, of equal or greater importance to its indirect capacity *via* respiration to yield utilizable energy.

4 STRATEGIES FOR LIFE IN THE ABSENCE OF CO_2

Most of what we know about the physiology and enzymology of acetogens comes from studies with glucose-cultivated cells of *Clostridium thermoaceticum*

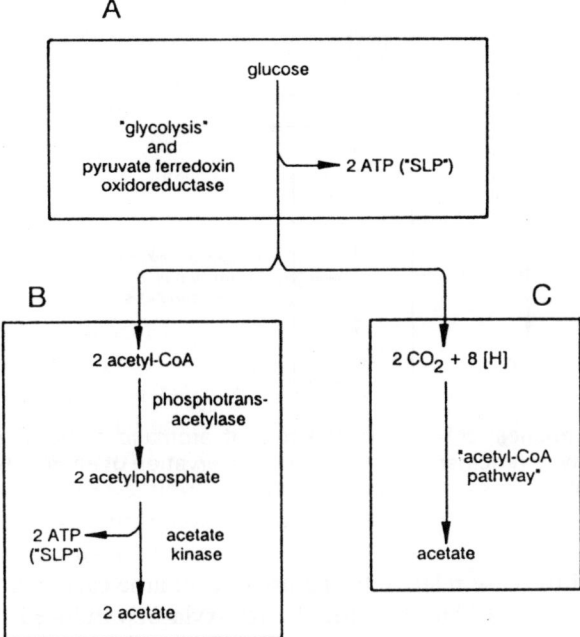

Figure 2 Overview of the 3 main processes of acetogens that collectively yield 3 molecules of acetate per molecule of glucose. A, illustrates the glycolytic conversion of glucose to acetyl-CoA, CO_2, and reductant ([H]); B, conversion of acetyl-CoA to acetate; C, conversion of CO_2 and reductant generated during the oxidation of glucose to acetate via the acetyl-CoA pathway (see Fig. 1). SLP, substrate-level phosphorylation.

and, for the most part, is restricted to the acetyl-CoA pathway. However, studies in several laboratories indicate that acetogens are metabolically one of the most versatile groups of anaerobes and harbor many diverse potentials that impinge on, or run parallel to, the acetyl-CoA pathway. In the following sections, a brief look at some of these potentials will serve to illustrate not only this versatility, but also point to the need to further explore these potentials relative to their biochemical features and ecological importance.

4.1 Alternative origins of CO_2 from lignin derivatives

Because CO_2 serves as an electron sink for acetogens, one might wonder if acetogens can do without it. In fact, growth is strictly dependent upon available CO_2 under certain conditions (e.g. with methanol as an energy substrate). It would stand to reason that acetogens might thus be able to generate CO_2 or CO_2 equivalents from specialized substrates to circumvent this limitation when necessary.

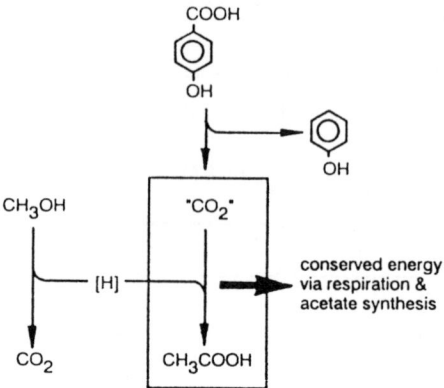

Figure 3 Hypothetical scheme for the role of aromatic carboxyl groups in the acetyl-CoA pathway (enclosed box) and the conservation of energy from methanol-derived reductant.

This is indeed the case relative to the use of aromatic carboxyl groups by *C. thermoaceticum* [15,16]. An aromatic decarboxylase is induced that activates the carboxyl group of certain carboxylated lignin derivatives and provides growth-supportive CO_2 equivalents under conditions that are deficient in CO_2 and thus non-growth supportive. In this case, the aromatic carboxyl group assumes the role of terminal electron acceptor. ^{14}C-studies demonstrated that acetate is totally synthesized from this CO_2 equivalent, thus proving that aromatic carboxyl groups indeed accept reductant and are utilized in acetogenesis. Figure 3 illustrates a scheme by which decarboxylation is coupled to the subsequent oxidation of methanol and provides the ultimate basis for acetogenesis and the conservation of energy. Under such conditions, growth is strictly dependent upon the utilization of the aromatic carboxyl group. Although little is known about the decarboxylase, its induction is dependent upon utilizable aromatic substrates (e.g. 4-hydroxybenzoate) and is not repressed by glucose.

4.2 Oxidization of reduced carriers via reduction of pyruvate

As shown in Fig. 2, when glucose is oxidized, reductant flows towards the acetyl-CoA pathway. Closer inspection of this process shows that, in the absence of supplemental CO_2, pyruvate, the immediate product of glycolysis, is unable to proceed towards acetate because the reductant generated in the oxidation of glucose has no CO_2 sink to flow towards. In other words, under such conditions, reduced electron carriers might be difficult to recycle. Indeed, in the absence of CO_2, glucose-dependent growth of *C. thermoaceticum* is severely impaired and often yields extensive lag phases and low yields. A simple way around this problem is illustrated by the capacity of *Peptostreptococcus productus* to form lactate. We have found (unpublished data) that this acetogen reduces significant quantities of pyruvate to lactate in the absence of CO_2, and thus recycles reductant carriers and provides the potential to decarboxylate pyruvate and generate CO_2 that is then utilized in the acetyl-CoA pathway and the additional conservation of energy. The overall reactions are (where X is the reductant carrier):

$$2\,C_6H_{12}O_6 + 4\,X \rightarrow 4\,CH_3COCOOH + 4\,XH_2$$
$$2\,CH_3COCOOH + 2\,XH_2 \rightarrow 2\,CH_3CHOHCOOH + 2\,X$$
$$2\,CH_3COCOOH + 2\,X + 2\,H_2O \rightarrow 2\,CH_3COOH + 2\,CO_2 + 2\,XH_2$$
$$2\,CO_2 + 4\,XH_2 \rightarrow CH_3COOH + 2\,H_2O + 4\,X$$

sum: 2 glucose → 2 lactate + 3 acetate

Significantly, in the presence of supplemental CO_2, lactate is not formed.

4.3 Alternatives to CO_2 as a terminal electron acceptor

As noted above, acetogens are able to conserve energy via respiratory mechanisms that are coupled to the reduction of CO_2 in the acetyl-CoA pathway. Additional avenues for the terminal routing of this reductant is also possible. One such mechanism involves the dismutation of fumarate to succinate and is utilized by *Clostridium formicoaceticum* and *Clostridium aceticum* [17–19]. These acetogens have two mechanisms for the consumption of reductant:

$$2\,CO_2 + 8[H] \rightarrow \text{acetate} + 2\,H_2O \quad \text{(acetogenic conditions)}$$

$$\text{fumarate} + 2[H] \rightarrow \text{succinate} \quad \text{(non-acetogenic conditions)}$$

In the absence of CO_2, these organisms have the potential both to generate and consume reductant and not to engage the acetyl-CoA pathway. When cultivated with both fumarate and methanol in the absence of CO_2, initial growth of *C. formicoaceticum* is independent of methanol utilization and appears to be strictly coupled to nonacetogenic conservation of energy (Fig. 4) [19]. However, when such cultures are provided with CO_2 and thus the option to engage acetogenesis, they do not, but instead continue to dismute fumarate, indicating that acetogenesis may not be preferred by these organisms under certain conditions. Significantly, acetyl-CoA synthase (CO dehydrogenase) activity levels from fumarate cultures are nearly 5-fold lower than those obtained with CO- or methanol-cultivated cells [19a].

Lignin derivatives containing aromatic acrylate groups can also be utilized by some acetogens as an energy-conserving terminal electron sink. This was first observed with CO_2-enriched cultures of *Acetobacterium woodii* [20,21]. Under these conditions, the cell would theoretically be able to use both the reduction of aromatic acrylates and CO_2 (the acetyl-CoA pathway). But can acrylates alone support growth when acetogenesis is no longer an option? To answer this question, the growth potentials of *P. productus* were examined [22]. This acetogen grew readily in the absence of CO_2 at the expense of methoxyl-derived reductant (see also 5.3 below) with caffeate as the sole terminal electron

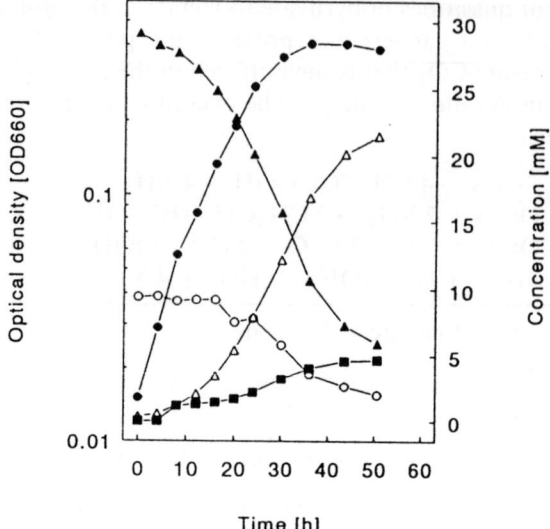

Figure 4 Methanol- and fumarate-derived product and growth profiles of *Clostridium formicoaceticum* ATCC 27076 under CO^2-limited conditions. Symbols: growth OD (●), acetate (■), fumarate (▲), succinate (△), methanol (○).

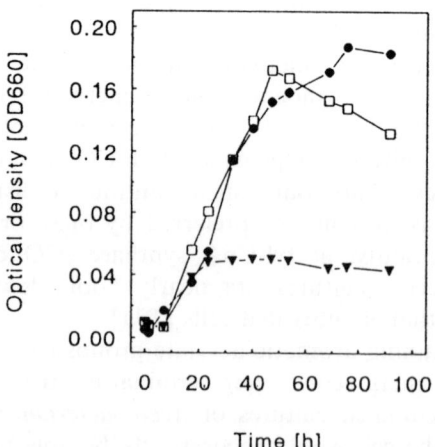

Figure 5 Effects of acrylate side chain of caffeate on the vanillate (methoxy)-dependent growth of *P. productus* ATCC 35244 under CO_2-limited conditions. Substrates: ▼, controls (no added substrates, or vanillate or caffeate alone); □, caffeate (10 mM) plus vanillate (2.2 mM); ●, ferulate (5 mM) (Has both acrylate and methoxyl groups present).

Table 1 Growth analysis of *P. productus* ATCC 35244 and *C. thermoaceticum* ATCC 39073

Substrates (mM)	CO_2 added	Acetate-to-biomass ratio[a]	2[H]-to-biomass ratio[b]
Peptostreptococcus productus			
hydroferulate (5)	yes	26	139 (300)[d]
ferulate (5)	yes	21	122 (110)
ferulate (5)	no	8	115 (45)
caffeate (10) + vanillate (2.2)[c]	no	3	57
Clostridium thermoaceticum			
oxalate (14)	yes	39	197
H_2 (0.8 atm.)	yes	276	1,298
CO (0.8 atm.)	yes	60	367
glucose (5)	yes	68	122

[a] mmol acetate formed (mg cell dry wt)$^{-1}$
[b] Reductant pairs (mM) consumed (mg cell dry wt)$^{-1}$
[c] Alone, caffeate, vanillate or hydroferulate were not growth supportive in the absence of CO_2.
[d] Numbers in brackets were obtained with *P. productus* strain Marburg.

acceptor (Fig. 5). Furthermore, during growth at the expense of caffeate plus vanillate, acetate production was reduced to near background levels, methoxyl-derived reductant going exclusively towards acrylate reduction. Significantly, the amount of reductant required to synthesize biomass was far less under these conditions (Table 1). *A. woodii* behaves similarly even with CO_2 present: methanol-derived reductant goes predominantly towards acrylates [20]. Thus, acrylate reduction appears to be preferred under some conditions. As with fumarate, the mechanism of energy conservation coupled to the reduction of aromatic acrylates is not known. However, this potential is independent of fumarate utilization and is induced by aromatic acrylates even when the cell is actively fermenting glucose [22].

Aromatic aldehydes are also reduced by certain acetogens [23]. This potential has not been examined in much detail, but unpublished studies in our group suggest that the reduction of aromatic aldehydes is not directly coupled to energy conservation. Instead, aromatic aldehydes might provide a reductant sink for the oxidation of methanol or aromatic methoxyl groups to CO_2 under CO_2-limited conditions, this methoxy-derived CO_2 being subsequently utilized in the acetyl-CoA pathway and the conservation of energy (Fig. 6). Under such conditions, growth would be obligately coupled to the initial reduction of aromatic aldehydes and the concomitant formation of CO_2.

Another interesting capacity of certain acetogens is the ability to reduce inorganic sulfur compounds. Inorganic sulfur compounds (e.g. sulfite, thiosulfate or sulfur) are (i) reduced to sulfide and (ii) used as growth-essential sources of sulfur by CO_2-enriched heterotrophic cultures of *Butyribacterium*

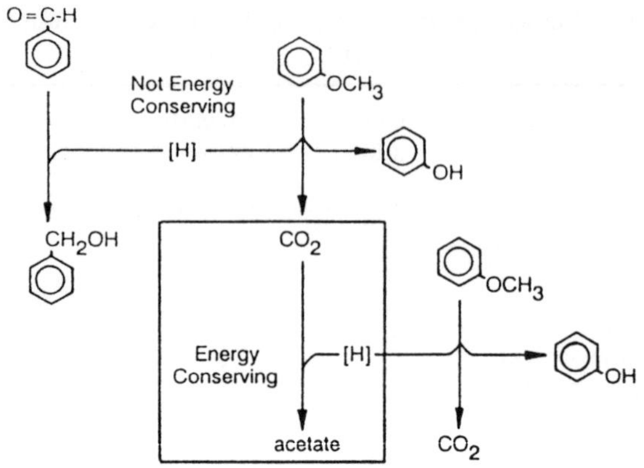

Figure 6 Hypothetical role of the reduction of aromatic aldehydes in the conservation of methoxyl-derived energy under CO_2-limited conditions. Enclosed box, acetyl-CoA pathway.

methylotrophicum and *A. woodii* [24]. Thus, the flow of reductant can be diverted away from CO_2. The potential to couple this reductant flow to the conservation of energy has been examined with *C. thermoaceticum* [25]. This acetogen has the capacity to replace CO_2 with thiosulfate or dimethylsulfoxide (DMSO), and derive growth-supportive energy during the 'respiration' of certain alcohols, including ethanol, propanol and butanol. Under such conditions, (i) thiosulfate and DMSO are reduced to sulfide and dimethylsulfide, respectively, (ii) alcohols are oxidized to their corresponding fatty acids, and (iii) the acetyl-CoA pathway is not engaged. Studies with vesicles indicate that a proton motive force is established in response to these CO_2-independent redox couples [25]; Lars Ljungdahl, personal communication).

5 ALTERNATIVE SOURCES OF REDUCTANT

In contrast to methanogens that have relatively limited capacity to oxidize substrates and are thus highly specialized, the capacity to extract reductant from very diverse substrates is a specialty of acetogens and is responsible for their environmental versatility. However, with the exception of the oxidation of glucose, very little is known of the other acetogenic oxidative processes and how they are coupled to the reduction of CO_2 (i.e., the acetyl-CoA pathway). Recent studies have identified several new oxidative metabolic potentials that may have important ramifications to the role acetogens play in the turnover of both carbon and halogenated compounds in nature.

5.1 Dechlorinations and oxidation of C_1 compounds

The growth-supportive oxidations/dechlorinations of methyl chloride (by strain MC) [26] and tetrachloromethane (by *A. woodii*) [27] illustrate that acetogens catalyze the anaerobic oxdiation of halogenated compounds. The pathways and catalysts are not resolved, but the oxidation of these compounds to the level of CO_2 are indicated by product stoichiometries. In the case of methyl chloride, the following overall reactions are predicted [26]:

$$\text{oxidative:} \quad CH_3Cl + 2\,H_2O \rightarrow [6H] + CO_2 + HCl$$
$$\text{reductive:} \quad [6H] + 3\,CH_3Cl + 3\,CO_2 \rightarrow 3\,CH_3COOH + 3\,HCl$$

sum of redox couple: $4\,CH_3Cl + 2\,CO_2 + 2\,H_2O \rightarrow 3\,CH_3COOH + 4\,HCl$

Thus, the flow of methyl chloride-derived reductant appears to be directed towards the acetyl-CoA pathway, acetate synthesis, and the conservation of energy. Significantly, growth yields with methyl chloride (7.9 g/mol) are greater than those with methanol (4.9 g/mol), suggesting dissimilar mechanisms of energy conservation, perhaps involving dechlorination [26].

5.2 Oxidation and generation of C_1 compounds from C_2 compounds

Surprisingly little is known of the potentials of acetogens to use C_2 compounds, or how such substrates are metabolized relative to the C_1-based acetyl-CoA pathway. Oxalate and glyoxylate were recently shown to be utilized as sole sources of carbon and energy by *C. thermoaceticum* according to the following stoichiometries [28]:

oxalate:
$$\text{oxidative:} \quad 4\,HOOC\text{—}COOH \rightarrow 8\,CO_2 + [8H]$$
$$\text{reductive:} \quad 2\,CO_2 + [8H] \rightarrow CH_3COOH + 2\,H_2O$$

sum of redox couple: $4\,HOOC\text{—}COOH \rightarrow CH_3COOH + 6\,CO_2 + 2\,H_2O$

glyoxylate:
$$\text{oxidative:} \quad 2\,HOOC\text{—}COH \rightarrow 2\,CO_2 + 2[CO] + [4H]$$
$$\text{oxidative:} \quad [CO] + H_2O \rightarrow CO_2 + [2H]$$
$$\text{reductive:} \quad CO_2 + [CO] + [6H] \rightarrow CH_3COOH + H_2O$$

sum of redox couple: $2\,HOOC\text{—}COH \rightarrow CH_3COOH + 2\,CO_2$

Coenzyme-A and formate-level intermediates are predicted for the oxidation of oxalate and glyoxylate. Both substrates are growth supportive in the absence of CO_2, and the amount of biomass formed per unit oxalate- or glyoxylate-derived reductant consumed is, with the exception of glucose, higher than that formed with other acetogenic substrates (Table 1).

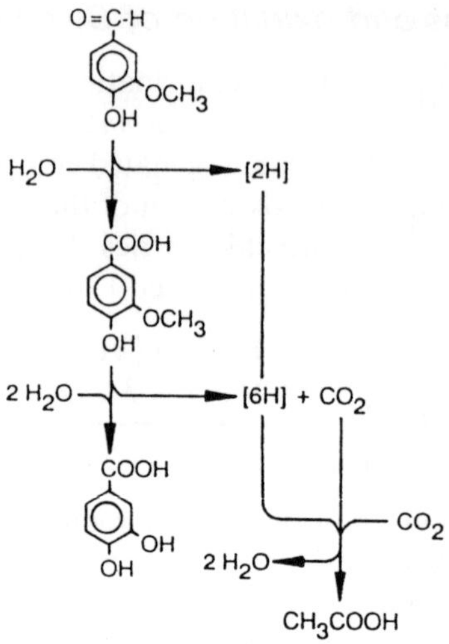

Figure 7 Stoichiometry for coupling between the oxidation of vanillin to protocatechuate and the reductive synthesis of acetate from CO_2.

5.3 Oxidation of aromatic substituent groups

A common feature of many acetogens is the ability to oxidize aromatic substituent groups. An unclassified acetogen, strain AmMan1, utilizes mandelate as follows ('ring' indicates the aromatic ring of mandelate and benzoate) [29]:

oxidative: $2 \text{ ring-CHOH—COOH} + 2 H_2O \rightarrow 2 \text{ ring-COOH} + 2 CO_2 + [8H]$
reductive: $2 CO_2 + [8H] \rightarrow CH_3COOH + 2 H_2O$

sum of
redox couple: $2 \text{ ring-CHOH—COOH} \rightarrow CH_3COOH + 2\text{ring-COOH}$

For each mandelate consumed, four reducing equivalents are generated, and benzoyl-CoA is proposed as an aromatic intermediate en route to benzoate [29]. Whether the energy of the benzoyl-CoA bond is conserved in the synthesis of ATP via SLP (during the sequential conversion to benzoylphosphate and benzoate) is unknown.

Acetogens also oxidize aromatic aldehydes and methoxyl groups [23,30–33]. Figure 7 illustrates how the oxidation of these two aromatic C_1-substituent groups of vanillin may provide reductant for the acetyl-CoA pathway. As suggested by this scheme, vanillin is not growth supportive without supplemental CO_2 (or alternative sink) since the oxidation of these two groups yields excess reductant. This has been confirmed in growth experiments. Virtually nothing is known of the oxidoreductases involved, but studies in our laboratory indicate that the oxidation of the aldehyde group by C. thermoaceticum and C. formicoaceticum is not coupled to the conservation of energy. This is in contrast to the oxidation of methoxyl groups that provide growth-essential reductant for the conservation of energy during the methoxyl-dependent synthesis of acetate. It has been proposed that THF-bound intermediates are involved in the O-demethylation and oxidation of methoxyl groups [33].

6 CONCLUSION

The above statements do not take into account many other intriguing potentials exhibited by both free-living and syntrophic acetogens. They nonetheless serve a single purpose: to illustrate that acetogenic bacteria harbor many unresolved catalytic systems. It is clear that these organisms have numerous mechanisms for funneling both carbon and reductant to the autotrophic acetyl-CoA pathway. They also have learned how to live without this pathway. Understanding the full magnitude of the roles acetogens play in nature will require further resolution of these processes and their regulation.

ACKNOWLEDGEMENTS

The author expresses appreciation to Lars Ljungdahl for communicating unpublished results, to Carola Matthies and Steven Daniel for review of the manuscript, and to Gabriele Diekert for the culture of P. productus strain Marburg. This work was supported in part by the German Ministry for Research and Technology.

REFERENCES

[1] Wood, H.G. (1991) Life with CO or CO_2 and H_2 as a source of carbon and energy. FASEB J. 5, 156–163.
[2] Wood, H.G. and Ljungdahl, L.G. (1991) Autotrophic character of the acetogenic bacteria. In: Variations in Autotrophic Life (Shively, J.M. and Barton, L.L., Eds), pp. 201–250. Academic Press, San Diego.
[3] Drake, H.L. (1992) Acetogenesis and acetogenic bacteria. In: Encyclopedia of Microbiology, Vol 1. (Lederberg, J., Ed.) Academic Press, San Diego.
[4] Ragsdale, S.W. (1991) Enzymology of the acetyl-CoA pathway of CO_2 fixation. Crit. Revs. Biochem. Mol. Biol. 26, 261–300.
[5] Fuchs, G. (1986) CO_2 fixation in acetogenic bacteria: variations on a theme. FEMS Microbiol. Revs. 39, 181-213.
[6] Diekert, G. (1992) The acetogenic bacteria. In: The Prokaryotes, 2nd ed. (Balows, A., et al., Eds), pp. 517–533, Springer-Verlag, New York.
[7] Meyer, O. (1988) Biology and biotechnology of aerobic carbon monoxide-oxidizing bacteria. In: Biotechnology Focus (Finn, R.K., et al., Eds.), pp. 3–31, Hanser Publisher, Munich.

[8]   Meyer, O., Frunzke, K., Gadkari, D., Jacobitz, S., Hugendieck, I. and Kraut, M. (1990) Utilization of carbon monoxide by aerobes: recent advances. *FEMS Microbiol. Revs.* **87**, 253–260.
[9]   Hugenholtz, J. and Ljungdahl, L. G. (1989) Electron transport and electrochemical proton gradient in membrane vesicles of *Clostridium thermoaceticum*. *J. Bacteriol.* **171**, 2873–2875.
[10] Terracciano, J.S., Schreurs, W.J.A. and Kashket, E.R. (1987) Membrane H^+ conductance of *Clostridium thermoaceticum* and *Clostridium acetobutylicum*: evidence for electrogenic Na^+/H^+ antiport in *Clostridium thermoaceticum*. *Appl. Environ. Microbiol.* **53**, 782–786.
[11] Heise, R., Reidlinger, J., Muller, V. and Gottschalk, G. (1991) A sodium-stimulated ATP synthase in the acetogenic bacterium *Acetobacterium woodii*. *FEBS Lett.* **295**, 119–122.
[12] Heise, R. and Müller, V. (1992) Sodium as the coupling ion for the ATPase of *Acetobacterium woodii*. Abstr. Ann. Meet. Am. Soc. Microbiol., Abst. K-19, p. 259.
[13] Geerligs, G., Schönheit, P. and Diekert, G. (1989) Sodium dependent acetate formation from CO_2 in *Peptostreptococcus productus* (strain Marburg). *FEMS Microbiol. Lett.* **57**, 253–258.
[14] Yang, H. and Drake, H.L. (1990) Differential effects of sodium on hydrogen- and glucose-dependent growth of the acetogenic bacterium *Acetogenium kivui*. *Appl. Environ. Microbiol.* **56**, 81–86.
[15] Hsu, T., Daniel, S.L., Lux, M.F. and Drake, H.L. (1990) Biotransformations of carboxylated aromatic compounds by the acetogen *Clostridium thermoaceticum*: generation of growth-supportive CO2 equivalents under CO_2-limited conditions. *J. Bacteriol.* **172**, 212–217.
[16] Hsu, T., Lux, M.F. and Drake, H.L. (1990) Expression of an aromatic-dependent decarboxylase which provides growth-essential CO_2 equivalents for the acetogenic (Wood) pathway of *Clostridium thermoaceticum*. *J. Bacteriol.* **172**, 5901–5907.
[17] Dorn, M., Andreesen, J. R. and Gottschalk, G. (1978) Fumarate reductase of *Clostridium formicoaceticum*. *Arch. Microbiol.* **119**, 7–11.
[18] Dorn, M., Andreesen, J.R. and Gottschalk, G. (1978) Fermentation of fumarate and L-malate by *Clostridium formicoaceticum*. *J. Bacteriol.* **133**, 26–32.
[19] Matthies, C., Schwarz, U., Freiberger, A. and Drake, H.L. (1992) Fumarate as an electron sink during aromatic methoxyl- and methanol-dependent growth by the acetogen *Clostridium formicoaceticum*. Abstr. 6th Intl. Symp. Microbial Ecol., Barcelona (in press, and manuscript in preparation).
[19a] Lux, M.F. and Drake, H.L. (1992) Reexamination of the metabolic potentials of the acetogens *Clostridium aceticum* and *Clostridium formicoaceticum*: chemolithoautotrophic and aromatic-dependent growth. *FEMS Microbiol. Lett.* **95**, 49–56.
[20] Tschech, A. and Pfennig, N. (1984) Growth yield increase linked to caffeate reduction in *Acetobacterium woodii*. *Arch. Microbiol.* **137**, 163–167.
[21] Hansen, B., Bokranz, M., Schönheit, P. and Kröger, A. (1988) ATP formation coupled to caffeate reduction by H_2 in *Acetobacterium woodii* NZva16. *Arch. Microbiol.* **150**, 447–451.
[22] Daniel, S.L., Misoph, M., Gößner, A. and Drake, H.L. (1992) Growth of acetogenic bacteria in the absence of autotrophic CO_2 fixation to acetate. Abstract 7th International Symposium on Microbiol Growth on C_1 Compounds. Abstract C133.
[23] Lux, M.F., Keith, E., Hsu, T. and Drake, H.L. (1990) Biotransformations of aromatic aldehydes by acetogenic bacteria. *FEMS Microbiol. Lett.* **67**, 73–78.
[24] Heijthuijsen, J.H.F.G. and Hansen, T.A. (1989) Selection of sulphur sources for

the growth of *Butyribacterium methylotrophicum* and *Acetobacterium woodii*. *Appl. Microbiol. Biotech.* **32**, 186–192.
[25] Beaty, P.S. and Ljungdahl, L.G. (1991) Growth of *Clostridium thermoaceticum* on methanol, ethanol, or dimethylsulfoxide. Abstr. Ann. Meet. Am. Soc. Microbiol., Abstr. K-131, p. 236.
[26] Traunecker, J., Preuß, A. and Diekert, G. (1991) Isolation and characterization of a methyl chloride utilizing, strictly anaerobic bacterium. *Arch. Microbiol.* **156**, 416–421.
[27] Egli, C., Tschan, T., Scholtz, R., Cook, A.M. and Leisinger, T. (1988) Transformation of tetrachloromethane to dichloromethane and carbon dioxide by *Acetobacterium woodii*. *Appl. Environ. Microbiol.* **54**, 2819–2824.
[28] Daniel, S.L. and Drake, H.L. (1991) Acetogenesis from two-carbon compounds by *Clostridium thermoaceticum*. Abstr. Ann. Meet. Am. Soc. Microbiol., Abstr. K-137, p.237.
[29] Dörner, C. and Schink, B. (1991) Fermentation of mandelate to benzoate and acetate by a homoacetogenic bacterium. *Arch. Microbiol.* **156**, 302–306.
[30] Bache, R. and Pfennig, N. (1981) Selective isolation of *Acetobacterium woodii* on methoxylated aromatic acids and determination of growth yields. *Arch. Microbiol.* **130**, 255–261.
[31] DeWeerd, K.A., Saxena, A., Nagle, D.P., Jr. and Suflita, J.M. (1988) Metabolism of the ^{18}O-methoxy substituent of 3-methoxylbenzoic acid and other unlabeled methoxybenzoic acids by anaerobic bacteria. *Appl. Environ. Microbiol.* **54**, 1237–1242.
[32] Daniel, S.L., Keith, E.S., Yang, H., Lin, Y.-S. and Drake, H.L. (1991) Utilization of methoxylated aromatic compounds by the acetogen *Clostridium thermoaceticum*: expression and specificity of the CO-dependent O-demethylating activity. *Biochem. Biophys. Res. Commun.* **180**, 416–422.
[33] Berman, M.H. and Frazer, A.C. (1992) Importance of tetrahydrofolate and ATP in the anaerobic O-demethylation reaction for phenylmethylethers. *Appl. Environ. Microbiol.* **58**, 925–931.

Species Index

Acetobacter sp. 262, 278, 279
   taxonomy 279, 281
Acetobacter aceti 229
Acetobacter methanolicus 222–4, 226, 406
Acetobacterium woodii 499, 501, 502, 503
Acidomonas sp. 262, 278, 279
   taxonomy 279, 281
Acidomonas methanolica 262, 279
Acinetobacter calcoaceticus 227, 229, 407
Agrobacterium sp. 262, 280
   taxonomy 279
Alcaligenes eutrophus 451, 452, 472, 481–2, 484–8
Alcaligenes eutrophus JMP 134
Alcaligenes hydrogenophilus 483
Alteromonas sp. 56
Aminobacter sp. 262
Aminobacter aganoensis 262
Aminobacter niigataensis 262
Amycolatopsis methanolica 245–6, 248–9, 329–34, 403, 406–7
Anabaena sp. 471–2
Ancylobacter sp. 262
Archaeoglobus sp. 136
Archaeoglobus fulgidus 152–6
Arthrobacter sp. 406
Asparagopsis nodosum 39
Asparagopsis taxiformis 38
Azospirillum sp. 262
Azotobacter sp., taxonomy 279, 281
Azotobacter vinelandii 451

Bacillus sp. 52, 211, 249, 262
Bacillus acidoterrestris 270
Bacillus alvei 270
Bacillus amyloliquefaciens 270
Bacillus amylolyticus 270
Bacillus aneurinolyticus 270
Bacillus anthracis 270
Bacillus atrophaeus 270
Bacillus azotofixans 270
Bacillus azotoformans 267, 269, 270, 273
Bacillus badius 270
Bacillus benzoevorans 270
Bacillus brevis 267, 268, 269, 270, 271, 273
Bacillus cereus 269, 270
Bacillus circulans 270
Bacillus coagulans 270
Bacillus cycloheptanicus 270
Bacillus fastidiosus 269, 270
Bacillus firmus 267, 269, 270, 271, 273
Bacillus fusiformis 270
Bacillus globisporus 270
Bacillus gordonae 270
Bacillus insolitus 270
Bacillus kaustophilus 270
Bacillus lantimorbus 270
Bacillus laterosporus 270
Bacillus lautus 270
Bacillus lavae 270
Bacillus licheniformis 270
Bacillus macerans 270
Bacillus macquariensis 270
Bacillus maroccanus 270
Bacillus medusa 270
Bacillus megaterium 269, 270, 271
Bacillus methanolicus 245, 246–9, 267, 273
Bacillus mycoides 270
Bacillus pabuli 270
Bacillus pasteurii 270
Bacillus polymyxa 270
Bacillus popillae 270
Bacillus psychosaccharolyticus 270
Bacillus psychrophilus 270
Bacillus pulvifaciens 270
Bacillus pumilus 269, 270
Bacillus schlegelii 439
Bacillus simplex 270
Bacillus smithii 270

510 Species Index

Bacillus sphaericus   270
Bacillus stearothermophilus   269, 270, 271
Bacillus subtilis   269, 270, 271
Bacillus tentus   270
Bacillus thermoglucosidasius   270
Bacillus thuringiensis   270
Bathymodiolus thermophilus   316–26
Blastobacter sp.   262, 278, 280
   taxonomy   279, 281
Blastobacter aggregatus   281
Blastobacter aminooxidans   280
Blastobacter capsulatus   281
Blastobacter denitrificans   281
Blastobacter viscosus   280
Bonnemaisonia hamifera   38
Bradyrhizobium sp., taxonomy   279
Bradyrhizobium japonicum   280, 451
Butyribacterium metholotrophicum
   501–2

Caulobacter sp.   281
Chromatium sp., taxonomy   281
Chromatium vinosum   471, 483, 487, 488
Clostridium aceticum   499
Clostridium acetobutylicum   247
Clostridium formicoaceticum   499, 500, 505
Clostridium thermoaceticum   165, 466, 496, 498, 501–3, 505
Comamonas testosteroni   407
Coriolus versicolor   42

Dehalobacter restrictus   354
Desulfovibrio sp., taxonomy   281
Drosophila sp.   247

Ectothiorhodospira, taxonomy   281
Emiliania huxleyi   18
Erythrobacter, taxonomy   281
Escherichia coli   56, 109, 111–14, 125, 134, 176, 185, 188, 229, 247, 270, 358, 386, 390–2, 444, 452, 466, 475, 486
   taxonomy   281

Fucus sargassum   37–8
Fucus vesiculosus   37

Gigartina stellata   37
Gluconobacter sp.   279

Hansenula polymorpha   193–204, 404, 415, 427–30
Hyphomicrobium sp.   22, 56, 59, 214–15, 224, 227, 246, 262, 280, 351–6, 358
   taxonomy   279, 281

Hyphomicrobium zavarzinii   401, 403, 406

Klebsiella sp.   475
Klebsiella aerogenes   403, 404, 407

Laminaria digitata   39
Laminaria hyperborea   39
Laminaria laminaria   38
Laminaria saccharina   39
Lemna minor   296
Lemna trisulca   296

Methanobacteriales   181–3, 187
Methanobacterium sp.   187
Methanobacterium formicicum   182, 444
Methanobacterium thermoautotrophicum
   138, 146, 152–8, 173, 177, 181–9, 444
Methanobacterium wolfei   152–4, 183
Methanococcus vannielii   157, 183, 184, 187
Methanococcus voltae   152, 157, 176, 189
Methanoculleus thermophilicum   136, 138
Methanogenium tationis   136, 138
Methanolobus tindarius   173, 174
Methanoplanus endosymbiosus   136
Methanopyrus kandleri   152–7
Methanosarcina sp.   58, sp. 164, 166, 173
Methanosarcina barkeri   136, 138, 146, 152–8, 165–7, 172, 174, 178, 444
Methanosarcina Göl   171–7
Methanosarcina mazei   166
Methanosarcina thermophila   157, 163–7
Methanosarcinaceae   182
Methanospirillum hungatei   136
Methanothermus sp.   187
Methanothermus fervidus   157, 181–9
Methanothrix sp.   164
Methanothrix soehngenii   157, 164, 166, 466
Methylobacillus flagellatum   292, 381, 382, 383, 384, 385, 386, 391, 392, 393, 394, 397
Methylobacillus fructoseoxidans   260
Methylobacillus glycogenes   224, 227, 260, 292
Methylobacillus sp.   211, 259, 278, 397
   taxonomy   277, 279, 281
Methylobacter sp.   255, 257, 258, 287, 297
   taxonomy   276, 281
Methylobacter capsulatus   111
Methylobacterium sp.   66, 259, 262, 278, 317, 355, 356, 357, 358, 359
   taxonomy   279, 281
Methylobacterium sp. D11   397
Methylobacterium sp. DM4   292

Species Index 511

Methylobacterium sp. M27   292
Methylobacterium sp. PK-1   292
Methylobacterium sp. PK-6   292
Methylobacterium sp. R14   397
Methylobacterium extorquens   212–13, 222–31, 278, 381–95, 397, 403, 415, 425–7
   AM1   292
Methylobacterium organophilum   214, 229, 311
   XX   292
Methylococcus sp.   257, 258
   taxonomy   276
Methylococcus agile   258
Methylococcus bovis   257
Methylococcus capsulatus   66–8, 93–105, 109–18, 255, 257–8, 286–7, 292, 297, 305
Methylococcus luteus   257
Methylococcus methanica   297
Methylococcus thermophilus   257, 258
Methylococcus ucrainicus   257
Methylococcus vinelandii   257
Methylococcus whittenburyi   257
Methylocystis sp.   255, 258, 297
   taxonomy   276, 281
Methylocystis parvus   113
   OBBP   292
Methylomicrobium sp.   262, 280
   taxonomy   281
Methylomonas sp.   255, 257, 258, 287, 383
   sp. A4   292
   sp. GJ6   121–31
   taxonomy   276, 277, 281
Methylomonas agile   111
Methylomonas albus   111, 113, 116, 257, 287, 305, 306
   BG8   292
Methylomonas aurantiaca   257, 258
Methylomonas capsulatus (Bath)   128–31
Methylomonas fodinarum   257, 258
Methylomonas gracilis   257, 258
Methylomonas luteus   287, 292
Methylomonas methanica   111, 214, 257, 258, 287, 292
Methylomonas methanolica   292
Methylomonas methylovora   292
Methylomonas pelagica   257
Methylomonas rubra   257, 258, 287, 292, 305, 306, 310, 323
Methylophaga sp.   259
Methylophaga marina   211–15, 224, 226–8, 260
Methylophaga thalassica   260
Methylophilus glucoseoxidans   260

Methylophilus methylotrophus   222–7, 260, 287, 365–76, 383
   AS1   292
Methylophilus sp.   211, 259, 356, 358, 359, 381–6, 389–95, 397
   taxonomy   277, 279, 281
Methylosinus sp.   111, 255, 258, 293
   taxonomy   276, 281
Methylosinus sp. B   292
Methylosinus sp. LAC   292
Methylosinus methanica   293
   (81Z)   292
Methylosinus sporium   66, 292, 293
Methylosinus trichosporium   66–78, 82, 109–13, 115, 121–31, 293, 305–6, 310, 339–40
   OB3b   292
Methylovarius sp.   257
Methylovorus sp.   259
   taxonomy   277, 279, 281
Methylovorus glucosotrophus   260
Microcoleous chthonoplastes   17, 18
Mycobacterium sp.   311
Mycobacterium convolutum   125, 131
Mycobacterium gastri   245, 246, 248, 249

Neurospora crassa   213
Nitrobacter sp.   262, 280
   taxonomy   279, 281
Nitrobacter hamburgensis   483
Nitrobacter winogradskyi   281
Nitrosomonas europaea   125
Nitszchia sp.   56
Nocardia opaca   483, 485

Ochromonas sp.   56

Paracoccus sp.   262
   taxonomy   279, 281
Paracoccus alcaliphilus   261, 262
Paracoccus aminophilus   261
Paracoccus aminovorans   261
Paracoccus denitrificans   212–13, 224, 227, 235–43, 261, 381–7, 389–94, 397, 403–4, 407, 449
Paracoccus kocurii   261
Peptostreptococcus productus   498–501, 505
Phaeocystis pouchetii   18
Phanerochaete chrysosporium   42
Phellinus pomaceus   41, 42, 352
Phlebia radiata   42
Polysiphonia fastigiata   16
Protaminobacter sp., taxonomy   277
Proteobacteria sp.   278, 281, 297

Proteus mirabilis 359
Pseudomonas sp. 58, 278, 280, 382, 403, 406, 415–25, 433–53
　taxonomy 277, 279, 281
Pseudomonas acidovorans 23
Pseudomonas aeruginosa 227, 394, 406, 444
Pseudomonas aminovorans 262, 280
Pseudomonas carboxydoflava 437, 440–6, 449–50, 461–6
Pseudomonas carboxydohydrogena 439–42, 444, 449, 450
Pseudomonas carboxydovorans 433–41, 444, 446–53, 461–6, 483
Pseudomonas cepacia G4 125
Pseudomonas sp. DM5 131
Pseudomonas facilis 483
Pseudomonas putida 54, 406, 444
Pseudomonas putida F1 125
Pseudomonas stutzeri 444
Pseudomonas thermocarboxydovorans 439–42, 452, 461–6

Rhizobiaceae 262
Rhizobium sp., 262, 280
　taxonomy 279
Rhizobium leguminosarum 451
Rhizobium–Agrobacterium complex 280
Rhodobacter sp. 281
　taxonomy 281
Rhodobacter capsulatus 471–2
Rhodobacter spheroides 444, 470–6, 483
Rhodococcus sp. 446
Rhodococcus erythropolis 406
Rhodococcus erythropolis JE77 125, 131
Rhodocyclus sp., taxonomy 281
Rhodocyclus gelatinosus 451
Rhodopseudomonas sp. 262
Rhodopseudomonas capsulatus 56
Rhodospirillum sp., taxonomy 281
Rhodospirillum rubrum 453, 471–6

Riccia sp. 296
Ricciocarpus sp. 296

Saccharomyces cerevisiae 247
Salmonella typhimurium 404
Serratia marcescens 359
Siboglinum poseidoni 316–26
Spirodela polyrhiza 296
Streptomyces sp. 437, 440
Streptomyces thermoautotrophicus 434–7, 440–2, 444, 445, 452
Sulfolobus acidocaldarius 176
Synechococcus sp. 17, 472, 475
Synechocystis sp. 17

Thiobacillus sp. 52, sp. 262
　taxonomy 281
Thiobacillus denitrificans 52
Thiobacillus ferrooxidans 483
Thiobacillus neapolitanus 52
Thiobacillus thioparus 22, 51, 52, 53
Thiobacillus versutus 381, 383–6, 390, 392–4, 397, 403
Thiocapsa roseopersicina 22, 23
Thiothrix ramosa 52
Thiovulum sp., taxonomy 281
Trichodesmium sp. 17

Ulva lactuca 37

Vibrio cholerae, taxonomy 281

Wofia sp. 296

Xanthobacter sp. 262, 280, 485
　taxonomy 279
Xanthobacter autotrophicus 280
Xanthobacter flavus 472, 483, 484, 487, 488

Zymomonas mobilis 247

Subject Index

Note: a full list of species mentioned in the text is to be found in the *Species Index*.

Acetate fermentation
  acetate kinase 164
  carbonic anhydrase 167-8
  CO dehydrogenase 165-6
  ferredoxin 167
  Methanosarcina thermophila 163-70
  methyl-coenzyme M methylreductase 166
  phosphotransacetylase 164
Acetobacter methanolicus, MDH docking with cytochrome c_L 222-31
Acetogenic culture, anaerobic utilization of dichloromethane 359-61
Acetyl-CoA pathway, autotrophs, CO_2 reductant 493-507
Acrylamides, degradation by *Methylophilus methylotrophus* 365-6
Alcaligenes eutrophicus
  Cfx genes 481-91
  Cfx operons 485-8
Aldehyde dehydrogenase, compared with carbon monoxide dehydrogenase 465-6
Algae, marine
  biomass, and production of dimethyl sulfide 18
  origination of methylated sulfur compounds 16-17
  see also Seaweeds
Amicyanin, as electron acceptor for MADHs 381-97
Amidase
  activity
    heat reactivation 373-5
    properties of switched-off amidase 372-3
    switch-off 372

  production by *Methylophilus methylotrophus* 365-81
    applications 370-2
    continuous culture 375-6
    large scale culture 367-70
    wild-type and mutant strains 377
    properties 367
Amines, oxidation *see* Methylamine dehydrogenase
Amycolatopsis methanolica
  methanol dehydrogenase, dye-linked 249
  methanol and glucose metabolism 333-4
  methanol oxidoreductases 248
  L-phenylalanine synthesis, RuMP cycle 329-34
Antarctic Ocean, halocarbons, production 37
Anthranilate synthase, L-tryptophan inhibition, *A. methanolica* mutants 332
Archaebacteria
  acetate fermentation 163-70
  methanogenesis 171-2
Arctic Ocean, halocarbons, production 36, 39-41
Aromatic substituent rings, mandelate and benzoate 504-5
Arrhenius equation, temperature effects 216
Atmosphere
  bromine compounds 35-45
  trace gases and climate change 59-60
ATP synthase, in methanogenesis 175-7
Autotrophs, acetyl-CoA pathway, CO_2 reductant 493-507

Bacillus methanolicus C1
  distance matrix tree 270
  methanol dehydrogenase 246–8
  morphology and physiology 272
Bacillus methanolicus PB1, taxonomy 267
Benzoate, aromatic substituent rings 504–5
Bioremediation
  chlorinated organics 337–50
    competition for active sites 339–40
    mass transfer limitations 340–3
    product toxicity 339
    reducing equivalents 340
  kinetic modeling 343–6
  Monod kinetics 343–6
Brillouin plot, *Methylococcus* membranes 101
Bromomethanes
  biosynthesis 37–9, 57–9
  bromoform in oceans 36–7
  list 58
  and ozone layer 35–6, 40
  production 39
  sources and sinks 35–45
    and atmospheric bromine 39–41

Calvin cycle, gene organization and regulation 470–2
Calvin–Benson cycle, ribulose-1,5-bisphosphate carboxylase 292, 319 209–11
Carbon dioxide
  absence, strategies for life 496–502
  chemoautotrophs, genetic regulation of assimilation 481–91
  methanogenesis, enzymes involved 151–61
  as reductant
    acetyl-CoA pathway, autotrophs 493–507
    alternative sources 502–5
  as terminal electron acceptor, alternatives 499–502
Carbon disulfide
  bacterial degradation 51–2
  microbiological sources 51
  utilization by novel bacteria 52
Carbon fixation
  phototrophic bacteria 469–79
  see also Cfx genes
Carbon monoxide, aerobic utilization 433–59

Carbon monoxide dehydrogenase
  acetate fermentation 165–6
  aerobic species 433–46
  Cut genes 452, 462–3, 464–7
  molecular genetics 461–7
  molecular weights of subunits 440
  molybdenum requirement 433, 442–5
  nitrate reductase 437
  pGB1 deletion analysis 464
  pGB1 sequence analysis 463–4
  reverse electron transfer 448–9
  selenium activation 445–6
  various sources, purification 440
Carbonic anhydrase, acetate fermentation 167–8
Carbonyl sulfide
  central role for 55
  pathway for degradation of thiocyanate 52
  production 53
Carboxidotrophic bacteria
  Cfx genes 452–3, 472, 481–91
  Cut genes 452, 462–7
  proton translocation 449–50
Cfx genes 452–3, 472, 481–91
  operons, regulation in *Alcaligenes eutrophus* 485–8
  physical organization 482–4
Chemoautotrophs, carbon dioxide assimilation, genetic regulation 481–91
Chlorinated aliphatic hydrocarbons, degradation by methanotrophs 121–33
Chlorinated methanes
  bioremediation 337–50
  biosynthesis 37–9, 57–9
  chloroform, degradation by methanotrophs 338–41
  dehalogenative reactions 352–5
    relevance to cell 354
  list 58
  sources 57–9
  as substrates 354–5
  see also specific named compounds
Clathrates, as natural sink for methane 7
CODH see Carbon monoxide dehydrogenase
CoM-S-S-HTP, exergonic reduction in methanogenic bacteria 171–80
Copper, pMMO activity 93–107
COS see Carbonyl sulfide
C_1 cycle, marine, and methylated sulfur compounds 26–8

C_1 substrates
  potential, in seawater 26
  S, N or halogen-containing,
    biogeochemical cycling 47–63
C_2 compounds, generation of C_1
    compounds 503–4
Cut genes 452, 462–7
Cyanate pathway, degradation of
    thiocyanate 52
Cyanobacteria, production of
    β-dimethylsulfoniopropionate
    17, 18
Cytochromes
  cytochrome b_{561} and b_{563} 448
  cytochrome c_H 221–2, 226–7
  cytochrome c_L
    interactions with methanol
      dehydrogenase 221–33
    and redox cycle of methanol
      dehydrogenase 214–16
  cytochrome d 446
  as electron acceptors for MADHs
    381–3, 387

Dehalogenation reactions, chlorinated
    methanes 352–4
Dehydroquinate dehydratase (DHQ), $A.$
    $methanolica$ mutants 332
1,2-Dichloroethane, degradation 122
$trans$-1,2-Dichloroethylene, degradation
    122
Dichloroethylene isomers, degradation
    by methanotrophs 337–46
Dichloromethane
  dehalogenase 355–9
  dehalogenation reactions 352–4
  as substrate 122, 355–9
  utilization
    aerobes 355–9
    anaerobes 359–61
Dimethyl sulfide
  oceans, microbial sources and sinks
    15–33
  degradation in seawater 15, 17–18,
    22–3
  sources and sinks, schema 25
Dimethyl sulfoxide
  distribution in seawater 24
  production by phototrophs and
    heterotrophs 23
  role in DMS cycle 23–4
Dimethylamine, microbiological role 56
5,5-Dimethylpyrroline-1-oxide, methane
    monooxygenase pathways 66–9

β-Dimethylsulfoniopropionate,
    production of dimethyl sulfide
    17–18
DMPO see 5,5-dimethylpyrroline-1-oxide
DMS see Dimethyl sulfide
DMSO see Dimethyl sulfoxide
DMSP see β-Dimethylsulfoniopropionate

Ecology of methanotrophs
  characterization 294–6
  fresh water habitat 3–70, 7, 294–7
  obligate methanotrophs 303–13
  see $also$ Oceans
Energy conservation, methanogenic
    bacteria 171–80
Energy transduction
  H_2 and CO_2 177
  methanol 177–8

Ferredoxin, acetate fermentation 167
Formaldehyde
  assimilation, ribulose monophosphate
    pathway 209–11, 292
  fixation, RuMP cycle 329–34
  intermediary of MSA degradation 49
  labelled, utilization 430
Formaldehyde dehydrogenase 406–7
Formyl transfer reactions,
    methanopterins 140–2
Formylmethanofuran dehydrogenase,
    activity 152–3
Formylmethanofuran H_4
    formyltransferase, activity 153
Fresh water
  acetate fermentation, $Methanosarcina$
    $thermophila$ 163–70
  characterization 294–6
  as natural sink for methane 7
  plants associated with methylotrophs
    296–7
$F_{420}H_2$-dependent heterodisulfide
    reductase system 173–5
Fungi, degradation of lignin, veratryl
    alcohol coupling 42

β–Galactosidase, mox gene expression
    239–42
Gas production see Natural gas
GBT see Glycine betaine
Global climate change, and
    halomethanes 59–60
Glucose, metabolism, $A.$ $methanolica$
    mutants 333–4

Glucosinolates, production of
    thiocyanate ion   51
Glycine betaine, interactions with DMS
    and DMSP   27–8
Gram-negative bacteria
    dichloromethane-utilizing   355–9
    methanol oxidation   209–20
Gram-positive bacteria
    methanol oxidation   211, 245–51
    taxonomy   267–73

Halogenated methanes *see*
    Bromomethanes; Chlorinated
    methanes; *specific named compounds*
Hansenula polymorpha
    methylamine utilization   427–300
    peroxisome biogenesis   193–207
Henry's coefficients   341, 342
Heterodisulfide reductase   158
    system   172–5
Hexulose phosphate synthase, ribulose
    monophosphate pathway   319
H_2-dependent heterodisulfide reductase
    system   172–5
Hydrates, as natural sink for methane   7
Hydrogenase, inducible uptake   450–2

Invertebrates, symbioses with
    methanotrophs, marine species
    315–28
Iodoform   58
Iodonitrotetrazolium chloride (INT)   436
Isotope studies, methane nutrition
    319–23

Lakes *see* Freshwater
Landfills
    as sink for methane   10
    see also Soils
Lignin, degradation, veratryl alcohol   42
Lignin derivatives, carbon dioxide origins
    497–8
Lysine residues, modification on MDH
    222–4

M2 bacterium, MSA metabolism   49–50
MADH *see* Methylamine dehydrogenases
Mandelate, aromatic substituent rings
    504–5
Marine environment *see* Oceans
Mau gene
    methylamine dehydrogenase   384–6
    mutants   386–9

Mau polypeptides, properties,
    comparisons, various methylotrophs
    389–95
MDH *see* Methanol dehydrogenase
3-Mercaptopropionate, production from
    3-methiolpropionate   21
Methane
    activation, dioxygen chemistry   104–6
    atmospheric, consumption by soils and
        MOB   304–6
    global budget   9
    global emission, consumption and
        production   9
    man-influenced sinks
        landfills   10
        natural gas production   8–9
        rice cultivation   8
    microbial oxidation, all sources,
        summary   74–5
    natural sinks
        atmospheric   304–6
        consumption by soils and MOB
            304–6
        freshwaters   7
        hydrates   7
        ocean   6
        soils   7–8, 304–8
        termites   6
        wetlands   3–6
    oxidation
        dioxygen chemistry   104–6
        direct measurements   3
        global estimate   3
        molecular biology   109–19
        rate measurements   2–3
        role of methylotrophy   1–13
        and water potential   308–10
        see also Methane monooxygenase;
            Methanotrophs
    production from carbon dioxide,
        enzymes involved   151–61
    soil methanotrophs, temperature
        responses   306–8
    see also Methanotrophs
Methane monooxygenase
    alkane oxidation, stereochemistry
        84–6
    colorimetric assay   290–1
    dehalogenation of chlorinated
        methanes   352–4
    homolytic/heterolytic pathways   70–5
    hydrogen peroxide/hydroxylase system
        65–80

Methylosinus trichosporium OB3b,
  pMMO or sMMO production
    122–33
  pMMO activity 93–107
    copper ions, effects 95–8
    copper ions, membrane-bound,
      magnetic susceptibility
      characterization 101–2
    EPR spectrum
      characterization 98–101
      and magnetic susceptibility data
        102–4
  sMMO catalytic cycle 65–80
    component interactions 86–90
    degradation of chlorinated
      hydrocarbons 121–33
    peroxide shunt of hydroxylase
      component 83–4
    proposed 83
    roles of protein component
      complexes 81–92
  sMMO genes
    cloning and sequencing 110–12
    expression 112–13
    transcriptional regulation 113–16
  sMMO and pMMO 65
  sMMO substrate radical adducts,
    hyperfine splitting constants 69
  spin trapping technique 66–70
  structural studies 75–8
  substrate and donor modulation 75
Methanesulfonic acid
  microbiological conversions 49–50
  natural sources 48–9
Methanol
  and glucose metabolism, *A.
    methanolica* mutants 333–4
  methanogenesis 171–80
    energy transduction 177–8
  oxidation
    in Gram-negative bacteria 209–20
    mutants 209
    Paracoccus denitrificans 235–43
Methanol dehydrogenase
  alpha subunit 224–7
  beta subunits, amino acid sequences
    225
  and cytochrome interactions 221–33
    electron transfer 215–16
    redox cycle of MDH 214–15
  dye-linked, *Amycolatopsis methanolica*
    249
  gamma subunit 224

gene probes 291–3
genetics of oxidation by *Paracoccus
  denitrificans* 235–43
in Gram-negative bacteria 209–20
in Gram-positive bacteria 245–51
inhibition by EDTA and other
  chelating agents 230–1
mox genes 235–43
  moxA, *K* and *L* genes 229–30
  regulatory *mox* mutants 240
  role of calcium 227–9
Methanol oxidase 210
Methanol:NDMA oxidoreductases
  (MNO) 245–51
Methanopterins 135–49
  biochemical specificity 136–40
  dissociation constants 139
  equilibrium constants, free energy
    changes and redox potentials 141
  formyl transfer reactions 140–2
  methenyl reduction 142–3
  methyl transfer reactions 145–6
  reactions at the formaldehyde state of
    reduction 143–5
  structural diversity 136
Methanosarcina thermophila
  acetate fermentation 163–70
  Göl strain, heterodisulfide reductase
    system 172
Methanotrophs
  kinetic modeling 343–6
  methanotroph-invertebrate symbioses,
    marine species 315–28
  obligate
    ecophysiological characteristics
      303–13
    taxonomy, Types I and II 255–8
  soil methanotrophs 304–10
  see also Methane; Methylotrophs
Methenyl reduction, methanopterins
  142–3
N-5,N-10-Methenyl-H_4MPT
  cyclohydrolase, activity 153–4
3-Methiolpropionate, anoxic
  production 21, 27
Methoxyphenazine methosulfate 436
Methyl chloride, production, sources and
  sinks 41–2
Methyl-coenzyme M methylreductase
  157
  acetate fermentation 166
N-5-Methyl-H_4MPT coenzyme M
  methyltransferase 156

Subject Index

Methylamine dehydrogenases
  comparisons, methylotrophs 383
  molecular biology and genetics
    381–400
    assembly model 395–7
  quinoproteins 403–4
Methylamine oxidase, quinoproteins
  404–6
Methylamines
  as substrates 56–7
  utilization, NMR in yeast and bacteria
    415–32
Methylated sulfur compounds see Sulfur
  compounds, methylated
N-5,N-10-Methylene-H_4MPT
  dehydrogenase
  co-enzyme F_{420}-reducing
    methylene-H_4MPT dehydrogenase
    155
  H_2-forming
    N-5,N-10-methylene-H_4MPT
    dehydrogenase 154
N-5,N-10-Methylene-H_4MPT reductase
  155–6
Methylobacterium extorquens
  MDH docking with cytochrome c_L
    222–31
  methylamine utilization 415, 425–6
Methylococcus capsulatus (Bath)
  sMMO genes, molecular biology
    109–19
  see also Methane monooxygenase,
    pMMO activity
Methylomonas spp.
  degradation of chlorinated
    hydrocarbons 123–6
  mutagenesis studies 116–18
Methylophilus methylotrophus
  amidase production 365–81
  MDH docking with cytochrome c_L
    222–31
  methylamine dehydrogenase 381–400
Methylosinus trichosporium, sMMO
  genes, molecular biology 109–19
Methylosinus trichosporium OB3b
  degradation of chlorinated
    hydrocarbons 123–6
  mutagenesis studies 116–18
  primer extension analysis 114
  promoter probe analysis 114–16
Methylosinus spp., see also Methane
  monooxygenase, sMMO
Methylotrophs
  16S rRNA sequencing 287–96

global methane budget 1–13
methylamine dehydrogenases 381–400
obligate methanotrophs 255–8
peroxisomes, biogenesis 193–207
phylogeny see Phylogeny of
  methylotrophs
plant associations 296–7
quinoproteins, biochemistry, new
  developments 401–13
taxonomy see Taxonomy of
  methylotrophs
see also Methanotrophs
Methyltransferases
  acetate fermentation 166
  methanopterins 145–6
Microbodies see Peroxisomes
MOB see Methanotrophs
Molluscs, symbioses with methanotrophs
  316–28
Molybdenum requirement, carbon
  monoxide dehydrogenase 433,
  442–5
Molybdopterin dinucleotides 442–5
Monomethylamine, microbiological role
  56
Monod kinetics, bioremediation 343–6
Mycobacterium gastri, methanol
  oxidoreductases 248
Mytilids, symbioses with methanotrophs
  316–28

Natural gas production, as sink for
  methane 8–9
Neurospora crassa, mitochondrial
  receptor proteins 198
Nitrate reductase, CO dehydroxygenase
  437
Nitroxide radicals, spin trapping
  technique 66–70
NMR, methylamine utilization in yeast
  and bacteria 415–32

Oceans
  bromomethanes 36–41
  methanotroph-invertebrate symbioses
    315–28
  methylated sulfur compounds 15–33
    and C_1 cycle 26–8
    DMS, DMSP and algae 16–28
    DMS fluxes 24–5
    role of DMSO 23–4
  as natural sink for methane 6
Ozone layer, destruction, and production
  of halocarbons 35–6, 40

Paracoccus denitrificans
  methylamine dehydrogenases 381–97
  oxidation of methanol 235–43
Pentosephosphate cycle 433
Peroxisomes
  biogenesis, *Hansenula polymorpha*
    193–207
  compartmentalization and function
    scheme 195
  deficient mutants 199–202
  degradation 198
  development 195–6
  metabolism, functional aspects 202–4
  protein targeting and assembly
    197–8
L-Phenylalanine, synthesis, RuMP cycle,
  A. methanolica 329–34
Phosphotransacetylase, acetate
  fermentation 164
Phototrophic bacteria, carbon fixation
  469–79
Phylogeny of methylotrophs 285–94
  16S rRNA sequence comparisons
    287–8
  colorimetric assay for MMO 290–1
  design of hybridization probes 289–90
  gene probes 291–3
Phytoplankton
  halocarbons, production 36
  production of dimethyl sulfide 17–18
  from β-dimethylsulfoniopropionate
    16–17, 18–20
  uptake of glycine betaine 28
Plants *see* Vascular plants
POBN *see*
  α-(4-pyridyl-1-oxide)-N-tert-
  butylnitrone
Pogonophora, symbioses with
  methanotrophs 316–28
Pore-forming protein, peroxisomes 197
Pseudomonas carboxydovorans, CO
  utilization 433–54
Pseudomonas MA, methylamine
  utilization 415–32
Pterins *see* Methanopterins;
  Molybdopterin; Sarcinapterin;
  Tatiopterins
α-(4-Pyridyl-1-oxide)-N-tert-butylnitrone,
  methane monooxygenase pathways
  66–9
Pyrrolo-quinoline quinone, MDH
  binding 209–17
Pyruvate, reduction, oxidation of reduced
  carriers 498–9

Quaternary amines, metabolism by
  marine organisms 56–7
Quinate, growth of *A. methanolica* 332
Quinoproteins
  biochemistry, new developments
    401–13
  putative PQQ-binding region 229
  see also Methanol dehydrogenase

Radioisotope studies, methane nutrition
  319–23
Rhodobacter sphaeroides, carbon fixation
  469–79
Rhodospirillum rubrum, carbon fixation
  469–79
Ribulose monophosphate pathway
  examples of methylotrophs and
    methanotrophs 290–1
  formaldehyde assimilation 209–11,
    292
  hexulose phosphate synthase 319
Ribulose-1,5-bisphosphate carboxylase,
  Calvin–Benson cycle 209–11, 292,
    319
Rice cultivation, as sink for methane 8
rRNA, 16S rRNA sequence comparisons,
  phylogeny of methylotrophs 287
RubisCO
  catalysis and regulation 472–6
  Cfx genes 472
RuMP cycle, formaldehyde fixation
  329–34

Sarcinapterin 136–7
SCN *see* Thiocyanate ion
Seas, seawater *see* Oceans
Seaweeds
  production of halocarbons 36–9
  see also Algae
Selenium, activation, carbon monoxide
  dehydrogenase 445–6
Serine pathway, examples of
  methylotrophs and methanotrophs
  290–1
Soils
  methanotrophs
    methane consumption 304–6
    temperature effects 306–8
  as natural sink for methane 7–8
  water potential and 308–10

Sulfur compounds, methylated
  exergonic reduction in methanogenic
    bacteria 171–80
  marine environment 15–33
  see also Dimethyl sulfide and sulfoxide

Tatiopterins 136–7
Taxonomy of methylotrophs 253–65
  acidophilic 279
  evolutionary aspects 282
  facultative methylotrophs 259–63,
    278–81
  Gram-negative
    facultative methylotrophs 278–81
    obligate methylotrophs 275–8
  Gram-positive, thermotolerant bacilli
    267–73
  groups of convenience 254
  obligate methylotrophs
    Gram-negative 275–8
    not utilizing methane 258–9
    utilizing methane 255–8
Temperature effects, Arrhenius equation
  216
Termites, as natural sinks for methane 6
Tetrachloromethane, dehalogenation
  reactions 352–4
Tetrahydrosarcinapterin 136–7
Thermopterin 136–7
Thermotolerant bacilli, taxonomy
  267–73
Thiocyanate ion
  oxidation, role of COS 55
  production 53–5
Topaquinone (TPQ) 401–9
Trace gases, atmosphere, and climate
  changee 59–60
Trichloroethane, degradation by
  methanotrophs 337–46
Trichloroethylene
  degradation by methanotrophs
    121–33

applications 337–46
continuous cultures of *M.
  trichosporium* OB3b 128–31
phylogeny 293–4
toxicity 126–8
Trichloromethane, dehalogenation
  reactions 352–4
Trimethylamine
  interactions with GBT, DMS and
    DMSP 28
  microbiological role 56
Trinitrobenzene sulphonate, modification
  of MDH 223
Tryptophan(yl) tryptophan(yl)quinone
  (TTQ) 381, 384, 401–9
L-Tyrosine, growth of *A. methanolica*
  332

Urothione, excretion 443–5

Vanillin, acetyl-CoA pathway 504
Vascular plants
  associated with methylotrophs 296–7
  methane transport 3–6
Veratryl alcohol, degradation of lignin
  42
Vinyl chloride, degradation by
  methanotrophs 338–41

Water potential, oxidation of methane
  308–10
Wetlands, as natural sinks 3–6

Yeasts see *Hansenula polymorpha*

Zooplankton, production of dimethyl
  sulfide from
  β-dimethylsulfoniopropionate
  16–17, 18–20